AF675175

Development Communication

Applied to Journalism & Mass Communication, Extension Education & Communication, Rural Development and Management Studies

Editors

V.K. Dubey

Ex. Head
Department of Extension Education
Institute of Agricultural Sciences
Banaras Hindu University
Varanasi (UP) India

K.N. Pandey

Ex. Head
Department of Extension Education
Institute of Agricultural Sciences
Banaras Hindu University
Varanasi (UP) India

Rajshekhar Pandey

National Council of Development Communication
Varanasi (UP) India

International Book Distributing Co.
(Publishing Division)

Published by
INTERNATIONAL BOOK DISTRIBUTING CO.
(Publishing Division)
Khushnuma Complex Basement
7, Meerabai Marg (Behind Jawahar Bhawan)
Lucknow-226 001, U.P. (INDIA)
Tel. : 91-522-2209542, 2209543, 2209544, 2209545
Fax : 0522-4045308
E-Mail : ibdco@airtelmail.in

First Edition 2009

ISBN

Composed & Designed at :
Panacea Computers
Agarwal Sabha Bhawan
Subhash Mohal,
Sadar Lucknow-226 002
Tel : 0522-2483312, 94522 95008, 93359 27082
E-mail : prasgupt@rediffmail.com, prasgupt@hotmail.com

Printed at:
Salasar Imaging Systems
C-7/5, Lawrence Road Industrial Area
Delhi - 110 035
Tel. : 011-27185653, 9810064311

Preface

Most of the publications available in the Development Communication area are of edited nature, generally as an outcome of seminars/ symposiums. Therefore, their focus is one sided. The present publication is an attempt to fill the gap of reading material, although a collection of Scholarly work, it has the capacity to bridge the gap which scholars are experiencing due to non availability of a quality text book in this field. It will address the Academic requirements of Post Graduate teaching in the field of Journalism and Mass Communication; Home Science Extension Education & Communication; Agricultural Extension Education; Rural Development; Social work; Rural and Agricultural Management.

Besides meeting the classroom curriculum needs, itis equally inclined to motivate the young researchers to explore the least travelled field of Development communication research. For that reason, it contains good number of articles which are full of imaginative ideas. Therefore, this collection will meet dual needs of academics not only in India but of developing world at large.

The Writings of Ms. Quebral; Cooney; Singhal and Rogers; Indu Bhushan, Kusum Singh; Beall; Melkote; Aunga; Shankaraiah; and Reddi; to mention few, are thoughtful and illuminating. It is difficult to find such useful Ideas of learned Scholars in one place. Therefore, this reading is of quality. It is expected that readers will find utility of their investment in this publication satisfying.

Developed world scholars were quite active in this field in 60s and 70s. Number of studies were undertaken, specially by U.S. scholars. But, interest started weaning from 90s onwards. However, the situation in the Developing world too has not shown significant change. More challenges have emerged. Poverty, hunger, diseases, Global warming still looms large on the humanistic horizon. These challenges need to be addressed in the light of modern Science and Technology of information science.

No single publication can answer these Himalayan questions. Therefore, we do not claim so, however, if this publication can serve to ignite interest in the field of development communication, it will be satisfying.

PROFESSOR V.K. DUBEY
Chief Editor
Journal of Communication Studies

Contents

Part I

Part II

Part III

Part I

Chapter 1

What is Development Communication?

Anura Goonasekera

Development communication is often looked upon, by writers on the subject, as a tool to bring about development in less developed countries. In this sense, what is emphasised is the use of communication media for development. Development communication within the present structures becomes subject to official control, and the high philosophical objectives of the right to communicate, free and balanced flow of information and concern for the identity of Third World countries, all recede further and further into the background

In a recently published Orientation and Resource Kit on Development Communication by UNESCO, Mayo and Servaes (1994) begin with observation that communication media are often called upon to support development programmes. They go on to state that, whereas development strategies could vary within and across countries, the communication activities associated with them are often quite similar. These activities are informing people about specific initiatives and urging their support for them. What I am going to discuss in this paper, however, are the much broader aspects of the intellectual discourse that underlie the concept of development communication. The practicalities of the use communication for development will be mentioned, only in so far as it is necessary to illustrate this intellectual tradition. Development communication as a field of academic inquiry has links with sociology, social psychology, economics and political theory. However, its immediate and direct links are with studies of modernisation of traditional societies conducted in the United States of America (USA) in the early 1950s. Daniel Lerner, Everett Rogers, Wilbur Schramm are among the most quoted authors in the Guru Gallery of Development communication. The origins of the development communication concept in the USA are not without controversy. Samarajiva (1987), who examined the origin of this concept, with special emphasis on the seminal work.

The passing of Traditional Society, by Daniel Lerner (1958), observes that the beginnings of this field of inquiry are rather murky. Relying in part upon original documents, he constructs the case that Lerner's Traditional Society was a spin off from a large and clandestine, audience research project conducted for the Voice of America by the Bureau of Applied Social Research. He notes the strong influence exerted by the demands of psychological warfare, in the context of the Cold War, on the early studies of development communication in the USA according to Samarajiva :

Exploratory work on the early period suggests the following pattern of net influence flows : marketing research to communication research; marketing and communication research to psychological warfare; from psychological warfare to communication and development (1987 : 17).

In the present analysis, I propose to first critically discuss two sociological paradigms of underdevelopment, of Third World societies as advanced by sociologists and social psychologists and then analyse their implications of a new information order. In order to illustrate the current intellectual discourses on communication and development, the following two paradigms of underdevelopment discussed are :

The paradigm of modernisation as enunciated by writers such as Lerner (1958), Rogers (1962), Schramm (1964), and Inkles and Smith (1976); and the dependency paradigm as advanced by Frank (1967), Wallerstein (1982), Portes (1982), Delacroix and Ragin (1978).

The Paradigm of Modernisation

The social scientific roots of the concept of development communication can be traced to certain early findings of sociologists and social psychologists working in the area of communication. It was these findings that later formed the basis of the modernisation paradigms. The following are the salient points of this paradigms :

1. Communication is a crucial element in the process of development of Third World societies (Schramm, 1964);
2. Communication, while spreading knowledge among the people, can also help create an attitude of mind and a type of personality which are necessary for traditional societies to become modern. This is quality of empathy which persons in traditional societies acquire by viewing mass media programmes that provide a vicarious universe for these persons to participate in (Lerner,

1958). Furthermore, the mass media, through advertisements of consumer durable creates new wants among the people which gives rise to 'a revolution of rising expectations'. The rulers can not resist fulfilling these new expectations of the people;

3. Communication through the mass media does not reach the masses directly. It is filtered through a variety of social layers. The message usually is first received by opinion leaders who spread it among their followers. This concept was popularly referred to as the 'two-step' flow of information (Lazarsfeld, 1948; Katz, 1960)

4. Communication is crucial to the adoption of innovation by a community. The process of adoption takes a specific pattern. The first to adopt new practices are the innovators who are usually the more educated and affluent members of society. If the innovation is successfully adopted, a large group of followers (compromising early adopters, early majority and late majority) will follow. The last to adopt a new practice are the laggards, who are generally the poorest of the poor in that community (Rogers, 1962);

5. The mass media are useful in spreading awareness but the final behavioural change will take place only after contact with persons of credibility at an interpersonal level;

6 Individual pay attention on a selective basis. There is, therefore, selective exposure and selective perception which are based on the already existing cognition's of the individual (Festinger, 1957; Lippit and White (1959);

7. Effective communication occurs when the source and the receiver have similar characteristics and interests (described as homophile). The most effective change agents are those that are most like the average client, except for his level of technical competence. Attention to feedback reaction from the audience is most, important in the continued success of a communicator (Rogers, 1962).

8. People pay attention to a message on the basis of uses & gratification derived form the message. The message should, therefore, be based on an understanding of the needs of the audience and its desires for gratification.

9. For better results, communication through the mass media should be combined with interpersonal channels of

communication. This is particularly crucial in rural areas of developing countries where interpersonal channels are still powerful and can sometimes negate message given by the mass media.

Development communication attempts to use these insights in one form or another in formulating communication campaigns in Third World countries.

Underlying Assumptions

There are several underlying assumptions in the concept of development communicators that need to be examined. First, is the assumption that the reasons for development and stagnation of societies are immanent, within or internal, to the societies themselves. Therefore, the rulers of these societies, by correctly identifying these immanent factors, can formulate policies that could bring about change in these societies. Second, is the assumption that communication, if used 'correctly' can set in motion the engines of development. Communication in this model is therefore, an independent variable. Further, it is assumed that what is lacking for development is communication. Third, is the assumption that people who live in developing countries have similar characteristics and will answer, in a predictable manner, to the call of the development communicator. Empirically, one could find many instances that could find many instances that could invalidate these assumptions. The assumption of the immanent of the causes of stagnation has been challenged by the world system theorists and dependency theorists. In addition, we cannot consider communication as an independent variable. The media of communication, in reality, has to operate within a social system with its values, stratification and political and economic conflicts. In such a situation, the media of communication can very well become the mouthpiece of dominant groups in society. Studies of the media in Latin American countries have observed that the mass media had little use for the peasants, as it met more the needs of the large farm owners. The media in certain Latin American countries appear to be deliberately oriented against aspects of social change which would be in favour of the peasants. The media is at the service of and owning elite's whose families sometimes own media organisations. In this situation, development communication alone will not help, for what is lacking may not be information but other resources, the mobilisation of which may require structural changes in society.

The problems of development are not always of a technical nature to be resolved by the experts. They are usually political ones. An important question is the extent to which government are willing to make political choices to bring about change. This is crucial for a meaningful assessment of the role of communication in development. Experience has also shown that development communication, when used without concomitant changes to the structure of society, could lead to the strengthening of the status quo or increasing and the widening of the gap between the rich and poor in developing countries. For instance, when new agricultural practices such as cash crops under irrigation were introduced into Third World villages, the first to adopt such practices were the more affluent and educated persons in the village. This is because such practices initially require the taking of risks and modest financial investments which are not within the reach of poor. The result is that the new practices give an opportunity for the rich persons in the village to become richer, thereby widening and strengthening the existing social disparities. Scarlet Epstein (1962), an anthropologist, reports that when irrigation was first introduced into the village of Wangala in South India, it raised the whole economy of the village to a higher level in one stroke. But the village remained wholly agriculture. It was unilinear development the employment structure and traditional role relations remained unchanged because the labour requirements for case cropping could be easily met under the traditional system :

The un-linear economic development in Wangala set up no incompatibility between the new wants and the old ways in the indigenous employment structure. The theoretical points are of interest here. Economic development may occur without any change in economic roles and relations, provided it does not result in re-allocation of resources or an increased range of economic relations (Epstein, 1962, p. 316).

While it is true that communication is a crucial element in the process of planned development in Third World societies, there are several other factors that one has to take into consideration in evaluating the role of communication in development. The development communicator, in giving primacy to communication in the process of development has tended, I think, to overstate his or her case.

The dependency paradigm

The dependency paradigm directs examination of international economic relations, within which any development of Third World societies has to take place. According to this theory, underdevelopment is the result of world economic relations which have created a centre and a periphery, with central economic relations with the peripheral states. This paradigm was advanced by sociologists of Third World countries, particularly Latin America. There are several variants of the dependency paradigm, but the more significant features are as follows.

1. Development of Third World countries is inimical to the dominant interest in advanced capitalistic exploitation the possibilities for development are limited. The only way out of this stranglehold is through political means (Frank, 1967).
2. Development and underdevelopment only in terms of the international system of economic relations, which in its present capitalist phase has divided the world into a group of core countries which are industrialised and rich (for example, USA and Japan) and a large number of peripheral countries which are poor and exploited (for example, Sri Lanka and Bangladesh). Bridging the core and the peripheral countries are the semi-peripheral countries which have developed a modicum of wealth & are crucial to keep the system in balance (Wallerstein and Hopkins, 1982).
3. Development and underdevelopment are, according to this theory, not two stages in the socio-economic progress of societies, but two concomitant parts of the same system. The underdeveloped periphery is the necessary corollary of the developed core countries. One cannot exist without the other.
4. State or government intervention is one way in which this situation can be changed in a significant way. Therefore, while this participation in the world economy impedes the economic development of poorer countries, this does not by itself shut off the possibilities of individual national development. A strong state can counter the deleterious effect of dependency. Delacroix and Ragin (1978) are advocates of the strong state theory. Some writers, for example, Wallerstein and Hopkins (1982), do not agree.
5. The situation of contemporary dependency is different from the situation of foreign colonialism, where the entire

administration was done by colonial instituted cadres. Dependency is not imposed by an occupying army. In the beginning, it is instituted by the willing participation of indigenous bourgeoisie and its effects are initially salutary to the peripheral countries because it brings in much needed technology and resources. The end result of dependency, however, is very similar to that of colonialism. It leads to exploitation, breeds social inequalities and leads to the miserisation of the dependent nations as a whole. The multinational enterprises are blamed for the bulk of this dismal outcome. Through the use of imported technology, these enterprises tie the dependent economy close to the centre. The cost of capital intensive technology and profit remittances abroad becomes constant drain on the foreign exchange resources of the peripheral countries. Finally, the activities of the multinational companies lead to the growth of a new privileged and exploitative group and the acceleration of social inequality.

The intellectual discourse on development communication has also highlighted issues relating to cultural implications of international communication particularly satellite broadcasting of television programmes. The early concern for the cultural implication of international communication stemmed from the large influx of media material form the developed countries (core) to the developing countries (peripheral). Varis (1993) in a UNESCO report observed That in a global perspective, imported television programmes developing countries, took up an average of one-third or more of the total time of programming in 1983. In certain countries such as Argentina, Brunei and Malaysia, it was over fifty per cent of the total programming time. This heavy bias in favour of developed countries is even more pronounced in the distribution of international news, where five giant international press agencies, all of them from the developed world, dominate the flow of international news. The result is that the dominant interest of the Western countries such as USA and Britain dominate the foreign news coverage by media of Third World countries. Da Costa in another UNESCO research report, states that :

Less than 20 percent of the news space in South American dailies for example, is given over to Latin America and the Senegalese press shows greater interest in a minor ministerial re-shuffle in France or the Federal Republic of Germany than in an election taking place in Gambia or the Ivory Coast (de Costa, et. al. 1980, p.5).

However, a recent study by Preben Sepstrup and Goonsekera (1994) has shown that the actual consumption of international television programmes are a minuscule part of what is made available by satellite broadcasters. While this finding is sobering, the implications of satellite broadcasting by Western nations on indigenous cultures is a serious one. Technological advances in communication appear to be worsening this imbalance of communication resources in favour of developed countries.

Let us consider two of these technological advances. One is in the field of satellite communication, which has opened up vast potentialities for easy, flow of information across national borders. With the development of Direct Broadcast Satellites (DBS), it is quite possible for a country that owns or controls a satellite to beam messages to a large number of countries in the world with or without the consent of such countries. The programmes may be considered wholly unsuitable for a variety of reasons cultural, linguistic, political or moral by dominant groups in these countries. For example, in a recent programme on Hong Kong based Star Television (Nikki Tonight), which is beamed to India, Mahatma Gandhi was referred to as a bastard bania (trader). This created an uproar in India where Gandhi is revered by a section of the population. A criminal defamation case was filed by a grandson of Gandhi and a warrant for the arrest of Rupert Murdoch, among others was issued in India. Such instances while rare, help to confirm the worst fears of critics of transferred television broadcasts as being deleterious, irresponsible and subversive (International Herald Tribune, July, 6, 1995).

Just as developments in DBS communication will result in the unrestricted flow of programmes and information over cultural and international boundaries, advances in computer technology and telecommunication industries have made it possible for the transfer of vast amounts of data electronically across national borders. This phenomenon is referred to as Transborder Data Flow (TDF). The central control of data which is made possible through TDF has far reaching implications for the economics of the countries that are caught within the system. It can lead to fundamental changes in the way in which goods and services are produced and distributed in these countries.

Transnational corporations are the prime users of TDF. It is both a commercial product and a management tool for these corporations. Developing countries participate in this international data market either as data suppliers or as buyers of processed data through experts or through parent multinational companies. In addition, the developing countries will also acquire the technology and equipment that is necessary for their own telematic sector.

In an information age, it may happen that developing countries are reduced to the export of data for processing to developed countries. The value added benefit will result at the processing stage and will accrue to the developed countries. Skilled jobs in data processing and system management will be retained by the affluent countries, while the developing countries will probably have armies of data entry personnel sending computer data via satellite for processing in the industrialized countries. This situation is similar to the export of raw material for processing in colonial times except that, in the modern information age, the export is raw data instead of raw materials. The result however is the same - exploitation, economic stagnation, and dependency.

A New Information Order

An issue basic to the discourse on development communication is the demand for a new information order which stems directly from this imbalance in the distribution of communication resources among developing and developed countries. The New International Information and Communication Order (NIICO) is seen as a way of balancing the distribution of information and connected resources in the world. At least, this is how the advocates of NIICO in the Third World see it. The debate on a NIICO started in the early 1960s, primarily as an ideological denunciation of imperialism and colonialism (Savio, 1982). The developed countries of the West rejected it as rhetoric designed to apply political pressure andnot as a genuine demand. But by the end of the 1970s, communication researchers advanced incontrovertible findings which indicated a profound imbalance between North and South in the field of information, and the southern dependence on communication structures of the developed Western nations for gaining access to information.

The world systems theorists argued that such a dependency of Third World countries was inherent in the present economic structure of the world. Poverty and stagnation were creations of the world economic system. While the debate on a NIICO began as a North South issue and continues to proceed on these lines, it is in fact much more than this. It is a question of developing a communication and information system which should ideally allow all persons to express themselves freely and choose the information that they wish to have. Viewed in this light, a NIICO addresses a basic question of the human right to communicate a question for everyone and not only for the South. The recognition of the concept of the right to communicate, it is believed, will create at the international level, a universal conscience necessary to convert freedom of information and the free and balanced flow of communication, which are basic to a NIICO, into a principle of international law (Cocca, 1982). At the national level, it will impose on the state, the obligation of creating the conditions under which the practical freedom and entitlements which derive from the right itself can be implemented. The opposition to the concept of the right to communicate and the demand for a NIICO stems from two main ideological standpoints. The Western nations distrust the concept of right to communicate because they see it as part of the New International Information and Communication Order of which they are suspicious. They fear that a NIICO will be used as a lever to impose restrictions on Western news agencies, on the marketing of Western films and television programmes in Third World countries and on exploitation of data transfer and other recent technology.

The opposition to the right to communicate stems from the fears, particularly among Third World countries, that it would be used to justify the continuation of the present massive imbalance in information flows and unrestricted importation of Western technology and information and consequently, of Western values. The concept of the right to communicate should be an important principle of a NIICO. Unfortunately, the debate to date has gone in other directions. While all sides involved in the debate accept the concept of the human right to communicate as a valid one, there is disagreement on the locus of this right. Some see it as a right pertaining primarily to the individual and only subsequently and secondarily to society; other see society or state as the primary locus with the state having powers to restrict the right in the public interest. There is also disagreement as to the content of the right. Some want the definition to include all the rights associated with communication, while others prefer a simple statement of the human right to communicate. The other aspects of communication freedom such as fairer sharing of resources, it is argued, should be left to a different forum.

A major problem identified by many is that the right to communicate is likely to remain a philosophical idea, unlikely to be implemented, because many societies do not have the economic capacity to put it into practice. Furthermore, the debate on the New International Information and Communication Order has taken the question of greater participation in the international exchange of information, more as a quantitative problem - the problem of a quantitative transfer of information capacity from North to South. Underlying this is the unrealistic idea of, the developing countries, sooner or later, taking over the transnational information system. The emphasis has been on production and distribution structures.

Accordingly, priority was given to the creation of national news agencies and the creation of regional agencies such as the Pan African News Agency, the Organisation of Asian News Agencies, the Federation of Arab News Agencies and the Accion Sistemas Informative Nacionales in Latin America. The limited financial assistance that was available has been used in the creation of this infrastructure.

These structures have shown little or no ability to compete with transnational systems. However, they have been able to give the appearance of competitiveness to regimes which are usually authoritarian and which will be often tempted to use new structures for controlling information instead of stimulating new and creative flows (Savio, 1982). Analyses of the news content of Third World news agencies how that it is concerned with events that stem from political and economic spheres of the urban centers and relate to the activities of the powers that be. In this, there is much similarity to the transnational news agencies of the West, except that the professional level is much lower. In both systems, the spot news dominates. The Western news agencies treat them professionally because the market pays for it. The Third World agencies treat news in a bureaucratic manner, probably because of government policies.

What can be done to correct the aforementioned situation?

The first step is to recognise the trends that development communication and the NIICO have taken in their practical applications. The emphasis on structures for a fairer distribution of news resulted in a number of regional and national organizations coming into existence. The model that has been followed is the Western model of news/film agencies. This has until now, not worked well in developing countries. From the emphasis on structures, it is necessary I think to look at the content of information.

- What is it that the Western news agencies do not provide, which development communicators can provide?
- Can development communication create a means of communication for small communities in the rural areas, in the cities, in the universities, in the plantations?
- Can it provide a forum for alternative content which is by passed by the mass media?
- Should not development communication emphasise the participation of small communities through their own media of communication such as newsletters, small radio stations and television stations?

The emphasis on cultural identity of Third World societies, while a counterpoint against the cultural 'invasion' by the media programmes of the developed countries, has had its own deleterious effects on the sub-cultures of the Third world. Most developing countries are plural societies with a dominant cultural group and a number of minority cultural groups. The concern for the preservation of cultural identity in the face of the 'invasion' by Western values has often resulted in an emphasis on the cultural values of the dominant groups in such societies. The majority cultures within such societies did not revive the same emphasis, partly because of the lack of financial and other resources. The latent effect of such a policy however, was to divide the communities further, particularly since the minority communities perceived the emphasis given through the mass media as an imposition of the dominant values of societies on them - cultural imperialism of a domestic nature. The mass media, under such a policy of cultural identity, instead of unifying the plural societies of the Third world tended to divide them further (Goonasekera, 1987). In such a context, the answer may be to decentralize the mass media by having smaller newspapers, television and radio stations that can cater to the interest of specific cultural and interest groups. This is a role of development communication. Cultural policies should aim at providing wider participation of people in the process of communication rather than emphasising, through a centralized mass media, the cultural identity and values of a dominant group.

In creating the New International Information and Communication Order, we should start from home. This means that the national communication policies and planning in Third World countries must take into account the following aspects :

The import of technology from developed countries;

Setting up of structure for both local and foreign dissemination of information;

Higher standard of training for media personnel (who are the gatekeeper); and

The introduction of 'laws' to facilitate the workings of the communication sector in keeping with the principles and spirit of NIICO.

One cannot outwardly demand a regulationof international flow of news form North to South unless inwardly the prerequisites for a fairer utilization of such facilities exist. At the Transborder level, the media organisations of Third World countries that have called for a new information order should initially create the means for communicating among themselves without bringing in the transnational agencies of the North. For instance, Asia vision Exchange is a step in the right direction. But much more remains to be done in this area. The communication links among Third World countries are still poor. Laws governing visas, foreign exchange, and accreditation of journalists all militate against the quick deployment of persons among these countries.

No doubt, the North-South dimension of the debate will remain the main challenge to the Third World countries. Many professionals and policy makers in the West now accept the argument of the Third World for adequate representation in the world information order and for self-depiction in the media of the North. But, this has remained mostly a theoretical gesture. Thus, most attempts at interstate regulation of the flow of media form North to South have failed. In this situation, the Third World will have to fall back more on their own resources and ingenuity to set in motion the conditions necessary for the New International Information and Communication Order which they have demanded. Hopefully, the Third World receive a modicum of unselfish media support form the developed countries of the North to create for them equal competitive chances in the international market for information.

References

Cocca, Aldo A. (1982) 'The right to communicate: an evolutive concept for a new personal and social dimension of fundamental human rights', *Evolving Perspectives.* pp22-37.

De Costa, A. L., et. al (1980) News Values and Principles of Crǿss Cultural Communication. Paris: UNESCO.

Delacroix, J and C Ragin (1978) 'Modernising institutions. mobilisation and Third World development: a cross national study', *American Journal of Sociology.* Vol. 84, pp123-150.

Epstein, S (1962) Economic Development and Social Change in South India. Manchester. *Manchester University Press.*

Festinger, L (1957) Theory of Cognitive Dissonance. New York: *Harper Row Publishers.*

Frank, Andre G (1967) Capitalism and Underdevelopment in Latin America. New York: *Monthly Review Press.*

Goonasekera, A (1990) 'Communication, culture and the growth of individual self in Third World societies', *Asian Journal of Communication.* Vol.1 No.1 pp34-52, Singapore; AMIC

Goonsekera, A and Lock Y K (1990) 'Violence on television in Asia', Asian Journal of Communication. Vol. No.1, pp136-146, Singapore; AMIC.

Goonasekera, A (1987) 'Third World cultures and new information order', In Jayaweera and Amunugama (eds) Rethinking Development Communication. Singapore; AMIC.

Haviwara Shigeru (1993) Mass Media and Cross Cultural Communication Japanese Study, *Interim Report. Singapore :* AMIC Collection.

Inkeles A and D Smith (1976) Becoming Modern : Individual Change in Six Developing Countries. Massachusetts : *Harvard University Press.*

Joshi, S R (1993) Mass Media and Cross Cultural Communication: Indian study, *Interim Report. Singapore:* AMIC Collection.

Karnik, Kiran S (1993) 'Socio-cultural and economic impact of transnational broadcasting and cable TV', *Publication of International Broadcasting,* Cable and Satellite Conference. Oct 25-28, New Delhi.

Karthigesu, R (1993) Mass Media and Cross Cultural Communication: Malaysian study , *Interim report.* AMIC Collection, Singapore.

Katz, E (1960) 'The two-step flow of communication', In Schramm (ed.) Mass Communication Illinion: *University of Illinion press.*

Katz, Elihu, et al. (1991) 'Neither here nor there: why Dallas" failed in Japan', *Communication 1991.* Vol.12 pp99-110.

Kodaira, Sachiko, I (1993) 'A review of research on TV violence in Japan', Paper presented *at ICC Annual Conference.* Sept. 21-23, 1993, Mexico City.

Lazarsfeld PF et. al (1948) The People' Choice. Columbia, New York *University Press.*

Lee, Paul S N (1991) The absorption and indigenisation of foreign media cultures : a study of cultural meeting point of the East and West Hong Kong', *Asian Journal of Communication.* Vol.1 No.2, pp52-72, Singapore: AMIC.

Lerner, D. (1958) The Passing of Traditional Society : Modernising the Middle East. Glencoe, Illinois: Free Press.

Lippit, R. and R. K. White (1959) 'An experimental study of leadership and group life', In EE Maccoby, et. al (eds.) *Readings in Social Psychology.* New York: Hold.

Mac Bride S. et. al (1980) Many Voices, One World. Paris : *UNESCO.*

Mayo J. and Servaes (1984) Approaches to Development Communication : An Orientation and Resource Kit' Paris : UNESCO.

Mc Anany, EG (1980) Communication in Rural Third World : The Role of Information in Development. Illinios : Praeger.

Nordenstreng, K. and T. Varis (1974) 'Television traffic a one way street? A survey and analysis of international flow of television programme *Material' Reports and Papers on Mass Communication.* No.70, Paris: UNESCO.

Portes, A. (1982) 'On the sociology of national development a critique', *American Journal of Sociology.* Vol.87, pp 247-279.

Samarajiva, R. (1987) 'The murky beginnings of communication and development field : voice of America and the passing of traditional society', In Jayaweera and Amungama (eds), *Rethinking Development Communication.* Singapore : AMIC.

Savio, R. (1982) 'Communication and development in the 80s', International Foundation for Development Alternatives. Dossier 32, Nov/Dec, pp75-79.

Schramm, W. (1964) Mass Media and National Development. Stanford, California : *Stanford University Press.*

Sepstrup, P. and A. Goonasekera (1994) TV Transnationalisation : Europe and Asia.. Paris: UNESCO.

Varis T. (1993) 'International flow of television programmes' *Reports and Papers on Mass Communication.* No.100, Paris : UNESCO.

Wallerstein I. and T. Hopkins(1982) 'World Systems Analysis'. London : Sage.

Chapter 2

Concepts and Practice of Development Communication

Nora C. Quebral

Since communication began to appear as courses in university catalogues and to affect extension programmes in agriculture and other areas, there has been some confusion about nomenclatures not to mention some doubts about the worth of the field among those in agricultural development work. To help clear up misconceptions, it is proposed that we take a quick look at the historical backgrounds of extension education, communication, agricultural communication and development communication. Our object is not so much to underscore differences as to trace conceptual roots as far as we are able in order for us to understand why and where their interests-converge at certain points.

It is a known fact that extension education as a service and as a field of study was first on the scene. It began in the 1800's in Europe, where it was first known as advisory work, but it was in the United States in the 1900's that it came into full flower. At that time the United States still had a primarily agricultural economy. There was a felt need for an organized service that would help the American farm family, as a socio-economic unit, obtain the information needed to improve its agricultural production, its home environment and its community life. We should note that as originally conceived, the service was to be "educational decentralized, cooperative and offered on a voluntary basis (Sanders & Maunder)." In the years following World War II the extension concept was transferred overseas to the developing countries, along with American aid.

Another American export in the 1950's was the practice and study of communication. Although it developed simultaneously in Europe, it was the U.S. variety that influenced Asian countries like India and the Philippines. The growth of the field was spurred by the advent of the communications revolution when radio, television and a successive array of communication hardware seemed to offer infinite possibilities for benefiting mankind. In the area of research, communication was initially linked to wartime persuasion and propaganda. Its behaviorisitic concepts inevitably gravitated after World War II to advertising and public relations where it was and still is used to sell products and images.

Meanwhile the same communication principles and techniques had also diffused to the American land grant colleges where they became an adjunct of extension education. Radio, TV, the print and audio-visual media comprised a manner of disseminating farm and home information that was collectively known as mass media methods. In agricultural curricula, courses were instituted on media use that went under the name of agricultural journalism and, later, agricultural communication. Further specialization soon led to their evolving into a separate major in the parent curriculum. From the U.S. land grant colleges, these courses moved in due time to the agricultural colleges of developing countries which were patterned after the former.

Also occurring in the 1950's was the confluence of factors that focused world attention on the phenomenon of "development," Poverty, food insufficiency, an alarming rate of population growth, degradation of the environment these and similar problems of the Third World became international and rather political issues in succeeding decades. Considering the volume of information needed to speed up the process of social change, the new communication technology seemed tailor-made for the job.

Since then perceptions about the causes of and remedies for under-development have altered and, along with them, the role that communication can play. Throughout the shifts, the systematic application of communication principles and techniques to move the development of poor countries forward has become known as development communication. It is to be noted that agricultural colleges, to whom the core problems of development are not new and which possess a long tradition of service to rural people, have taken the lead in formalizing these development and communication concepts into an academic discipline.

Our thumbnail histories of extension education, communication, agricultural communication and development communication are meant to show that each grew out of some need that information was expected to solve at some place and time. That they overlap at certain points does not change the fact that they already-co-exist.

While parallel, they each have certain specialties that render somewhat academic such questions as "Why should agricultural universities institute agricultural or development communication programmes when they already have extension education programmes?"

Instead of vying with each other, the thing to do seems to be for extension education and development communication to combine their strengths against common problems. Since the concerns of developing countries today are predominantly agricultural and rural, it follows that at this moment development communication, like extension education, centers on agricultural and rural development. The question then is not whether development workers should get training in extension or in communication. Both types of expertise are needed; both streams feed into and can give added depth to a common pool. Development communication has borrowed liberally from extension education, but the process has not been entirely one-way. Many of the concepts of the former have already crept unobtrusively into the later.

What is development communication?

It has been the custom to define it by explaining what development is, then explaining what communication is and then putting the two meanings together. This is still a valid way of doing it as long as we allow for the synergism between the two that may lead to new ideas that are not inherent in either one separately. As an aggregate concept, however, development communication has drawn freely not only from its parents but also from extension education, from non-formal education from the basic and applied social sciences, from the humanities and from wherever fresh insights might be gained. Rather than attempt one inclusive definition of development communication, therefore, it might be more useful for us to say what it is and is not.

We can begin by saying that development communication uses the communication media to help bring about the socio-economic upliftment of people in ways that are compatible with its humanistic goals. In this respect it is both purposive and interventionist.

It has come to mean more than just communicating information on agriculture, family planning, nutrition, environmentalism and the like. It has become associated with a type of communication planning that is systematic and cyclical, that begins with objectives based on user needs and that is periodically renewed by formative and summative evaluation.

Development communication is a circular, not a linear, process. In the process the users are just as active as the sources and affect the outcome of communication just as much if not more. The accent of development communication is not on the dissemination of information but on its reception and utilization. The users of the communication media are seen as participants in the communication process rather than as audiences.

It views people not as independent atoms but as connected to each other in networks where a good many decisions are made by clusters of individuals. Many communication strategies, then, especially at the village level, are built on this assumption.

It can operate on several planes for community, sectoral, national and international development. Insights generated at one level move easily to and enrich strategies at the other levels. From a macro viewpoint it deals with information imbalances, with visible and invisible policies that affect communication systems, which in turn affect national and international relationships that promote or impede development. Over the long haul, however, equalizing these world relationships become meaningful only insofar as they raise the quality of lives in the villages.

It accepts that there is no one model of development. It recognizes that countries may take different paths to self-realization, depending on their histories, values, resources and aspirations.

It supports development that is participatory and decentralized in planning, execution and benefit, sharing, as indeed espoused by extension education but which was eclipsed in implementation when the latter was adopted by the developing countries. The people for whom development projects are meant have a valid right to be involved even in such project activities as research and evaluation.

It is more multi-directional than the top-down movement implied in the diffusion approach to rural development. It advocates a development process where decisions, to be lasting, are made as well by village people.

Development communication is primarily educating people, not simply passing along information to them. The latter may suffice for progressive farmers who are already disposed to seek and use pertinent information. But motivation may still be needed for the small farmer and his family who have little education and few options and for whom the risks of unwise decisions are greater. Motivation has recently fallen out of favour in some development circles because it smacks of persuasion and, hence, of manipulation. It is considered an integral part of learning, however. So long as the learner is apprised of the facts all along the way and so long as motivation means demonstrating to people the pros and cons of alternative procedures so that in the end they make the final choice, then motivation need not be the ogre that it is sometimes made out to be.

Development communication is not merely publicity for a project and its executors. While the public should know the whys and wherefores of a project in order to be able to participate in it, building images, particularly false ones, does nothing for anyone's real development.

While linked with the mass media, rarely does it depend on the mass media only, simply because the latter have not yet become part of the warp and woof of village life. The mass media, in any case, are but extensions of face-to-face communication and should not substitute for the real thing unless warranted. Many times development communication means attaching the mass media to interpersonal channels instead of the other way around. It does not advocate the exclusive use of the mass media, but neither does it rely mainly on personal channels as extension education normally does in the developing countries. It views a communication situation as multi-faceted, requiring the right medium or media mix for the user of communication at a given time and place.

It sees the causes of underdevelopment as present in people themselves, but also as present in their environment in the community, in the nation, in international relations. The solutions to underdevelopment, therefore, require changes in people's knowledge, attitudes and practices, but also changes in social and economic structures, institutions and policies which exist both in the countryside and in the cities.

It is not exclusively or even mainly government communication. The development communicator is accountable to his clients, not primarily to government or its agencies. He acts in the people's best interests. One of his jobs is to encourage the flow of communication from them to government and its planning and implementing bodies as well as among the people themselves.

To a great extent it is transfer of technology to the end users, but it assumed that several steps have happened before the actual transfer. The most important ones are the identification of needs by the end user, with the help of the field agent, and the search for the technology that will most appropriately answer the need. The transfer process thus begins with the user of the technology, not with the extension service or the university.

Technology that arises from research is probably the chief commodity transferred at this time, but technology that is suited to the socio-cultural circumstances of the user. It therefore has to be location and culture-specific, as demonstrated time and again in agriculture.

While much of development communication deals with innovation or what is new, it has ample room for the old, whether in terms of cultural values, of sound folk practices, of traditional but proven channels to and from people.

For all practical purposes it is today focused on agricultural and rural development, but it does not preclude urban development for two reasons. One is that the urban poor also require the services of the development communicator. Another is that urban development and rural development may not be separated from each other. They are two faces of the same coin. It is true, however, that the majority to be helped today are in the rural areas.

While it may begin with raising agricultural productivity and incomes, it looks beyond agricultural or even economic development. Its goal is a higher quality of life for everyone where the indicators are greater equality, increased self-reliance larger freedom for the individual.

Because the trickle-down approach to development, or working through the more progressive farmers and communities, has not appreciably improved lives in the countryside, development communication must now bring information directly to its chief users. It must bias information deliberately for the benefit of those in the villages.

Because its beneficiaries are in the villages and because development should be decentralized, it follows that the most pertinent communication media are the localized media like small radio stations, community newspapers, video facilities, whose control lies more with the community residents rather that with vast, centralized communication networks. The latter can be supportive of the local media by making people aware of certain events, by providing the agenda for national discussions, by holding up examples to be emulated, by doing the things that their bigness and longer reach qualify them to do best. But for teaching, for promoting dialogue, for influencing decisions, the local media may work better. And because development should be participatory, interactive communication media that encourage query and response can also play a major role.

It should be clear by now that even while we acknowledge the electric nature of development communication, its present meaning is actually shaped most strongly by the prevailing views on development. These views will likely continue to change, as they have done in the past.

Perhaps the next question waiting to be answered is what may be expected of the development communicator. If he is worth his sail, he should have the following capabilities :

1. An understanding of people and the society in which they live.
2. A working knowledge of development problems and issues.
3. A working knowledge of the subject matter that is to be communicated.
4. The skills and social commitment to utilize the communication media for developmental objectives.
5. The ability to plan, executes, monitor, evaluate and document multi-media strategies.

In planning strategies, the development communicator has a gamut of communication technology to choose from. The most advanced are not necessarily the most cogent, since they should match user needs and conditions. Five general categories of media are now being utilized :

1. Group and organizational media such as associations, formal and informal groups, organizations.
2. Folk and traditional media such as theatre, puppetry, fairs.
3. Print media such as newspapers and publications.
4. Broadcast media such as radio and television.
5. Audio-visual media such as film, slides, video and audio cassettes.

The last three are particularly useful for three communicative purposes :

1. For amplification and range. The ability to transmit the same message rapidly to more people over longer distances is perhaps the best recognized attribute of communication technology.
2. For more permanent records. Earlier a characteristic only of the print media, permanence is now also made possible by discs, tapes and films.
3. For fidelity of message. This third function still has to be fully explored especially in agriculture where recommendations, say, on input proportions and on planting and fertilizing intervals, depths and spacing have become so precise that a small farmer with undeveloped cognitive abilities may be apt to forget or misinterpret directions without the support of communication technology.

At the risk of stressing the obvious, let me point out that with his knowledge and skills, the development communicator can be useful to the following people :

1. To extension workers, teachers and trainers as a producer of teaching aids and of update materials.
2. To development administrators in the preparation of reports and organizational literature.
3. To researchers in the interpretation and dissemination of the results of their investigations.
4. And finally to the farmer and his family in his capacity as teacher, interpreter and middleman through the communication media.

Because the development communicator is supportive of the first three groups in many instances, there is a real danger that he will lose sight of the fact that his true clientele is the farmer and his family. Hence, it is important that he is in touch with them at all times by getting out to the field as often as he can.

The international aspects of development communication pose a similar danger to its students and practitioners : their being carried away by global issues and forgetting those closer to home, of being too general at the expense of specificity.

It can be closed with a home truth that communication technology, in spite of its great capabilities, can never totally substitute for the field worker. The latter's role might change since technology continually produces new artifacts that should make his job easier to do. And that is how the extension worker should regard communication technology as tools to be used and development communication as an additional perspective from which to view his job.

Reference

H.C. **SANDERS AND** A.M. MAUNDER, "Why an Extension Service Today?" *The Cooperative Extension Service,* H.C. Sanders (ed.), N.J., Prentice Hall, 1066. p. 6.

Note :

- Based on a paper delivered at a Workshop on Agricultural and Development Communication held at the Centre of Advanced Studies in Agricultural Communication, G. B. Pant University of Agriculture and Technology, Pantnagar, Uttar Pradesh.

Chapter **3**

Bottom-Sideways Communication for Bottom-up Democracy

Kusum Singh and Bertram Gross

It is essential to develop a bottom-sideways strategy of working with hundreds of millions of illiterate peasants, workers and tradesmen and spread out - across the country by using all available means of interpersonal, organizational and political communication to transcend the divisions imposed by religion, caste, language, gender and social role.

In the technocratic world of information management, the idea of "democracy" is an unwelcome stranger. It suggests the intrusion into top-level decision making of untutored minds that have not yet learned the difference between a bit and a byte or plumbed the essence of front-end and friendly systems.

With enough money and "top-top" teamwork, however, information managers can program any formal system of representative democracy to sidetrack unfriendly intrusions. The unexpert, unrich, unwhite and even unmale multitudes can be held at bay by selective cooptation, divide-and-conquer strategies and well-polished imagery. Thus can *representative* democracy, with its constitutional machinery for the bottom-up selection of top leaders, be converted into *image* democracy? Elected officials may then represent a minority of self-serving elites who finance electoral and lobbying campaigns while feeding the majority of people symbolic representations of selfless service for everybody.

When this occurs, the fault does not lie in the formal machinery of representative democracy, one of the great institutional inventions in human history. The difficulty, rather, is that the majority have too rarely developed the many means (shall we label them "software"?) of using this machinery.

What are these means?

The Software of Democracy : Uses and Misuses

The most obvious are standard devices for deconcentrating government power. Separations between executive, legislative and judicial branches may restrain the power of any member in this customary constitutional trinity. A federal structure, also embodied in many constitutions, provides states with certain residual - or even countervailing-authority.The states themselves may be countervailed by the authority of their subdivisions : countries, cities, townships and villages. Outside of constitutions, political parties and factions may compete with each other in nominating - and trying to elect - candidates for public office. In trying to "throw the rascals out," minority parties may unveil the unsavory actions of the "in's." They, in turn, may portray the evils that could result if the "out's" win seats of formal authority. With constitutionally guaranteed freedoms of the press, these conflicting views may enlighten the general public. With constitutionally and legally protected freedoms to organize, non-government groups may articulate and advance their interests through all channels and modes of communication.They may thus win a "say" in the actions of almost any official, legislator or judge. To prevent secret manipulation by invisible hands, procedures may be established for "open decisions openly arrived at" - from public notice, meetings and hearings to consultative arrangements, advisory boards, representative councils and other methods of "maximum feasible participation." These procedures may be strengthened by freedom of information laws that restrict government privacy by establishing public rights to know. Each of these devices has now become an essential component in the complex machinery of constitutional democracy.

For most people, however, each of these devices is to use computers - "user unfriendly." To appeal an executive decision by going to a court or legislature takes lots of time, money and special skills. State and local authorities are often as much insulated from popular control as a federal government and in some circumstances even more so. Political parties that may compete savagely in dividing the spoils and handling minor issues may be tacitly united in defense of institutionalized privilege. Political campaigns are increasingly dominated by the big money used to finance competing image-creation strategies of media consulting firms. Outside of government, the freedom to organize collectively has been used most effectively by giant transnational corporations, officially defined as "persons" under the present interpretations of the 14th amendment in the United States and similar edicts in other countries. The largest of these collectivities (along with their trade associations and multitudinous advisers) have the power to win entitlements from local, state, national and foreign governments by operating through, alongside or over - not merely under - them. Through their global reach, they have effectively reduced the counter-vailing power of labor unions, all of which act nationally or occasionally, on a continental basis. In many countries, including the United States, they have curbed the spread of blue collar or white collar unionism. Public and private bureaucrats have mastered the ritualistic arts of giving other people an illusory "sense of participation," while seeing to it that their own power remains undiminished. Many representatives of vested interests establish monopolies of seats on so called "representative" boards. Freedom of information laws tend to be used mainly by business representatives seeking to uncover the plans of private competitors or public controllers.

Thus, has it come about that the "Have Most" in the world do the most in using these remarkable devices. The "Have Littles" and "Have Nots" usually do the least. Indeed, some of their self-designated spokes-people preach abstentionism from entrapment in what they label "bourgeois democracy." In so doing, they unwittingly go along with the dominant logic in most countries of constitutional capitalism : the drift toward a new corporatism of interwoven power in the hands of Big Government and Big Business. This drift is openly encouraged if not even legitimated by the many technocrats and academic elitists who feel that ordinary people could never understand the facts of life on such complex subjects as war, unemployment, inflation, crime and prejudice. Their own difficulties in handling these matters rarely interfere with their willingness to be annointed as media gurus.

Democratic Currents

On the other hand, anyone who looks carefully can also find an enormous variety of democratic currents. Most of these take form outside of government in that diverse domain of civic society encompassed by "mediating organizations." Many voluntary organizations now assume functions of self-service and consciousness-raising, as well as advocacy. Through groups of this type, senior citizens assert their ability to give more to society rather than accept consignment to the junk heaps of profiteering nursing homes. Racial and ethnic minorities win freedom from the most blatant forms of prejudice, while preparing new attacks on respectable racism. Women achieve important bits of freedom from entrenched patriarchy, while moving ahead in the more difficult battle against subtle sexism. Millions of people in churches, synagogues and temples transcend old confinements to Sunday and Sabbath morality with courageous attention to the morality of behavior during the rest of the week. As part of a large but nebulous anti-war movement, they have exposed much of the suicidal behavior of the more aggressive militarists. An environmental movement has created a widespread consciousness of the need to conserve the land, water and air resources of the planet. Civil liberties organizations have defended the Bill of Rights against repeated attacks on constitutional protections against government invasions of personal freedoms. In some areas particularly among teachers and public employees -labor organization has spread. Part of the labor movement has moved beyond bread-and-butter unionism and concerned itself with issues of workplace democracy, the fate of the unemployed and the larger issues of social and economic policy. In low-income and middle-income neighborhoods, community groups have often started to rival the remarkable levels of civic action heretofore found only in high-income neighborhoods. Called "the backyard revolution" in Harry Boyte's book under that title, this neighborhood movement has brought people from all parts of the "Right-Middle-Left" political spectrum together on local issues of common concern.

In most of these democratic currents, moreover, there is a new awareness that on most issues the Left is bereft, the Middle muddled and the Right wrong. If this one-dimensional classification had some meaning when first used almost 200 years ago in the French assembly, it means much less in today's multi-dimensional world. To identify the position of a person or a group (let alone a party or candidate), many dimensions are needed not only attitudes towards more or less government, but also approaches to civil liberties, violence, sexism, racism, crime and various combinations of decentralization and centralization. A still more important set of dimensions is provided by the many people who - while divided on other issues - are fed up with the elitism, snobbery and overdone hierarchy in most private and public bureaucracies and are inventing more democratic and non-charismatic styles of leadership. Most of these people like to think that by practicing "bottom-up" decision-making and communication they can counteract "top-down" elitism.

The Limitations of "Bottom-Up"

We demur. Bottom-up efforts usually imply a tacit acceptance of verticality that is, a Jacob's ladder going up from the "lower" or "inferior" classes or "levels" of people to those who are regarded as "higher" or "superior" beings. One of us was born into the "ruler" (or "warrior") caste in India. She learned at an early age that many people of superior intelligence, morality and aesthetic taste were to be found not only among Kshatriyas or Brahmins but also among the Vaishas (tradesmen), Sudras (menial workers) and Harijans (outcastes or untouchables). The other worked many years in the Upper House of the U.S. Congress, the President's executive office, and local government. He learned that many of the best legislators are in the Lower House, that people in the White House may lack the most rudimentary culture or morality and that local government may often represent the highest form of public service. Both of us have learned that in the halls of so-called higher (really tertiary) education, office workers and custodial staff may have more common sense and moral judgment those tenured professors. We are trying to expunge from our minds the very idea of lower and higher levels of government, with all their invidious connotations. After all, the real difference between cities, countries state and the federal government is the size of their geographical areas. While small may not always be beautiful, larger does not necessarily imply brainier or better.

Also, the senders of messages from the bottom up often look upwards in breathless anticipation ofcrumbs, compassion (the tribute that superiors may charitably pay to inferiors) or cooperation. Instead of looking sideways toward other people with common interests, they accept the societal fragmentation imposed by bureaucratic specialization. They speak the language of *single* issues, closing their eyes to the interweaving among most important issues. In neighborhood or urban affairs, when extending their consciousness to *many* issues, they usually confine themselves to a single small area. This breeds a parochial approach to many problems that are global and national as well as local.

The Importance of Bottom Sideways

The essential alternative to top-down, in our judgment, is "bottom-sideways." When Mohandas Gandhi confronted the top

British rulers of India, he developed a bottom-sideways strategy of working with hundreds of millions of illiterate peasants, workers and tradesmen. He and his associates spread out across the country and without access to any mass media (all controlled by the rulers at the tip of the imperial pyramid) achieved mass communication. This meant trying to use all available means of interpersonal, organizational and political communication to transcend the divisions imposed by religion, caste, language, gender and social role. To the extent that this bottom-sideways outreach was successful, Gandhi's messages up the ladder of the British monarchy were no longer humble petitions from submissive subjects; they had power behind them.

Similarly, in any country of constitutional democracy, the bottom-up machinery of formal representations can yield more genuine representation when people and groups reach out to each other across the many boundaries that divide us. This may yet prove to be the only way to forestall that crystalization of top-top communication and cohesion among the Have Most which would be the essence of a new corporatism dominating the machinery of democracy.

References

This phrase is Woodrow Wilson's famous "open covenants openly arrived at" adjusted to include decisions other than treaties and express more fully recent approaches to "goldfish howl" democracy.

This new corporatism has been referred to as "apple pie authoritarianism" by Kevin Phillips *(Post - Conservative America,* New York; Random House, 1982) and "friendly fascism" or "the repressive conjunction of Big Business and Big Government" by Bertram Gross *(Friendly Fascism,* Boston: South End Press, 1982).

This view is more fully articulated in "People Against Charisma" by Kusum Singh *(Communicator,* Quarterly Journal of the Indian Institute of Mass Communication, New Delhi, India, October 1981).

KUSUM SINGH, "Gandhi and Mao as Mass Communicators," *Journal of Communication,* Summer 1979.

Chapter **4**

Development Support Communication : A brief review of the concept

B. S. S. Rao

Communication cannot be seen in isolation, but as a part of the total development system. Thus Communication cannot be dealt with by itself but only in relation to the political and socio-economic structures and cultural pattern. It is important to avoid a simplistic relation between Communication and the development effort.

The origin of developmental communication dates back to the application of agricultural extension education during 1945-50, launched in most of the developing countries. Initially it was popularised as agricultural communication. Later, when the scope of extension effort was enlarged to include other areas like industry, health, social education etc., it became popular as rural communication. With the emergence of development as a distinct concept in the international arena, communication applied for initiating and accelerating the process of development, branched off as a serious study under the discipline of development support communication (DSC) or developmental communication.

The 1948 UN Conference on Freedom of Information which described free and adequate information as the touchstone of all freedoms had sown the seeds of developmental communication. Developing countries had set their goals to build infrastructures for free and adequate flow of information without which political articulation, economic growth, and social justice cannot be achieved.

The subsequent UN General Assembly resolutions of1953 which emphasized the role of information media in education, economic and social progress, provided a fillip for a faster growth of developmental communication. The idea that development is dependent on effectiveness of communication, and it is conditioned to large extent by the character of communication, gained momentum, it was widely appreciated that; (a) wherever communication existed on a planned level, development has been accelerated, and (b) major developmental programmes will not reach their targets without adequate and planned communication inputs.

Media and development

Based on his studies during 1950-51 in the middle East, (covering Iran, Egypt, Turkey, Syria, Lebanon, and Jordan), Daniel Lerner (1958) pointed out that "mass communication serves as the great multiplier in development, the device that can spread the requisite knowledge and attitudes immeasurably more quickly and widely than ever before". Lerner's analysis revealed that there is high correlation between media growth and national growth. He argued that the person who changes in a developing society is usually "a mobile personality" - one who has high capacity to identify with new aspects of his environment, and who can take in stride the new demands upon himself that are out of his habitual experience. According to Lerner, the first element in the social dynamic of development is the infusion of modern or mobile personality the second element is the mobile multiplier: the mass media.

Lerner analysed data for 54 countries and showed the high correlation between the four measures : urbanization, literacy, media participation, and political participation. Schramm and Carter (1959) similarly found correlations of 72 and 74 respectively between the scale of economic development and other scales representing the development of mass media receiving and propagating systems, in 100 countries.

Leonard Doob's study in Africa (1961) and YVL Rao's study in India (1966) are generally cited as further works corroborating Learner's hypotheses on communication and development. Both the studies explain how extensive exposure to mass media helps bring about change in the traditional society.

Rao, who studied two villages in Andhra Pradesh in two different setting, was convinced that "where the channels of information are varied and broad based like newspapers and radio, etc., the changes resulting from economic, social or political ideas creeping into the society are smooth. Where the channels of information are tight and controlled by few, changes are difficult to make and often lead to a worsening of the factions". He argued that the amount of information available and the wideness of its distribution are key factors determining the speed and smoothness of development, While according to Lerner mass communication is the great multiplier of ideas and information, according to Rao it is the great smoothener of transition. Rao concluded that it is futile to attempt to treat the interaction between the communication and development as casual relationship because the interaction is constant and cumulative.

Wilbur Schramm's study of the role of information in the developing countries (1964) is the first major treatise on the subject. The analysis in this report provides the backdrop for the evolution of a distinct philosophy and content for developmental communication. Hence, some of the broad conclusions of the study are reproduced below :

1. Wherever new skills and new attitudes are to be learned, wherever people need to decide new things, wherever traditional society trembles on the verge of modernizing, there a more efficient form of communication can be helpful..... Therefore, the efficiency of communication is a consideration in determining the speed and smoothness of national development.
2. The flow of information is sadly inadequate. The situation is particularly cruel for the countries that are in early state of development..... It greatly thins out as it moves from cities towards villages, and still more as it feeds-back from the villages to the centre.
3. The whole information system, not a particular channel or a combination of channels, needs to be planned to fight the battle of development.
4. Investment in information, in a developing country, is investment in the most essential social and economic changes which make up national development.

Pradipto Roy, Rogers*et. al* (1969) conducted a comparative study on the impact of communication on development in Costa Rica and India. Communication treatment with Radio Rural Forums, Literacy-cum-Reading Forums, and Animation Leadership Training was given to 20 selected innovations in agriculture, health and social education. The study revealed significant correlation between the degree of participation and exposure to the Radio Forum treatment and amount of change in the important variables among the forum participants.

The Radio Forum treatment not only accelerated the growth of knowledge and adoption, but also discouraged discontinuance of innovations.

Kivlin, Fliegal *et. al* (1971) conducted another study on the spread of modern agriculture and health practices in India under a three nation project and concluded, among others, that villages with greater access to and contact with mass media channels were more likely to score high on agricultural adoption. They also pointed out that mass media channels were widely used and that promotional effort via them were likely to be successful.

Communication tasks in development

Nora C. Quebral (AMIC : 1976) expressed the opinion that development communication is : "the art and science of human communication applied to the speedy transformation of a country and mass of its people from poverty to a dynamic state of economic growth that makes possible greater social equality and the greater fulfilment of the human potentialities". This provides a broader perspective of the communication tasks in development. But the specific tasks need to be analysed from the practical life situations of developing countries.

Schramm who is considered the father of developmental communication distinguished three kinds of tasks for communication in national development : (1) people must be given information about national development, their attention must be focused on the need for change, the methods and means of change, and if possible their aspirations must be raised; (2) there must be an opportunity to participate meaningfully in the decision making process the leaders must have an opportunity to lead and the common people to be heard; (3) information flow must be both up and down the hierarchy; and (4) the needed skills must be taught, farmers must learn the new methods of farming, adults must be educated, and extension workers must be trained.

Developmental communication is not un-dimensional. It envisages multi-dimensional linkages for ensuring free and adequate flow of information, and building interaction between the sources and the users of the knowledge. It implies linkages between all agencies involved in development work, such as planners, administrators, subject specialists, field workers, local leaders, media representatives, researchers and individuals who constitute the final consumers of the information. Thus, the routes of communication envisaged are not only vertical from top to bottom and bottom to top, but also horizontal between the institutions and personnel connected with developmental work. The networking also presupposes constant monitoring and redrawing of communication strategies to eliminate seepage in the effort.

John L. Woods (1976) conceived a triangular nexus with three points : knowledge generators, political leaders and development knowledge users as shown in the figure on the next page.

Wood's model of developmental communication

The role of developmental communication, according to Wood, is to link all the three elements in the development linkage triangle *plus* all the intermediate audience groups. He desired emphasis not on pushing the information towards the target audience, but on taking into accounts the information seeking patterns of the target audience, and integrating them into the development planning process.

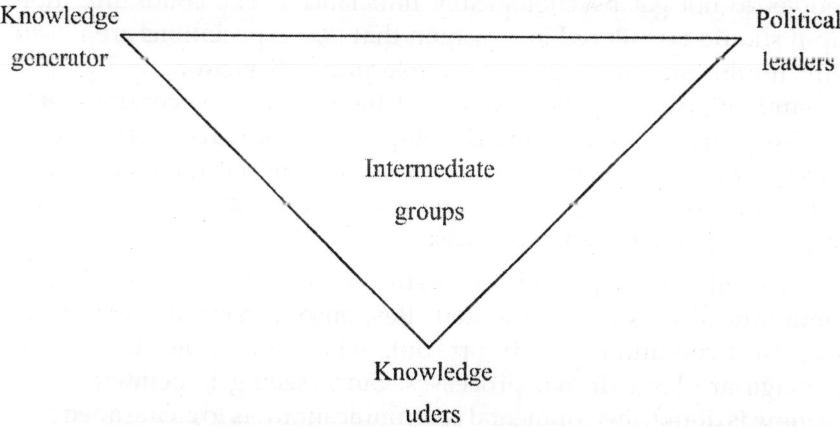

One essential prerequisite of developmental communication is that it should not be tied to any specific channel or communication technique. It must be able to exploit fruitfully informal and institutionalized networks, for promoting interaction through interpersonal means by word of mouth, by folk and traditional channels, as well as mass media. The strategy of integrated communication is gaining popularity. According to this concept integration of communication inputs is envisaged at three levels : (a) interpersonal; (b) mass media; and (c) inter-agency and within agency communication. Even though the decisive role in development is played by the interpersonal channels interacting at the point of delivery, the philosophy in this strategy is to evolve coordinated system to avoid isolated efforts, and divergent messages that are likely to create confusion. The system also envisages selection of right type of a communication channel or channels in combination (which is now popular as media mix) for specific informational input at the right moment.

The Seminar on Developmental Communication organized by the Asian Mass Communication Research and Information Centre (AMIC) identified two prerequisites : (1) shift from dissatisfaction to satisfaction; and (2) psychology of completeness (AMIC : 1976). The first prerequisite stresses that information delivery or transfer is only a prelude. It is the motivation which is the basic citade of development communication. Without adequate motivation and persuasion it would be impossible to make people move along certain desired lines. Therefore, it is incumbent on development communication to affect the information transfer and behavioural change in a way that the people do not get psychologically imbalanced. The communication input should be tailored in a fashion that : *(a)* expectations are raised with minimum frustration and maximum satisfaction; and *(b)* the dynamic equilibrium is maintained between socio-economic and psycho-cultural components. Development effort raises aspirations, calls for new adjustments and implies discarding of traditional modes. Communication effort needs to be geared to minimise the irritants and smoothen the change process.

The other presupposition lays stress on the results. Development communication is result-oriented. Response is the indicator of the success of communication effort. But, action and application of the message are long drawn processes. Since seeing is believing and learning is doing developmental communication has to create adequate opportunities for thinking, seeing, and learning new innovations. It is, therefore, essential to take up only such projects which have the backing of the infrastructure and launch a coordinated strategy to achieve the results to the fullest satisfaction of all.

The Asian Regional Seminar (1979) on the theme of Rural Development and Communication policies jointly sponsored by the Indian Institute of Mass Communication and the International Institute of Communications, concluded : "communication cannot be seen in isolation, but as a part of the total development system. Thus, communication cannot be dealt with by itself but only in relation to the political and socio-economic structures and cultural patterns. It is important to avoid a simplistic relation between communication and the development effort."

To conclude, it may be stated that development is a dynamic process. As such communication intervention to stimulate change and development has to be equally dynamic. There cannot be a set pattern of fixed formulae. The means of communication have to change with the change in perspectives of development.

References

AMIC, Study Seminar on Development Communication, Ed. Guy-do-Fontagaland, Asian Mass Communication Research and Information Centre, Singapore, 1976.

DOOB, LEONARD. W., "Communication in Africa : A search for boundaries," Yale University Press, New Haven, 1961.

KIVLIN. JOSEPH et. al., Innovation in Rural India, Bowling Green University, Ohio, 1971.

IIMC, Rural Development and Communication Policies, papers and proceedings of Seminar organised by IIMC and the International Institute *of* Communication, Indian Institute of Mass Communication, New Delhi, 1980.

LERNER, DANIEL, The Passing of Traditional Society, Free Press. Glenco, 1958.

RAO, YVL, Communication and Development, University of Minesota Press, Mineapolis, 1966.

ROY, PRADIPTO *el. al.* "The Impact of Communication on Development," National Institute of Community Development, Hyderabad, 1969.

SCHRAMM, W1LBUR, "Mass Media and National Development," UNESCO, Paris, 1964.

SCHRAMM, W. & CARTER, R. F., "Scales for describing National Communication Systems", Institute for Communication Research, Stanford, 1959.

WOODS, JOHN, L., "Report on the International Conference on Integrated Communication for Rural Development." East-West Centre, Honolulu, Hawaii, (1976).

Chapter **5**

Development Communication : A Historical and Contemporary Perspective

Binod C. Agrawal

In more than one way, my presentation to-day is an exercise in self indulgence of what I have witnessed in last two decades in the field of development communication which is also not more than 20 years old. I have attempted a synoptic view rather than a synthesis of what has happened in national and global perspective. The novel of "development communication" has beencondensed into a short story without loosing the essence. The aim is to bring out the main features and the diversities in the multilinear evolution of the emergent field of development communication within the discipline of communication.

In India, communication applications and research in its broadest sense for development is a post Independence phenomena. Hence, its growth must be seen in response to several converging needs, and meeting points of multiple divergent theoretical and methodological approaches for improving socio-economic life of the poor and weak. The phrase now is "better quality of life". The canvas is very large from adoption of non-conventional energy to contraceptives, and awareness of environment to consumer education. The preview that follows is also selective to be pointed and penetrating. In doing so, I must not be accused of being negligent, this being a review exercise that too a purposive one.

Multiple Meaning and Use :

Few weeks ago, I received a beautifully done report of great national concern and international implications. It relates to National AIDS Control Programme in India. The report has been brought out by National AIDS Control Organization in April 1993. The report provides a detailed status report on the alarming rate of spread of Human Immunodeficiency Virus (HIV) in India and the national level efforts to check the spread through Information Education and Communication (IEC) approach. The report provides four levels of strategies to combat AIDS HIV and STD. While outlining the communication strategy, the authors recognize "Communicating about AIDS/HIV and STDs is extremely difficult as such communications necessitate discussion of sexual practices, a topic many people in many cultures would rather leave undiscussed" (National AIDS Control Programme 1993 : 28).

Availability and procurement of drinking water is another national problem requiring multi-disciplinary efforts to combat. Among many disciplines involved in solving water problem, communication has been considered as one of the critical element in appropriate utilization of water for health and sanitation. Recently, Government of Karnataka initiated an Integrated Rural Water Supply and Environmental Sanitation project with the assistance of World Bank. Strategic planning and implementation in health communications has been thought essential for improving the health and sanitation of the people.

In 1990, as a part of National Literacy Mission, an adult literacy programme was initiated which utilized radio to accelerate the process of adult learning. This effort was named as Project in Radio Education for Adult Literacy (PREAL). The PREAL covered four Hindi speaking States of India. Spread over 3600 villages, the target group for the adult literacy programme were rural women.

These three development projects can easily be seen as development communication projects in the present context of India and in the world. In all three projects, externally generated information would be disseminated to bring about desired change in knowledge and behaviour. For this purpose, variety of means and methods of communications would be utilized to convey the message to reach to the target group.

As a matter of fact, where I live in Ahmedabad, a Master degree is awarded in development communication at the Centre for Development Communication, Gujarat University. A similar centre, with an identical name, has been in existence for more than a decade, at Hyderabad, assisting a number of national and international agencies to reach to unapproachable, unreachable mass of humanity through various means of communications for their development. I can continue giving similar illustrations from across the world. My intention here is to indicate the scope of development communication.

In a variety of cultural contexts, communication has been considered as a critical element in achieving higher agricultural productivity, acceptance of birth control methods, health and hygiene practices. Discussion around these ideas has been for more than four decades and especially after the Second World War on communication and development horizon. All these have been referred to by a variety of names which has now been broadly accepted as development communication.

Some of the Assumptions :

What are some of the basic assumptions in the use of communication for development? Let me identify a few, before I move further.

Human behaviour has built in plasticity which can be molded, modified and changed through media intervention by persuasion or fear or by creating conducive conditions. They can also be advocated for participation.

It respects individual freedom, free choice and creates congenial conditions in which new ideas could be experimented or accepted.

Exceptions apart, societies in which such external information interventions have been attempted are heterogeneous in its social composition. Some sort of social stratification and built in inequality also exists in these societies.

Not all members of a given society are having the same level of information and awareness about any issue.

Increased communication would lead to change in behaviour :

With some of these broad assumptions in view most communication efforts have been attempted to bring about change in the human condition. Let me illustrate from a beautiful publication, which came out from this college Studies in Communication (1988). It contains abstracts of "over two decades of research and extension in rural and urban communities" (Anandalakshmy 1988 : v), Broadly based on the above, assumptions these studies bring out some of the salient aspects of development communication.

Evolution and Continuing Confusion :

Credit to coin the concept of development communication, to best of my knowledge, goes to an Asian Professor Nora C. Quebral. Professor Quebral in 1980 told me that she wrote the first article on the concept of development communication, after practicing it for few years in Media Asia (see Quebral 1975). Actually this article is based on the material she had earlier presented in a course at Singapore in 1973 (see Quebral 1973). About ten years later she wrote another article on the concept of development communication in an Indian journal Interaction (Quebral 1983). This article contains elaboration and refinement of her thoughts on her earlier work. I must mention in passing that very few Euro-American workers and practitioners have recognized the contributions of Quebral in this field.

I personally became aware of the concept of development communication in late seventies. It is an interesting little-story. I prepared a training proposal for development communication researchers in 1979. My superior after reading the proposal thought that I had made a typing error and had typed "development" instead of "developmental". It took an effort to convince that the term development communication was used most appropriately in the proposal. This terminological confusion has continued even to-day to a large extent as later events have shown. The word developmental was used by National Council of Development Communication (NCDC) until 1990, and Gujarat University Diploma had similar title also during the same period. Several research workers continued to use the term in some what loose manner (see for example Rao 1992). It is the external academic pressure thatmade many in India to accept the concept of development communication to confirm with the rest of the world. It must be mentioned in passing that even to-day these two terms are being used interchangeably and find place in the literature of development communication. Another term often used is development support communication, though the same is being questioned.

In the context of India, the largest number of papers on development communication can be found in the journal Interaction published by National Council of Development Communication, Varanasi. Though the journal started as another agriculture education publication it has assumed the status of a multi-focal development communication clearing house of national and international researches. An older journal in this area is the Indian Journal of Extension Education published since 1965.

Agriculture and Development Communication :

Extension Education as an academic discipline in its many incarnations, has the credit to highlight the role of communication in development. No doubt the entire extension education in agriculture, home science, animal husbandry, fisheries, horticulture and poultry was a direct American import in the post independent India to help increase production. The initial theoretical framework for this purpose was directly borrowed. Of course the Indian researches mostly in the agriculture and home science colleges and universities spent a great deal of their time, in disproving, modifying and refining the "adoption process" model originally postulated by a committee of rural sociologists of United States of America in 1955 (Rogers and Shoemaker 1971). Literally many thousands Masters and Ph. D. thesis have been written using adoption process model in last four decades all over world in which India has contributed a substantial number. I must mention in passing that this model while rejected by its founders, in India, a significant number of researchers continue to use this model to assess communication effects not only in agriculture but in several other development programmes (for detail discussion see Singh 1984 and Chu 1987).

While turning the leaf of communication and agriculture history, I was most intrigued that although the concept of development communication was conceived by an extension education person like Nora C. Quebral seldom the same has been referred by the development agents, scholars and planners of agricultural communication. One of the recent books where a great deal of recent communication research approaches for agricultural and rural development have been discussed is entitled Communication and Indian Agriculture (see Ostman ed. 1989) in which no mention has been made of Quebral.

In spite of this remark, I must say Ostman (1989) provides valuable insights in two areas of development communication. First the American original assumption about the key role that communication can play in the rural development has been reaffirmed. Due to recent advances in the communication technologies there is a renewed vigor in the use of communication for development. Second, a great deal of agricultural researchers continue to follow older framework for their development activities.

In the context of India, Colle (1989) has tried to outline some of the newer areas of development communication to be explored in

Twenty First Century. The key issue of development communication in agriculture will continue to be higher agricultural production with an added caution on the use of chemicals, injurious to humans and animals alike.

IEC for Health and Family Planning :

Information, Education and Communication (IEC) remained the corner stone of development communication in family planning and health since the family programme was initiated way back in 1952, though it took a visible shape in 1966. While discussing communication and family planning, Kakar (1987) saw communication in relation to development. "By and large it means dissemination of information in a manner as has the potential to influence people's thought process and to change their behaviour". (Kakar 1987 : xvii). The basic assumptions about human responses to externally generated information on family planning have not been questioned or altered as seen in the illustration given in case of AIDS. Instead a great deal of emphasis has been laid on the nature and content of the messages and delivery system or media. The entire family planning programme, largely supported by the Central Government, followed "all media" approach to create awareness about the importance of family planning and economic and social benefits in limiting the family size. At another level, the communication thrust has been in providing knowledge about methods of family planning followed by adoption of family planning methods both by men and women. IEC approach represents three distinct but interrelated steps in development communication. This can be clearly observed in a communication study carried at national level to measure effectiveness of media by National Institute of Health and Family Welfare, New Delhi (see Technical Report 9, 1985). While little change in Philosophy and approach of communication has been observed, departure from "family planning" to "family welfare" is note worthy thereby enlarging the scope of development communication for the welfare of the family.

While literature is available on Integrated Child Development Services (ICDS) in which IEC approach has been followed to reach to millions of children and their mothers, relatively fewer studies are reported on development communication (see Punhani and Mahajan (1989). Johti (1989 : 147) has recommended the use of mass media in ICDS programme. Rewal and Sadana (1989 : 2) concluded that "the combination of a mass educational campaign and face-to-face communication has been successful in creating awareness and adoption of a novel practice in the management of diarrhoea "This was reported from Gujarat where oral dehydration therapy **(ORT)** was promoted in 1236 ICDS programme villages. The interesting part is that IEC approach was by and large followed in this programme also.

Development Communication : Women and Children :

Two publications recently appeared in India where there is direct reference to communication for the development of women and children). (Thakur and Agrawal 1989 and Agrawal and Sinha 1986).

Both these Publications emerged out of UNICEF supported seminars held in Ahmedabad and Delhi.

Similar efforts were made by National Institute of Public Co-operation and Child Development, New Delhi in mid eighties where use of media was discussed in the development programmes related to children. *Marketing Women's Health Through Television* (1987), focused especially on women for family planning programme (see Shah 1987). Further, most of the Masters thesis written in the department of Home Science Extension Education has covered one aspect or other of women and children. All these have contributed in the understanding of the role of communication in development. Most of the analyses have been carried out using adoption process model discussed earlier.

A book on development communication related to women appeared in 1988 (See Agrawal and Rai 1988). The authors observed that "Communication for development in most cases, assumes that the major burden for bringing about desirable changes in rural life remains man's responsibility, though women equally participate in it" (Agrawal and Rai 1988 : 84). It w as observed "that the communication boundaries, in most cases, are separate for men and women" (Agrawal and Rai 1988 : 91). Based on an in depth study the authors concluded that the information needs of rural women were comparatively much more than urban women and men. Hence, development communication for women should be exclusive and separate (Agrawal and Rai 1988 : 91).

These findings support the views expressed in the report *Shramshakti* (1988) where nineteen studies have been reported related to women communication and development. The report observes that "A crucial shortcoming in the existing machinery relates to the lack of any effective mechanism for dissemination of information among women regarding women's development programme (Shramshakti 1988 : 15). SEWA members have used video as an aid to reflect women's problems and as a tool for development.

Compared to development communication in agriculture, health and family planning, little is know about women and children. It would be one of the most important areas to be explored for organizing women to empower them for total human development (Ramiraz 1987).

Development as Communication :

An insightful critique of development communication was brought out by Narula and Pearce (1986). The authors while analyzing development efforts of government in Delhi areas included a variety of actions as part of communication. Narula and Pearce (1986 : 5) suggested "communication perspective" which included "all of the actions taken by the government legislation, public works, and the development bureaucracy itself as well as the IEC program are powerful forms of communication." Hence, they have recommended that the concept of development communication must "include explicit statements about relationship between government and people" (Narula and Pearce 1986 : 206). Dialogue Action Strategy has been suggested to achieve the goals of development even though this strategy might interrupt the existingpatterns of development process. This excellent research work has not found much acceptance among the workers of development communication in the last decade in the country.

Development Communication : The Chinese Experience :

In 1976, a group of communication and development experts met in Honolulu and discussed in detail the China's experience with development communication. The question these scholars raised : How transferable was development communication experience of China for other countries?

"It has become almost trail to say that China's experience in development communication is impressive and intriguing but how instructive it is to developing countries remains unclear. (Schramm, Chu and Yu 1976 : 87). The authors indicated that Chinese basically depended upon interpersonal communication face to face talk, the discussion the decision making group and the meeting (1976 : 90). The important element in the communication approach was nearly 20 million cadres who acted as chief channel to inform and guide. Responsibility for development was given to the local community.

The "engine" that drove development was social pressure and moral generated by local communication and cooperation. Mass media though not neglected mostly used for reaching to cadres to hear policy among other things. In the opinion of the authors, it was not possible to replicate China's approach of development communication in other countries.

Seventeen years later the same question was examined by a Chinese student of development communication Ling Chen (1991) who used "multiplicity paradigm" for this purpose. Chen (1991 : 113) concluded that "The multi-channel approach to communication (as advocated by the multiplicity paradigm) so far has worked well for rural development in China. Mass media and traditional face-to-face contact have functioned in complement of each other. No large-scale investment in expressive communication technologies is undertaken beyond current need and financial ability."

In spite of historical changes, shift in ideology, opening up of China to outside world and reform movement, after 1979, the development communication approach by and large remained same. Development communication, if examined, it emphasizes the China's unique characteristics, a long cultural tradition and socialist political system that yielded development results. These conditions are hard to come by in many countries so is Chinese development communication approach (Chu 1987. 98).

Development Communication : The Indian Perspective

The development communication experiences of China has been narrated to compare the Indian situation for bringing out the diversity created by political and cultural history. For this purpose I have chosen three recent books published in last eight years related to development communication. All of them have been written by social anthropologists who now are working as development Communication experts (see Sinha 1985, Ambekar [Yadav] 1992; and Rao 1992). The choice of these books are only to discuss the current state in development communication in India. All the three studies have been carried out in rural geographic regions of India namely Bihar (Sinha 1985), Karnataka (Ambekar 1992) and Maharashtra (Rao 1992). In each case television was used as a primary source of communication for development. Each study has analyzed long terms effects of television on rural development. The areas of development include agriculture, animal husbandry, health nutrition and family planning.

While Sinha (1985) and Ambekar (1992) have made several visits to their respective villages over a span of ten years, Rao (1992) collected his data in eighties from five villages of Maharashtra. All of them were involved in Satellite Instructional Television Experiment (SITE) and the inspiration to work in development communication came from SITE. Sinha (1985) and others visualized communication as process hence, carried an in depth qualitative study.

What kinds of new insights have this researcher provided? Some in depth analytical description and spread of externally generated television messages They concluded that it helped peasants adopt innovations in agriculture and animal husbandry, health and nutrition and family planning. Each of these authors clearly brought out that messages did not move in a linear direction. Process of communication was intricately interlinked and interwoven within the village social structure. There were multiple foci of sources of information in which, different human sources, acted as subject specific experts. Another view that emerged from these analyses relate to the notion of "communication alliance". (See Agrawal 1982) within the family, jati and village.

Development communication as a concept is yet to be fully internalized both by scholars and practitioners. As late as 1988 while talking about development communication, Eapen (1988 : 67) refers to development support communication. His diagnosis and analysis certainly opens up the pandora of problems related to the inherent conflicts and contradictions between the stated policy of the state regarding electronic media and actual use for development. "Indian development strategy emphasizes attitudinal change, motivation, and participation. Communication is regarded as a mobilization mechanism," (Dube. 1976 : 315) Further Karnik (1992) added that communication in the context of India should be used for agenda setting and for creating an 'ambience'. Development communication as it stands to day is multidisciplinary in approach, content and goal in which almost all disciplines of social sciences have contributed which has anything to do with, development and change.

Development communication in India has yet to come out of the old paradigm. The privatization of communication, rapid changes in communication technologies and increasing influence of the concept of social marketing might change the direction of development communication. That will be an unfortunate event for all those who are committed to development.

References :

Agrawal, Binod C. (1982) *Communication Structures in India* : A Case Study of Television. Communicator Vol. 17 No. 4.

Agrawal, Binod C. and Kunkum Rai (1988) *Women Television and Rural Development*. Delhi : National Publishing House.

Agrawal, Binod C. and Arbind Sinha (1986) *SITE TO INSAT*. New Delhi : Concept Publishing Co.

Ambekar, (Yadav) J. B. (1992) *Communication and Rural Development*. New Delhi : Mittal Publications.

Anandalakshmy, S, (1988) Preface. *In Studies in Communication*. Mridula Seth ed., pp.v. New Delhi : Lady Irwin College.

Chen, Ling. *The Door Opens to a Thousand Bloosoms* : A Preliminary Study of Communication and Rural Development in China (1970-88). Asian Journal of Communication Vol. 1 No. 2 pp. 103-121.

Chu, Godwin C. (1987) *Development Communication in the Year 200*. Future Trends and Directions. In Rethinking Development Communication. N. Jayaweera and S. Amunugama, eds., pp. 95-107. Singapore : The Asian Mass Communication Research and Information Centre (AM1C).

Colle, Royal D. (1989) Preparing for the Twenty-First Century in Indian Communication Research. *In Communication and Indian Agriculture*, Ronald E. Ostman ed., pp. 271-291. New Delhi : Sage Publications.

Dube, S. C. (1976) *Development Change and Communication in India*. In Communication and Change : The last Ten Years-And The Next. W. Schramm and D. Lerner eds., pp. 98-118, Honolulu : East-West Center.

Eapen, K. E. (1988) *Development Communication* : A Country Cousin Syndrome. ICCTR Journal Vol. 1 No. 1, pp. 67-74.

Johri, A.C. (1989) *Changing Communication and Media Technology in IODS Projects in India*. In Research on ICDS : An Overview, by Rita Punhari and Rachna Mahajan, pp. 446-447. New Delhi : NIPCD.

Kakar, V.N. (1987) *Communication in Family Planning* : India's Experience. New Delhi : National Institute of Health and Family Welfare.

Karnik, Kiran (1992) *Communication, Technologies and Population*. In Science Population and Development, Vasant Gowariker ed., pp. 308-327. Pune : Unmesh Communications.

Narula, Uma and W. Barnett Pearce (1986) *Development as Communication*. A Perspective on India. Carbondale : Southern Illinois University Press.

National AIDS Control Programme India (1993) *Country Scenario* : An Update. New Delhi : National AIDS Control Organizations, Ministry of Health

& Family Planning, Government of India.

Ostman, Ronald E ed, (1989) *Communication and Indian Agriculture.* New Delhi : Sage Publications.

Punhari, Rita and Rachna Mahajan (1988) *Research on ICDS* : An Overview volume I (1975-85). New Delhi : National Institute of Public Cooperation and Child Development.

Quebral, Nora C (1973) *Teaching Communication for Development.* In AM1C Refresher Course On Mass Communication Research and Information Centre (AMIC).

(1975) *Development Communication :* Where does it Stand Today. Media Asia Vol. 2 No. 4.

(1983) *Concepts and Practices of Development Communication* Interaction Vol. No. 2 pp. 97-114.

Rao, B.S.S. (1992) *Television for Rural Development.* New Delhi : Concept Publishing Company.

Ramiraz, Mina (1987) *Devcom and Community Organizing Process - A* Philippine Experience. In Rethinking Development Communications, N. Jayweera and S. Amunugama eds., pp. 227-238. Singapore : AMIC.

Rewal, Sneh and Rajeev Sadana (1989) *Promotion of Oral Rehydration Therapy, An Educational Strategy.* New Delhi : CARE India.

Regers, Everett M with F Floyd Shosemaker (1991) *Communication of Innovations* : A Cross Cultural Approach. New York : The Free Press (Second Edition).

Chapter **6**

New Strategy in Communication and Development

Indu Bhusan Singh

This paper is an attempt to point out the main fallacies in Western thinking about the role of communication in development. The article proposes strategies related to new conceptual framework for constructing models of development communication and new communication research designs that are methodologically rigorous and ideologically relevant. An in depth and logical analysis of development communication is highlighted with the help of exhaustive reviews to support and substantiate the various ideas indicated in the article.

A general optimism of the past two decades that mass communication can foster rapid development in the Third World countries has ended with the beginningof a new decade of the 1980's. The all powerful mass media and their predicted influence have failed to "multiply" the process of development "magically." With the exception of the impact of radio broadcasting, which is worth noting, Western communication scholars are finally settling for a subdued influence of the mass media, in development.

The explanation for this seeming failure of the mass media in development does not lie in the nature of the technologies per se, but in predictions about the use of these technologies. The main cause of failure may lie in the flaws related to the conceptual framework of Western development communication models.

This paper is an attempt to point out the main fallacies in Western thinking about the role of communication in development. More specifically, it subscribes to the following criticisms : (a) The Western models of development and development communication are inadequate and sometimes irrelevant for the Third World countries, and (b) development communication research based on these models have been misconceptualized and misdirected.

This article proposes two new strategies : (a) new conceptual frameworks for constructing models of development and development communication, and (b) new communication research designs that are methodologically rigorous andideologically relevant.

What is Development?

The old paradigm of development, which was functional until the mid-seventies was based on a single major criterion of economic growth. Most economists and development theorists attempted to define and measure development in terms of the gross national product (GNP) and per capita income.

Many definitions of development are also closely linked to the principles of Western liberal democracy and capitalism. "Democracy, training, modern factories, more money - these words sum up the major development policies of the Western democracy," Rogers has summarized four academic, and historical influences on the old paradigm of development. They are : (1) economic growth through industrialisation and accompanying urbanization; (2) capital-intensive, labor-extensive technology mainly imported from more developed nations; (3) centralized planning, mainly by economists and bankers in order to guide and speed up the process of development; and (4) the causes of underdevelopment within the developing nation, rather than in their external relationships with other countries.

The emerging new paradigm of development has largely been the result of two major publications : P. Meadows et. al. The Limits to Growth, and E. F. Schumacker Small is Beautiful. Meadows et. al. challenged for the first time the concept of infinite-growth; whereas, Schumacker questioned the relevancy of capital intensive, labour extensive industrialization route to development in Third World countries. Schumacker suggested development planning that was based on intermediate technology, was labour intensive and was oriented toward establishing small industries in the rural-based population centers of the Third World countries. Whereas, the old paradigm of development was based on the rate of economic growth (a concept that didn't include economic equality) the proposed new paradigm is based on two major concepts "quality of life" and "integrated rural development." By the early 1970s, it was realized that the fruits of economic growth in the Third World countries were not being shared by the masses. The trend was that the rich became richer and the poor became poorer. For example, the Green Revolution that increased wheat production in Asian countries manifold, also contributed significantly to the widening gap between the rich and the poor farmers. The concept of "integrated rural development" is emerging as the main element of new paradigm of development. This concept means appropriate technology, labour intensive planning, small industries, agricultural development, health development, income distribution, and so forth. This is considered as a comprehensive approach to enhancing the quality of life of the rural population.

The new concept of development is considered nation-specific, where each country might pursue somewhat different pathways to development. It rejects any global definition of development, although it does not undermine the significance of supranational factors in the national development process. Development is simply a powerful change towards the kind of social and economic system that a country decides it needs". Inayatullah writes, "development as change toward patterns of society that allow better realization of human values, that allow a society greater control over its environment and over its own political destiny, and that enables its individuals to gain increased control over themselves."

My own definition of development is based on the broad concept of "quality of life." Development, thus, is a process of planned social change to bring about a higher quality of life through economic advancement and social equality for the masses, with their participation in and control of their own destiny.

This new definition of development also evokes the idea for a new model of development for the Third World countries. The following discussion sheds some light onthe problems of the Western model of development that has been imposed on the Third World countries in the past, and presents a new model for consideration in the future.

Models of Development

The controversy about the relevance of the Western model to the developing countries and the ethnocentric predicament of the Third World intellectual elites in rejecting it has continued for over two decades. And one finds no end to this controversy in the coming decades; if anything, it is only more likely to intensify during this period. There could be several reasons for this the partial success of the Western model of development in non-Western societies, growing political power of the Third World nations in the United Nations, the widening gap between the industrialized and non-industrialized nations, along with the increasing demands by the latter for greater equality among the two, and an enhanced international competition for obtaining the shrinking natural resources of the world.

As pointed out earlier, the Western model of development has been based on the rate of economic growth through industrialization. Since Industrial Revolution was the key to Western economic growth and materialistically oriented life style this model was advocated for the Third World countries uncritically. The model looks like the one below:

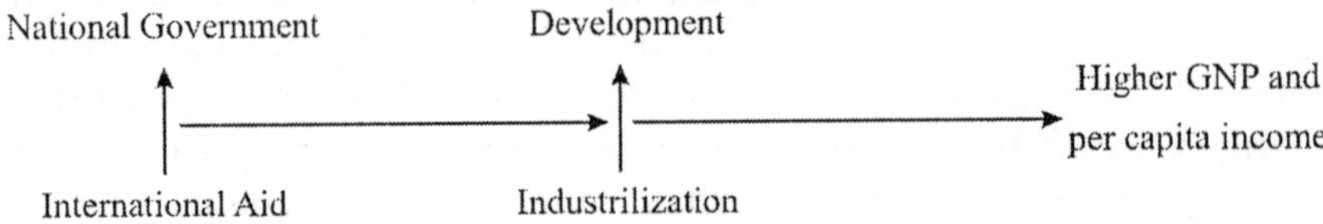

The question arises, if this model was not right for the Third World countries, why did they accept it in the first place. The acceptance of this model can be attributed to several factors. But the most pertinent of these has been explained by Inayatullah : "Asian students, scholars, and bureaucrates continued to look at development through intellectual lenses that blurred and distorted the true image of their own societies. They were still fired with zeal to transform them into replicas of affluent and powerful Western societies, and saturated with bias for a development model that was not always relevant to the societal conditions." Missing from this comment is the fact that the rules of development were, and still are, controlled by the industrialized nations of the world.

In any event, the relevance of the Western model of development has been questioned on the following basis :

1. The Western model of development is socio-historical in approach; it simply describes what has happened in Western countries. It, thus, cannot be used as a predictor of change in non-Western countries.
2. The model focuses on how the cultural heritage of non-Western societies has blocked their path to modernity, and how adaptation of Western institutions could rescue them from

grinding poverty.

3. The model seeks explanation of underdevelopment in the psychic domain of individuals lack of "need achievement", poor "empathy", "traditionalism", "fatalism", etc. among the poor masses of the Third World. It limits or virtually ignores the influence of other structural constraints for explaining the cause of underdevelopment.
4. The model assumes that the main causes of underdevelopment are intrinsic to the Third World nations rather than they are both intrinsic as well as extrinsic.
5. The model presents a mono-cultural view that depicts Western developed nations as the ideal which the Third World countries should aspire.
6. The model is based on monolithic theorizing of the development process through high rate of economic growth. It ignores the concept of "quality of life" obtained through a combination of economic growth and social equality for the masses.

Since the late 1970s, there have been some sporadic attempts by a few Western development and communication development scholars to question the relevancy of their models for non-Western societies. It was facile to forget that India, China, Persia and Egypt were old, old centers of civilization, that their rich cultures in fact were now poor (in a cash sense); and even their family life displayed a warmer intimacy and their artistic triumphs were greater, that was not development.

There is another evidence for the increased appreciation and understanding of non-Western cultures and traditions by Western development theorists. They now espouse the idea of preserving diversity of world culture, and the need for adaptation rather than adoption of the Western model. Despite these seeming progress of thought, one finds no significant change in the essential core of their development doctrine - the traditional Western model for the Third World countries prevails, at least in their minds.

New Framework for Conceptualizing the Model of Development

In view of the apparent inadequacies of the Western model of development, what would be an appropriate approach to conceptualizing a model that fits the development climate of the Third World countries? The model on the following page is being proposed for this purpose.

Basic assumptions of the model

1. Successful development efforts in the Third World countries must begin at the centre of the model and gradually expand towards the periphery of the national and international boundaries. This implies that development projects should begin at the grass-roots level, with wide participation in and control of the process, means, and outcome of development efforts by the majority of the people.
2. Development process involves at least four major sub-systems - economic, political, technological and cultural. Although not all four of these will be equally important at one point in time, the social system must be open to adjust and readjust these sub-systems in order to foster rapid development. Changes brought about in one of the sub-systems due to any reason may initiate change in other sub-systems. The role of communication at this juncture would be to create climate propitious for development and serve the linkage function among the sub-systems, local, regional and national governments. The flow of communication must be dichotomous in nature.
3. Unlike the traditional Western model of development that relies heavily on an asymmetrical orientation leaning towards the changes in economic sub-system, this proposes a symmetrical development in all four sub-systems to obtain a better quality of life for the masses.
4. Since the development process in a nation is no longer dependent on internal variables alone, this model incorporates the importance of external variables such as international technical assistance, international monetary assistance, and foreign aid etc. But the planning for such assistance too must involve the grass-roots level.

In sum, this model can be used as a heuristic tool for conceptualizing the development process in the Third World countries. The three main features of this model make it uniquely suitable for the development climate of the Third World countries.

The process of development must begin at the grass-roots level involving a wide participation by the masses. In the past, development planning has been top-down from national government to the grass-roots level. The process also requires simultaneous adjustments in the various sub-systems to achieve planned development. It does not consider changes in economic sub-system as a sufficient condition, as suggested by the traditional Western model.

The means to bring about planned development must take into account all of the available local resources, including human resources. The role of the internal as well as external variables in development must be examined constantly. For example, the role of national and supranational agencies should be of facilitators rather than controllers of the development process.

The outcome of development must be planned and determined with the active involvement of the local populace. National as well as international agencies should only serve as facilitators. Enhancement of the quality of life should be the key variable towards which all needed efforts should be directed. Then only, the dreams of development can become the reality of progress in the Third World countries.

Communication in Development

The recent changes in conceptualization of development have caused similar changes in thinking about the role of communication in development. Ever since the publication of two influential books on communication in development by Daniel Lerner. The Passing of Traditional Society and Wilbur Schramm Mass Media and National Development, Western communication researchers, especially in the United States, have exaggerated the power of the mass media in fostering rapid development in the Third World countries.

Based on Lerner's study of modernization in the Middle East, the mass media were called "magic multiplier" for development in the Third World countries. Pye wrote, "It was the pressure of communications which brought about the downfall of traditional societies."

With this over-optimism regarding the mass media role in development Western communication researchers embarked on a series of empirical studies in the Third World countries (Frey; Rogers; Rogers and Sven-ning; Gruning; Kincaid; McNelly and Molina; Inkeles and Smith).

Basically, two kinds of studies were conducted co-relational and field experimental.

Co-relational studies on mass media and modernization demonstrated a certain degree of covariance between the two sets of

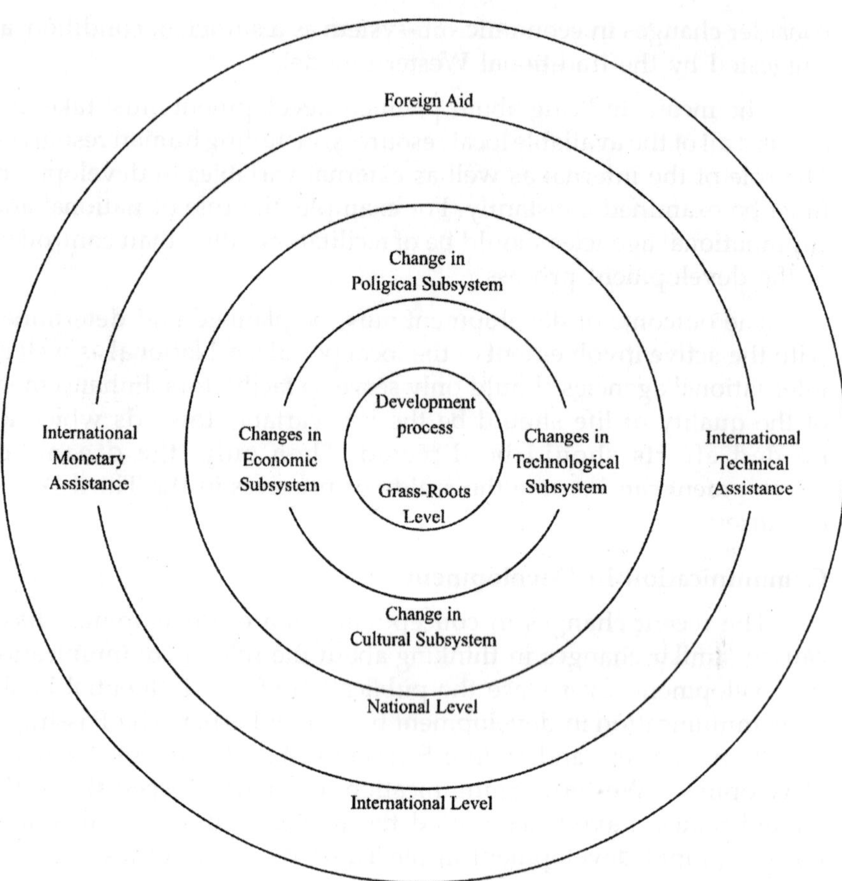

variables but didn't prove any cause-and-effect relationship. On the other hand, some field experimental studies demonstrated more direct effects of the mass media on the modernization process.

Despite the fact that some studies confirmed that mass media exposure was highly correlated with individual modernization variable (Frey) and "the mass media were in the front rank, along with the school and the factory, as inculcators of individual modernization" (Inkeles and Smith) the powerful role of the mass media was more often assumed than empirically established. This trend in the West, unfortunately, continues.

Criticism of Development Communication Research

Development communication research in the past has suffered from two basic pitfalls : (1) Since the models of development communication were originated in the United States and Europe, its underlying assumptions hinged upon the role of information in Western liberal democracy and the free enterprise system. It is thus, culture bound. (2) Asymmetrical orientation of these models towards behaviorism and operationalism has provided a weak conceptual framework to analyze development communication processes in non-Western countries. Behavioral orientation specifically forced Western communication scholars to design studies based on hypodermic-needle model of media effects.

In a thorough review of the past communication research efforts in Latin America, Beltran concluded that "Latin American communication research has followed conceptual and methodological orientations established by researchers in Europe and the United States, some of which proved inadequate and others obsolete. Noting further on this issue, the Latin American scholar remarked, the methodology itself limitsa critical comprehension of the social system in which communication research operates. Analysis of this methodology reveals its pro-status quo bias, in that it never considers the alternative of the creation of a new system but rather presents 'functional' adjustments to the old". Beltran's observation in this regard actually applies to the entire Third World countries.

Everett Rogers, who has conducted numerous studies in Asia and Latin America, conceded as early as 1969 that most of the findings of development communication research carried out in the underdeveloped countries may be misdirected and incomplete because of "the inappropriate use of culture-bound research methods developed largely in the United States."

Lately, Rogers has been much more critical of his diffusion studies conducted in Asia and Latin America for the past fifteen years or so. He writes, "the innovation is usually a technological idea, and thus one can see that past diffusion research fits well with the dominant paradigm's focus on technology and its top-down communication to the public." Other critiques or Western approach to development communication have also surfaced from time to lime. They mostly commented on the assumptions and direction of diffusion research in the Third World countries (Marcea; Gruning; Golding; Haven; Beltran). They all claim a pro-innovation bias in diffusion research along with the rejection of the decisive influence of the social structure on the individual decisions involved in adoption or rejection of innovations. Dependency of diffusion research on the linear model of Western communication has been pointed out repeatedly.

Development communication studies so far have overemphasized the significance of individual variables and ignored the role of social system variables. This may have been the result of the following factors : (a) most mass communication researchers came from individual psychology background. They were prone to applying individual level analysis in their research designs; (b) consumer oriented mass communication research was considered of utmost importance for the benefit of the vast advertising empire of the West and especially in the United States; (c) the very nature of Western culture which is individualistically oriented.

As Rogers pointed out, and very wisely so, the emphasis on individual variables tended to define social problems studied in terms of "person-blame" rather than "system-blame."

McNelly pointed out the excessive preoccupation with attitude in U.S. communication research and attributed this to the persuasion orientation of the research. "Much attitude research has tended to deal with relatively trivial or shallow preferences in purchasing or voting." One finds no relief from this built-in biasness in U.S. communication research, even after over a decade of criticism.

Nordenstreng's observation of the direction of U.S. communication research centers perhaps sums up the general feeling on this issue. After a tour of 20 U.S. communication research centers, he concluded : "American communication research has grown up in an atmosphere of behaviourism and operationalism, which has made it correct in technical methodology, but poor in conceptual productivity."

If the past can he prelude to the future, the scope of development communication model does not seem very bright. I, for one, believe that there is still some hope for developing a realistic conceptual framework for development communication model. It is the essence of this optimism that is captured in the following discussion.

Development Communication Model : A General System Approach

Conceptual changes proposed earlier for the model of development implies that corresponding changes must be proposed for the model of development communication. A comprehensive model of development communication based on general systems approach is therefore proposed.

The general systems approach emerged out of developments in biology and mathematics. Two modern developments are considered its integral part : Cybernetics and information theory. The feedback concept based on cybernetics "showed how the life likeness of living and social, as well as mechanical, systems depended on feedback...." (Klapp, 1973). The information theory idea, on the other hand, proposed that information is a central part of the organization of any system, increasing the order of that system. Living systems survive and grow through the input and use of information.

In a general systems approach to development communication, social systems are considered a dynamic living system. The principles of general systems approach that apply to the living organisms can also apply to the realm of social systems. Social systems survive and grow through the use of information, although their absolute survival may not be directly, linked to the availability of informational input.

Two general systems characteristics are essential to understanding its application in development communication model. First, the social system in which the model will be operational is considered an open system, a changing system, and a continually adapting system. Second, the concept of "systems" is considered to be some whole form in terms of structure, operation and functions. The whole is composed of several parts that are united and integrated. For example, a social system is composed of several subsystems such as political, economic, cultural and technical. These subsystems are integrated through some interactive forces of the system as a whole. As a result, the social system provides constant interactions among its subsystems. Based on this general systems approach, a new conceptual framework for development communication model has been proposed on the following page.

Basic Assumption of the Model

The existing models of development communication are only tangentially applicable to the Third World conditions. This is because of the following weaknesses of these models : (a) an information based hypothesis that assumes informational input as the key to rapid development; (b) the assumption of need for a well-developed information system as a precondition to rapid development; (c) a linear mode of thinking that increased informational input can lead to rapid development; and (d) mass media are more important to development than the dynamics of interpersonal sources.

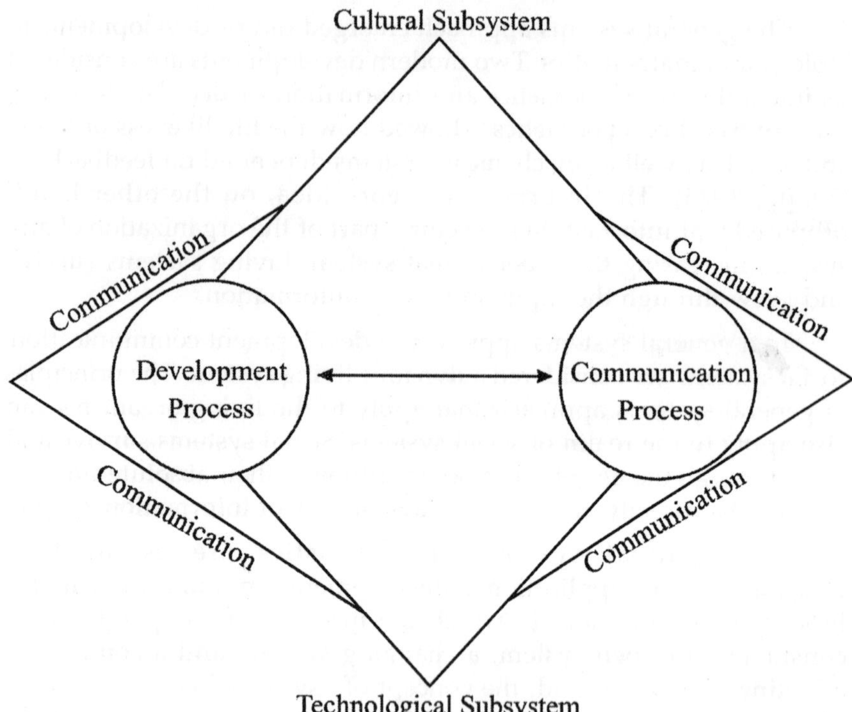

In contrast with these, the proposed model is based on the following assumptions : (a) in the development process, communication systems are an integral part and the role of both mass communication as well as interpersonal communication is omnipresent; (b) continued development in a social system brings about simultaneous changes in its communication systems, creating no need for a separate, well-developed system of information as a precondition to rapid development; (c) appropriate communication inputs are important to planned development, but standing alone, they can't foster rapid development; and (d) in development process, communication is an agent that interlinks the various subsystems such as political, economic, cultural and technical in order to achieve rapid and balanced form of development. Communication stimulates simultaneous changes in subsystems and creates conducive climate for development. It helps subsystems constantly adapt to new environments created by the development process. Just as there are various pathways to development, so are there various ways of utilizing communication in development. Different societies emphasize different sub-systems to achieve national development objectives. For example, Communist systems are most likely to start development process by modifying their political subsystem first. Capitalistic systems, on the other hand, rely on changes in the economic sub-system. Other systems can initiate changes in two or more sub-systems simultaneously. In the case of Japan, a rapid industrial development was accelerated by changes in the economic and technical sub-systems simultaneously.

In order to bring about rapid development, the Third World countries must initiate focused change. A focused change is one in which certain aspects of development are emphasized but the necessary changes are brought about in all of the sub-systems.

In a focused change process, communication is utilized in a planned way. For example, to develop economic sector through improved agricultural practices, informational inputs regarding farming, marketing, etc. are vital. But the mere availability of information will not contribute significantly to focused change unless essential changes are made in other sub-systems. For this purpose, communication systems become a vital link among all the sub-systems as depicted by the model.

Research Implications of the Model

This model can be used as a heuristic tool as well as an empirical model for analyzing development communication in the Third World. There can be two basic empirical utilizations of this model based on the following postulates : First, in order to study the role of communication in development, researchers must study the entire communication systems, i.e., who controls the mass media, how policy and programming decisions are made, the feedback system, the interaction between the mass media and interpersonal sources, etc. The previous studies in development communication have ignored the importance of these variables. Second, the effects of communication systems cannot be studied comprehensively in absence of its interaction with the political, economic technological and cultural sub-systems.

The general systems approach, as explained before, assumes that the various sub-systems of a social system are interdependent, and they operate through some interacting forces present in that system. In view of this, the role of communication in development must be studied in relation with all of the sub-systems.

I propose that development communication scholars of the Third World countries must embark on empirical studies that are methodologically rigorous and ideologically relevant to their local conditions.

References

OWENS, E. & SHAW, R. (1974). *Development Reconsidered: Bridging the Gap Between Government and the People.* Lexington. Mass, Lcxington Books.

ROGERS, E.M. (1976), "Communication and Development : The Passing of the Dominant Paradigm." *Communication Research,* 3 : 213-237.

MEADOWS, P. *et. al.* (1972). *The Limits to Growth,* Cambridge : MIT Press.

SCHUMACKER, E.F. (1973). *Small is Beautiful : Economics as if People Mattered,* New York : Harper & Row.

SCHRAMM, W. & LERNER, D. (1976). *Communication and Change in Developing Countries : Ten Yean After.* Honolulu : University of Hawaii/ East-West Centre Press.

1NAYATULLAH (1976). "Western, Asian and Global Models of Development," In Wilbur Schramm and Daniel Lerner (eds.) *Communication and Change : The Last Ten Years - And the Next.* Honolulu : University of Hawaii/East-West Center Press.

ROGER, E.M. (1976). "The Passing of the Dominant Paradigm - Reflections of Diffusion Research." In Wilbur Schramm and Daniel Lerner (eds.) *Communication and Change : The Last Ten Years - And The Next.* Honolulu : University of Hawaii East-West Center Press.

LERNER, D. (1958). *The Passing of Traditional Society : Modernizing the Middle East.* New York: Free Press.

SCHRAMM, W. (1964). *Mass Media and National Development.* Stanford, Calif : Stanford University Press.

PYE, L. (1963). *Communication and Political Development.* Princeton, N.J.: Princeton University Press.

FREY, F.W. (1964). *The Mass Media and Rural Development in Turkey.* Cambridge : MIT Center for International Studies, Rural Development, Research Report 3.

ROGERS, E.M. (1965). "Mass Media Exposure and Modernization Among Colombian Peasant." *Public Opinion Quarterly,* 29 : 614-625.

ROGER, E.M. & SVENNING, L. (1965), *Modernization Among Peasants : The Impact of Communication.* New York : Holt, Rinehart and Winston.

GRUNING, J. E. (1971). "Communication and the Economic Decision-Making Processes of Colombian Peasants." *Economic Development and Cultural Change.* 18 : 580-597.

KINCADI, D. L. et al. (1973). Mother's Club and Family Planning in Rural Korea : The Case of Oryu of Li Honolulu : East-West Communication Institute Report.

Mc NELLY, J. T. & MOLINA, J.R. (1972). "Communication, Stratification, and International Affairs Information in a Developing Urban Society." *Journalism Quarterly*, 49: 316-326.

INKELES, A. & SMITH, D.H. (1974). *Becoming Modern : Individual Change in Six Developing Countries.* Cambridge : Harvard University Press.

NEURATH, P.M. (1962). "Radio Rural Forum as a Tool of Change in Indian Villages." *Economic Development and Cultural Change.* 10 : 275-283.

BROWN, M. (1970). "Communication and Agricultural Development : A Field Experiment." *Journalism Quarterly*, 47: 725-734.

Ibid.

BELTRAN, S.L.R. (1975). "Research Ideologies in Conflict." *Journal of Communication, 25* : 187-193,

Ibid.

Ibid.

MARCEAU F. J. (1972). "Communication and Development : A Reconsideration." *Public Opinion Quarterly*, 36 : 235-245.

Ibid.

GOLDING, P. (1974). "Media Role in National Development : Critique of a Theoretical Orthodoxy." *Journal of Communication*, 24 : 39-53.

HAVENS, A.E. (1972). "Methodological Issues in the Study of Development," *Sociologia Ruralis*, 12 : 252-272,

Ibid.

Ibid.

Ibid.

NORDENSTRENG, K, (1968). "Communication Research in the United States : A Critical Perspective." *Gazette*, 14: 207-216.

KLAPP, O.E. (1973). *Model of Social Order.* Palo Alto: National Press Books, p. 287.

Chapter 7

Small is necessary : New Information Technology and Local Community Development

Dr. Sean Cooney

It was long expected that the "Ideal World Order" would result from a Development process which led from the Traditional Small, local economy to the big economy. The supporters of both the capitalist and socialist methods have agreed that material progress is the main purpose and aim of Development.

While this "Big Economy" solution has been promoted heavily by the rich countries, convinced of the merit of their own system, it has been widely accepted as a goal by the less developed, seeking to "modernize". Following the colonial pattern, the "developed" countries have acted on the premise that "Development" is something which they have, which they can give to others less fortunate than themselves - much as in former centuries conquerors regarded the conquered as being at the receiving end of so-called "civilization".

In seeking to help less developed countries to improve their life styles, it was assumed that what worked for the rich countries in the past could be made to work for others. The way to achieve progress in the Third World, it was said, lay in accelerating economic development by communicating thetechnology used in rich countries to them.

The so-called "less developed" nations have accepted the rich as teachers, and themselves as learners, because the western nations demonstrate in their rich life style the benefits of technology, promising similar rapid material benefits to other nations which will follow the same path.

This view was once put by Jawaharlal Nehru as follows "It is science and technology which has made other countries wealthy and prosperous and it is only through the growth of technology that India shall become a wealthy and prosperous nation." My own country, (Ireland), another former British colony, which is now a member of the European Community, is no less affected by this philosophy than is India.

The Linear Theory of Progress

An important foundation of the BIG theory of Development believe in the Developed countries, is the Linear theory of Innovation, Economic Development and Social Progress. This is fully discussed and illustrated in my book "Work for All" and in my study for the European Commission "Productivity progress and Innovation".

According to the Linear Theory, the primary source of progress throughout human history has been the improvement of our ways of meeting our survival needs that is, technology. There is, it is claimed, a step by step or linear, cause-effect relationship between the amount of money invested in the invention and communication of technological innovation, rising productivity (measured in money), economic development (also measured in money) and general social progress. This view is shared by all types of economic philosophy in western and third world and even in socialist countries.

There is much evidence to support the view of advocates of the BIG Linear theory of Development about the benefits of communicating new technology in gene. Illustration I show's the way new technology has always increased the productivity of labour throughout pre-history and history.

Changes in Communication technology proceed in the same one way manner and have similar effects to other Technological changes. The present Information Revolution was not the only information revolution nor the first (3) (Illustration 2).

Information Revolutions

Changes in Communication technology proceed in the same one way manner and have similar effects to other Technological changes. Information technology, like the larger field of technology as a whole, is adopted because it yields more results for less labour. It is not a cyclical phenomenon. The Information Revolution about which we hear so much today was not the only information revolution nor the first.

The role of computerized information retrieval systems in promoting the widespread dissemination of scientific and technical

How the Work Input to Meet a Given Need has Fallen Over Time

3 Million Years	Work Required	Need Met
10,000 Yrs. Ago	Agriculture	
7,000 Yrs. Ago	Writing	
4,000 Yrs. Ago	Metals	
1000 Yrs. Ago	Water Power	
1650 A.D.	Scientific Rev.	
1750 A.D.	Industi. Rev.	
1950 A.D.	Computing	
1975 A.D.	Micro Electr.	

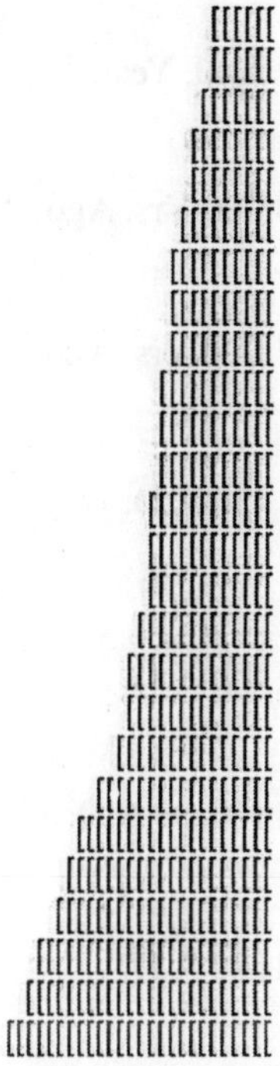

information has also been a major factor in speeding the development and adoption of new technology in other fields of industrial and social activity. Twenty five years ago, technical information for development was transmitted by meetings between people, by publications, and correspondence with a significant but not overwhelming role for the telephone. These remain and will remain important means of communication. But others have now assumed importance as well.

Today we have been through several stages of development (the KWIC index, computerised bibliographies, word processors, internationally accessible data bases, and now satellites and videotex systems) while the technology of printing and of telecommunications have both been revolutionized. We are well into the era of network communication and on the verge of the time when any computer will be able to communicate with any other, using integrated software that can access any type of file.

Information Revolutions

	Work Required	Need Met
1 Mill. Years?	Speech	
35,00 Yrs. Ago	Calendar	
7,000 Yrs. Ago	Writing	
1450 A.D.	Printing	
1800 A.D.	Powered Printing	
1860-80 A.D.	Telecommunications	
1680 A.D.	Computing	
1975 A.D.	Micro Electr.	

What Really Happens?

While the Big Linear Model of Development mode 1 appeared for a generation or two to be working, it no longer seems 50 (1, 2, 4). The promise of "development" through technology and industrialization has never really been fulfilled. Some of the most significant real effects of technological communication during the past decade and a half even in the richer economics have been to create irreversible unemployment, to threaten the natural environment on which we depend for survival and human health, and to make it more and more difficult to maintain conventional economic growth. This failure of the development process is affecting not only the countries in the "market economy", with its capitalist bias, but also the socialist countries.

But if these are the problems experienced by the rich-countries of the north, the story in the third world is considerably worse. Even in the years of greatest optimism about world development, when economic growth was rapid, in the 1960s and 1970s, the benefits of new technology and of economic development, took longest to reach the poorer communities and contributed least to improving the quality of their lives. But recently "Linear" development appears to have brought almost to a standstill the previous all too slow development process in poorer regions.

What is of particular concern is that countries of the south are actually slipping backward or experiencing negative development. Far from developing towards prosperity they are becoming relatively poorer and more deprived as economic power is increasingly centralized. They are losing their ancient cultures and their craft traditions on which rest their capability for self-sufficiency, without gaining the promised prosperity (5, 6),

An excellent example of "Good and Bad" effects of Development Communication based on technology in the Third World is the Green Revolution, whose impacts would seem to be (7) :

(a) Nutrition and Food : it made many countries self-sufficient in food, countries which previously were food importers and often were dependent on food aid from the rich.

(b) Economics : Farmers had to borrow from banks to buy fertilizers, seed and pest control chemicals and machinery. They also produced substantial food surpluses which had to be sold through the market. But because food surpluses replaced deficits, prices received by farmers for grains often fell, so that they were not able to repay the banks.

(c) Politics : The availability of adequate food has reinforced the regimes in power and weakened opposition movements in several countries. It is also said to have strengthened the hold of landowners and weakened the pressure for agrarian reform.

(d) Population : The availability of more plentiful food, together with modern medical technology, reduced the death rate in all age groups. This socially welcome result also increased the pressure on urban settlements, and on the ecological system, since cities grew faster and more of the wild places were harnessed to food production.

(e) Regional Effects : New crops created were first produced to suit regions inhabited by large populations from major cultural groups. Minority cultural groups inhabiting less favoured regions were left at a disadvantage.

(f) Culture : The Green Revolution changed cultural patterns. New crops, instead of being planted according to the seasons, are non photo periodic and can be planted almost at any time. Life styles and working patterns based on seasons, involving a whole rhythm of living, are altered in this way.

(g) Ecology : Many of the above impacts have ecological consequences. The use of new, uniform strains of crops over wide areas encouraged the spread of pests and diseases, which developed immunity to the control chemicals. Chemicals also have a serious effect on less robust and often beneficial insect and plant populations. The demand for more land to produce more food has also had a serious effect on forest resources.

It is therefore clear that the Green Revolution has started, and continues to stimulate, a complex chain reaction of consequences spreading outwards in space and time to affect all aspects of society and the environment effects whose final outcome is as yet quite unpredictable. The Bhopal tragedy is just one example.

To sum up: Instead of an economy of growing abundance, most societies are faced today with increasing scarcity. Instead of rapid growth and plenty of jobs, there has been persistent and repeated economic recession, severe and rising unemployment and accelerating damage to the natural environment. Business profits have been eroded by both recession and competition. Governments are experiencing very high demands on their financial revenues, which are at the same time being eroded by the same rising unemployment and by falling tax revenues (Illustration 3).

The failure to achieve greater welfare has led to social disruption and crime, while cultural decline has not been rewarded by increased leisure. Meantime, pollution of the environment becomes an ever greater hazard to human health, to natural resources and, in some places, to human survival.

The Attraction of the Small

But if the goal of developing for third world communities the material living standards of the Americans, Europeans and Japanese is failing, this is not the only option. There have always been other

Illustration 3

Expectations	The Reality
Plenty of jobs	Unemployment
Continued Growth	Recession
Greater Welfare	Social Disruption
Civil Order	Crime
Leisure, Culture	Culture Decline
Clean Environment	Pollution

concepts of Development, based more on the achievement of local aspirations than on copying the rich, and placing greater emphasis on qualitative progress than on the accumulation of money.

There are, if we simplify the matter, two main approaches to the question of how to handle the fragile planet we inhabit, two attitudes towards life and the survival of our species. One derives from the assumption that the fast return based on conquering nature is all that matters, since there will always be new lands to rape. The other is based on the superstitions dread of offending the spirits of the forest by taking from them more than tradition dictates, or, put another way, recommends sound use of the environment and the limited resources available to us because it is aware of the inter-relationship of all life.

The second or Small attitude has never been put so well as by the North American Red Indian Seattle in 1851 (I have edited this from Hanbury Tenison :

"The white man does not understand our ways. He is a stranger who comes in the night and takes from the land whatever he needs. The earth is not his brother but his enemy and when he has conquered it, he moves on. He leaves his father's grave behind, and he does not care. He kidnaps the earth from his children and he does not care. His appetite will devour the earth and leave only a desert.

Every part of the earth is sacred to my people. Our dead never forget the country of their birth when they go to walk among the stars. The perfumed flowers are our sisters, the deer, the horse, the great eagle are our brothers. The rivers quench our thirst. The rivers carry our canoes and feed our children.

This we know; the earth does not belong to man; man belongs to the earth. If we sell our land you must teach your children, as we have taught ours that the earth is our mother. If men spit upon the ground they spit upon themselves."

There are some 200 millions of people in tribal societies still isolated from the international Big Economy. We may take these as the extreme example of the Small Societies, which are resisting, being absorbed and wiped out by the Big international system. The American Indian is the best known example, but there are small societies in every continent in Africa, in South East Asia and so on which, are trying to survive.

Small is Necessary

This second great model of Development, the small, largely self sufficient local economy, which was typical of all societies in the past, has been in decline and decay for several generations. It will again become attractive to people everywhere who are disillusioned with failure of the Big Economy. As the failure of the Linear Theory and Big development begins to bite, the changing balance of the world economy will create pressures for a return to greater local self sufficiency.

In this area of local self sufficiency and crafts the less developed countries are not at so great a disadvantage compared to the rich, as they are in relation to big industry. The strong rural craft heritage of the poorer countries, a heritage which has become weak in the developed countries, will be what the poor have to depend on for their own survival and to offer the rich.

But the small economy is also attracting attention in developed countries. There is a revival of interest in craft skills and ways of becoming self sufficient on a local basis, as problems of unemployment and sluggish development bite into the quality of life.

Some advocates of the Small Economy see a new type of world coming, after Capitalism and Socialism have both failed. It will arrive out of awareness that we have over-exploited our world and that further exploitation will only bring about our own destruction.

Among the forerunners of these ideas, Mahatma Gandhi is notable, while in recent years the idea of an "Alternative" or "Ecological" approach and the use of "Intermediate Technology" have become more prominent. This view has been enunciated by many people, in many ways, notably by E. F. Schumacher (who developed his ideas first for India). But the idea which I am presenting here goes further than Schumacher (who said "Small is Beautiful") or his deputy, George MacRobie, (who said "Small is Possible"), the new idea might be described as. "Small is Necessary".

The Multiple Economies

New technology, properly used, can provide mankind with a steadily improving life style. New information technology can perform the crucial role of transferring technology to families and local communities who need it, anywhere in the world. In seeking to make use of this potential, must we then choose between the Big and the Small economics? Is there no other alternative?

Recognition of the existence of two opposed views leads to three important conclusions :

First, it is urgently necessary to find a better understanding of the full impacts of communicating new technology. This will require the adoption of new, more holistic and qualitative methods of forecasting, and a new concept of Development, as discussed more fully in my book.

Second, while waiting for these new approaches to bear fruit, it will be necessary for those in both rich and third world countries, for whom the Big Economy can no longer be relied on to provide a secure living, to turn to local, self-sufficient solutions. Information technology can play a useful role in this adjustment, both in improved forecasting, and in supporting local community initiatives. In India such ideas have been long advocated by eminent leaders such as Mahatma Gandhi and Tagore. This self-sufficient philosophy is also developed in my book.

Thirdly, we need to reconcile the big and small and not have them in conflict with each other. The new economy of the future, in both rich and third world countries, will combine elements of the traditional small, self-sufficient economy with big capitalist and big socialist policies. This synthesis will reconcile the capitalist, socialist and small traditional systems which have been in mutual competition up to the present.

A new strategy for Development, which may be called the Multiple Economy, will arise as a result of a synthesis between the Big and the Small economics. The role of communication in the Multiple Economy will be to promote progress in the life style of families and local communities by using either the Big or the Small economy or mixtures of the two as is found suitable. There will be some big enterprises, both private and state run, there will be some welfare by the state, but there will be small enterprises and a lot of self-sufficiency and voluntary welfare recognized as such. This will replace dogmatic adherence to Big Economy, top down leadership, whether of the Capitalist or Socialist kind.

The increasing use of new, small technology may be the only way in which peripheral communities can maintain and improve their life quality and protect themselves from the inexorable effects of dependence on the larger market economy, in which they are ill-equipped to compete. It may be the way in which everyone everywhere may enjoy the full benefits of new technology without its social costs in unemployment, pollution and economic instability.

Communication in the multiple Economy

Therefore, it is not just a question of how to get people to move from the small traditional to the big capitalist or socialist mode, as is assumed in conventional or "Linear" models of Development. There is an alternative - the small economy - and there is also the Multiple economy, in which Big and Small are reconciled and the scope of "Development" broadened.

It is important and necessary that people have the choice at the level of individuals, families and local communities, to decide the proportion of big to small in their society and the power to see that it turns out that way. This will depend on their having also the necessary information to enable them to decide between the different options.

The First characteristic of communication in the multiple economies is that it will not be confined to vertical communication downward, as in the Big Economy, but will also include horizontal communication between local communities and individuals.

The Second characteristic of communication of technology in the Multiple Economy is that it will take place, not from north to south alone as in the Big economy, but also from south to north and most importantly - from south to south. Up to now the assumption has been that the less developed will simply learn from the more developed. Now that the concept of development is being forced to change by the failure of the big systems, there is room for two way exchange.

Priority Strategies

My final section deals briefly with the priority strategies for development communication in the Multiple Economy. The following scale of priorities would appear to offer best strategy for meeting information needs quickly and effectively :

1. Technological Gatekeepers :

The first priority should be given to promoting meetings between people who have information and those who need it, that is, to developing the "Technological Gatekeeper" function. The main thrust of information science has been towards the improvement of documentary sources and libraries and the development of computer and telecommunication techniques. During the same period, research has shown that this emphasis is probably wrong because direct face-to-face contacts are generally more effective than documents or computer services.

2. Training Materials for Local Use :

The second priority is to provide local communities and families with a range of training materials for local use in solving locally identified problems. These can be based today on videotape, backed by print materials and instructions on the best methods of imparting the information contained in the package. Local users can make their own choice among the packages available and therefore not be limited to those materials which those at the Top wish them to use (as in the case with radio and TV today).

3. Literature Services :

The third priority is to make current periodical literature and books more easily and quickly accessible to users in developing countries. It is important that there be good, large collections of current and past journals in a country. Given these, it is possible to use photocopying technology to spread this material around to those who need it. The details of operation of the literature service and its advantages have been discussed in some detail in earlier papers.

4. Computers, Telecommunications :

The fourth priority Computer information systems and the telecommunication networks which make them widely available - have expanded with great rapidity in recent years, and greatly improved in quality. They have been married to satellite technology in which India is well advanced which will enable them to reach into the villages.

Computer systems cover a much larger proportion of the world's recorded information than other systems, and can be searched quickly. Narasimhanao sees information technology as a way to bridge the enormous geographical distances between communities and to combat illiteracy. Provision of education, health care and other services to most of the 575,000 villages in India is out of the question by conventional means with the resources available. He proposes radically innovative solutions using information technology.

However, from the viewpoint of LDCs, computer services have still a number of disadvantages. They are not yet available to the majority of users in rural areas. They are controlled by the Big Economy and, so far, have not been adapted to solve the problems perceived as important by small local communities, or to respond to the demands of these communities rather than those of the Big Economy.

In spite of these reservations, I would like to conclude by drawing attention to a certain potential of the Videotex system. Most of the options offered by Videotex are "vertical" and consist of information offered by the big economy (such as market information, management, calendar, and so on). But the MAILBOX option will allow different users such as villages to send messages to each other over the system. If Small should, as I have suggested, prove Necessary, the Videotex Mailbox facility could prove an important means of small-to-small, horizontal communication of information between local communities, quite independently of the big system. Similarly, the use on the Videotex network of Expert Systems, in which the user determines the questions and answers independently of the central manager could become useful in due course to local-communities.

References

J. Cooney, S. Work for All. Inforecast Ltd, Kilmacanogue, Bray, Ireland 1985 (IrE5).

Cooney, S. Productivity, *Progress and Innovation*. Report of a Study for the FAST Technology Forecasting Programme of the European Commission, 1982.

Condey, S. Maximizing the Benefits from New Technology. IFIP Conference *"Can Information Technology result in Benevolent Bureaucracy?"* Namur, Belgium, January 1985.

Rahman, I. A., *Concluding Address to the Dublin Conference on* "Informatics and Industrial Development" (March 1981). Tycooly publishers Dublin 1982.

Cooney, S., "Ireland and the Third World's Food Problems.", *World Food Day Seminar,* organized by Gorta, Dublin, October 16, 1981.

Cooney, S., Information for Progress: Revising our Priorities. Dublin Conference on *"Informatics and Industrial Development"* (see 4 above).

Cooney, S., "Progress Through Technology : need for a new assessment." *Science Policy II* (1) February 1984.

Hanbury Tenison, R, *Worlds Apart. Granada* 1984.

Cooney, S., "Self-Sufficiency as a possible strategy for New Technology.", In *"Information Society for Richer, for poorer"* (Bjorn-Andersen, Earl, Holst, Mumford) North : Holland, 1982). Proceedings of a conference supported by the EEC FAST Programme and held in London in January 1982.

Cooney, S., "A Two-way Transfer Self Sufficient Technology." *Interaction-1* (2) 1983.*Proposal to the Youth Employment Agency for a programme to train the young unemployed in self sufficiency skills.* Muinin Teo., Dublin, 1983.

Schumacher, E. F., "Small is Beautiful", *Abacus* 1974.

McRobie, G., "Small is Posssible", *Abacus* 1982.

Alien, T. J., J. M. Piepmeier and S. Cooney., "The International Technological Gatekeeper", *Technology Review,* 17(37-44) 1971.

Cooney, S. and T. J. Allen., "The Technological Gatekeeper and Policies for National and International Transfer of Information.", *R & D Management, 5* : 1 "October 1974.

Cooney, S., "Criteria for a User-oriented, Cost-effective information service : An Analysis of the East African Literature Service." *Special Libraries,* December 1974.

Cooney, S., "The East African Scientific Literature Service A Manual." *An Foras Taluntais,* Dublin, 1968.

Kaungamno, E.E., "Review of developments of library and documentation services in anglophone countries since the Kampala meeting 1970." *Tanzania Library Service, occasional paper No. 1,* 1976.

Cooney, S., "Information for Agricultural Development the role of Literature Services." *VII World Congress of IAALD,* Ottawa, June 1985.

Narasimhan, R., "The Socio-Economic Significance of Information Technology for Developing Countries, in "Informatics and Industrial Development"; *Proceedings of the International Conference on Policies for Information Processing for Developing Countries.* Dublin, March 1981 (Tycooly Dublin 1982).

Chapter 8

Research on Communication for Development Squinting - A Mirage or a Myth

Nalini Vittal

This paper has a limited objective : to show that in the field of communication research, the fault lies as much in ourselves as in our stars that we are underlings. This it does by analyzing one piece of research, which is a good example of the kind of research prevalent in our country that claims to be relevant to both communication and development. The problems highlighted in the example typify those in communication research as a whole. The dangers in it are common to most work. I have seen in our country with its prolific writing on communication.

The writing has grown in response to a variety of impulses. Where once university departments of journalism conducted the odd survey on newspaper readership, we have research units of advertising agencies, audience research wings of media organizations, full-fledged media and market research bureaux, departments of extension and communication in universities, voluntary agencies, experimental media groups, individually-funded investigators and a variety of academic organizations with at least a tangential interest in 'communication'. Alongside and fueling this interest in communication among the lay and specialized elite has grown the physical reach of media. The last decade or more has seen much public investment in electronic media, and private investment in the press, especially magazines newer thrusts, also by government, in non-media 'development sectors' such as adult education and by the international development lobby have given rise to a new rhetoric (such as calls for popular participation), and the bureaucracy has joined the ranks of people with an interest in the communication arena.

As for communication research, like the young Nelson, fear is unknown. There is no field it dare not tread : family planning and I am grateful to Graham Murdock, Gaston Roberge and that Late Saulat Rahman for ideas I have taken from them.

The paper was presented in a Seminar on Communication Research Trends & Priorities of India, organized by the IIMC, New Delhi.

Population education, advertising and market research, political propaganda, organizational behaviour, agricultural extension, semiotics, voter behaviour, international news flows, business correspondence, education, popular culture, and linguistics all are deemed fit to accommodate it.

In each of these fields of work/study, communication is seen to play a central or supportive role. For example, advertising is often delinked from the (presumably crass) commercial purpose, concerning itself with the communication of ideas and aspirations, making possible the "delivery of a higher standard of living". In agriculture, communication is often supposed to be the essence of technology dissemination, with wideranging applications from farmer motivation to extension/transfer of the ideational (as opposed to the material) component of technological practice. Women, singly and in groups, have expressed concern at popular images that reinforce stereotypes and distort, as well as inhibit, emancipation.

At this point, it is useful to distinguish between studies of media and those loosely concerned with all forms of communication. For this, it is worth considering the different institutional streams that feed into present day communication research. While advertising agencies and their support services have had much to contribute to status reports on media use, and national media organizations regularly conduct recall and impact studies of their services, the more recent impetus for smaller studies on various aspects of media has come from university departments and other teaching centers of journalism and communication. These, for the most part, may have begun with a purely journalism orientation but have since expanded to include other forms of mass communication and, later, more dispersed themes in the broad area of interpersonal or group communication, or the role of each of these within particular contexts. Agricultural universities, on the other hand, began from the classical agricultural extension approach to educating farmers and moved through a diffusion of innovations phase to the present concern with transfer of technology. Simultaneously in these and other universities, communication studies have come to claim for itself a distinct status. A different perspective has come from management theorists, who have been interested in communication processes as they relate to individuals and the organization. Similarly, academics from the social sciences have been interested in subjects such as group dynamics or the mass media as a sociologically explicable phenomenon.

Thus, Indian communication research exhibits a diversity some studies are media centered, others, not. Much of it, however, has been concerned with some end-use or purpose which media/ communication are seen to serve. One end-purpose that has gained much sympathy in the recent past, particularly since SITE in 1975 and the Janata Government in 1977 is rural development. It is seen as among the prime contexts within which communication must work. It has urgent demands, which, in the larger interests of society; only communication can meet. The actual process is seen as (ideally) two-way, with planners at one end and the public at the other. This two way communication is essential if the people (i.e. potential or actual participants in development) are to (a) be informed of developed programmes designed for them, (b) act in ways which maximizes benefit from these programmes, and (c) convey to planners their own reactions in order that programmes be modified to meet their needs. Communication, then, further the objectives of development programmes and helps achieve national goals.

Much writing on communication for rural development has these objective-furthering functions of communication at heart. Typical of such writing is Development Communication : Process and Impact, (3) which calls itself "a micro-study" of a given rural district. This is an extended report of research conducted between 1978 and 79 by a national academic organization. The reason for analyzing it here is that it is illustrative of the mass of present-day communication research, and captures neatly the current malaise in the field. The date of publication is less important than the overall style of the book and what it has to say.

Development Communication begins with a seemingly unexceptionable statement : The importance of communication drives from the fact that it is for the people that development is designed. On closer examination, the statement appears less innocuous: two assumptions are implicitly in it One, that development is a given, and that communication acts within this frame work to support/subvert it. That development, by being at the base of the discussion, is an established agreed-upon concept, and that the meaning of the word is set. The second assumption is to do with the primacy of the state :

development is "designed". Planners presumably know their job and people are expected to know theirs, i. e., to participate in development. Besides, knowledge, competence and power are vested with the designers of development, and this is natural and inevitable. The quarrel that this paper has with this opening and with most communication research we produce is its complete ignorance of what constitutes development and the mechanics of its operation. For one thing, the entry of the word 'development' into academic parlance is relatively recent, given the history of disciplines such as economics and sociology. It is much disputed, and often used for want of more adequate term, and then with suitable reservation and qualifying riders. Treating it as a given, particularly when it forms the context within which a process (communication) is explored, is both presumptuous and misleading. Second, the casual description of development betrays on unquestioning uncritical acceptance of the nature of things, and thereby, a particular ideological stance : the state is seen as an objective mechanism for reconciling conflicting interests.

What makes these objections relevant? After all, the animal under dissection is communication, not development. The answer is simple : the obsession with intricacies of communication has made us lose sight of the big picture, as a result, we are doing a great disservice to contemporary and future communication research, theory, teaching, and practice. That development communication is as poorly defined as development is more to do with our notions of development than our notions of communication. Once we try to understand how development whatever it is operates, we will find it easier to define development communication. We may even decide to discard the concept altogether.

It is precisely this concept which constitutes the second stumbling-block in Development Communication. Communication is seen at once as "an organizational process and a resource in development". One can only sympathize with this kind of desperation. On the one hand, a clear management perspective is adopted : information is transmitted through the hierarchy of development administration, the organization is itself a skeleton on which are arranged an assortment of interpersonal relationships, and it is through this bureaucratic-human framework that information on development reaches the people. The administrative structure, then, not only delivers development but also mediates it by mediating its information.

Distinct from this is the second perspective adopted by the study on hand : that of extension education, Supporting development but concerned with "education rather than publicity". This kind of education is to do not only with awareness of development activities, but also the interalization of knowledge and skills. This makes for more efficient and effective manpower in the countryside, and development efforts are thus rendered more productive. Thus, communication is oriented to human resource development which in turn serves the national interest.

A conceptual fuzziness is evident in the attempt to see communication as both "an organizational process and a resource in development". While Development Communication points out that it is not bound by this twin perspective, its later analysis seems to adopt it. It seems as though the authors have taken the word 'Communication' interpreted it in ways which appear relevant to their research, and then tried to see a relationship between these different approaches. This is part of the problem in this and most other communication research in the country. Communication has been vested with a magical ability to relate the unrelated, merely because it is used to describe exchange and interaction of any kind. It is a strange and sorry situation that we have allowed ourselves to be trapped by the limitations of our language. It is part accident, part linguistic inadequacy, that one word can describe different processes, from the enjoyment of a piece of music to conversation to education. We have, then taken this semantic convenience to mean an objective unity.

This insistence on the linking function of communication also informs the objectives and analysis of Development Communication. To begin with, it wants to (i) explore mass and interpersonal communication media, their distribution and efficiency; (ii) assess the extent of public awareness and information utilization, and (iii) study the status of inter-agency integrations as well as of "communication planning". These seem to use an information-centered management perspective, where there is an admitted disparity between popular knowledge and its application which is attributable to various factors such as lack of resources or skills, where planning and co-ordination can ensure that development goods are delivered.

Later, when the book turns to interpreting data from the field, it seems to reach an impasse. Several interesting points are made, such as the relationship between levels of awareness and traceable-sources of information. However, the report seems to find it difficult to derive other conclusions relevant to communication. For example, its final conclusions look at communication as a neutral process, with no relationship to society save that "different groups have different access to media" or that "the field situation is fraught with problems unanticipated by communication agencies". Ultimately, it sees no conflict or convergence of interest among segments of society which may be articulated in forms of communication. Neither does it seek to analyze why media are the way they are, though much of the focus is on media in the rural milieu. The approach is very instrumentalist, closely woven with it" and ending with the Importance of "the need for integration between infrastructure and communication."

And no wonder. The book does not, as pointed out earlier, question the concept of development, much less that of communication 'support' to development. It does not discuss what communication may be, or consider functions other than information transfer it may perform, such as reinforcing social inequality or legitimizing the given structure. Neither does it seem able to explain, for example, why dissimilar villages do not show any great difference in overall awareness of development programmes.

The point is these shortcomings arise as a result of being communication bound, ratherthan despite it. A betterway to conduct this and much other research on communication would have been to begin from the context of communication in this case, development. This must involve grappling with the concept of development in the first place. For example, it may be seen as a specific (interventionist) variant of social change. If society is full of pulls in different directions, if in a 'developing' country the bureaucracy is an importantinstrument that facilitates or inhibits change, if access to services and benefits are functions of access to power and wealth, then surely a study should begin here. Then, for example, media can be shown as true reflections or distortions of the structure. Whatever the analysis, it must begin from a view of social structure and not from information or one of the many other faces of communication.

This analysis, however, must make explicit its suppositions. The problem with most communication research is that if its several ideas are thought through, it does seem to embody some idea of society of social change. There is some kind of perspective which seems to inform the research but even the writers do not seem aware of it.

Awareness of the historical bias of one's work is as important as an awareness of the present conditions in which our work is set. After all the theme of communication and social change has not just happened to us; there is a definite history which has had its influence upon our thinking. We may have had our peculiar stimuli such as the overriding public allegiance 'development' has seen in our society. At the same time, we are also products of a wider informal society, that of international academe. As Rosengren (8) illustrates in the following matrix, there have been differing viewpoints on whether there is any kind of relationship at all between culture (which includes the institution of mass media and social structure.

		Social structure YES	**Influence culture NO**
Culture influence	YES	Interdependent	Idealist
Social structure	NO	Materialist	Autonomous

Most people now a days would agree on an interdependence between cultural institutions and social structure. If we do, too, then it is up to us to make the connections between what we see as 'communication' and out particular past and present, our ancient and persistent social organization, our recent emergence as an independent nation, our grinding poverty and immense potential for genuine development.

Having considered these conditions, we might decide as Prokop (6) does, that what we describe today as mass communication is precisely the absence of any meaningful communication between individuals and groups. Prokop argues that in a formal sense mass media reach more individuals than before, but that "the quality and real effect of public discourse and rational communication of social aims are left untouched". In a sense, the efforts of our experimental media groups and occasional calls for deprofessionalization of the media echo the same concern. Yet, sweeping reviews (9) of the current status do no often question the idea that media perform the normative communication function.

However, in the context of development some small beginnings have been made by communication researchers. One analysis that may be useful at this point comes from earlier work I have done (10). Two approaches have been identified as being relevant to research in communication : the 'extension' and 'sociological' approaches. The former refers to the information-deterministicview discussed at length in this paper, and forms the bulk of our research. The (sociological view an admittedly loose term, but a beginning), sadly underrepresented, sees all communication as "an articulation of the social structure". Perhaps both are relevant, but I see a need for greater attention to the latter view, to context, which is completely lacking. Occasional genuflections to "people's needs" or "developmental goals" do not make for a 'sociological' analysis. Neither should this be taken as an attempt to silly inject 'outside' disciplines (such as sociology) into the alleged purity of our own. I even challenge the idea of communication studies as a distinct discipline. I suggest that the field resembles development studies or women's studies, which I once described (11) as "an area of popular and academic concern, a junction of several disciplinary tracks".

As I see it, the need in research on communication in general, and in the context of development in particular, is for some sound theory. The need is for us to be aware of our never-stated perspectives, to challenge the concept of development and recognize that one human being's development may be another's irreversible alienation, to find better ways of studying this welter of confusion. The danger if we do not is that we will carry our sins into the future.

References :

JUAN E. Diaz BORDENAVE : (1977), COMMUNICATION AND RURAL DEVELOPMENT, UNESCO, Paris.

COLIN CHERRY (1955), COMMUNICATION THEORY AND HUMAN BEHAVIOUR STUDIES IN COMMUNICATION, Seeker and Warburg, London,

N. K. JAISWAL, N. VITTAL & G. RAMACHADRAIAH (1981), Development Communication : Process and Impact. A micro Study in Mahbubnagar District. Andhra Pradesh, National Institute of Rural Development, Hyderabad.

ROBIN, McCRON (1977), CHANGING PERSPECTIVES IN THE STUDY OF MASS MEDIA AND SOCIALIZATION. International Association of Mass Communication Research, Leeds.

GRAHAM MURCOCK and Peter GOLDING

DIETER PROKOP :

GASTON ROBERGE

ER1K ROSENGREN

NALINI VITTAL

NALINI VITTAL

NALINI VITTAL

(1978), THEORIES OF COMMUNICATION AND THEORIES OF SOCIETY, COMMUNICATION RESEARCH Vol. 5, No. 3. July

(1983), PROBLEMS OF PRODUCTION AND CONSUMPTION IN THE MASS MEDIA. MEDIA, CULTURE & SOCIETY, Vol. 5, No. 1, January,

(1982), MEDIA ASIA,

(1980), MASS MEDIA AND SOCIAL CHANGE : SOME CURRENT APPROACHES. Wilhoit and de Bock (eds.) MASS COMMUNICATION REVIEW YEAR BOOK. Volume 1, Sage.

(1979-81), COMMUNICATION FOR RURAL DEVELOPMENT IN INDIA: SOME FACETS National Institute of Rural Development, Hyderabad.

(1982), EFFECTIVENESS OF COMMUNICATION TO RURAL POOR : A STUDY IN AN IRDP DISTRICT, JOURNAL OF RURAL DEVELOPMENT, Vol. 1., No. 4, July.

(1983), Untitled note (mimeo) to the WORKSHOP ON WOMEN'S STUDIES, Institute of Social Studies Trust. Bangalore.

Chapter **9**

Perspectives on Communication Research in Asia

Usha Vyasulu Reddi

Not a single year has gone by in the past four decades when, somewhere in Asia, there has not been a workshop, a seminar, or an article published on the state of communication theory and research in this continent. Each New Year has also seen a demand for the rejection of western concepts and models and a call to Asian scholars to develop indigenous ways of looking at the development problems of this vast and varied continent. Yet for all of this activity, communication theory and research remain very much on the starting block today. It is not that there has been no research activity. Quite the contrary, research is flourishing and projects are undertaken, articles published, reports written and filed away, and action taken on the results. But time and again, we end up asking the same questions : what and where are the research issues which communication scholars in Asia must address.

I do not propose in this article to either berate or applaud Western models or to provide an alternative Asian 'perspective'. It would be presumptions on my part to try in this brief space while others have failed. Nor do I propose to undertake a 'survey of research' in Asia. Still further, I do not wish to prescribe norms and procedures for Asian scholars to follow. What I would like to do is to write as a researcher in Asia who is searching for the questions I must answer while faced with the task of undertaking action research to solve operational problems on a day to day basis. To do all of this, I seek the reader's indulgence because I, like many others, have been trained in Western conceptual Models which I have used, modified, discarded, accused of functional irrelevance and unlearned. And because, instinctively, I react to communication problems by using Western paradigms.

It is not that there is something wrong about the Western conceptual models. There are communication behaviour patterns which are common to both the East and the West and there by can be subjected to a common theoretical framework. At the same time there are behavioural patterns which exist in the two hemispheres and which cannot be explained by common frameworks. These differences are both real and relative; and it is these differences, largely in value orientations which turn communication paradigms upside down and which form the basis for the worldwide discussion on the relevance of western models in the East.

Two fundamental questions must be answered here. First, what, if any, are the premises upon which this discussion is based, and second what is the structure of this paper.

The Premises and the Structure

The Premises : The first fundamental premise of this paper is theory and research share a symbiotic relationship, each is part of the process as also the result of the other.

A second fundamental premise is that, despite all our efforts, value orientations do intrude on the process of research and it is difficult, perhaps impossible for any researcher, to conduct value free research. A third premise is that there are both commonalities and differences in Western models; it is necessary to understand these thoroughly before applying or discarding any There can be no blind acceptance or rejection of any model.

A fourth premise is that the philosophical bases of Western and Asian cultures are very different and may be there is truth in the Mark Twain saying "East is East and West is West; and Ne'er the Twain shall meet".

The Structure

To undertake this discussion, this paper shall proceed through a description of Eastern and Western perspectives vis a vis issues of research to a look at the kind of research commonly carried out in Asia today. This will be followed by a discussion of the methods and their use in Asian contexts and an evaluation of their relevance and application.

This is not a rigid structure because there are both overlapping and mutually exclusive areas involved, and hence repetition of key issues and concepts may exist.

The Context of the Cultures

The difference between Asia and the West is not merely geographical. It is historical, philosophical, and societal and the very notion of 'shared meaning' which forms the basis of communication has different connotations. Nor is the difference merely dichotomous, there are similarities as well as differences all operating at the same time between Asian civilisations and the Greco-Roman civilisations on the other. There are also great diversities and disparities within Asia forming a rich tapestry. But there are few common strings and it is these in terms of their influence on research with which I am concerned here.

To the Asian, born as he is and governed by an intangible scheme of things, the spiritual is more important than the temporal, the past and future converge with the present to form a continuum which cannot be explained in isolation. The collective unit of the family or community is more important than the individual, duty more sacred than rights the whole greater than its constituent parts. There is a creative union between man and nature; what appears to be on the surface is but the tip of the iceberg and action is in no way reflective of either expression or thought. There is both contradiction and reconciliation of thought and action within the same individual, and when we communicate with him/her, we can never be sure if the response we are getting is in deed the accurate one.

What Chu (1988) said in his article about Chinese culture is relevant for all Asian cultures :

'...authority is respected and harmony is cherished. This respect for authority calls for the observance of specific rituals in communication between elders and their juniors and between superiors and their subordinates. Choice of words and salutations so as to show respect and to maintain the hierarchy of somial order can be very tedious and difficult. Respect for elders and superiors has also influenced the expression of opinion in public. To avoid offending elders or superiors, dissenting opinions are more often than not withheld.

Among peers, harmony takes precedence over respect for authority. To maintain harmony, conversations are not only carefully worded but also calculated for this purpose......

'...A deeper root of this concern with respect for authority and harmony can be found in the importance given to 'face' by the Chinese.

The concern with 'having face' or 'losing face' in the presence of outsiders has led to the exhibition of contradictory behaviour in public and private.'

The traditional modes of communication in Asia were and remain today, highly interpersonal. At a more mundane level, the very identification of an individual through his name is indicative of his role, status, economic function, geographical origin and community identity in the country. This placed invisible bonds and strong restrictions on the nature of communication between individuals and communities.

Methods of communication, mostly oral, ensured the unquestioned acceptance of the somial order and of the rigid rules of social stratification and conduct which gave legitimacy to the ruling order. All communication took place within a common frame of reference and was asymmetrical, i.e. downward from the higher to the lower status individuals.

These forms of communication are in direct contrast to the communication traditions of the west, where a great emphasis is placed upon the individual as an entity distinct but in a somial relationship with the society in which he/she lives. There are few, if any, rigid distinctions of class and community, greater social and geographical mobility, and it is the individual upon whose rights and responsibilities the society is based rather than the converse.

It is therefore, interesting that the successful communicators of twentieth century Asia, i.e. Mahatma Gandhi and Mao Zedona were leaders who drew their communication styles from the ethos of the societies in which they lived and succeeded in transforming the two major giants of Asia. It is also worth noting that the Western societies which were lined up in opposition against these two leaders never got the full measure of what these two leaders achieved in mobilising vast populations until it was far too late for their governments to do anything but accept the fait accompli presented by the two master communicators.

In logical sequence, it follows that the successful methodologies of research will be those which emerge from the cultures drawing their inspiration and techniques from the living traditions of Asia.

While we are looking at the cultures of Asia, it is also essential that we examine the culture of communication within the Asian context. Despite variations within Asia, the dominant culture of the communicators in the post World War II era, i.e. 1950s and 1960s, has been that of using the scientific method to the solution of developmental problems. If they had only been a little more introspective they would have come across a practice which they themselves followed and which greatly influences communication.

The notion of 'shared meaning' and of 'face' are two concepts which have greatly influenced both communication and research in Asia. In its essence, the notion of shared meaning is an attitude, concerned with not only the meaning of a message, but also with the context of meaning in any given set of messages, rather than messages as discrete entitles and understandable only in relation to each other. What is implied here is that it is necessary to delve into the philosophical bases of culture and tradition to explain the notion of shared meaning, because the premises of meaning are rooted in the many philosophical debates and because the stress is not only on meaning and interpretation, but also on who is the source of the message, what are his/her motives, and what is the context in which the message is embedded.

The concept of 'face' has even more relevance for the communication ethos in Asia 'Face' influences the choice of communication modes (Chu, 1988). To avoid losing face, for both the communicator and the receiver, a critical message is conveyed through writing rather than in a face to face situation. At the observable level, the tone will be both (deceptively) polite and cordial, while being extremely tough and harsh in writing, exactly the contrary of the communication situation in the West.

'Face' leads to a preference for indirect forms of communication, and changes the semantic and social meaning of communication content. In the interest of external harmony, superficial unity is often found in the content of communication and it is very difficult to 'interpret' the message until both the source and the receiver are aware of all the hidden contexts of the message.

The implication for 'face' for communication scholars is that we cannot take responses to questionnaires and interviews at 'face value' without a clear and thorough understanding of the context in which the message is prepared, the source of the message, the mental frame of the source and the receiver, the social and cultural conditions which govern the message and the source. A task very difficult to do unless and until the researcher is himself/herself steeped in the ethos or is fully conversant with the nuances which cultures will impose upon a message.

And that is precisely the point that "I am trying to make here. While in their own personal lives, the Asian researchers participated in all the traditional systems of their individual cultures, when it came to research they choose to opt for the only models of research which their educational training provided them with : the Euro-American theories and methods, forgetting their own cultural roots. We have upon wave of Asian scholars sought and obtained communication training in the universities of the West, and began to analyse their own systems through the eyes of their mentors.

With great enthusiasm, experiments in communication were planned and executed, statistics on the status of communication gathered. Underlying all the experiments was the assumption that the problems were due to a lack of information, ignorance and poverty, and change was possible if only the backward and poor people were uplifted by the information provided. What information was needed and what was provided, from the research data were two opposites.

The lessons of the 1960s were learned in the 1970s when experiment after experiment failed to yield answers to the questions the policy makers were asking. But it was, in my opinion, the strong influence and complete acceptance of Western models by Asian scholars (who ignored their own social backgrounds) which are to be blamed for the failure of Asian scholars to build up viable concepts and methods suited to Asia.

But beyond this broad condemnation of the scholars and the methods that they choose, there are reasons why these conceptual approaches 'failed'. The assumption that we can look at the individual as a separate entity from the social system in which he/she lives tends to blind us from the beginning. We are limited by the approaches we choose to use, by the tools which we fashion to gather the information and by our assumption that we as researchers, are outsiders to the ethos. And that is why we have fallen behind.

Kinds of Research in Asia

While communication research elsewhere in the World is currently at a very critical juncture with intense debate about changing paradigms and polemics, communication research in Asia is very much as it was inherited from the West two or three decades ago. Kishore (on-going) has argued that communication researchers in Asia do not focus on the development of theoretical concepts, models, research methodologies appropriate to the local way of life and social systems, borrowing blindly from Western concepts, etc.

Part of the reason for the unscrupulous use of Western models came from the historical growth of development research. The late 1950s and the 1960s witnessed a mass migration of researchers from Western countries to Asian and other developing countries with the mission of educating their counterpart's ensitu. The researcher would come gather his data according to previously developed conceptual frameworks, methodologies, and tools, and return home to complete his analysis. At most these visitors left behind a partial understanding of the research methodology.

A parallel migration of scholars from Asian to American universities to study communication took place. Although only a handful of these scholars returned to home countries, those who did brought back with them the dominant research paradigms behavioural in orientation and empirical in methodology.

The experience of foreign trained scholars was such that the pure and rigorous application of their learning a broad was impossible. Further, Asian scholars began to raise research issues which they perceived as more locally significant. But, because of the disparity between the needs and the methods and more so between the scholar's training in behavioural techniques and the more urgent search for solutions, the gap between the ideal and the actual situation has never been wider than it is today.

The urgent need in the Asian countries soon after independence was rapid development. National developmental projects and international technical assistance to developing countries burgonet. Yet most of these projects were launched without adequate planning and even less evaluation. Many of these projects failed either because of a lack of scientific guidance or because evaluation was done through improperly designed.

In the belief that without the contribution of social sciences, the process of development is slower, more painful more arbitrary, and more unequal (poolm 1966), social researchers began to undertake large and small surveys. The purpose of these surveys was both to build a database of basic information as also to identify critical development indicators. Soon the most popular method of research became the sample survey, followed by the experiment. A study conducted by Alfonso and Dissanayake (1981) showed that in communication research conducted in eight Asian countries between 1971 and 1981, 83.8 per cent were empirical in nature. Most of these, no doubt, were surveys. Similar findings emerged from a study conducted by Kishore (on going). Case studies, policy research, and basic research were noticeable by the paucity of studies in these areas. Yadava (1984) also reported similar findings.

As UNESCO and other aid agencies promoted the communication approach to development, much of the practices and research in this field became to examine the role of media in nation building and thus, much research became main stream.

Two events in 1970s led to a questioning and to a breakaway form this thinking. First the 'development communication' paradigm as it emerged in the work of Asian scholars which questions old theories and argued for an entirely new look at the relationship between communication and development (challenging the existing definitions).

Second, the SITE experiment in India ushered in a new look at research with a holistic approach replacing the earlier broad based survey. The SITE experiment criticised previous experiments which relied solely on surveys and argued for qualitative anthropological validation of the process of communication through observation and other techniques (Agrawal, 1981). Such an approach sought to avoid a static ethnographic description of the culture of the audiences. The success in using this method was such that after SITE, it became the dominant research paradigm in Asian studies relating to television's role in development. A similar study was subsequently carried out in Indonesia (Alfian and Chu, 1979).

Next to the surveys, the experiments carried out in Asia in the post independence years demonstrated a desire to identify key variables affecting the pace of modernization, communication being one of many. From the controlled experiment to the field experiment which sought to manipulate variables within a real social context, researchers hoped for cues and for an in depth understanding which the survey method lacked. But one aspect became clear the guaging of long term results from experiments became a problem.

What resulted then was a dis-satisfaction with the available and established research and an un-happiness over the concept construction, methodology used, and the resulting applicability of findings. To understand this feeling of despair, we have to examine the methodologies themselves to identify the conditions in which they are either ideally suited or totally inappropriate.

Using surveys, for instance, in situations of great variability limits the method itself. For what kind of a survey can be designed for the heterogeneity of languages, styles of life, attitudes, economic and ethnic backgrounds and other characteristics which are too many to even categorized? And that, too, in the absence of baseline information to help simplify sampling procedures. And what if the little available baseline information is inaccurate and unreliable? Micro level surveys do not yield us the depth of information we require; micro level studies cannot serve bases for generalizations.

A major consequence of these characteristics is that it is nearly impossible to provide the 'development planner' with a set of thumb rules by which to operate. Coupled with the weakness of the survey as a method, is the social distance separating the researcher and the respondent, where the former is an elite schooled in Western institution and the latter is a farmer. Each is in the unfamiliar terrain of the other.

The overuse of structured vast surveys by Asian scholar (Dissanayake and Belton, 1983, Kishore, 1989) has led to the irrelevance of this method.

Looking in greater depth at survey as research methods we have to distinguish between the unit of response and the unit of analysis. The source of information from whom we obtain the data is the unit of response. The unit of analysis is the entity toward which these data are related as the statistical bases of analysis (Roy, et. al., 1976). Conceptually these are different but often in Asian conditions, they are not and analysis can easily become confused if the source of information is also the unit of analysis, e.g. a family or community response to an interview meant for the individual.

The basic data gathering tool in a survey is the self administered questionnaire or the interview. Inadequate demographic data, population lists, physical communications such as telephones, posts and telegraphs, literacy levels, and other logistic facilities are sometimes non-existent, making the questionnaire useless as an instrument for data collection in Asia.

As far as the interview is concerned, it has become a major research tool. This approach is conceptually best suited for developing countries provided that the interviewers are thoroughly trained and manage to avoid the collection of stereotyped, irrelevant, and invalid responses, a situation which could easily occur if the respondents do not wish to lose 'face' before the interviewer.

Other techniques for gathering data, viz. mail questionnaires or telephone interviews are never used, given the vagaries of telecommunications. Roy and others (1976) have also pointed to other disadvantages of the survey method when they have cited suspicion, interviewer respondent differences, lack of time, non-involvement, and a typical conversational patterns which inhibit in-depth probing into culturally sensitive areas.

In the search for cause effect relationships, field experiments have been used quite extensively in Asia. The field experiment tries to incorporate the advantages of laboratory conditions with the benefits of the field surveys and case studies in a naturalistic setting. The achievement of such conditions is both difficult and expensive. Further, the results of such experimental prototypes are very difficult to replicate and base generalizations upon. The most they can do is to throw up a set of pointers and establish a short term cause effect relationship. Logistics and other extraneous factors also intervene so that compromises in research design lead to reduced reliability of results.

Finally, we cannot afford to bypass the issues of sampling, measurement and data analysis, one area where researchers encounter innumerable problems. It is in these areas that maximum concessions to reality must be made.

If a sample design is the theoretical basis and the practical means by which we infer the characteristics of the population, the sampling frame from which samples are drawn must ideally be a complete enumeration of all units of the population. The reality in most parts of Asia is that we have to draw samples form incomplete, inadequate and inaccurate sampling frames. Thus, faithful adherence to the strict rules of sampling is both expensive and difficult to achieve. Our objective is usually to draw a 'scientific sample' one which yields the desired information with a tolerable error and a realistic compromise between the rigours of theory and practical inhibitions of time, money, staff, materials, and logistics. Ultimately, because of cultural, economic, and policy constraints, we end up doing the best we can under very trying circumstances. To summarize, sampling designs in Asian countries often have to settle for incomparable, often crude, statistical approaches.

Similar problems arise with data measurement and analysis. At the outset, for any sample for which the statistical probability is unknown, inferences from probability statistics are invalid. Again, yardsticks for measurement which work well in the West make no sense in Asia. For example, it is nearly impossible to correctly ascertain the age of a rural or poor respondent in India. Another measure which poses a problem is that of wealth. Is wealth measured as individual, house-hold or annual income, visible or hidden income, land holdings, gold, or material possessions. The definition of wealth is culture bound in Asian societies making this a difficult yardstick by which to measure economic status.

Further, what kind of statistical analysis can be undertaken? Parametric statistics have to be ruled out, given the sampling procedures followed. And what will non-parametric statistics tell us beyond suggesting associations and correlations, which may be neither consistent nor replicable over time and location.

All these issues point toward the central question of the application value of the methodologies and relevance of subsequent findings. Are we merely stoically carrying on despite all these constraints?

Relevance and Application of Findings

Time and again, scholars have criticized the dominant research methodologies for their irrelevance in Asian conditions, Such a criticism comes from a failure to apply the results of research, Lozare (1983) has called this a 'failure to root analysis in social realities' while Dissanayake and Belton (1983) have blamed researchers who persist with empirical designs when a critical approach is needed.

I consider the question of applicability of research findings to be a complex one which cannot merely be treated as a result of either researcher bias or method limitations. The many aspects discussed earlier must surely give the reader a clue to some of these complexities.

Initially, I believe that it is necessary to establish priorities for communication research in Asia. What are our long term goals and our short term strategies? The building of a coherent or cohesive body of knowledge on research methodologies is a long term goal.

This calls for basic research into methodological issues.

However, what appears obvious is that if communication research must serve as an input to problem solving and policy formulation, the social scientist must actively advocate more serious problem solving studies aimed at untangling social and economic issues. The social researcher must provide not esoteric theorizing but concrete information which will help the decision-maker and the practitioner alike.

Hursh-Cesar (1976) has outlined a setting of priorities which I believe can serve as a guideline. He speaks of providing knowledge to serve as programme inputs, delivery outputs, and service outcomes.

The establishment of priorities will bring us to the next part of research the identification of research objectives in terms of relationships which are realistic and which arise out of the socio-cultural ethos itself. This of course, implies the development of accurate database on the social processes, of institutions, both social and political which influence the life of each individual in the system and which determine how the society operates. Dissanayake and Belton (1983) have argued that if social researchers of Asia are to cope with more pertinent methodologies, they must move away from the functionalist perspective which governs most of mainstream communication research in the West, because this approach has problems, both pragmatic and metaphysical, in Asian conditions.

To tackle this third issue, I would call for more participatory research where the social distance between the researcher and the respondent is narrowed and the latter participates in the process of information gathering. This would increase the extent to which the researchers can flexibly respond to local conditions and make the findings of his work relevant and reliable. The local condition may in turn dictate the kind of methodology the researcher should adopt rather than the present top down approach.

Further, I would not totally rule out the application of Western research techniques under Asian conditions. This is because there are pockets of Asia which resemble the West and also because methodologies have strengths and weaknesses which are irrespective of the cultural situation in which they are applied. Its inappropriateness as a method may be due to its improper use rather than its lack of relevance. Proper use of a method means that researchers must have a thorough understanding of research techniques. My own experience in Asia has been that many communication scholars are not conversant with the research techniques and are equally poor in understanding of the social conditions which necessitate on the spot modifications of tools of data gathering.

In the ultimate analysis, what I see as the major debates over relevance is really a consequence of the distance between the techniques we are familiar with and the real and fault needs of practitioners who have to apply the findings of research. The debate is between the kinds of research which we would like to undertake vs the kind of research which is urgently needed. To throw out the Western methodologies outright is simply to throw the baby out with the bathwater. What we have to do in Asia is, sit down, identify priorities, set objectives, time frames, share experiences on what has worked and what has not and why and come with some kind of a guidance we can give to upcoming researchers. We have to strengthen our foundations of knowledge about our societies and the methods and look towards an interdisciplinary approach to research itself instead of as communication researchers working in isolation from the mainstream. We have to keep abreast of the developments in research techniques while also keeping our feet rooted to the ground.

We do not loose 'face' by saying that we do not know the answers. I feel we will lose face in not searching for the answers both as part of our daily work and also in the long term.

References

Agrawal, B.C. (1981) SITE Social Evaluation : Results, Experiences, and Implications. Ahmedabad; Space Application Centre.

Alfian and Chum Godwin (1979) Satellite Television in Indonesia. Hawai : East West Communication Institute.

Chu, Leonard (1988) 'Mass Communication Theory : A Chinese Perspective' in Dissanayake, W. ed. Communication Theory : The Asian perspective. Singapore : AMIC p. 126-138.

Dissanayake, W. and Belton, B. K. (1983). 'Reflections on Critical Theory and Communication Research' in Dissanayake, W. and Mohd. Saod eds. Communication Research and Cultural Values. Singapore : AMIC. P. 127-139.

Felicianao, Gloria D. and Lozare, Benjamin W (1974) 'Doing Research in Rural Asia : problems of Methods' Media Asia. Vol. 1, no. 1.

Hursh-Cesar, Gerald (1976) 'Problems in Applying Research' in prodipo Roy and G. Hursh-Cesareds Third World Surveys, Survey Research in Developing Countries. New Delhi : Macmillan Company of India. p. 399-458.

Kishore, Devesh (on-going) Future Trends : Strategies for communication Research. Unpublished project Report.

Lozare, B.W. (1983) 'Problems of Conducting Communication Research in Less Developed Countries : The Need for Innovation and Adaptation' in Dissanayake, W. and Mohd Said, eds. Communication Research and Cultural Values. Singapore : AMIC. p. 55-83.

Pool, thiel de Sola (1966) 'The Necessity for Social Scientists Doing Research for Governments' Background, Vol. 10, p. 111-122.

Reddi, Usha V. (1988) 'Communication Theory : An Indian Perspective' in Dissanayake, W. ed. Communication Theory : The Asian Perspective. Singapore AMIC p. 6978.

Reddi, Usha V (1989) 'Media and Culture in Indian Society; Conflict or Cooperation'. Media Culture and Society. Vol. II. P. 395-414.

Roy, Prodipto, et. al (1976) 'The Survey Siting in p. Roy and G. Hurs-Cesar, eds. Third World Surveys : Survey Research in Developing Countries. New Delhi : The Macmillan Company of India p. 42-90.

Vasudeva, S. and Chakravarthy, p. (1986) 'The Epistemology of Indian Mass Communication Research' Media, Culture and Society, Vol. II, p. 415-433.

Yadava, J. S. (1934) 'Communication Research : Some Reflections' paper presented at the National Seminar on Communication Research Trends and priorities, March 8-9, 1984. New Delhi.

Chapter **10**

Indian Concepts about Human Communication : A Pilot Study on Tolkappiyar, the Ancient Tamil Grammarian

T. Manonmani

This paper tries to extract the basic strand's of thoughts of Tolkappiyar, concerning human interaction through language both verbal and non-verbal from his grammatical work viz., Tolkappiyam generally believed to be written around 4th Century B.C. with a view to evolve human communication theory in the Indian context. This study would stimulate the communication scholars in India belonging to various sub-cultural, religious and language groups. This kind of interior, intra and inter-cultural studies would be an appropriate and needed research at present.

As far as the subject is concerned, it may be asked why should one go back to 2000 years in a modern scientific age when there are enough of Western theories. Here it becomes necessary to understand the various reasons and the need to pay attention to formulate communication theories and models in consonance with the Traditional Indian Society and its cultural ethos and norms.

The emergence of Development Communication in the Third World Nations in 1970s, demands appropriate communication strategies to spread developmental schemes. As Development communication is more of interior, local and cultural in nature it need not be based on outside models only. One of the main reasons for the failure of the developmental schemes is the adoption of communication strategies of North American origin. These theories were developed to help the modern, industrial and urban society and not for the traditional, agricultural and rural society like India. So, the communication strategies carried out in India are largely irrelevant to the society.

Secondly, Indian culture is society oriented unlike its counterpart viz., the western culture which relies heavily on individuals. An Indian may accept modernizations by his rational outlooks, but during adoption he considers largely the views of his society rather than his own. Indian cultural norms are contradictory and diversified in nature and inexplicable to Westerners.

Indian scholars have to come up with alternate theories, paradigm in the Indian context for better results. So normally there should be a revival of literatures in the areas of traditional classical and folk arts both verbal and non-verbal as they print to the philosophy of human communication of our predecessors.

This paper tries to do an experimental study on Tolkappiyam, an ancient Tamil cultural grammar generally written by Tolkkappiyar and ascribed to 4th century B.C. Although it was done in the 4th Century B.C., its main strands of thoughts are similar to the contemporary rings and more relevant to communication research as it deals with language which is a primary channel or human communication and links people in a social network. Moreover, he deals not only with the structure of the language but also discusses the behaviour of that language group. Thus Tolkappiyar's concentration lies on the combined theme of language and behaviour, which would help to conceptualize the communicating behaviour of that group. So, the study on Tolkappiyar would be useful in the theory building research.

Tolkappiyam :

Tolkappiyam has been divided into three main sections, Eluttu atikaram (Section on letters or phonology), Collati-karam (Section on morphology & syntax), and porul atikaram (section on meaning or subject matter). Each section comprises of Nine chapters and totally there are 1610 sutras.

The first section of Eluttu (letter), thoroughly discusses the phonology of Tamil language i.e., phonetics, phonemics and morphophonemics. The section on col (word or morpheme) deals with the morphology of Tamil language. He explains the morphemes, semantics and dialects in brief. He also speaks about the syntactical elements. In the first two sections Tolkappiyar completely deals with lingual variables.

The third section on Porul (meaning) needs elaborate study because of its uniqueness. In this section he talks about not only the units of meaning but also he tries to picture the entire social structure of Tamil land of his period.

The first four chapters of Porul such as Ahattinai lyal (Family life), Purattinailyal (life out of family i.e., social and political life), Kalaviyal (premarital love), Karpiyal (marital love) discuss the socio-cultural variables i.e., nonlingual variables. The fifth chapter "Poruliyal" starts to speak about meaning i.e., lingual variable, in the First and last sutras.

One can infer from the contents and the placements of the main three sections, his principal objectives that is to establish the intimate relationship between language structure and meaning. Language and meaning are basics to human communication. Because the success of communication depends on the tasks of reflecting the intended meaning and of understanding or grasping the reflected meaning. Hence Tolkappiyar gives more importance to the "meaning" and the variables involved in meaning viz. culture, cognition and codification. Moreover, as he combines language with human behaviour, this paper tries to evaluate his philosophy of human communication in the interpersonal level.

Culture :

Each community develops its own language and own methods of language behaviour to share and express ideas that relate to the life. Tolkappiyar elaborately discusses socio-cultural variables in the major portion of his work. As mentioned earlier, the sections on Eluttu and Col deal with lingual variables such as phonology and morphology. The first two sections appear to be compact and their verses are knitted together one by one, the subject of each chapter being dealt with cogently. Therefore the third section of porul (meaning) should normally be dealt with the meaning or semantics. If it is so, there will be an order. But the major portion of this section is devoted to elaborate discussion about society, court and culture. For example, Ahatinai lyal the first chapter of third section speaks about the geography of Tamil land, ecological variations which are termed as 'mutal' (Space and Time), and 'karu'. He explains the flora and fauna and the things peculiar to different ecological regions such as Gods, food, animals, trees, birds, occupations, musical instruments etc., (Porul, Ahan), then he explains the term "Uri" in which he discusses the emotions and feelings of the people. He explains the various routine day to day incidents in his society which are termed as Thinai and Thurai based on 'Uri'. In the second chapter, Purattinai lyal he completely pictures the political scene, especially the critical description of war.

In the above geographical domain he introduces human beings. With in this environment people make interaction with themselves through language for various reasons in various occasions. In the third and fourth chapters as Kala-viyal and Karpiyal, he does a case study with the two selective units of male (kilavan) and female (kilatti) both the adolescents. He pictures how the two units gradually develop their communicating behaviour and by that, how they establish their relationships shaped by various social factors such as family (mother, foster mother) peer groups (Tholi, Pangan), Opinion leaders of the group such as Brahmin (Parpar), seers (aravor), Intellectuals (arivar) and other change agents like artists (kuttar, panar) etc., and also the cultural norms. After these, he begins the description of meaning or message in the fifth chapter. Here one logical question is raised. Why does Tolkappiyar discuss society. Court and culture while he discusses about the structure of the language as Eluttu (letter), Col (ward), and Porul (meaning)? Why does he link lingual and non-lingual variables? Why does he mix language with human behaviour?

Tolkappiyar believes that language is a cultural code. Language reflects the relationships in a culture which in turn gives names to the things or objects. Only the members of that cultural group can understand the meaning of the word representing the objects. Moreover the connotative meaning of a world is nothing but its cultural meaning. Thus in any language, words are associated with the meanings prevailing in that culture. As the cognitive psychologists say that the intentions or messages of the speaker are transformed into signals in the culturally accepted codes (encode) and whereby those signals are received, transformed and interpreted by the receivers in the same way by using the same culturally accepted code (decode). According to Tolkappiyar's view culture is one of the important determinant factors in figurating the meaning in the process of communication.

Tolkappiyar conceptualizes language and culture always goes together. He pictures culture is the nucleus of his society and hence he stresses the importance of cultural conventions in many occasions in figurating meaning or message. He says that "if the conventions are changed the meaning will be distorted." - (Porul, Marapiyal, 93)

From the above discussion, it can be viewed that the individual's communicating behaviour and skills shaped not only by innate abilities and learned behaviour of individuals as pointed out by Noam Chomsky but also by geographical situations. Tolkappiyar views that language is not an external instrument which facilitates the exchange of information, ideas and opinions but as constitutive part of the communication environment itself in the geographical situation.

Cognitions :

The common denominator for all these language spoken, written and understanding i.e., sending, receiving, understanding and knowing is the unique-capacity of human being to think. In human communication how meanings or messages are learnt and how they relate to internal and external behaviour are the central point.

Tolkappiyar believes that the thinking capacity of human being plays a role in encoding and decoding process. He is of the opinion that the individual's mind is the centre of thought process. When he classifies the beings or species of the world on the basis of sensations such as touch, smell, etc., he points out that the human being has the sixth sense i.e., 'mind' which is unique to mankind (Porul, Marapiyal, 17). And also he has the concept about head (Eluttu, Pirappiyal, 1) or black box as mentioned by modern psychologists. He acknowledges all the five senses as potential information channels of human being. It indicates that these stimulating factors are common for both sources and receivers as conceptualized by David K. Berlo But Berlo does not deal with human mind, the sixth sense, the main element involve in coding the meaning.

Mental Process :

Sensing or perceiving the meaning takes place in the thinking process. One cannot look at a tree or flower without the intention to see. 'Looking' is sensation and identifying the image of the tree by past experience is Perception. Perception always involves 'sensation' within itself. These two are names of different cognitive functions but they are parts of the same sort of mental process. Perception is understanding instantaneously and has to have thought flow for mental visualizations of the message. Mental image or visualizations for sensing or understanding is the central) point in human communication. Hence, he distinguishes the 'word' as the form of the word (Sonmai), and the sense of the *word* (Porunmai) since he understands the role of visualizing process *in* communication. According to his view, the meaning of a word cannot be understood by taking merely its form. The form is only like a Jacket or skeleton.

When one visualizes the sense of the word or phrase; the meaning or message is understood.

Again, he says that it is difficult to explain why one meaning stands for a particular word (Col, uriiyal, 98), and one cannot say when and how a meaning is assigned to a particular word (Col-Uriiyal, 98). One cannot explain the meaning for the words which stand for cultural values such as chastity, love, kindness etc., and human feelings such as fear, anger, beauty etc., by showing an object as in other cases. The meaning of these words can be understood purely by mind through social conventions i.e., through the beliefs, attitudes and culture which one acquires through his experiences in his social environments. Tolkappiyam says :

"Likeness, fear, abundance
Chastity, growth, beauty
these are all abstract ideas to be understood by mind through the conventions of the world and not
to be shown to the eyes say so
the scholars".

(Poru, Poruliyal, 51).

From the above sutra, it is concluded that the meaning or message is perceived by a thought process which is mentioned as cognitive process by modern psychologists. He conceptualizes like modern linguist Benjamin Whorf that the culture, cognition and meaning or message go together in human communication both verbal and non verbal. These cannot be separated since the function of the human communication depends upon these three acting together.

According to Tolkappiyar, geographical situation, culture, cognition and attitude are all seen as important to understand the way communication operates. He also stresses the idea that "meanings are in people, not in words".

Tolkappiyar says that when the meaning is understood by wisdom, the thought process becomes very stable and the meaning or message could be retained or set in memory. Retention is another aspect in human communication. Thus the real meaning or truth is perceived while the serious thinking is in operation. The following sutra explains the concept;

"If taught clearly, there will be no
confusion in knowing the meaning of the word,"

(Col, Uriyiyal, 96)

"Knowing by wisdom depends on the

skill of the receiver."

(Col, Uriyiyal, 96)

He views that the above task is possible only for an active receiver. Here the perception implies, complete attention; the nerves, the brain, the ears, the eyes, the heart every thing at their highest reception. Perception includes seeing, hearing and reacting or responding. There is no gap between sensation and perception. Hence he conveys in 'Meyppattiyal' or non verbal communication, the receiver should have the motivation or intention to understand by mind whatever he receives through eyes and ears :

"It is difficult to understand

the significance of 'body-language' for

the people other than those who have the intention and capacity

of understanding well by means of eyes and ears".

(Porul, Meyppattiyal, 27)

He opines, "When mind is made active the reflected meanings are perceived and interpreted".

Tolkappiyar emphasizes the importance of meaning attached to the message connected with both speaker and receiver. His concept has shift away from views of Westerner's models which emphasize the transmission rather than the inter-pretation of information.

From the above discussions it can be inferred that Tolkappiyar identifies two aspects in the levels of understanding meaning. They are : Perception level and cognition level (knowing level). Cognition level is the level in which the receiver arrives at the complete, real meaning or truth at the end of the cognitive process. There are different stages in the entire process of verbal and non verbal cognition such as Perception, recognition, imagery, judging decision making and conception. Assuming meaning in the first stage i.e., in the perception stage, does not mean that one understands the reflected meaning perfectly at all times. For example, the suggestive meanings or the meaning in the deep structure can be understood only after passing all or some of these stages in the cognition process. In this context assuming meaning in the perception stage is equivalent to 'mistaking a rope for a serpant'. Here the mind has to look at everything with complete attention, totally and not just a part of it.

Tolkappiyar's concept of knowing real meaning by wisdom is developed later by a philosopher-poet called Thiruvalluvar in Tamil Nadu, He says,

"To descern the truth in everything from whomsoever it may be heard is wisdom".

Mental lexican :

When the participants or the individuals develop the behaviour of realising or knowing the exact meaning or truth, the human communication would be perfect. The task of retaining experiences (theoretical and practical knowledge, mental lexican etc.) can shape the above behaviour. Tolkappiyar believes that perfect communication can perhaps happen when the speaker and the receiver have-good mental lexican which helps the speaker to give the intended meaning clearly and the receiver to perceive the reflected meaning. He says that the speaker should, explain simply and clearly by using unambiguos words, by whichever possible ways. While using unfamiliar words, the speaker should be able to explain the meaning with the help of familiar words. He says,

1. *"One who wants to express his ideas clearly must express it in an unambiguous terms".*

(Col. Kilaviyakkam, 55)

2. "One word stands for many senses or many words stand for one sense the *unfamiliar words being instructed combined With that which is often used each in it's traditional form",*

(Col. Uriiyal, 1).

A better vocabulary cultivation only can help to express and understand the meaning of unfamiliar words. Above all, he gives suggestion to develop mental lexican since he realises its important role. He says that along with the traditional words one must accept the words which undergo changes and gain currency in course of time and also the new words. Such words are not to be discarded in communication. This view is explicated in the sutra given below. "No word is to be discarded if it has gained currency in course of time".

(Col, Eccha lyal, 56)

It is of Tolkappiyar's concept that if words are retained in the mental lexican in course of time, the communication would not be distorted.

Cognitive balance :

The completion of communication both verbal and non-verbal is based on the commonly shared system of interpretation or balanced system of cognition between the communicator and the receiver. He speaks this concept at the end of the chapter of Meyppattiyal (body language) since he realizes the need of cognitive balance to understand the message especially in non-verbal communication. He says that the lady love's cognitive system must be balanced in various aspects with the lover's cognition to observe the message of love sent through the body language.

"Nativity, family trait, strength age the balance feelings of lust, chastity, graciousness, understanding and wealth these being said in order are the kinds of similarity between the lovers."

(Porul, Meyppattiyal, 25)

According to Tolkappiyar's concept, the cognitive balance includes different variables. As mentioned in the above sutra Nativity, family trait, chastity, wealth, status are the socio-cultural variables. Strength, age, balanced feelings of love and lust, understanding capacity is the sentimental and physiological variables. While these variables are balanced, the female can understand the message and will respond to the male. His view can be compared with the view of modern cognitivist Fritz Heider, "A positive unit relation occurs when a person is tied to another person of thing through any of a number of sociological, physiological or physical variables such as causation, ownership, membership and proximity".

Tolkappiyar's theory of cognitive balance is not only based on the sentiment variables as stated by Theodore Newcomb, the modern communicologist such as the degree of likings or disliking between the participants in communication, or the degree of knowledge or frame of reference as advocated by other communication theoreticians, but also it is based on the socio-cultural variables such as social values, culture, beliefs and attitudes etc. Here it is highlighted that the concept of Newcomb and other's poses that the communication is taking place in a social vaccum. Moreover, in intercultural communication the balance only in sentimental variables can not help even to the occurrence of communication. Perhaps if it occurs the communication would not be completed.

Tolkappiyar views that there is no communication between the two if their cognitive system is not being balanced. In his case study he explains, when a male unit expresses his desire to love, the female unit does not respond to male since she is not possessing the commonly shared system of interpretation with the male. Even though the male implies all the persuasive techniques to make her love him by attributing good and evils to her, she does not respond to him. It is evident in the following sutra :

"Unreciprocated love occurs in the act of having
pleasure in addressing young girl who is not
matured of love making . . .
attributing good and evil to himself and to her without having any reply from her".

(Porul, Ahattinai iyal, 50)

From the point of view of communication it is conceptualized that "Persuation" is not a main metre for attitude change always. If persuasion can change the attitudes, the immatured girl also must have been persuaded. It is realised that human communication is not based mainly on persuasion or manipulation as believed in Aristotlean based Western models. As Osgood says, Western models explain the communication there is one system a source influences another destination by manipulation". But Indian human communication is reciprocal. Its Philosophy is based on knowing the truth or reality as mentioned earlier.

Codification :

In human communication the communicator has certain image in his mind and he intends to create the same image in the minds of his audience. Encoding and decoding are the basic functions in the communication process.

How does a listener make sense of the stream of sounds uttered by the speaker? or how the sense of a word is codified by receiver? To clear the doubts, he speaks about the role of syllables in a word or phrase. The very first sutra in the chapter of 'poruliyal' pronounces this verbal cognition.

If change the sound, the meaning may be conveyed But should not misspell or distort the organization of the syllables say so the scholars - (Porul, Poruliyal, 1).

He believes that the syllables are the organs of a word or phrase and the organisation of the syllables are the pivotal factor in meaning giving function. If the organisation of syllables are distorted the phrase does not elevate meaning. Take for example, the sentence "Sappadu podap padum" (meals will be given). The syllables in an organisation give the meaning. But it should not be uttered or written as Sappa duppo dappadum. Here the phrase, does not figure any meaning. Thus, Tolkappiyar conceptualizes that the exploded sounds as such in the organisation of syllables of a word are accounted for in the decoding process. The different sounds of the syllables in the organisation do not fuse together but come one after another in order in which the different syllables are uttered.

Secondly, he views that the syllables are purely in terms of relative duration. He fixes the length of time (Mattirai) taken for utterences of long and short vowels and consonants which constitute syllables (Eluttu, Nolmarapu, 4). His theory of unit of time is not only to make the speaker to utter easily and convenientil but also to make easy to decode to the receiver who receives the syllables one by one and reunites them. Thus, Tolkappiyar has made an attempt to study the act of communication from the receiver's point of view. He is of the opinion that human communication is essentially transactional, with the receiver playing an active role in the course of any communication process : and does not flow directly from one sender to one receiver.

Coding Process :

When a listener hears the speaker's language, the explosion of sound pattern of each syllable gradually received one by one and again reunited in the left hemisphere of the frontal lobe of the brain to form the thought content. It is purely a neurological and psychological process.

The initial aspect of decoding is taking place when by sound pressures and frequencies pass on to the ear or eye and are transformed into electro chemical impulses by neurons in the nerve fibers. The impulses are relayed to the audio or video section of cortex portion of the brain which controls the thinking process.

Then the sensory impressions are made one by one and the receiver reassembles the codes and interprets and finally understands the message. For example the word animal can be demonstrated as follows to show how the word 'animal' is decoded.

When a person is about to speak, the hearer is quite ignorant of what he will say and the hearer has only a blank expectancy that allows no room even for a guess. But as soon as the speaker utters, the hearer hears the sound of his first syllable 'a', his imagination or guess begins to work. He identifies 'a' sounds within his memory and rules out of all words not beginning with 'a'. When he utters the next syllable 'ni' the hearer's expectation is further defined and limited to the words that have 'ani' and there is possibilities to remove the words which are not having the first two syllables are as 'ani'. When the last syllable 'mal' is uttered all the unknowns are solved and the result is completely defined to consciousness as a meaning. When the process going on, the hearer visualizes himself the image of an animal with four legs and two horns as he has experienced in his society. Thus Tolkappiyar thinks the utility of all different syllables in an organisation. At the same time he understands the gradual revelations of syllables through the gradual understanding of the word symbol.

Process of Human Communication :

Tolkappiyar in his combined theme elaborately discusses the geographical units of Tamil Community, the sub-cultural group in India. In his concern geographical units play a predominant role in shaping communication behaviour of humans as mentioned earlier. Geographical data or information links individuals together in a social network. Hence, this is the process oriented way of thinking about human communication and not the static view advocated by the most of the western linear theoreticians. He insights the evolution of human relationships and the integration of community on the foundations of constant, consistent interaction of people that is interpersonal communication through language both verbal and non-verbal.

Evolution of Relationship :

According to Tolkappiyar language is not merely a channel of communication, but it is essentially a constitutive part of communication itself which helps to establish mutual understanding and rapport and it finally develops relationships with a result to generate commonality attitude. Denis Mcquail says "Communication creates commonality".

Tolkappiyar has demonstrated the above process with his two selective units viz., Kilavan (male) and Kilatti (female). In the chapters of Kalaviyal (Premarital love) Karpiyal (marital love) and Meyppattiyal (body language) he pictures the gradual development of relationship between the units by describing how the two individuals indicate, develop, grow and generate their relationships. He carefully unites trie occurrence of accidental meeting of the two units (Porul, Kalaviyal, 2) recognising in the next meeting (Porul Kalaviyal, 5) and initiating data transmission by body language (Porul, Kalaviyal, 6). They are attracting each other (Porul, kalaviyal, 7) and developing constant interaction, creating mutual understanding and relationship in course of time (Porul, Kalaviyal,). Then the formal bonding or wedding as symbolic acknowledgement takes place (Porul, Karpiyal, 1). They often try to avoid or resist dissonances in their life in order to continue their relationship (Porul, Karpiyal, 14) and live together with children and relatives until their death (Porul, Karpiyal, 51).

Although he discusses the process with two different sexes, their love affairs and family communication, the basic elements involved in the process are common and hence it can be extended to the interpersonal communication in general.

He believes the process takes place on the foundations of love and harmony which are referred to as the primary values in the society (Porul, Kariyal, 1 and 2). It is a clue from Tolkappiyar that the relationship is formed whenever a reciprocal data processing (interpersonal interaction with messages) occurs that is two or more individuals mutually take into account one another's verbal or non-verbal activities in order to satisfy their goals and needs.

Consistency :

When the actions and reactions of individuals in a relationship are consistent with their goals and needs, the relationship progresses on a positive tone with continual increases in the level of harmony. Tolkapp iyar's major attention is focused on the factors which are to impact the maintenance of positive, progressive relationships whenever the situation affects the dynamics of interaction between his two units, for example, when the male unit delays to marry the female (Varaivu Neettittal) or when the parents of the female make arrangements to get her married with another person or when the male goes for prostitution. Tolkappiyar considers two types of factors in resisting the change. The factors are : 1 Participants themselves 2. The social factors like primary and secondary groups and the cultural norms.

Participants :

Whenever the interaction is distorted due to the difference of opinion, one of his units mostly female engages in making harmony by adjusting with its attitude because of its likings towards male, or its drive to fulfill its needs and goals or due to the pressure of the cultural norms. As every individual is highly influenced by society and its norms he gives priority to them and hence instructs female unit to compromise with male mostly for common cause (Porul, Karpiyal, 21, & 26). In his opinion the male also tries to adjust with female for his own reasons by showing some obedience to female (Porul, Karpiyal. 19). Hence, it may be concluded that either of the participants engage in resisting dissonances which obstruct to the continuity of communication. His view is very close to the view of Theodre Newcomb, but Newcomb considers only the participants and their sentiments such as likings or disliking but not the cultural norms and primary and secondary groups.

Newconb's consideration isolates the participants from the social environment.

Social Factors :

In moulding individual's communication behaviour and in shaping relationship, the author refers to the roles played by the primary and secondary group viz., vayilkal. These groups are the dominant influencing factors than the participants themselves and their likings or disliking. Whenever distortions in communication arise, these groups influence the participants. The Primary groups like foster mother (Sevili) and peer groups (Tholi and pangan) are the intimates to the units or the individuals. The sutra given below indicates the influence of the group :

> *They say that it is the duty of the foster mother to instruct of good things and to prevent bad ones in accordance with the past, future and present.*
>
> (Porul, Karpiyal, 12)

All these two occurrences of 'Pulttal' (feigned, dislike) and ideal discard the maid friend will have the right of making expression, interfering (Porul, Karpiyal, 16).

The maid friend will have the right of saying "you are wicked at the face of the lover for preventing him from going astray". (Porul, Karpiyal, 14).

Since primary groups are very close to the units or individual they interfere and avoid dislikings.

Interpersonal communication within relationship is also influenced by the structure and the distribution of power in the society. The personals of social institutions or the so called opinion leaders or change agents of the group often serve as the preservers of the culture. They guide the public to lead a successful life. In Tolkappiyar's society, Brahmin (Parpar), Seers (Aravor) and intellectuals (Arivar) have been emerged as the opinion leaders. They are the well-wishers of every individual without discrimination (Porul, Karpiyal, 37). At the time of confusion or crisis among the individuals or among the groups in the society the secondary group or opinion leaders have the right to interfere. They advice or even command the participants to maintain harmony for the sake of communal unity preserving cultural norms. His sutra explains :

"To make them behave with in the moral code by admonition belongs to them, for the husband and wife stand by their advice"

(Porul, Karpiyal, 15)

From the concept concerning the influence of primary and secondary groups, it is generalised that human communication is more mediated than directive as Katz and Lasar-field and John W-Riley believed. But it is highlighted that these groups are in their turn influenced by culture. Tolkappiyar's stress on the direct and indirect influence of cultural norms requires special attention. This view indicates that no individual family in Tamils can think of independent identity exclusive of community since the culture is group oriented.

Community Integration :

When the participants often adjust in interaction there is a possibility of the emergence of tolerance which is considered as the primary value in the society. Human communication becomes continuous and progressive spiral in nature when the individuals desire to overcome the dissonances and to continue interaction.

The reciprocal message processing in persons involved, contributes positively and progressively to one another, leading to a sense of "positiveness" in their experiences. The satisfaction derived from others and from environment buildsand contributes at the same time to the satisfaction of other's drive. The result is an integrated community which is a source of growing pleasure and value for the participants. This philosophical view is the sole aim of Tolkap-piyar's grammatical work. His entire theme is conceived in one sutra at the end of the chapter Karpiyal. It enriches the purpose of human life as to achieve great things like "unity" by maintaining harmonious relationship with fellow beings. His philosophy can be elucidated from the following sutra : "After the enjoyment of love is fulfilled *being surrounded by the pleasure giving children and having the virtuous relatives with them the husband and the wife must perform what is best for the world and die for it* that is the way of reaping benefit of this life."

(Porul, Karpiyal, 51)

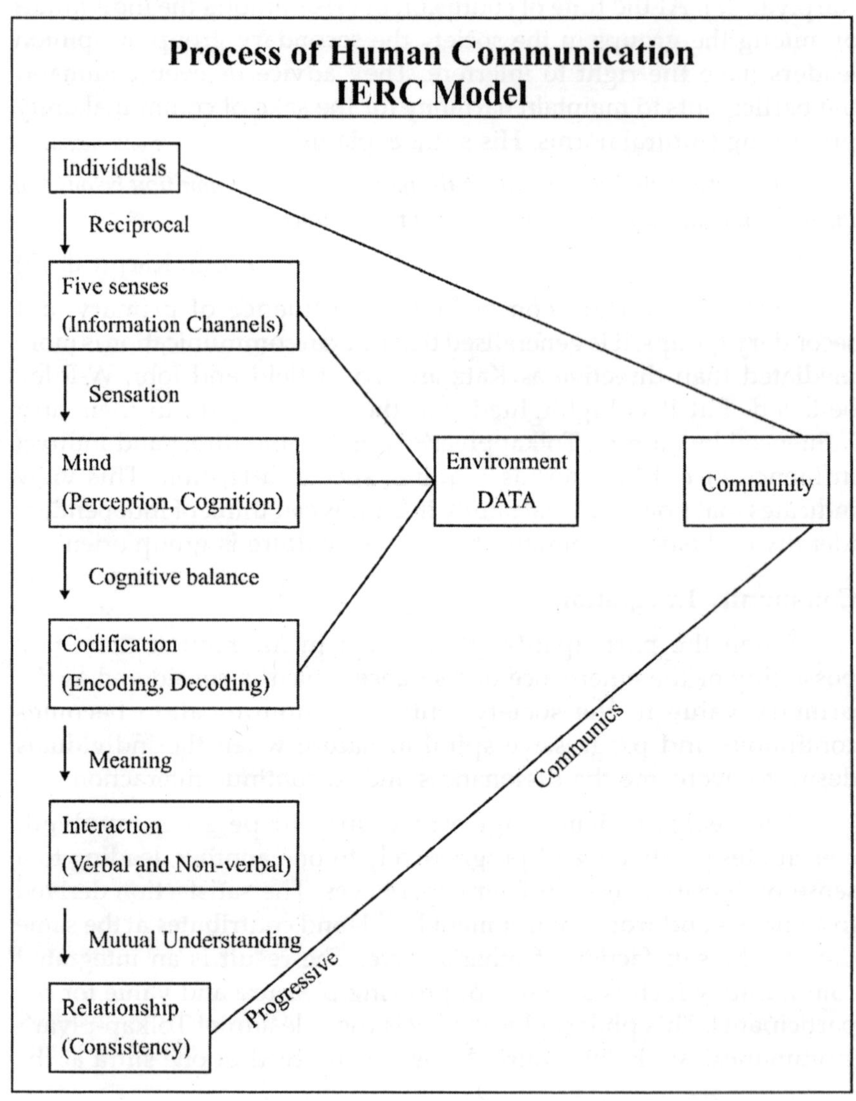

The above sutra can be explained as while every individual performing life with children in central and relatives in-peripherals, the resultant network becomes an integrated community. This philosophy of Tholkappiyar is developed in sangam classics as follows in later period. "All are our nations and all are our kiths and Kins".

The understanding of this slogan is necessary for every Indian in the present context.

Communication scholars in India may consider the view of Tolkappiyar as Human Communication is an essential life process through which individual interact with data in order to establish human relationship and the integrated human society which is necessary to carry out the life activities. From his view point one can realise the meaning of Latin termed "communis". Tolkappiyar's concepts can shake most of the western views about Human Communication as an intact transmission of messages or an interaction from a source to a receiver.

According to Tolkappiyar the various elements involved in human communication process are : Individuals, Five Senses, Head, Mind-perception and Cognition, Geographical situations-data, Codification-encoding and decoding, Meaning-reality. Interaction-mutual understanding, Relationship-consistency, Commonality attitude and community. With these the researcher here tries to bring out the following graphical representation.

References:

Sitaram, K. S. (1991), Culture and communication; A world View, India, Associated printers, p. 160

2. Osgood, Charles E., et. al (eds) (1964), Psycho Linguistics, London, Indian Press, p. 192.

Little John, Stephen W. (1989), Theories of Human communication, California, Wordsworth Publishing company, p. 74.

Berlo, David K. (1960), The Process of Communication, Newyork, Holt. pp. 23-28.

Brant, Ruben D. (1975) Communication and Behaviour, New York, Macmillan Publishing Company, p. 98.

International Encyclopedia of communication, New York, Oxford University Press, Vol. II. p. 317.

Dissanyake, Wimal, (1988) "The need for Asian approaches to communication", Communication Theory : Asian Perspectives,

Singapore, AMIC Publications, p. 5.

Sitaram, K. S. op. cit., p. 172

Ibid.

Datta, D. M., (1972) The six ways of knowing. India, University of Calcutta, p. 259.

Mcquail, Denis, (1975), Towards a Sociology of Communication, New York, Collier Macmillan, p. 75.

Brent, Ruben D., op. cit., p. 160.

Ibid.

Ibid.

For Further Readings :

Aubery Fisher, B., (1978), Perspectives on Human Communication, New York, macmillan Publishing Company.

Branland, Dean, (1960), Interpersonal Communication, New York, Holt.

Bettinghaus, Erwin P., (1968), Persuasive Communication, New York, Holt.

Ernest, Borman G., (1980), Communication Theory, New York - Rinehant and Winston.

Hockett, Charless, E. (1958), A course in Modern Linguistics, New Delhi, Oxford and IBA Publishing Company.

Ithiel de solapool, et.al., (eds), (1973), Hand book of Communication, Chicago, Chicago, Mcnally.

Mcquail, Denis, (1988). Mass Communication Theory : An Introduction, London, Sage Publication.

Srinivasan, T. M., et. al. (eds), (1969), Sense perception (Visayapratyaksam) in Science and Sastras. India, Sri Sharada Trust.

Rogers, Everette, M. and Lawrence Kincaid, D. (1981), Communication Networks, New York, Free Press.

Tan, Alexis S., (1985), Mass Communication Theories and Research, New York, John willy and Sons.

Chapter **11**

Action Research Methodology for Developing Information Packages and Training Modules -Participative Message Development Case Studies from Australia

Shankaraiah Chamla

In this paper an attempts is made to describe briefly how communication technologies and research methodologies are used by commercial and development sectors in the information age and their consequences for sustainable development Two case studies of action research methodologies used in developing information packages and training modules for agricultural development are presented with a view to promote the use of communication research methodologies. This hopefully helps to achieve balanced access to all types of development information.

We are in an information age. Technological developments in communication enabled to convert us into one global village. Information generated in one country is traveling with a lightening speed to another country through satellites, electronic media like T. V., Radio, Videos and various print media. Computers can connect to various data bases through video text, electronic mail and many other wonderful means and modes of communication to any corner of the country you choose to communicate.

Use of information technologies and various communication research methodologies is not uniform in commercial and development sectors. Commercial sector has innovatively and effectively used various communication research methodologies to understand the target audience, develop messages but the development sector such as agriculture, health, education and environment has not utilized these methodologies in designing their messages. Most of the investments and huge budgets are spent in research and developing products or innovations in agriculture, health and less money is spent in designing, testing and modifying the messages to create greatest possible impact on the target audience.

Use of Communication Research Methodologies in Commercial Sector :

Big advertisement budgets for promoting products, consumer goods and the competition in the market place has influenced the various media agencies to understand and reach the target audience, they employ various communication research methodologies in the name of market research, audience research or field testing or field testing of messages and revising them to draw their attention and motivate to use them or buy a particular product. This information from communication research is used to develop precise and well developed messages in the form of 30 second ads on TV, radio, films, home movie videos or short and sharp visual/written ads in newspapers, magazines, posters, etc. To illustrate this point I like you to study in detail the methodologies by ad agencies especially fast foods, health products (drugs/chemicals) or any consumer goods in the world. For example in Australia Forex (XXXX) beer, or Multinational Corporation like Coca Cola spend millions of dollars to produce 30 second ad. The message or jingle is developed based on their target audience research by communication experts and well tested in many locations and countries to create greatest possible impact on the audience. When such messages are presented the target audience responds like a 'PAVALOVIAN DOG' to the powerful stimulus produced by the ad. This S-R (stimulus respond) is not limited to the uneducated poor masses but well educated and higher socio-economic target audience also fall for such ads. This is the power of well researched, professionally prepared communication messages. This approach is also used by commercial agencies such as fertilizer, machinery, pest and disease control and seed companies in all developed countries. The consequences of this approach produced both positive and negative results as stated briefly below.

Positive Results/Negative Effects

- Increased food production
- Reproduced child mortality

- Controlling and epidemic diseases
- Increasing longivity
- Better life style in general

Negative effects

- Environmental degradation
- Excess use of chemicals 'pill-popping' syndrome to improve health
- Excess weight increase, heart attack, cancer, etc.
- Increased population
- Unsustainable development and life styles.

Most of the negative effects are emanating because the development agencies don't use these effective communication methodologies to provide a balanced view. Let us, examine how these are used in the development sector.

Use of Communication Research Methodologies in Development Sector :

Government development agencies such as agriculture, forestry, irrigation, health, education, environmental departments have recently realised the need to invest more on communication. In developed countries these agencies are using advanced communication research methodologies in designing, testing and refining the messages professionally before their widespread use. In developing countries government agencies involving in family planning, forestry media and agriculture have modestly started using the action research methodologies.

When I worked in All India Radio & T. V., as audience research officer in early seventies we used several communication research methods such as surveys, content analysis, impact studies (pre & post testing) of agricultural programs. Agricultural Extension as a profession provided this communication research methodology but their use was limited due to lack of funds by administrators and technical researches in investing money on message development treatment. As departments lagged behind the commercial agencies who employ communication research methodologies and provide budget support to implement professional programmes.

Participative Message Development : Case Studies From Australia :

If we want our clients or target audience to pay attention, understand, accept, and act upon the communication messages it is imperative that we understand the; target audience and make them relevant, professionally test and modify them. We call this process a Participative Message Development (PMD) which uses several communication research methodologies in an action research mode. The PMD process can be divided into the following major stages/ steps. STAGE 1 : Understanding client needs, socio-economic context and communication behaviour.

In this stage communication research methods such as surveys and case studies are conducted using quantitative and qualitative research methods.

STAGE 2 : Developing relevant messages of information packages.

Using various communication principles and clients information from the first stage information packages are developed by professionals.

STAGE 3 : Testing these messages and information packages.

These messages are tested with the sample of target audience for clarity of message (written & audio-visual) and it's intended impact. Trial meetings are conducted to test these materials for their use in groups. Questionnaires and quantitative methods are employed in testing these messages.

STAGE 4 : Fine tuning messages by incorporating the feedback from the testing stage.

Based on the results from the testing process, the materials are fine-tuned by incorporating the feedback.

This is a brief description of Participative Message Development which incorporates various communication and survey research methodologies in a collaborative mode with action orientation.

What is Action Research-a brief explanation :

Before the explanation of research methodology used in these two case studies, it is important to understand the concept of action research. It is not attempted to provide exhaustive discussion on action research definition but help recapitulate it's meaning in the context of communication research methodology in designing development messages.

According to Kemmis and Mc Taggart (1988) 'Action Research is a form of *collective* self-reflective enquiry undertaken by participants in social situations in order to improve the rationality and justice of their own social or educational practices, as well as their understanding of these practices and the situations in which these practices are carried out.' They further emphasize that 'The approach is only action research when it is collaborative, though it is important to realise that the action research of the group is achieved through the critically examined action of individual group members.'

According to Holley (1990) 'Action Research is seen as a process of innovation, change and problem solving'. He identified three developmental phases of action research span, in British education system.

Phase 1 Teaching level : When action research was individualised and conducted with the help of outside experts between 1960s to 1984.

Phase 2 Learning Level 1 : When the emphasis was on the excellence of learning i.e. participative cooperative learning during 1984 to 1988.

Phase 3 'The Learning level 2' : When collaborative school has become a community school based on partnership of all involved with shared values and a collaborative culture. Since 1988 and will continue to develop further.

First Case Study of Developing Conservation Information Package:

The first case study deals with development of Conservation Information Packages consisting of Two Videos, one pamphlet and an extension officers' (users) guide.

Stage 1 : *Understanding clients information needs and communication behaviour :*

In 1982 a detailed study was conducted (Chamala, Conghenour and Keith 1983) to understand the Conservation Cropping on the Darling Downs. It was a collaborative study involving University of Queensland, visiting Professor from University of Kentucky, U.S.A. and Queensland Dept. of Primary Industries (Q.D.P.I.) and farming community. Survey research methodology was used in general but the following new elements were incorporated in the survey research design.

- A farming systems research methodology is used in developing questionnaire to understand clients behaviour during cropping as well as follow periods, thus getting the total picture of their behaviours and actions.
- Field extension officers collected the data. They were trained and supervised during the data collection. University Researchers also conducted some interviews.
- A staff survey was conducted in April 1982 to understand the departmental extension staff opinions on why farmers did not adopt recommended soil conservation practices and if there were any organizational constraints to developing an effective coordinated programme on agronomic soil conservation practices.
- A workshop involving all departmental extension research and management staff was held in June 1982 to present preliminary results of both farmer and staff survey and to define the forces constraining/promoting the adoption of conservation cropping.

The results of this first phase study provided quantitative data for extension programme, planning and implementation. It was also felt that there is a need, inter alia, for conservation cropping information package. Thus, from collaborative research sets of actions started following. It was a collaborative research project with action orientation.

Stage 2 : *Developing Conservation Information Package.*

Based on the results of first phase study a team of extension & communication personnel from O.D.P.I. and University staff were formed to develop conservation information package consisting of two videos, one pamphlet and extension officers' guide.

Two video's The content of the Message is similar but the Presentation was different :

Lectures vs farmers experiences, an extension Officers guide. University graphic artists, television unit were used in designing these packages. Two videos message was similar in content but the treatment and presentation was markedly different. One video was presented in a logical sequence in the expert lecture type while second video was presented interviewing the farmers who implemented the conservation cropping practices.

Pamphlet : The pamphlet cover design was important to all part the all children of the farmer and make him read and use. Therefore, two cover designs were developed for field testing. One more uniqueness of this pamphlet was to include information on the concerns related to conservation cropping such as disease and weed control, dangers of chemicals use etc. In the first phase, farmers expressed several concerns about the conservation cropping. Instead of usually presenting one sided positive information on conservation cropping, common concerns were identified and discussed. It also had innovative farmers' experiences of adopting this technology.

Extension Officers User's Guide :

It was found necessary, based on staff survey to develop extension officers' users' guide. It gave a brief introduction on conservation cropping, role of videos and pamphlet in the overall extension strategy to get farmers adopt the conservation cropping technology. How to use these package in groups and what to do before the groups view these videos and what type of follow-up activities to be undertaken by extension staff in order to accelerate the adoption process etc.

Stage 3 & 4 : Testing and fine tuning the information packages

These packages were designed by specialists in extension, communication (artists, cameramen, Journalist trained in video communication and subject matter specialists) to field test them for the designed impact as well as any defects/groups or concerns in the package.

The methodology used was very comprehensive and showed these videos to farmers, young high school children studying agriculture and extension staff in Queensland. The samples were spread out in the entire state. Meetings were held by extension staff and questionnaires were developed to test them. The results were very interesting and few are briefly described here :

Video :

The background music used in the videos (selected by the team in good faith) was hated by farmers. It distracted them and had a negative effect. Farmers got most from the video where the message was presented in the form of questions and farmers answering while high school liked the expert lecture type. The sample of farmers from Central Queensland felt that the practices recommended were good thought it could be applicable Only on the Darling Down region as all the farmers interviewed were from that region. We knew it is applicable to that region but did not include a farmer endorsing from that region. There were several other minor aspects. We changed the background music, included another farmer from Central Queensland and made several minor modifications based on the pre-testing results.

Pamphlet :

The cover designed tested were not able to draw the audience attention. Consequently a new cover design based on the results was prepared. Another farmers (photo) endorsement from Central Queensland was included to make it more applicable. Similarly, several concerns of conservation cropping and information on weed control and machinery were included.

Extension Officers' Guide :

Conservation cropping agronomic recommendations were the domain of agriculture branch while structural aspects (contour banks and water ways etc.) were promoted by Soil conservation branch. Survey indicated problem of domain in order to overcome the domain problem the endorsement from Directors of both the branches. Several other aspects of planning group meetings and follow up activity details were incorporated.

This is how the information package on conservation cropping was developed using truly collaborative action research approach. Several communication and survey research methodologies were combined to achieve the results required for the action orientation.

Second Case Study of Developing training Modules :

The second case study deals with designing, testing and fine-tuning, of Training Package for Land care group extension personnel and community leaders. This consisted of two publications namely.

1. *Working Together Land : Group Management skills and strategies.*
2. *Group Management skills for Landcare : A Trainers guide.*

The methodology used in designing and developing the above learning (Training) modules for development agents is again described in the stages.

Stage 1 :Understanding the clients information and communication behaviour:

In 1985, a national study (Chamala and Maurer) of District soil conservation. Advisory committees were conducted to understand the extent of community participation and its impact on soil & water conservation and their information and training needs.

The study identified several constraints which groups experienced in trying to contribute more effectively to land and water management. The need for training in the areas of planning programmes, mobilizing financial/human resources, communicating with individuals and the public, managing conflicts and understanding government budgets. In 1987-88, a National Advisory Committee was set-up. The training needs were classified into four categories.

- Management of groups involved in community participation.
- Communication skills required for Land and water management programmes.
- Planning.
- Financial and human resources.

Stage 2 : The first book design was based on the needs analysis. A market research methodology was used in identifying the needs first, developing the contents. First four learning modules were developed based on literature & similar publications in U.S.A.

Stage 3 & 4 : Testing and Fine tuning of training modules :

These four modules were pre-tested in July 1988 by conducting workshops in Queensland, Western Australia, South Australia, and Victoria and Tasmania. Participants in these workshops clearly indicated the need for smaller learning units with each unit addressing a specific topic. They also suggested additional topics. This resulted in fourteen chapters, each devoted to a particular learning topic.

Then fourteen small learning modules (chapters) were designed by the specialists and they, were sent to all the States Representatives who were on the National Advisory Committee for pre-testing. We had a mass of feedback from these pre-testing from all over the country, with extension staff and community leaders.

The major problem was with learning modules.

Title : Citizen Participation in Soil and water Conservation :

Several respondents didn't like this title as it brought fears and concerns among the professional extension agency administrators and community leaders. There was a feeling that non-farming citizens such as urban environmental groups (Greenies) will start dictating to farmers and others involved in the programmes. Some language used was complex, few sentences were too long, some cartoons produced unintented reactions and several qualitative suggestions were made so that they could be more relevant to the intended audience.

We modified the title into Working Together for Landcare Group Management Skills and Strategies and made changes to language, cartoons, length of the sentences and incorporated several other practical activities. Thus, in 1990 two types of publications were brought. The entire 15 chapters were put together in a book as well as each chapter was published as separate Learning Unit for use by landcare groups. In 1991, A trainer's guide was published using the similar methodology. As a result of this approach the training modules developed were more relevant to target audience and they are sold out, hopefully, used in Australia, parts of Canada where similar problems exist, parts of South Asia.

Conclusions :

We strongly believe the need for use of communication research methodology in designing and testing development messages, information packages, or training/learning modules. These methodologies are already used in commercial sectors with a tremendous impact. When audience are presented with development messages without testing their use is retarted or limited and causes environmental degradations. These two case studies provide as examples for participative Message Development research methodologies.

References :

Chamala S., Conghenour, C. M.; and Keith K. J. (1993) 'Study of Conservation Cropping on the Darling Downs : A basis for extension programming' Brisbane : Dept. of Agriculture, University of Queensland, p. 100.

Chamala, S.; Graham, D.; Hejoworth, L; Pitkeathley, A.; Bateman, R.; and Keith, K.J. (1984) 'Developing and Testing a Conservation Cropping Information Package' Brisbane : Dept. of Agriculture, University of Queensland.

Chamala, S., and Maurer, R.C. (1985) 'National Study of District Soil Conservation Advisory Committees' Dept. of Agriculture, University of Queensland, p. 85.

Chamala, S. and Maurer, R.D. (1986) 'Community Involvement crucial for Conservation in Australia-Summary and Recommendations from the National Study of District Soil Conservation Advisory Committees : Brisbane : Dept. of Agriculture, University of Queensland p. 23.

Chamala, S. and Mortros. P.D. (1990) 'Working Together' for Land care : Group Management Skills and strategies Brisbane : Aust. Academic Press, P. 368.

Mortirs, P. D. and Chamala, S. (1991) 'Group Management skills for Land care - A Trainers Guide Brisbane : Aust. Academic Press.

Kemmis, S. and MC Taggart (1988) The Action Research Planner, (3rd ed) Geelong, Victoria : Deakin University Press.

Holly, Peter (1999) 'From Action Research to Collaborative Enquiry; The Process of and innovation' in Zuber - Skerrit, 0. (ed) Action Research for Change and Development, Brisbane : Centre for the Advancement of Learning and Teaching. Goriffth University.

Chapter **12**

The Study of Popular Communication and the Theory of Play

James Combs

Students of popular communication should cultivate the concept of play. Several different strands of thought recognized that people occupy worlds of play. Play is a tricky but useful concept, and at best should be understood as part of a larger theory of human action. Play intrudes into real-worlds such as politics, and explains how such a "serious" arena of action as politics can become an object of play. Popular communication is an autonomous cultural field of play-learning in which people acquire fantastic ideas and images which they may "transport" to other areas of society as a way of understanding and acting towards that arena.

It would be worthwhile for popular communicators to cultivate the theory of play. Rolf Meyerson has recently argued that the subject matter of popular communication "falls within the sociology of play." If this is the case, then it behaves us to understand the concept of play. Here, an attempt has been made to discuss the concept, and how it might be usefully integrated with other concepts and approaches in popular communication studies.

Play is an idea that has had considerable impact on academic disciplines in recent years. Much of the *impetus* for this intellectual innovation came from the work of Johan Huizinga. Who discovered in his historical researches the importance of the "play element" in culture. The literature that followed in the wake of Huizinga modified and enriched his formulation, but largely agreed that play was a multifaceted phenomenon which informed an astonishingly wide variety of human activities. We are now faced with explaining man not only in terms of reason *(Homo Sapiens)* and work (*Homo Faber)* but also play *(Homo Ludens)*.

Although distinguishable in origin and development, there were several other intellectual strands which were to converge on the contemporary play theory. Firstly, there was the work of social theorists such as Simmel and the symbolic interactionist school, especially George Herbert Mead, who wrote about the role of play in society. Mead in particular, anticipated the work of Piaget and others, concerning the importance of play in childhood learning. The girl here seems to be that play, and the form of interpersonal play in rough and then regularized *games,* is a major way in which people learn to think and act socially.

Secondly, there was the development of the idea of society as drama. Here, the metaphor was drawn from theatre, with the argument that "real-life" social action can be best understood from the perspective of the "play-life" of drama. The "social play" is the true source of the fictitious and condensed play seen on the stage. As Kenneth Burke insisted, "The difference between the symbolic drama and the drama of living is a difference between imaginary obstacles and real obstacles." The "drama of living," the play of society, is the source of the created theatre, the play on the stage. But the theory recognises the centrality of play, of the drama of human play enacted in both the "real" and the threatrical worlds.

From a somewhat different source came the formal theory of games. Game theory had mathematical and economic modeling origins from such figures as John von Neumann. But it soon was recognised that the concept of game might serve as a model for social behaviour. The analogy of game highlighted the "interplay" of actors. Economic models of market behaviour could predict the nature of "rational" behaviour of individuals in a system. Similar models have had an impact on strategic thinking among international relations experts. Yet like the study of childhood play and social play, game theory stemmed from the fundamental root metaphor of people playing, although the game of international politics might be more complicated and consequential than marbles, or dating.

The power of the game metaphor was to be such that it would be applied to a variety of social contexts. Eric Berne wrote about the various everyday "games people play." Adam Smith wrote about "The Money Game." Stephen Potter wrote about "Gamesmanship." Such works articulated the notion that there was "play" in the various games one may participate in; *i.e.,* that strategic "moves" and outcomes were uncertain but manipulable. Like Machiavelli's "Prince," the player may be able to, through foresight and adroit performance, to control the action. On the other hand, the "play" may be out of his control; Machiavellian *fortuna* (luck) may sway the game against him. Games have patterns and twists beyond the control of even the most skilled player.

Although there is much more work we could mention (such as the study of sports) this should give us a rough idea of the range of uses of the concept of play. But what do they have in common? A recognition I believe, that people occupy worlds of play, in both a subjective and an objective sense. In the realm of play, we enter, either actually vicariously, arenas of action where we participate in plays, dramas, or games which we perceive to be worth the time and make effort for the enjoyment we hope to derive from them. The spirit of play asserts indeed that "the play's the thing," whether it be in a work or play setting. Work is play if it is approached as such; play may be work in similar fashion. A salesman may enjoy the sale for the play in it; the football player may see the game as pure drudgery. But the buyer may think the sales pitch a big bore, while the football fan may immerse himself in the excitement of the game the player grinds through. The game does not guarantee play it only provides the context of play. As Csikszentmihalyi and Bennett note, "play stops when it becomes boredom."

Play is autonomous. It is an interlude in the mundane, an episode that "flows," that means something over and above the usual arid habitual. It is even useful to think of the play-experience as if one were in a stage play. One is separated from everyday life by heightened awareness of being "in a play," with action, opportunity, skill, risk, all the elements of "interplay". The autonomy of play makes it a moment out of time, a play-period in a play-pen. It is a "recess" from that which matters, a retreat into the pointless, a play on reality. This is illustrated by Geertz's famous analysis of the Balinese cockfight. The "deep play" of the cockfight in a very real sense meant nothing. It changed no one's status or wealth, and the outcomes were soon forgotten. But the play-experience was important in an *aesthetic* sense. The cockfight became a context for a kind of symbolic rendering of the existential features of life. It made "every experience comprehensible by presenting it in terms of acts and objects which have had their practical consequences removed and have been reduced (or, if you prefer, raised) to the level of sheer appearances, where their meaning can be more powerfully articulated and more exactly perceived, it catches up these themes death, masculinity, rage, pride, loss, beneficence, chance and, ordering them into an encompassing structure, presents them in such a way as to throw into relief a particular view of their essential nature."

The symbolic and aesthetic experimental qualities of play have been echoed by William Stephenson in his "play theory of mass communication." "Playing is pretending," he writes, "a stepping outside the world of duty and responsibility. Play is an interlude in the day. It is not ordinary or real. It is voluntary and not a task or moral duty. It is in some sense disinterested, providing a temporary satisfaction." On the one hand, he argues, we occupy the world of reality, concerning ourselves with what is useful, and adjusting ourselves with the necessity of painful work in order to make do. But there is another world we occupy. The world not of "reality-testing" but rather "fantasy-making," with little concern for utility, and taking ourselves out of the world of work into a world of pleasure. For the consumer of mass communication, this means that he or she prefers and seeks a world of "communication-pleasure," a play-world of news and features which one experiences as part of a fantasy world of play-time.

Now, if we contend that play may even invade the work-world, and that arenas of action such as politics are plays, we are entering the realm of ontology, or theories of action, which the concept of play in itself is not adequate to encompass. For my own part, I think play should be integrated into dramaturgical theory as an inclusive perspective on human action. For the most part here, we would focus on those zones of social action we must associate with play, most specifically popular communication. In that respect, let me now describe a bit how my study of the relationship between politics and popular communication was enriched, and complicated, by reflection on play theory.

It is clear from the above discussion that plays can be either active or passive. The active participant in play likes the action, enjoying the playing, the competition, teamwork, agonic achievement, risk, or whatever for its own sake. The passive participant in play enjoys the vicarious pleasure he or she derives from attending the action. In either case, they participate in has if it were a play-world, even though the stakes and the consequences may be quite real. Such autonomous play-worlds may range from chess to war, but the unifying feature is the attitude of the participant in the play. It is here that the distinction between "real-worlds" and "play-worlds" break down, and gives us a clue as to how play intrudes into real worlds. Further, this intrusion gives us some hints as to how popular communication affects "serious" worlds, which I will treat below.

Politics, I came to see, is one of those arenas of social action in which plays (1) a large role. Stephenson has noted that politics is "from the public viewpoint, play. The diplomats and politicians do the work; the public merely has something given to it to talk about, to give them communication-pleasure." Stephenson's view has been supported by much research. "For the individuals who comprise the mass," writes Dan Nimmo, "politics is a playful, expressive drama by which they air needs to be with others, needs to be liked, self-acceptance and self-doubts, aggressions, love, and hate." Most people thus experience politics as passive play, which they know through the mass media and play-talk. It is part of the vicarious world "out there" beyond their immediate and palpable experience, but which they "know" through the fantastic construction of a dramatic, aesthetically pleasing world. For a variety of reasons, politics is for most people a play-world of the imagination.

But what is astonishing on reflection is that in many ways politics is play for the actor's tool. Not only do they have imaginative "constructs" as to what the political world is like, they often enjoy the play for the action too. Politics is hard work, of course, but many politicians and political activists enjoy the play itself. An obvious example is the campaigner who likes "pressing the flesh", and the campaign worker who exults at the hard work-play of the protracted contest. (It is instructive to go back and read Machiavelli, with his somewhat whimsical treatment of the intense political games people play).

The concept of play permits us to understand better the "linkage" between popular communication and "serious" worlds such as politics. The relationship between popular communication and other worlds that people occupy is subtle, long-term, and "covert", but examination reveals that it is nonetheless evident and important. Even though popular culture is an autonomous field of social play, it has impact on people because, as a form of play, people *learn* from it. Figures such as Mead, Piaget, and Erikson have understood, that play is not neutral, harmless, or frivolous, even though it may seem so. What could be more innocent than children playing in a sandbox? But not so as Erikson writes, "child's play is the infantile form of the human ability to deal with experience by creating model situations and to master reality by experiment and planning." Since social learning is a lifelong process of coping with temporal change, then people seek and find cues for thought and action in subsequent play contexts. Even though popular communication may be produced as "mere entertainment," since the object of play created deals, like the Balinese cockfight, with existential and social questions; it has the potential of being a major source of social learning.

This became clear to me during the course of my own study of politics and popular communication. The overwhelming proliferation and attention to popular communication should have convinced us that it is a major source of political learning, although political scientists have not devoted much time to it. Simply the overt depiction of politics and politicians alone should have some impact. But perhaps much more important are the non-political "covert" messages which originate in popular communication but which have subtle and long-term consequences. Popular communication is inextricably linked to the "symbolic universe" of the larger culture, and to its identifiable parts, such as the political culture. There is a reciprocal, understandable flow of ideas and images to and from these different play-worlds. Politics affects popular communication, and popular communication affects politics, each giving the other something to play with. The actors, events, values, and processes in the political drama have their impact on the creations of popular communication. And, conversely, the impact of popular communication indirectly on mass audiences and directly on the political order can be traced.

For example, when one studies the phenomenon of political heroism in contemporary America, failure to consider the enormous impact of popular communication on conceptions of heroism makes full understanding impossible, because, we are all party to the vast array of heroic figures icons, celebrities, superheroes, fad figures, and so on which parade across our consciousness. It should not surprise us, then, that political heroism, although their to a rich tradition, should be affected by changing tides of popular conceptions of heroism. Indeed, contemporary political heroes appear to be consciously styled combinations of the traditional and the contemporary, and shape their appeals around such popular expectations. Carter "plays" as a kind of John-Boy Walton who brings traditional down-home values and good sense, although also being an engineer and administrator : Kennedy is a sort of "royal redeemer" drawn from the Family imbued with *noblesse oblige*, but also as the subject of eternal and lurid interest in the women's magazines and tabloids; Brown is in the mold of the maverick reformer from the West, but also moves among the beautiful Beverely Hills and safaris with Ronstadt; Reagan is a "citizen politician" whose celebrity originated as a pop hero in the movies; Glenn a contender for 1984 is equally a celebrity (astronaut) who will give us, no doubt, "the right stuff."

The point is that the "play" of such political figures is not exclusively a fiction created from symbols of the political culture, but also draws upon symbols of popular communication. Similarly, our expectations as to what political heroes can and should do are affected by popular communication. Consider the following quote from an expert on terrorism at a conference held before the abortive raid on Tehran :

> We want 'cinemaphotogenic solutions'. We want solutions in the way John Wayne and Errol Flynn produced them. The cavalry rides in people like this because it is like the movies. If that is inapplicable, we have a sense of national frustration because we can't play Wild West there. Tehran is a big city, close to the U.S.S.R. We can't just jump in and say, 'We'll show those bastards whose mythology should prevail'.

I suspect that both elite and mass fantasies have been "fed" by the heroic depictions of such successful daring raids as the Israeli raid at Entebbe. That such commando raids are risky seems subdued by the dramatic glory of "pulling it off" just like in the popular account. The play stirs play-fantasies about repeating the play in another "real life" context that somehow the play of political life can and should live up to the dramatic standards of art.

Popular communication, then, is an autonomous cultural field of play learning in which people acquire fantastic ideas and images which they may "transport" to other arenas of society as a way of understanding and acting towards that arena.

There are some well-known ways in which the play of popular communication may instruct us. As a mode of knowledge popular communication provides us with the ideas and images which help shape our attitudes towards the wider social world. In the long run, popular communication may either reinforce or undermine views we previously held. But there seems little doubt that both the conservative and subversive possibilities of popular communication are quite real.

As a mode of action, popular communication conjures role models and dramatic scenarios which shape our behaviour.

Perhaps all this is no more than to say that popular communication is a source of identity. Klapp has argued that popular heroes take us on "identity voyages" in which we learn various stances towards the world. Zweig has remarked that the "play-heroes of popular culture, crude and repetitive as they may be, remind us with an unexpected realism of who we are, and what we do." As George Hervert Mead long ago understood, the "social self" is a creative and dynamic process of self-identification and adaptation. It is surely one of the major items on the agenda of inquiry for scholars to specify the relationship between social identity and the play of popular communication.

What is argued is the development of a ludenic theory of popular communication. "All play," wrote Huizinga, "means something." It is our task to interpret the meaningful "webs of significance" that popular communication both shapes and reflects. But as Geertz understands, it is not an easy task, since it involves the daring pursuit of what he calls "thick description." The meanings of play are multifaceted and changeable for those that create the objects of play, those that seek uses and gratifications in them, and even those who don't want to play. All those shifting groups "out there" which "attend to" popular communication in some sense are the subject matter of our inquiry. Why do they play? What do they play with? Why do they play with one thing at one time and something else at another time? And even, who do popular communicators play with inquiring into the "whys" and "wherefores" of popular communication?

It is also clear that playing with ludenic theory is not the ultimate game that will preclude other popular playing. For those imbued with other play-forms mythic, formulaic, Freudian, Marxian, cultural, or aesthetic play theory can enrich their own approach to popular communications. Whether one is interested in the motives, the social structure, the aesthetic form, the mythic meanings, or whatever popular communication, play theory might be of some help. Although theoretical integration incorporating a variety of such approaches is unlikely in the near future, certainly play theory is a rich set of ideas that some future Einstein of popular communication (or even a future Kenneth Burke) can play with.

But there is one final caveat, what might be called the paradox of seriousness. Both in the conduct of play, and in the study of play, there is the danger that play can become too serious, and that seriousness can become too playful. In the first instance, if playing with popular communication, both for the actors and the inquirers, becomes too solemn, play loses the very playfulness which is its virtue. If, for example, the Popular Communication Caucus takes on the academic seriousness of, say, the American Political Science Association, then not only will it become dull and dreary, but will also lose that vivacity and daring which is the glory of play. Similarly, if the creators and consumers of popular communication in society reduce it to "serious business," then whatever verve or stimulation it might have had, will be lost. It is believed that the major American professional sports of baseball, basketball, and football are in danger of doing just that, collapsing under the tedium of money, legalities, over organization, and ideological seriousness.

The second instance seems to be less dangerous, and indeed if one is going to err, the side of play is the one to take. Play theorists tend to argue that people are at their best when they don't take themselves and the world, too seriously; in other words, when they cultivate the arts of play. Seriousness, after all, is just another form of play, and indeed considerable attention should be given to the study (playfully, I hope) of "serious play." And here, of course, we come to the nexus of our paradox, for this very paper is an instance of serious play, written playfully. Some will take it seriously, and thus overlook the spirit of play which inspired it, and the very insight which might contribute to our understanding of popular communication. Yet, it will be urged, such study is serious, and play must be eliminated from study for it to be conducted properly. If it is so then the play-element in the play of popular communication is likely to be missed, thus eliminating from the inquiry, precisely, what is important in popular communication. But if popular play is approached playfully, then one may be accused of being frivolous. And indeed if one is too frivolous, one may well eliminate the concentration which is necessary to do proper inquiry into popular communication. And no doubt, if one is self-conscious about one's serious task to be playful, then play is destroyed by the seriousness of the effort to be playful. Yet "spontaneous" play-inquiry which eliminates seriousness may omit the attention to rationality and academic standards which are the core of any serious discipline. Play after all is Dionysian, and perhaps that is what makes the figure of *Homo* Ludens so elusive, for both the serious actor and the serious inquirer. In any case, one cannot not play, and how one should play, both in action and in inquiry, is the existential core of the paradox. This of course takes us into the higher reaches of the meaning of the whole play, which is, one might suggest, beyond the scope of this paper.

References :

ROLF MEYERSON (1951). "The Sociology of Popular Culture : Looking Backwards and Forwards," *Communications Research,* 5; 4 (July 1978), p. 388; sec also Martha Wolfenstein, "The Emergence of Fun Morality," *Journal of Social Issues, 7 :* 4 pp. 10-15.

JOHAN HUIZINGA (1956). *Homo Ludens : A Study of the Play Element in Culture* (Boston : Beacon Press), and *The Waning of the Middle Ages* (New York; Double-day).

See much varied works as David L. Miller, *Gods and Games* (New York : World Publishing Co., 1970); Jacques Ehrmanr (ed.). *Game, play. Literature* (Boston; Beacon Press, 1971); Riger Cailloids, *Man, Play, and Games* (Glencoe 111.: Free Press, 1961).

HUIZINGA, *Homo Ludens,* foreward.

For Georg Simmel, See in particular his section, "Sociability as the Autonomous Form, or Play-Form, of Sociation," in Jurt H. Wolff (Ed.), *The Sociology of George Simmel* (New York : The Free Press of Glencoe, 1964), pp. 43-57; and the classic, George Herbert Mead, *Mind, Self, and Society* (Chicago : University of Chicago Press, 1934); see also the excellent interpretation of Mead, David L. Miller, *George Herbert Mead : Self, Language, mid the World* (Austin : University of Texas Press, 1973).

See Jean Piaget, *Play, Dreams, and Imitation in Childhood* (New York : Norton, 1951); Erikson, *Toys and Reasons: Stages in the Ritualization of Experience* (New York; Norton. 1977).

See the collection in James E. Combs and Michael W. Mansfield (eds.), *Drama in Life : The Uses of Communication in Society* (New York; Hastings House, 1976) : Dennis Brisset and Charles Edgley (eds.). *Life as Theater : A Dramaturgical Source book* (Chicago : Aldine Publishing Co., 1975); Stanford M. Lyman and Marvin B. Scott, *The Drama of Social Reality (New* York : Oxford University Press, 1975); and James E. Combs, *Dimensions of Political Drama* (Santa Monica, California : Goodyear Publishing Co., 1980); more directly, see James Combs, *Polpop : Politics and Popular Culture in America* (Brunswick, Ohio : King's Court, 1982).

Quoted in Michael Feeham, "Kenneth Burke's Discovery of Dramalism," *Quarterly Journal of Speech,* 65 (1979), p. 410.

See the discussion of von Neymann in J. Bronowski. "The Long Childhood," in his *Ascent Man* (Boston : Little, Brown, 1973), pp. 432-436; see also the delightful book, John McDonald *Strategy in Poker, Business, and* War (New York : Norton Library, 1963).

KENNETH E. BOULDING (1962). *Conflict and Defense* (New York : Harper and Brothers).

ERIC BERNE (1964), *Games People Play* (New York : Grove Press).

ADAM SMITH (1968), *The Money Game* (New York : Random House).

MIHALY CSIKSZENTMIHALYI & STITH BENNETT," An Exploratory Model of Play," *American Anthropologist,* 73 : 1 (February 1971), p. 56.

See Mhaly Csikszentmihalyi, *Beyond Boredom and Anxiety : The Experience of* Play in Work and Games, (San Francisco : Jossey-Bessa, 1975).

CLIFFORD GEERTZ, "Deep Play. Notes on the Balinese Cockfight," in *The Inter-pretation of Cultures* (New York : Basic Books, 1973), p. 443.

WILLIAM STEPHENSON, *The Play Theory of Mass Communication* (Chicago : University of Chicago Press, 1967), p. 46.

Ibid., pp. 57-59.

See "A Concluding Remark : Existence, Communications, and the Dramaturgical Perspective," in Combs and Mansfield, *Drama in Life,* op. cit., pp. 429-431.

EDWIN CADY, *The Big Game : College Sports and American Life :* (Konxville : University *of* Tennessee Press, 1978), p. 52.

STEPHENSON, Play Theory, op. cit., p. 60.

DAN NIMMO (1974). Popular Images of Politics (Englewood Cliffs, N.J.; Prentice-Hall), p. 143.

ERNEST G. BORMANN (1972). "Fantasy and Rhetorical Vision : The Rhetorical Criticism of Social Reality," *Quarterly Journal of Speech,* 58, 396-407.

See the data gathered in D. E. Berlyne, "Laughter, Humor, and Play," *The Handbook of Social Psychology,* 2nd ed., edited by Gardner Lindzey and Elliot Aronson, Vol. 3 (Reading, Mass.; Addison-Wesley, 1969), pp. 795852.

ERIKSON, Toys and Reasons, *op. cit.,* p. 54.

BERNARD BOWRON, LEO MARX & ARNOLD ROSE (1954). "Literature and Covert Culture," *American Quarterly,* IX (Winter 1957), pp. 377-386; Arnold Rose "Popular Logic in the Study of Covert Culture," Chapter XXI of *Theory and Method* in the Social Sciences (Minneapolis : University of Minnesota Press, 1954). pp. 320-326.

DAVID HACKER, quoted in Rogers Worthington, "Terrorism : Mindless Violence or New International Politics," Chicago *Tribune,* November

29, 1979, section 2, pp. 1, 3.

ORRIN E. KLAPP (1969). *Collective Search for Identity* (New York; Holt, Rinehart, and Winston), pp. 211-256.

PAUL ZWEIG (1978). "The Hero in Literature" *Saturday Review*, December, p. 35.

GEORGE HERBERT MEAD (1913). "The Social Self," *Journal of Philosophy*, 10; pp. 374-380.

HUIZINGA. *Homo Ludens, op. cit.*, p. I.

GEERTZ. *Interpretation of Cultures, op. cit.*, p. 6.

Chapter **13**

Priorities in Development Communication Research in India - Some Salient Issues

Uma Narula

In the present paper, I have presented an analytical review of the conceptual development communication perspective; the government development efforts and the ensuing development communication situations. The critique for the two dimensions is based on the, Gandhian plan of rural reconstruction, two development experiments of Etawah (1948) and Nilokheri (1950) and the six National Development plans. The assessment of the periodic development Communication situation is based on periodic appraisals, both by government and independent institutions; research studies and forums of communication scholars and practitioners.

My major concern is the communication research priorities in the development communication perspective perse which may solve the puzzle of current development disappointments. The puzzle is that development progress has not commensurate with the facts so much efforts and resources have been put in the development programmes.

I have assessed the issue from three perspectives : (1) a conceptual point of view; (2) the problematic node in the current development as an outcome of a specific communication perspective-some unanticipated and unwanted consequences; (3) the solution-proposed communication strategies to handle these unanticipated and unwanted consequences.

The thesis of this paper is : the concept of communication needs redefining. The new definition suggests that communication is not just a discreet act but a perspective from which a set of acts, technologies, and institutions are assessed. If viewed from this perspective; many aspects of development efforts may seem problematic which were not seen as such before or they were not even assessed as a form of 'Communication'. The result is some unintended messages, unanticipated and unwanted consequences.

The paper was presented in a Seminar on 'Communication Research Trends & Priorities' organised by IIMC, New Delhi is *a* Research Officer, Department of Communication Research, IIMC, New Delhi need certain communication strategies in order to counteract these consequences; and to succeed in our development efforts. Priorities should be given to the proposed communication strategies.

Analytical Review of Indian Development Communication :

Development in India has not occurred haphazardly. Since independence, development efforts have been based on a conceptual scheme, their effectiveness evaluated and the programs revised as deemed appropriate. The governing documents have been a series of national development plans, usually designated for a five year time period, which articulate the rationale, objectives and means of development during that period. The review of literature suggests that each development decade operated upon a specific development and development communication model. Some of these models, like community development and Panchayat Raj were unique to Indian Society and were the result of much experimentation in Indian Development Planning. Others were international models, such as the Dominant paradigm, borrowed to fit the Indian Development scenario.

A review of Indian development perspective viz, conceptual perspective in development communication; government efforts; and evaluation of development communication situation is presented.

We accepted Gandhi meta-model as a conceptual model for development communication. This stressed the village development through participatory communication, development of human beings through agriculture and rural industrialization. The development situation at this time was that people were not ready to accept development. They did not have faith in the government development plans. Development procedures were unsystematic and development administration was not oriented to development tasks.

The development experiments of Etawah (1948-52) and Nilokheri (1950) identified three conceptual dimensions of the development perspective that effective development depends on the co-ordination of active village participation development administration and strong village leadership. The government made efforts to operationalise these three dimensions through various five year plans.

In the first plan : the development perspective was to create development awareness among the people; develop-mass-media, interpersonal and traditional channels and encourages effective public participation. The government efforts were to use extension education approach; develop extension agencies for interpersonal communication links; develop radio technology and promote traditional media. People participated by contributing in cash, labour and kind.

In the second plan : the conceptual development perspective was: Panchayat Raj was initiated for decentralized rural administration; land tenure reforms for structural change; structured leadership and structured citizen participation were introduced. The development administration was reoriented to development tasks. For the first time, efforts were made to initiate development communication programs for women. The government implemented all except structured citizen participation and partially implemented the development communication programs for women.

In the third plan : the conceptual development perspective was to give adequate information to the masses to understand the development messages : to change their altitude and behaviour towards development; encourage active people's participation in the program implementation; assess people's felt needs and give them feedback through interpersonal and media channels particularly through organised media feedback programmes, and to develop traditional media for social communication. The government made efforts to implement these through extension agents and media channels of radio, T.V., film and print. But the development of traditional media only remained a commitment. The evaluation of the development situation indicated that only radio was effective, the press neglected the rural scene. People adopted the development programs. The program specialists were effective in assessing the felt needs of the people. The people did not give feedback either through media or the interpersonal channels.

During the fourth plan :

The conceptual development perspective was to change attitude and behaviour; motivate people for adoption. This was the first time the interpersonal (Personal) channels of family, friends' opinion leaders were used as development communication links. The government made intensive and extensive campaigns using media mix and interpersonal channels. But the development situation indicated serious development information imbalances between urban rural progressive regions and backward urban/rural regions. Development personnel did not have adequate development skills. There was lack of co-ordination between communication agencies and development agencies the interpersonal (personal) channels proved ineffective. Development community felt that there was information glut but in reality there was information imbalance and blockage by the 'haves' and the development bureaucracy.

In the fifth plan :

The conceptual development perspective was to initiate more of development awareness, information and adoption; to remove structural constraints by legislation/education; to adopt a different development communication model for the urban (urban slum) communities; to formulate a national media and communication policy. The government made efforts to achieve these ends. The evaluation of the development communication situation indicated, that, there was satiation effect in development was ineffective due to communication problems with the people media, messages and development bureaucracy. The people had inadequate access to the media, interpersonal (impersonal) channels. They lacked aspiration and motivation and were unable to articulate their needs and grievances. The rural elites were opposing change and progress. On the media scene; the existing media was inadequate and underutilized. There was lack of co-ordination among different media. The messages were not oriented to the weaker sections these were overloaded with information, inconsistent, untimely and were poor in treatment. The development bureaucracy had low awareness of rural problems they had inadequate communication tools, and inadequate facilities to train rural communicators. They also paid inadequate attention to feedback given by the people. Over and above, there was credibility gap between the development bureaucracy and the masses.

The sixth plan :

The sixth plan (current) has emphasized the support of media in development communication. It has stressed communication action programs for meaningful participation of people. It has advocated on integrated approach to development communication as well development programs. These government efforts are at an experimental stage. (for detailed description of the analytical review please refer Narula Uma, 1984).

At present, we are faced with two major questions how is that more development had not occurred since so many interim goals of development programs have been achieved? Are the disappointing results of development due to unintended messages and unanticipated features of the existing communication model? The critiques of the present development communication models suggest why development is producing disfunctional effects.

The present Indian Development Communication model is rational and block recursive model. The major blocks being : development demands, efficacy of the masses, learned dependency, and the efficacy of the development functionaries. The recursively lies in the unwanted repetitive patterns (urps) of these blocks. The review of development literature and findings of empirical studies have shown that URPS of these "Blocks" have resulted in development disappointments.

A critique of the development literature and the three field studies done in rural community and urban/ urbanslum community done in the year 1982-1983. (Narula & Dhawan 1982, Narula 1983, Narula & Pearce 1983) has suggested that there is a lack of coordination between the three agents of developments viz. planners, change agents and the masses. This lack of co-ordination has resulted in such consequences as (1) discounted and perceived helplessness; (2) Learned dependency and the neo-fatalism; ethic of proportional efforts; (3) and the counter productive effects. These consequences are inter-related and a reunanticipated and unwanted but are URPS. The interaction analysis among three agents of change points that URPS are resulting in development disappointments (Narula Pearce 1984).

Taking a Communication Perspective :

The currently operating development communication perspective in Indian Society is specific to Indian Development community, masses and Indian situation. The present Indian situation endorses the 'New paradigm' which aims at promoting participation for development. But review of development literature and field studies imply that development is not effective as a suitable Communication perspective is lacking. These conclusions indicate the needs for redefining the concept of development communication. The international development community has defined communication as information, education and communication. But what they have not considered is the Communication perspective in which these functions are performed.

Pearce (1983) has defined 'communication' as a perspective from which to assess acts, technology and institutions. From a communication perspective the particular acts, events and objects are examined in terms of the way they express the social reality or the way they Create or maintain the social reality. Therefore whatever one does or does not do CAN BE LOOKED AT from a communication perspective.

Taking a communication perspective has several consequences. One of them is that events and objects which were not clearly seen as forms of communication; clearly appears as functioning communicatively. The scope of the issue being considered and the variables involved is increased as certain acts, technology and institutions are not defined irrelevant by the narrow concept of communication. Another consequence is that particular event or object may be seen as unexpectedly problematic. For example an act may be required as an EXPRESSION of an existing social order, but have the EFFECT of undercutting the legitimacy of that social order or producing unintended effects which are unwanted.

The combinations of these advantages make possible a radical rethinking of some of the situations to which the professionals apply their efforts. One such situation is development communication in India.

Development Situation :

The current development situation is that development does not commensurate with the efforts and resources put into it. The receipt of messages, awareness and even adoption is not highly related to participation the people are getting more and more dissatisfied. The people are not taking the kind of initiative that government wants them to take if development has to occur. This warrants rethinking of the whole structure of how development occurs.

The conceptual move that in-forms this paper is that 'communication' is a way of creating social reality. From this perspective, it is not sufficient to examine the contentof the messages, or frequencies of behaviour and to correlate it with the effects, but equally necessary is to look at how the patterns of communication, including but not limited to government development/ communication, operate.

One way of looking at is from the dominant mode of thinking and working in India fatalism. The government has tried to convert fatalism to a kind of modernity; but has been successful in some but not all of its attempts to change the people. An analysis in the later section will indicate that these government attempts have made people 'dependent' on it (government) I have labeled it as "Learned Dependency" or Neo Fatalism.

Narula (1982) describes the sundrome of learned dependency not only as the absence of 'development motivation' but a cluster of beliefs and attitudes and folkways which are an undesirably alternative to it. I suggest that the behavioural cues for dependency are lack of initiative, shifting responsibility, exaggerated perception of obstacles, seeking directions and solutions and problems, seeking favour of a superior and lack of individual choice and overconfirmism. Figure 1, the Schematic categorisation of 'Learned dependency'.

Learned dependency is the interplay of such factors as : fatalism present both in individual and the social structure (2) government perception of fatalism as a development problem; (3) government intentions for development; and (4) the development efforts of the government. This assertion is based on the data based analysis of rural, urban slum population of India, and development, communication.

The fatalism in Indian people suggests the traditional characteristics of contentment, individual dependency, ignorance of modern innovations, exploitations by exploitative linkages. The fatalism in social structure suggests participation by social-hierarchical dependent relationships.

These posed development problems for the government. Development communication was operationalised to change this fatalism to modernity.

Through its planning agencies, the government identified four of these characteristics of traditional Indian life as problems (all except the 'individual dependency') and specified the alternatives they would strive to create the combination of which is many developers concept of 'modernity'. The objectives of the development plans were an improved quality of life, a better informed populace who were emancipated from the exploitation of the local elites, and an individualism which would enable persons to improve themselves, independently of the existing social hierarchies (Refer figure 1).

Learned dependence not only inhibits development, but it does so in a manner which is relatively impervious to government control. The motivation of the people cannot be rechanneled as straightforwardly as the flow of traffic by building the road or the flow of water by building a dam. Motivation exists within the 'social reality'; the meanings created and maintained by the people in their interactions with each other and in the context of their physical and cultural environments.

How did learned dependency happen? And what can be done to transform it into 'development motivation'?

These two questions are obviously related. One hypothesis is that; learned dependency is the unfortunate and uncorrectable result of poor communication between development functionaries and the masses. The second, more fundamental is that learned dependency is the manifestation of social meanings which make 'active participation' paradoxical.

The experience of those engaged in planned change, ranging from individual psychological therapist to the international developers, have shown that 'interventions' are assimilated into the system in which they occur, and often have results unanticipated and often not wanted by the change agents. These literatures suggest that attempts to introduce change be based on the best available analysis of the situation.

Learned Dependency :

Communication Structure of development :

The government development efforts consisted of three components : (1) direct action to provide infrastructure (e.g. paving roads, draining ponds, installing sewage drains etc.); (2) legislation removing social obstacles to development (e.g. setting maximum amount of agriculture land that can be owned by any one person; removing the exploitative linkages of money lender bonded labour etc. in the rural areas (3) development communication designed to inculcate discontent with things as they are and information about modern improvements. Only the latter of these was seen as 'Communication' though in retrospect all of them should have been. The administrative hierarchy consisting of multiple levels of government officials who are in contact with the masses is 'invisible' not seen as a component of the communication process but simply as the 'overhead' or 'delivery system' of goods and services. This traditional perspective identifies 'communication' as one, necessary but not central, component of much larger program. The concept that communication is a discrete class of acts, distinguishable from other classes of acts, has been a major impediment to the understanding of human interaction. From the communication perspective, the legislation and direct action by the government were strong means of communication-perhaps stronger than the radio programmes and educational packets which were thought as development communication.

The 'development Communication' very effectively informed the masses about modern alternatives and successfully instilled discontent. However, by its legislation and direct action policies, the government loudly nominated itself as a surrogate parental figure who accepts responsibility for making development work, which controls the resources by which the benefits of development may be delivered to the masses. The direct action, legislation and creation of immense development administrative machinery have affected the social reality of the masses. Rather than replacing the local social hierarchy with an individualistic and community orientation by the masses, the government announced itself as the top of the social hierarchy. The 'control mechanisms' or 'recepies' for dependent behaviour in traditional Indian culture have been perpetuated by the Government assuming the role of paternalistic provider. The content of messages which advocated individual and community initiative were unintentionally and powerfully contradicted by legislation and government action which communicated a very different relationship between the masses and the government. The government clearly nominates itself as the provider of benefits and guarantor of liberation from the exploitative linkages. This is clear and effective and people response is 'Learned dependence' in which they ask, "why should I (for what is needed"?) Since it is the government responsibility.

Ironically, these problems as indicated by our data are the result of the efficacy of the development problems and not dueto the failure of the development programmes.

The social reality of the people has changed to one of learned-dependency on the government. Traditionally dependent on local, social and hierarchy and on fate they have transferred that dependency to the government as a very appropriate response to the multiple messages which the government gave; that they were ready to take responsibility for everything ranging from economics to sewers to birth control.

They were persuaded by the content of the development communication efforts that they should not be contented with what they have, that they deserve more,and that they should communicate their grievances to the government. They do, but seldom get a response, because the government simply is not capable of filling all of the requests they receive. This creates a paradoxical situation and produces a counter productive situation. The effects of "Counter productive" situation will be discussed in a later section of the paper.

In the two communities, that I studied the 'fatalism' had been replaced not with modernity as had been hoped but with a 'Neo-fatalism' in which people were not content but had learned to depend upon the government to make things better. Learned dependency is less of a novelty than a shift from dependency on the village social structure to depending on the government.

The above explanation of 'learned dependency', views it as the unfortunate result of the unintended-communication messages by the government and the misinterpretation of other messages by the government and the people. Again, generalizing, development motivation was blocked by four perceptions by the masses. (1) They perceive the government as the perpetual provider of the benefits of modernity. (2) They perceive the government as ineffective and inefficient in the delivery of the benefits and resources of development. (3) They perceive the government as uninterested in their own suggestions or initiations in their development programmes. (4) They perceive themselves as ineffective in accomplishing development objectives. These perceptions are reasonable, and they give rise to three perceptions on the part of the development functionaries that the masses : (1) limit their participation in development programmes to passive acceptance of what the government did for them and complains about what it did not do; (2) have no useful suggestions for development; (3) do not maintain or repair what the government gives them. The resulting dialogue between these groups implements frustration on the part of development functionaries and increased dependency on the part of the masses.

If this explanation is correct, the problem stems from a remediable mistake. A continued use of one way, top down messages would be ineffectual, but a better form of communication may be achieved by other means. The 'Development-Dialogue-Actionstrategy section of this describes an action project designed to test the effectiveness of one situation to this problem. The creation of this strategy might suffice to counteract previous dysfunctional & Communication patterns by enabling development functionaries and the masses to replace erroneous with correct perceptions of each other.

The second explanation views learned dependency as a manifestation of a pattern of meanings which make 'active participation paradoxical'. If this explanation is correct the action project described not is effective. The problem stems from the interface of concepts of self, development and the government. Specifically, development is seen by the masses as something for which government has resumed responsibility and in this government has assumed a parental role, 'Self is seen as 'one of the masses', a complex concept which reflects the lack of what Pareek (1963) called 'extension motive', and the presence of what we designate as 'ethic of proportional effort'.

Ethic of Proportional Efforts :

This concept requests that masses are not willing to put more effort for development in kind, labour or cash; if the development benefits accrued are not in the same proportion. There are two consequences of such a move; (1) the haves; of the society may have more development resources but they are not willing to contribute those resources for the simple reason that both they and the have nots will be equally benefitted. The have not's also reciprocate the same attitude. But such an attitude of the masses is not conducive to active participation of the masses. (2) The masses are not willing to put more development effort since they feel the government should put more development effort, as it is the govt. responsibility to deliver development both adequately and efficiently and why should they (masses) participate actively.

The hypothesis of an ethic of proportional effort in the social reality of the masses is significant if true, because such an ethic is the antithesis of the ethic implied in the concept of "democratic socialism".

Counter-productive situation :

The activities of the development functionaries themselves create a counter-productive situation by being unreceptive to the people's participation. There is research evidence that general attitude of the development functionaries towards public participation and that of the official particularly associated with the development programmes affect the development awareness, adoption and approval of the programmes.

At present, in the 1980's with high level of awareness and information, people are toady to accept and participate in development but their participation is ignored by the development functionaries. The significant fact is that by not approving people's participation they are encouraging people's dependency for development.

The implications are that these development functionaries have failed people in doing development effectively by creating counterproductive situations. Moreover, what bothers people most is the inadequate feedback loops among the various agents of change. The planners, change agents and the masses. The inadequacy of feedback loops does affect the adequacy and efficiency of the delivery system of socio-economic facilities and the active development participation of the masses.

Counter-productive situation is : development functionaries demand participation by the people, but clearly communicate their disapproval of the form and content of the feedback given by the people, so people stop giving feedback. Moreover, the development functionaries are unreceptive to people when they (people) contact them personally for various development programmes, so people stop communicating to them personally. 'No contact' with the change agents and no development feedback only lead to passive participation of the people. People loose initiative to participate.

It is further alleged that development functionaries express disapproval of the public participation because of their own 'trained incapacity' to value the conventional wisdom and accumulated experience of the people in contributing to development. The development functionaries have the biased conviction that people are incapable of handling their own problems due to illiteracy tradition and not having media and technological exposure. The end result is that development functionaries do not encourage public participation. Thus, public assessment of development bureaucracy is low and the existing redress channels are inadequate and inefficient.

Narula and Pearce (1984) have advocated certain suggestions, for reducing and coping feelings of discontent learned dependency and show that masses are NOT helpless. They have posed recommendations at three levels;

(a) broad strategic considerations;

(b) some matters involving the administration of development;

(c) some aspects of communication infrastructure;

(d) development - Dialogue-Action strategy;

(a) Broad Strategic Considerations :

To counteract learned dependency, we recommend that development communication include explicit statements about the relationship between the government and the people, as well as information and exhortation about the particular development projects. It demands carefully articulated statements. In this context, the development communication requires two features : educating the masses to see development as an interaction process loading to partnership with the government. (2) It should acknowledge the contradictory messages the masses have received and provided a way of interpreting them so that does not lead to learned dependency and confusion among the masses.

We recommend a programme of social education; first for the development bureaucracy and then for the masses. To show eligibility for this programme the community would have to demonstrate 'active participation'. A community participating would submit its own descriptions of needs give a priority ranking to them, match government contributions to its own contributions in labour, cash or kind.

We recommend that the decision whether a community is ready for participation be made by the community itself. This recommendation is the key that makes the whole package work. Given the history of the 'interaction', virtually any thing that government tries to say to the masses would be interpreted as "experts telling us what to do". 'An interpretation which would increase dependency'.

(b) Administration of Development :

The development experience as a whole is a process of communication, therefore the organization and functioning of the administrative bureaucracy is a part of the construction of social reality of development. As such, specific attention should be given to the administrative matters in terms of the way it functions as a participant in the development interaction.

(c) Communication Infrastructure :

The data base analysis indicates that the development interaction has taken some unfortunate patterns because of the deficiencies of the communication infrastructure.These deficiencies have substituted infrequent and inadequate "feedback" for the "dialogue" which should occur.

The infrastructure is well developed in terms of getting messages to the masses from the government, and has some capacity for getting 'feedback' from the masses to the government. However, 'feedback' is a unindirectional message flow from the masses to the government. The masses are not at all sure that they have been heard; and if heard, understood; and if understood taken seriously; and if taken seriously, become the cause of government action. The most frequent responses to their feedback are disclaimers from the change agents (who say that they have passed the message through the proper channels and do not know what is happening to it now). And inferences from "general" messages produced by the government which are usually not specifically related to the complaint or suggestion. Having one's messages vanish without trace into the bureaucracy is an insurmountable obstacle to the development of a sense of efficacy and involvement.

The concept of participatory development, as well as democratic socialism suggests that 'dialogue' should replace feedback; not only the communication patterns be 'two-way' but they should be interactive. Obviously genuine dialogue cannot occur everywhere, since planning commission cannot sit with residents of every village. But dialogue can be productively approximated as carefully planned activity.

We recommend that DIALOGUE BE CREATED BY COMMUNITY ACTION PROJECTS, which incorporates a variety of procedures adapted to the specific characteristics of the communities in which they occur.

One procedure is the development of 'interpersonal' contact-points in communities where they do not exist. Our data show that rural communities have ample interpersonal channels whereas the urban communities do not have. A second procedure is the two-way use of media. Radio forums and teleclubs have proven a viable way of approximating dialogue in rural areas. These procedures should be explored for new applications and introduction to new areas.

(d) Development-Dialogue-Action Strategy :

Development communication strategy identified as DEVELOPMENT-DIALOGUE-ACTION-STRATEGY is suggested to counteract the low development motivation among the masses. Such a development communication strategy may focus on community involvement and involvement of the development functionaries. The assumption is that such a strategy can focus on self-participation, self-reliance and self-management of the socio-economic facilities provided by the government. The strategy suggests the active involvement of both the development functionaries and the community as a whole (not just the group leaders) in a mutual search for solutions. The active involvement of both the groups can reduce the low perceived efficacy of the development functionaries and their own actions. This can result in effective development delivery system. The decisions whether that community is ready for participation and development interaction are made by that community itself.

Methodology :

Development-Dialogue-Action Strategy is a step wise involvement, of people in general and the development functionaries in group-discussion meetings.

Step1. A voluntary group meeting with the development, functionaries. Both the groups express their discontent, expectations, about on going development activities. The group should not consist of more than 25 persons at a time. The group of development functionaries remains the same for all the groups. They are the concerned functionaries for the ongoing development facilities.

Step2. In the second discussion meeting: both the groups may be more precise in their discontent/expectations. This sets the process of what can be done by both the groups.

Step3. In the third discussion meeting both the groups search for the solutions. This makes the group realize in the process the feasibility and shortcomings of the solutions proposed.

Step4. The fourth discussion group meeting recommends action to be taken. What who and in what way?

The assumption is that these four steps "dialogue" meetings in quick succession with a short gap should make both the groups ready for development action and mutual cooperation for achieving the results.

Results Expected :

1. Both the groups get a forum to express their discontent and expectations for each group. (The research evidence is that both the groups blame each other for development not progressing so well. And each group does not realize what the specific complaint is).
2. Both the groups will mutually search for solutions. (Research evidence is; that a solution suggested by the development functionaries may not be acceptable to the masses as it may not be need based. Solution suggested by the masses may not be acceptable to the development functionaries because of limited resources of funds, man-power, etc.).
3. Both the groups act on the action recommendation in cooperation and are in better positron to counteract the practical difficulties.
4. The secondary results may sharpen the ensuing causes of development discontent. If the second explanation is correct; that learned dependency is the manifestation of a pattern of meanings which make active participation paradoxical, some 'interventions' much more subtle than described in the action project is required. At this point, I am not prepared to describe this strategy although it will be based on analogues in three literature : (1) small farmers programmes of the food and Agriculture organization (FAO) which uses a strategy of 'misdirection'. (2) Family therapy using the Milan model in which 'counterparadoxes' are employed and the Palo Alto group using the 'Paradoxical injunctions' (Palazolli, et. al. 1978) Waltzwick and Weakland, 1974) as therapeautic injunctions. The basic idea of the three is to disrupt the logic of a system by encouraging a person to do that which he has been striving to avoid. Could some comparable form of "counter-paradox" be used in a nationwide development process? Could it be targeted on the unwanted parental role of the government; on the ethic of proportional efforts and the counter productive effects produced?

The answer to these questions are by no means simple, but the analysis of the dialogue between government and the masses about development suggests that they are important ones to consider.

The contributions of this scholarly exercise to knowledge and process of communication are :

(1) the importance of both INTENDED AND UNINTENDED messages for development. Any analysis and evaluation of development programmes should include both.

(2) Taking a communication perspective of the analysis of the interaction itself should be the content of the development communication. In this way we may be able to achieve the goal of 'interpretative' research : that of increasing the possibility of communication among the participants of social interaction.

(3) The lessons learnt are communication experiences and necessary for the significant changes in the on going Indian development programmes towards modernity and to bring second order changes of modernity.

(4) Narula & Pearce (1984) have developed a new research technique labeled as 'CONTEXTUALISING' for studying the Interaction analysis for development. Such a technique studies interaction at different levels of context-from broader to narrower contexts of development.

Learned Dependency – Unexpected Result of Development Communication in India

Fatalism	Govt. Perception of the problem	Government intentions	Development efforts	Actual results
Contentment	-	Improved quality of life	Make them discontented with existing conditions	Very discontented
Individual dependency	-	-	Reinforced dependency people substitute Govt. as source of everything (first order change of modernity)	Very dependent on Govt.
Ignorance of modern innovation	-	Inform	Devcom-IEC functions motivation discontent intervening (Patenalistic-attitutde)	Not so ignorant
Being exploited	-	Emanicipate	Legislation Devcom (Patenalistic attitude)	Well informed about personal rights
Participation by social hierarchy dependent relationships	-	Individual development	Individual development Government hierarchy for traditional hierarchy	Learned Dependency lot of initiative about personal rights

Summary : From Fatalism to Neo-fatalism rather than to modernity.

: Leaned-Dependency.

: Categorization based on Data-based analysis of rural and urban slum population of India.

Footnotes :

1. This paper is informed by the 'co-ordinated management of meaning (CMM) theory. For a fuller description of this **theory of** social interaction, see Cearce & Cronon 1980.
2. **The currently** vogue 'hermaneutic' move of treating actions as 'texts' demanding interpretation is the basis; but the distinctive character of the present study consists of its focus on "interaction" and "reflexivity".
3. A full explanation of the rationale for and the use of the procedure of **CONTEXTUAL I SING FOR analyzing** interaction is found in Narula & Pearce 1984 **chapter.**

References :

Ensminger, Douglas. (1972), RURAL INDIA IN TRANSITION. Faridabad Thompson Press.

Meyer, Albert & Marriott Mackim, et. al., (1958) PILOT PROJECT INDIA, Berkeley, University of California Press.

Narula Uma & Dhawan R. K. (1984), Indian Development Communication Models: an Analytical review of literature. IN DYNAMICS OF DEVELOPMENT

COMMUNICATION AWARENESS MOTIVATION, PARTICIPATION, Chap. 3. University of Microfilm International. University of Massachusetts, Amherest. USA.

(1982 a.), PUBLIC PARTICIPATION IN DEVELOPMENT PROGRAMS - an empirical study, Indian Institute of Public Administration, Delhi Chapter, New Delhi.

(1982 b.), LEARNED DEPENDENCY - AN UNEXPECTED CONSEQUENT OF DEVELOPMENT IN INDIA. Papers presented to the Eastern Communication Association (ECA), Amherst, USA.

(1983), LEARNED DEPENDENCY - UNEXPECTED CONSEQUENT OF DEVELOPMENT IN INDIA. A CMM. & Pearce W. B.

Pearce W. B. & Narula Uma.

...& Narula Uma & Cronen Vernon

Pareek Udai.

Planning Commission

Planning Commission

Pradhan : P.

Pearce W, B.

PERSPECTIVE. Paper presented to the 5th International Conference

Cluture and Communication Philadelphia.

In Press. DYNAMICS OF DEVELOPMENT : THE INTERACTION AMONG PLANNERS, CHANGE AGENTS AND MASSES. Sage Publications, California.

In press. LEARNED DEPENDENCY AND CONFUSION IN THE DEVELOPMENT DIALOGUE IN INDIA. In Communication in Third World Countries, Concept Publishers, New Delhi, Ed. Sharma, S. P.

(1983), Communication-Action Plan. World Bank, Washington.

(1960), COMMUNICATION,ACTION AND MEANING. Newyork, Praeger Press.

(1968). MOTIVATIONAL PATTERNS AND PLANNED SOCIAL CHANGE.

International Social Science Journal; 20(1) 465-473.

Government of India. Sixth Five Year Plan (1980-'85), A FRAMEWORK, New Delhi. Government of India Press.

FIVE YEAR PLANS. First, Second, Third, Fourth and Fifth, New Delhi. Government of India Press.

(1980), THE SOCIALIST THOUGHT OF MAHATMA GANDHI, Vol. 1, DELHI GDK Publishers.

(1983), SCIENTIFIC RESEARCH METHODS IN COMMUNICATION STUDIES AND THEIR IMPLICATIONS IN THEORY AND KNOWLEDGE. In Speech Communication in 20th Century, Carbondale, Report of International Conference.

Report of the working group

Schramm, Wilbur, AND Porter, William (Eds.)..,.

Seminar

Singh, Kusum Jitendra

Southern University Press. Ed, Tom Benson,

Report of ASIAN REGIONAL SEMINAR ON RURAL COMMUNICATION, 1977 Pg. 2, Hyderabad, organized, by Osmartia University, AMICand ICSSR.

February, (1979) Delhi. RURAL DEVELOPMENT AND COMMUNICATION POLICIES. Organised by Indian Institute of Mass Communication, New Delhi.

Report of the COMMITTEE FOR EVALUATION AND PUBLIC PARTICIPATION (1959) P. 10 Luck-now, UP Government. Report of the COMMITTEE ON PANCHAYAT RAJ INSTITUTIONS. (1978) Ministry of Agriculture, Irrigation and Rural Development, New Delhi. Government of India.

On BLOCK LEVEL PLANNING (1978) New Delhi, Planning Commission, Government of India.

MEN, WOMEN AND MEDIA UNDERSTANDING HUMAN COMMUNICATION. Honolulu, University of Hawai Press, 1972 revised 1982.

(1979), Feb. 21-24- RURAL DEVELOPMENT AND COMMUNICATION POLICIES. Hyderabad, organized by the Indian Institute of Mass Communication, New Delhi and International Institute of Communication, London.

(1976), GANDHI AND MAO AS; COMMUNICATORS, A COMPARATIVE STUDY OF PRACTICE AND THEORY Unpublished dissertation. University of Philadelphia.

Chapter **14**

Connecting Culture, Listening and Storytelling

Melissa L. Beall

Culture may even be described simply as that which makes life worth living.

T.S. Eliot

I grew up in a first-generation immigrant farm family in Nebraska, in the middle of the United States of America. My father and aunts and uncles could speak only Danish when they went to school. The teacher spoke only English and told her students that English was the only language they could use while in school. They learned English quickly, despite living in homes where only the Danish language was spoken. My grandmother (Besta Mor) and her friends often told me stories and shared memories about things from the "old country." Besta Mor wanted us to know about Denmark and our Danish heritage, but she also told us that we were "Americans and we needed to follow American ways." I grew up listening to stories, learning about "the old country" and fascinated with what I heard. As an adult I am still fascinated by stories, particularly stories that tell me something about the culture : clothing, food, candles, rituals, celebrations, and traditions. Everywhere I travel, I seek out the stories of that culture. I've come to believe that if everyone listened to more cultural stories, we would learn to appreciate all the similarities and differences the world holds for us.

Storytelling, Listening and Identity :

Besta Mor told me stories about how and where she lived and the conditions in Denmark when she was young. She told her grandchildren more than her children, perhaps because we were better listeners, or were more curious and didn't ask so many personal questions. From the stories she told, I knew that her days in Denmark were not the happiest ones, and that money was very scarce. She shared things with me, but always reminded me that she came to the United States to find a better life. Even so, she left behind brothers and sisters whom she missed, and whom she never saw again after her departure from Denmark as a young woman. Interestingly enough, her stories remain clear in my mind. So much so, in fact, that when I first visited Denmark, I recognized her village, even though none of us had remembered to write or consciously note the name. We were driving through Jutland, the western most part of Denmark that is the peninsula rising out of Germany and into the North Sea. When we went there, I said that I did not know the name of the village. But, when we drove into the village of Faster and I saw the name, the Church, and a few old buildings, there was a stirring of memory. I stopped the car, turned around, and went back to the Church. In the cemetery, I found my great-grandmother's tombstone and knew that I had at last found what my family had searched for years earlier. They wanted to find family but could not. Recollection of Grandma's stories led me to stop at the right place. And, when we had all but given up hope of finding family, suddenly we had an ending and a starting place. Family members there provided insights we had never before known.

Each successive trip to Denmark has provided insights into whose I am as well as my ancestry. And, each trip and each new piece of data has provided an impetus to learn more about the culture that makes me who I am as well as the many stories and many cultures that make up this diverse and interesting world.

Story telling, Listening, and Identity :

My own interest in intercultural listening began when a curious child listened to Besta Mor's stories about the old country. She was a tiny, tiny woman who had survived many difficulties. She left Denmark on her own, moved to the USA, worked in a can manufacturing factory in Chicago (and lost the tips of two fingers while she worked there), eventually moved to Nebraska, met and married my grandfather, bore four children, lost her husband when my mother was one-year-old, managed to hang on to the farm and her children despite the odds, and lived to the age of 96. I never heard her complain but her stories made me appreciate what she did. Grandma was a survivor, and despite her tiny stature, she possessed a daunting determination to overcome obstacles. Her story was not that much different from her family members who remained in Denmark, but her journey created differences. And that journey created more stories for her family to learn and integrate into their own lives. I often think about what I would have missed had I not sensed the need to be quiet and listen to her.

My view of my ancestry is not much different from that of many immigrants from around the world. Here in Iowa where we live, for example, there are large numbers of second, third and fourth generation-immigrants from such places as the Czech Republic, Denmark, England, Germany, Ireland, the Netherlands, Norway, Slovakia, and Sweden. In addition, there are new immigrants from at least 50 different cultures who reside within a 100 mile radius of my own home. Each of these immigrant families brings their culture in the form of celebrations, clothing, customs, food, language, rituals, stories, and traditions. And, when we take the time to learn about these unique aspects of these cultures, we are certainly richer for it. Collins and Cooper (1997) suggest that among the values of storytelling are that stories "improve listening skills" and listening to stories "helps [people] understand their own and others' cultural heritage."

In many parts of the world, March 17 is St. Patrick's Day. The stories we hear about St. Patrick and Ireland are based on both truth and embellishment. Historian Thomas G. Ryan (2004) says that 10.9 percent of U.S. citizens claim Irish ancestry, and local members of the Ancient Order of Hibernian societies keep the stories and the traditions alive with parades and other celebrations. And, while we are not all Irish, and not all the stories are true, the stories we hear help us to know something about the culture that is the origin of the story and the customs (http://www.historychannel.com/exhibits/stpatricksday/ *history/*.

Holidays and special events are often at the center of cultural storytelling. Christmas as it is currently celebrated around the world is an amalgam of many different countries additions to the holiday originally celebrated as a religious holiday. While it is still a religious holiday, it is often celebrated and stories are told that have little to do with the original religious connotations. Denmark's Julknisse has little resemblance to Sinter Klass in Holland or Santa Claus in the United States. But, the stories we hear about these characters and their deeds have much in common. And the differences clearly show us each culture's unique characteristics as well. Food is an important part of the holiday traditions and customs in the Christmas celebrations. In Denmark, freshly baked goods are always available during the holiday season and homes are decorated with candles and flowers then and at all times of the year. Because the Danes love life and light, they use more candles, per capita, than any other country in the world (Danish Immigrant Museum, Elk Horn, Iowa). The Danes believe if a guest leaves a home during the holidays without being fed the guest may carry the Yule spirit from the home. One rarely enters a Danish home without being offered coffee and "a bite to eat" at any time of the year, holiday or not. Many cultures bring candles, flowers, and food into their celebrations. The African-American holiday, Kwanzaa, is a celebration of Pan-African family, community, and culture. The symbols include food, corn and other symbols of the harvest, and a candleholder with seven candles, all representing some aspect of African culture and traditions (Kwanzaa website, http://www.officialkwanzaawebsite.org/. Chinese New Year celebrations have become more prevalent with ever-increasing numbers of Chinese moving from their native land to other cultures. "Probably more food is consumed during the Chinese New Year celebrations than at any other time of year." The Chinese New Year is also celebrated with lanterns and candles and celebrates the ancestors and their contributions to family [www.educ.uvic.ca/faculty/mroth/438/CHINA/chinese_new_year.html.]. The traditions and customs accompanying different religions and cultures have become better known to many of us. Why? Because we are curious and because the stories people share with each other help us to understand our similarities and differences. If we could get all government leaders to listen to the stories of other cultures, we would likely learn that we share more than we differ.

Images of a Culture :

In many cultures people welcome others into their homes with food in an environment where certain items are a necessary touch. An invited guest to any home in many parts of the world, for example, is likely to find flowers and lighted candles gracing the room. Food and drink, taking many different forms and ranging from biscuits to cake, cookies, or pastries are accompanied by tea or coffee. Hospitality means that one offers one's guests food and drink. And, that visual, physical display of hospitality can often be found in the stories or images of the culture. The colours we use, the ways we decorate our homes, yards, and gardens, the shape and form of the artwork are all subtle images of culture. University of Northern Iowa art graduate student, Bounnak Thammavong says his father's stories and his Laotion heritage are reflected in all of his art. "I don't always notice it, but there is a cultural influence that shows up in my work, sometimes in an abstract way."(Waterloo/Cedar Falls Courier, March 28, 2004, El). And, for those of us who are one or two generations removed from our ancestral homes, that culture is often conveyed to us through the stories we learn.

Culture, Listening and Story telling :

Are there geographical differences in listening? Yes. There are certainly differences from country to country and culture to culture. There are even differences within specific areas, but for the most part, western Europeans and citizens of the United States are generally not good listeners. We spend too much time talking or thinking about talking, and too little time in allowing the other to talk so we can listen to them and their stories. Are today's Danes and American-Danes good listeners? My research tells me that they are much better story tellers than they are listeners. A Danish Communication professor said when he interviewed new immigrants about their experiences in Denmark, they told him they didn't think their new neighbours were good listeners and wondered why. He responded that Danes are not good listeners because they "love to hold on to the word." (Jensen, personal email, March 2004) When I interviewed Danish citizens, and international students from Denmark, all were surprised that I would spend time researching listening in cultures. Most indicated that "listening is expected in school, but listening is not an important area of study. Listening is just something you do." Teachers in what we in the US call middle or high school asked for teaching activities or suggestions for reaching their students, because, they said, "Danish students are not good listeners." (Personal conversations with teachers in Denmark, August 2003) Colleagues from other northern European countries tell me that communication is now an area of study, but few teach or research listening. And yet, listening scholars around the world believe that if we all were better listeners, we would have a better chance of achieving peace and understanding in the world. I often receive emails from Dr. Om Kharbanda, a retired Chemistry professor in India. He is a frequent contributor to newspapers in his own country, and writes to many people around the globe about the value of listening. Dr. Kharbanda believes that we all need to learn to be better listeners and that can only be accomplished with a specific teaching program.

Colleagues in Asia, especially China, Korea, and Japan, decry the attitude change in the younger generation. These colleagues tell me that today's young people "do not listen." International students from those countries tell me that the expectations are there, and they are taught mostly through stories and in the verbalization of clear expectations from family members and teachers, that listening is important. The students told me story after story about what they learned from listening. But, they also indicate that listening is not as important to them, personally, as it is to their elders, especially when they interact with their friends. Even so, the Chinese, Japanese, Korean and Thai students suggested that listening is the way they learn the stories of their culture, the way they learn what is important in the classroom, and what is important for their families. The Chinese character for "to listen" suggests that listening is only effective when it totally involves a person: the eyes, the ears, and the heart.

Every interview with an Asian student reinforced the view that listening is both directly and indirectly taught in their countries. And, while the students may not practice their listening as much when with their friends, listening is an expectation they must meet at home and in the classroom. And, listening to them is more than just hearing and remembering things. Listening truly requires total, undivided attention on the part of the listener. Listening is an active process that requires the listener to work hard at listening. One of the international students told me she always had to listen to what the speaker was NOT saying. Another told me that listening must especially come from the heart. She talked about how when we meet an American acquaintance in the United States, the American will say, "Hi, how are you?" as they keep walking. But, in her country, that would be considered rude behaviour, and unless the one who asks the question waits for an answer, there would be no response because one was not expected. If the person asking does not wait, the question is not perceived as genuine, and the event is viewed as a "strange" kind of greeting. She said it was difficult to get used to that concept because in her country you would never ask that without stopping to wait for a response, even if you were in a hurry. Listening cannot (and should not) be hurried in her culture.

Another student told me that in Taiwan she was punished in the following way by her mother for talking and not listening. One day she talked back to her mother and failed to listen to her. The next day her mother visited the school and told her teachers that she was a disobedient child and did not listen, so they, too, should punish her for not listening and acting as a child should act. (She was a teenager at the time.) Others told me that there were more subtle ways of punishing students and family members who do not listen, but listening is a clear expectation. One of the ways Asian students learn what it means to be a good listener is through the stories they are told. The same could probably be said for all of us. Thus, storytelling and listening are closely linked in some Asian cultures, at least. And, several Japanese interviewees stated that each of us "should listen with one's heart," because only then will one be able to find the connections and similarities that people share, and sharing is important if we are to work well with others in the world.

Students from South America also stated that their American colleagues are not good listeners, and they don't seem to know the stories of their cultures. In many South and Central American countries, young people are taught to listen to their elders and their teachers, and even as they become adults they are expected to show respect by listening. Their listening stories, too, indicate that stories as well as directions and modeling are the ways they learn to listen, and the students all felt that listening was important, was expected, and they would be chastised in some way if they did not listen well. The graduate students indicated that listening was even more important in the workplace in South America than it seems to be in the United States.

Asking three questions about listening in another person's culture has provided me with greater insights into the cultures of the world. The interviews provide stories that help me get to know more about the world from which my international students and colleagues come. The interviews reinforce my view that competent listening may be an ideal goal that is hard to achieve, but well worth the effort to at least try to accomplish because it is the one way we can learn to understand others, increase sensitivity and awareness, and learn the stories that show us our similarities and differences and make us appreciate others.

Story telling and Listening : Some Research Results :

At this point in my own research I have interviewed thirty people. The following generalizations may be drawn from my interviews.

1. Asian students are taught to listen. There are very clear expectations about what constitutes a good listener and there are punishments if one does not listen. Teachers and families teach listening, not so much by talking about guidelines and definitions, but rather by examples in stories and in the clear communication of expectations.

2. European students, teachers/professors, and other adults don't teach listening, although teachers for young people wish they could do something to improve their students' listening behaviours. Teachers for those age groups are willing to try to teach listening. Many of us fit the description of Professor Jensen from the University of Roskilde, Denmark : we like to "hold on to the word."
3. People in South and Central America tend to be better listeners, and their culture places an importance on listening. Students, family members, and workers are taught to listen. There is a politeness and sense of courtesy that interviewees from Ecuador and Argentina found missing in the United States. Interviewees had trouble understanding how U.S. students learn anything about themselves or their culture without being better listeners. Many college professors have the same question. All of the stories I've heard lead me to believe that there are some guidelines we need to know and use if we hope to be better intercultural listeners. The following guidelines provide at least a starting point.

Guidelines for Intercultural Listening :

Competent Intercultural Listeners Should :

1. Be prepared to listen.
2. Learn to control internal and external distractions
3. Behave as a good listener. Stop talking; let others have a chance. Don't interrupt. Concentrate on what is said, not who is saying it, or what the speaker is doing.
4. Good listeners maintain eye contact with speakers, if it is a part of their culture.
5. Good listeners learn the "rules" ol the culture.
6. Good listeners ask questions at appropriate times and maintain flexibility as they carefully listen to the speaker's views.
7. Remember that dialects, accents, and "different" vocal dynamics can but should not distract. Learn to focus on the message.
8. Learn to ask appropriate questions.

We should remember that all cultures and co-cultures have both similarities and differences. We should not focus on differences. Instead, we should emulate our Asian and South American colleagues : listen with undivided attention with our ears, our eyes, and, most especially, our hearts.

References :

Beall, Melissa L., Interviews with International Students and Teachers, March 2003-2004.

Chinese New Year traditions, found on the following website : www.educ.uvic.ca/faculty/mroth/438/CHINA/chinese_new_year.html

Collins, Rives, and Pamela J. Cooper (1997), The Power of Story : Teaching through Story telling (2nd ed.), pp. 13, 17, Boston : Allyn and Bacon.

Danish Immigrant Museum, Elk Horn, Iowa, [This is an "international cultural center dedicated to preserving the culture and history of the Danish people who immigrated to America." The museum has extensive collections of artifacts, special annual and traveling exhibits and displays of Danish cultural practices and traditions. The town has an authentic Danish windmill, shipped piece by piece from Denmark and re-constructed in this western Iowa village].

Jensen, Iben, personal email, March 18, 2004.

Kwanzaa website, http://www.officialkwanzaawebsite.org/

Personal conversations with teachers in Denmark, August 2003, Endeslev and Horne, Denmark.

Ryan, Thomas G., Professor emeritus of History, University of Northern Iowa, "Iowa's Irish Celebrating Today," Waterloo/Cedar Falls Courier, A6, March 17, 2004.

St. Patrick's Day website, (http://www.historychannel.com/exhibits/stpatricksday/history/

Chapter **15**

Use of Stories in University Teaching

Cherryl Garner

Stories are "more true" than facts because stories are multidimensional. A story incorporates when and who lasting generations, or minutes (Simmons, 2001) and the saga continues. A generational or "native rhythm" story may be understood in terms of what Carl Jung (1980) called the magical images stored in our primitive brain. A minute-by-minute, "here and now" or "circadian rhythm" story points to a level of truthfulness or what Sartre (1985) called our authentic being. MacIntyre (1984) argued that the body revealed the stories of the soul. According to Callahan (2001) when the body experienced a planned sequence of events a particular story, "algorithm story" will be released. When in university teaching we are moved to utilize narration or called upon to foster story-telling environments, it becomes our civic duty, our role as "good citizens" to promote values-based stories - stories that teach tolerance, inclusion, acceptance and respect of self and others. The following will be addressed to discuss the need for values-based stories in university teaching : Native Rhythm : He Would Spin Me A Story; Circadian Rhythm : And Still, They Persisted! Algorithm : Places in the Body Where Stories are Stored.

Native Rhythm : He Would Spin Me a Story :

It was August 1991. My psychology clinical internship at an ethnically diverse, major teaching institution was ending. Most importantly, closure was being broached in clinical supervisions. A primary tool for teaching and developing therapeutic skill, supervision also provided an avenue to bring depth and meaning to my therapeutic work. One of my supervisors, the chairman of the department of psychology, in delivering his parting comments to me expressed his amazement at our patient's continued progress and asked for the first time, "are you sure she is a patient"? How disconcerting! He had University of South Carolina Beaufort supervised her treatment every week for the past eleven months. Where was this question coming from? His expertise was instrumental in my understanding of our patient's psyche. Much to my chagrin, just as the supervisory hour was ending, with stern voice and demeanor he added, "It is only because of your native rhythm you are able to work with her." His "native rhythm" remark left me confused. Its intent was not flattering. I had no desire to burst into song or dance or to celebrate the challenges and complexities of our clinical work. In our eleven months of supervision, the term "native rhythm" was never mentioned. Our supervisory sessions spoke to my relationship with our patient in light of the evolving therapeutic process and therapeutic fit, appropriate and or inappropriate mirroring, building trust and my struggles with our patient's narcissism. It is not unusual for the final scene of supervision to be experienced by the supervisor as a sacred rite of passage. Lane (1990) suggested as an end goal of supervision, the supervisor should welcome supervise as a colleague, one who would continue the craft. Supervises are, to some degree, extensions of their supervisors. That was the last time I would meet with my supervisor.

Not knowing whether shame or anger was the more powerful emotion, I took comfort in my fellow in terms shared perception of my supervisor's insensitivity. Neither the five psychology in terms nor the chairman's colleagues interpreted his native rhythm comment as positive. Little did my supervisor know, he would spin me a story.

Our patient was young, attractive, verbal and intelligent (a member of MENSA). Successful treatment was, on the one hand, due to her tenacity to work through some painful issues by employing the gifts with which she entered the therapeutic arena. On the other hand it was, I thought, due to the dynamic learning process of supervision. Culturally, our patient and I met on bumpy terrain but in the spirit of that landscape, her story came alive. She was a Brahmin (of honourable caste). Unflinchingly, she described me as a displaced Indian (born in Trinidad, West Indies), product of East Indian indenturement, untouchable at best and the Indian that India rejects. Our patient's mother was a psychiatrist whose bias about psychologists was glaring. She demanded the patient seek treatment from a "real doctor," a psychiatrist not a psychologist, and most definitely a "real Indian" before she would agree to absorb treatment expenses. It was our patient's feudal and inconsistent relationship with her mother that enabled her to choose the very therapist her mother had condemned.

As our patient exposed her narrative, her stormy confessional echoed struggles with abandonment, envy, anger and prejudices. Many sessions were filled with Hindu mythological concepts. We explored the meaning of her goddess name and preference to indulge its flagrant nature. Recognizing the weightiness of her struggles, I related the Indian folk tale of how two frogs, unaware of their behavior, landed in a pail of milk. One frog, overwhelmed by the task, decided it was useless to keep kicking and eventually died. The other frog persisted tirelessly. The milk eventually turned into butter and the frog leapt back into the world transformed, resilient, with a newfound respect for itself and for life. Therein, she heard the analogy of courage her courage. The transferential and countertransferential experiences were epiphanic and the therapeutic alliance became more and more evident. Through her descent to self-discovery, she was growing her new story by embracing her old, literally and metaphorically, spinning her toxins into gold. Her truths had grown her and positive manifestations of her insights became palpable.

A somewhat secret symmetry, supervision links therapist, patient and supervisor. Rife with cerebral interpretations of unconscious projections on the part of my patient and myself, and uncovering layers of pent-up aggression were our supervisory hours. The very reasons, I believe, native rhythm was never given any attention in the supervisory milieu. Despite the fact that supervision lends itself to progress, it can fuel unconscious hostilities. Furthermore, the very issues that are discussed between patient and therapist can manifest as parallel processes in the supervisory arena. I perceived my supervisor's comment about my native rhythm as hostile, discriminating and certainly ill-timed. If my native rhythm were instrumental, ultimately, my supervisor's native rhythm must have penetrated the therapeutic triangle. Or, was it simply a pas de deux? Had he abandoned the process much like our patient was abandoned and as I too, was about to abandon her at the close of my internship? Was he envious about my work much like our patient struggled with issues of envy and as I too envied her "real Indian" experience? Howard (1999) suggested "We can't teach what we don't know" and by the same token we can't share what we don't have. Why was the tenor of native rhythm discriminating and primitive? I believe that was the intent. Furthermore, my skewed idea of self-knowledge at that time could only afford me a negative interpretation. I was unable to discern the good in the bad, to locate native rhythm on the same cognitive map that would explain psychological prowess. It never occurred to me to process the comment as one art (native rhythm) enhancing the other (therapy).

Could the sterile boundaries of some teaching institutions, some pedagogical styles, some professors, foster denial or shame about one's native rhythm? Over the years, in relating the story to professors from different universities some of them interpret my supervisor's comment as an attempt to undermine my cognitive abilities. Others understand it as an attempt to compliment the cognitive-affective abilities, the experiential, emotional components that fueled my creativity. In the years that followed, my supervisor's comment will be transformed and a veritable paean will ensue.

The power of my native rhythm was revealed while exploring concepts such as : "Racial Memory" (Freud, 1939), "Animism" (Freud, 1950), "Inner Culture" (Harding, 1970), "Primordial Images" (Jacobi, 1973) and "Archetypes and the Collective Unconscious" (Jung, 1980). Primitive as defined by Harding (1970) is a union with nature. It is only through the emergence of ego development when consciousness comes to the fore, that the union with nature is broken. Predating these theorists were Socrates who suggested "know thyself" (Lavine, 1984) and Aristotle who argued that "every realm of nature is marvelous" (Lear 1998). Many researchers have borrowed from Socrates. For example, Elbaz (1983) believed, because teachers survive in a world of stories, self-knowledge is a necessary tool. We ground ourselves in being, in balance and in harmony with our nature when we can experience the rhythms that express our deepest values. By fully embracing our existence, we automatically connect to others. Five years ago, I created an abstract design on canvas, An art appraiser, artists, as well as people who are not involved in the art world have made the following comments : "it looks like a wolf woman", "wolves playing with a woman" and "wolves are in the design". Three years ago, I stumbled upon Estes' (1995) Women Who Run With The Wolves. Not only did her celebration of native rhythm (wild woman) furnish me with the title La Loba (The wolf woman) for my artwork, but it made salient, my supervisor's envy, his incompleteness and disconnection to the home within himself. A colleague recently suggested that awesome experiences find me because of my "karma". The law of Karma is the law of cause and effect wherein action and reaction are interdependent; deeds of the past life affect the present and deeds of present life affect the future. A variation on the theme of native rhythm, his words struck a harmonious chord.

The concept of native rhythm is as deep as it is broad. It tells a story about the metamorphic self and collaborative thoughtfulness that influence us over time. My native rhythm story is not one of blame. For, as Simmons suggested, a blame story neither fleshes out the problem nor invites resolution. A blame story is the problem and it always goes wrong. Furthermore, hope is usually absent from a blame story. As I continue to mine my story, its treasure becomes increasingly abundant.

Along with art, I utilize poetry as a medium for enjoying the unconscious. One exercise employed in my classroom is to ask students to create stories in poetry form or create poetry out of stories. They are asked to conjure up a word that crystallizes a story and then create a descriptive verse to explain the story as it relates to them. Here is my rendition of Native Rhythm :

Something primitive something new, stories of me, stories of you?

Each** tale **bears** witness **to pleasure pain and drama

And miracles unfold in variegated karma.

Something to give, something to receive?

Opulent gifts our thoughts we may weave.

Nature unencumbered! Cosmic dance!

Stored collective wisdom, future, present and past

Another creative exercise uses visuals, drawings or paintings to define a story. Students' artistic interpretations then serve as grist to explore here and now stories.

Circadian Rhythm : And Still, They Persisted :

According to Surya Das (1998) the "nowness," the "right now" or what is often referred to as the "here and now" should be fully lived because in this space we can become emotionally untangled, uncoiled, unknotted and experience a curing level of truth. Present moment-to-moment awareness reminds us of our limitlessness and our freedom to express and is the keystone of emotional intelligence. For Sartre (1985), becoming self-aware means taking full responsibility not just for ourselves but for all mankind. In this action-oriented here and now position, we may exist fully and be authentic. Our here and now authenticness is affected by what we hear in the now and the idea of "entertainment" suggests that different rhythms of the body, circadian rhythms to be exact, may be changed through sound.

Goldman (1995) offers the approach that synchronicity takes place when the more powerful rhythmic vibrations of one object changes the less powerful rhythmic vibrations of another object and causes them to synchronize their rhythms. In this newness, the rhythm that emerges can be used to heal and to transform.

I experienced that awesome level of synchronicity when in 1998. I met the Brazilian Fulbright Scholar, Idea Dos Santos. Santos had traveled extensively throughoutIndia, Africa, Guyana and the United States and was expert in African and African-Brazilian cultures. Furthermore, the authority with which she discussed African-American history was enough to intimidate educators at the historically black university where she was scholar in residence. Santos stood 5 feet 3 inches and was nothing short of interesting. She carried a slender stature, olive complexion and wore her straight, black hair in a short, unisex style. Her indigenous-looking high cheek bones, equine nose, slanted eyes and long ears conjured up a Spock-like aura. It was ludicrous to affiliate Santos with any particular era or any particular ethnic group because in her being she was timeless, universal, multi-racial, multi-cultural and multidimensional. An evolved soul, tough, assertive but definitely influenced by her experiences, Santos shoulders would bend in the midst of adversity like a tree in a storm. She smoked about four packs of cigarettes daily, and her discoloured teeth and fingers revealed her battle with anxiety. Materialistically unattached, clothing served a utilitarian purpose for this non-conformist intellectual, and it was not unusual to find her walking around campus in bare feet. It didn't matter that her attire usually evoked raised brows from the elitists who couldn't fully appreciate her anti-bourgeoisie views. She questioned me in her thick accent, about the first wave of East Indians who were brought to Trinidad and Guyana in the 1800s as indentured laborers and then informed me of the wave of African Candomble slaves who were living in Bahia, Brazil in the 1800s. It was astonishing to discover these two seemingly different religious and ethnic groups shared such striking similarities. Santos' mission to link cultures to the grasslands of Africa and to diminish racial divides appealed to my rebellious, caste-bashing and ethnocentric-intolerant nature. I felt a rush of adrenaline as her otherwise even-tempered manner was replaced by exhilaration to proclaim the inter and intra-relatedness of culture. With synchronizing rhythms vibrating in a defining moment, Santos and I were chanting the same mantra. "And still, they persisted!"

Narrating the likeness between these two displaced ethnic groups who did not borrow from each other, became for us, an obsession. We spent hours sitting on the floor of her university residence, eating our favourite Indian and Brazilian foods with our hands and planning our delivery. It was Black History month and the entire school was invited to a historical, story telling hour.

The lines on Santos' face resembled riverbeds into which the waters of life had dug deep. She wore a crisp, white cotton dress symbolic of a priestess. I was draped in a red, silk, sari, reflecting pain and passion. Humbled to be in her presence, I was at the same time, energized at a visceral level. Intuitively, from childhood this information was stored in my primitive brain. The heart and logic of the African, Candomble story, Santos took her place on the left, posterior side of the lecture hall. Synthesizing her truths with the comparable Hindu message, I was situated on the right, posterior side of the hall. Our movement from the back of the darkened hall to the front where lights shone brightly was symbolic of our people's plight, their journey, their upward mobility. We intended to affect the audience at auditory, visual and motor levels, to stir them, have them away from left to right be moved by what they heard. Santos' deep, smoky voice harmonizing with my alto tones, we opened in rhythmic dialogue the central theme that resonated from both sides of the hall.

Santos & Garner : And still, they persisted! In the 1800s the Hindu indentured labourers who were taken to Trinidad and Guyana and the African slaves who were taken to Bahia, Brazil arrived in their new environments with little material goods. What these parallel traditions tenaciously clung to, were their precious religious and cultural identities. Both religions are so ancient there is no known founder and both believe in a supreme God.

(Presenters now in the middle of the hall, the audience swayed from left to right augmenting the melodic exchange.)

(S) : Olorum (Yoruba), Nazambi (Bantu), Mawa (Ewe/fon)

(G) : Brahma,Vishnu, Shiva

S&G : Between God and men there are intermediate, anthropomorphic deities possessing good and evil, destructive and constructive abilities.

S : Orishas (Yoruba), Voduns (Ewe/Fon), Inkisis (Bantu)

G : Narasimha, Hanuman, Ganesh

S&G : Every god has a female counterpart.

G : Saraswati for Brahma, Lakshmi for Vishnu and kali for Shiva.

S : The same holds true for several African gods. Shango for Oya, Obatala for Nana and Oshossi for Oshum.

S&G : Both religions are demonstrated through poetry.

S : In Candomble, the If a Poems and Secret Prayers are recited and the legend of African poetry is orated through generations.

G : Among the Hindus the Mahabharata and Ramayana are proclaimed poems, literary masterpieces.

S&G : Both religions are passionate about life and love, and sexuality is not repressed but almost glorified.

G : The Hindus gave to the world the Kama Sutra.

S : The Africans have the Orikis.

S&G : Neither group proselytizes.

S : In Candomble it is said "who is outside shall not enter and who is inside shall not leave."

G : Hinduism is a birthright

(Presenters face the audience symbolizing resistance to oppression)

S&G : Both cultures engage in rituals of walking through fire, making offerings of fruits, food and flowers and demonstrating reverence for the banana and other fruit trees. Both cultures use triangular shaped flags for certain forms of celebration. The East Indian kali and the African Orisha cults are both forms of folk culture. (Pause) And still they persisted.

S : The East Indians in Trinidad and Guyana and the Africans in Bahia resisted oppression in manners reflecting hostile and docile temperaments. Notwithstanding their nature, religion was at the core of their social and cultural resistance. They were attacked for their attempts to maintain social and cultural identities that were not befitting a eurocentric style of existence. In Brazil, slavery was abolished 66 years after independence was gained. Brazilian authorities intended to build a European country that would carefully conceal any African heritage. Priests and priestesses were arrested and persecuted. Shrines were invaded and sacred objects confiscated as evidence of criminal offences.

G : British rulers in Trinidad and Guyana ridiculed the East Indians. Since Christianity was the only acceptable form of worship, there were attempts to have Hindus adopt Christian names and convert to Christianity.

(Santos arched her back and grew taller with emphasis. Garenr's eyes widen with pride.)

S&G : Ignorant of the sophisticated rituals and complexities of Hinduism and Camdomble, the oppressors behaved with contempt. That contempt was met with resistance as the Hindus and Africans learned to disguise their beliefs. And still, they persisted!

S : In Bahia, syncretism was utilized as every African deity was assimilated to a Catholic Saint. For example, Ogum, Yoruba god of Iron and war was affiliated with Saint Anthony.

G : Hindus attended Catholic-church but secretly worshiped at their temples. They used Christian names when in public and Hindu names in private.

S&G : Despite the jeering about their food, both cultures maintained culinary ties to their motherland. Yes, they did persist!

S : Bahian foods : palm oil smoked and dried shrimps, black peas, yams and plantains are notably African.

G : Curries, roti, dahl and vegetarianism remained staple diets for the Hindus. Even though they were forced to live in sugar regions instead of geographical areas that produced the vegetable protein so crucial for fortification, The exacerbated malnutrition did not result in the consumption of meat.

S&G : And still, they persisted! Both groups continued to wear their native attire and embrace theirmusic, art, language, and culture. Religious resistance and cultural persistence resulted in victory.

S : In Bahia, festivals such as Yemoja Day is celebrated by Brazilians of all religions, races and classes. It is largely publicised and Brazilian presidents show reverence to the old Yalorishas (Candomble priestesses).

G : In Guyana and Trinidad, Divali (Hindu festival of lights) is a public holiday celebrated by all races, religions and cultures.

S : African dignitaries who visit Bahia are always struck by the preservation of the African culture.

G : In the same vein, visitors from India marvel at the extension and continuance of Hindu tradition in Trinidad and Guyana.

S : Heroic preservation of the ancestors religion means honoring the linkage with mother India and mother Africa. It means sustaining cultural identity and human dignity. It means to escape being made into an object or thing, to escape being made nameless and selfless spirit less. It means that as a result of their struggles the enslaved and the indentured never forgot that their ancestors were kings and queens, hunters and warriors, healers and artists, despite racial and economic superiority and hierarchy.

S&G : Like the Africans who were taken to Bahia, Brazil, the East Indians in Trinidad and Guyana continue to be powerful and to some extent overwhelming. They did not merely infiltrate but have left legacies on virtually every walk of life.

We were thanked profusely, but more importantly, symbolic of blessings, were the questions with which we were showered. African-American students who heard for the first time, the Candomble people's victory against oppression were insatiable of the message the scholar had so powerfully imposed on their minds. Most East Indians who were born in India are uninformed about the diaspora of their people and have little regard for East Indians born in the West Indies. Furthermore, East Indians regardless of where we are born Burma, Malaysia, Africa or the West Indies are not very receptive of the idea that Africans share with us similar customs and values.

Santos and I left the lecture hall. There was no need to elaborate on the fact that the clashing of civilizations, racial discord and racial segregation among slaves and indentured labourers were the result of the "British divide-and-rule techniques" (Jagan, 1966). As with kindred spirits whose hearts beat in sync, an unsung message unfolded. Santos was wise beyond her years and that wisdom seemed to have settled between her shoulder blades in the kyphotic posture she carried like a vessel. I could almost imagine her joys, her burdens, her stories safely stored, therein.

Algorithm : Places in the Body Where Stories are Stored :

In 1893, Sigmund Freud and Joseph Breuer's seminar paper on understanding hysteria identified a relationship between trauma and memory. They were perplexed by the tenacity of past traumatic memories and believed those memories were powerful enough to haunt the present. They suggested expressions such as acts of revenge allowed for physical reactions or releases that eventually led to alleviation of emotional distress. Furthermore, they believed when a traumatic memory was associated with other stored memories the psychological space that occupied knowledge of the traumatic memory expanded and eventually caused the intensity of the trauma to become diffused. In his work with hysterics, Freud discovered his patients' bodies bad "the strange habit of converting insults, criticisms," and wishes unacceptable to society into physiological symptoms. For example, Freud treated patients suffering with bodily afflictions like facial tics, painful legs, outstretched arms and hallucinations with what he termed the "talking cure" (Gay, 1988). Contemporary theorist Roger Callahan devised a cutting edge modality called Thought Field Therapy for tuning in to and tapping out the body's stories.

Callahan employed the mathematical concept of algorithm or common solution to a problem, to heal the body of its traumatic stories. A general meaning of algorithm suggests that a planned sequence of intentions will result in solving a problem. There is always one start and one end point. There may be alternate pathways through an algorithm, but each step instructs as to what should come next. In the medical field, algorithms are used for making diagnoses as well as treating and managing diseases.

Thought Field Therapy is totally non-invasive. Depending on the nature of the trauma, particular points of the body, when tapped in a specific sequence while the patient is tuned in to the disturbing thought, will produce a shift in cognition and an ultimate release of a crippling image.

Louise Hay (1987) has been fruitful in linking mental causes with bodily stories. She argued, "Ingrown Toenails may tell stories concerning : worry and guilt about your right to move forward. Loss of Speech: So mad you can't speak. Fear of speaking up. Resentment of authority. Diabetes : Longing for what might have been. A great need to control. Deep sorrow. No sweetness left."

Van der Kolk (1994) believed the traumatic "imprints" that steer peoples' bodies can be overcome or certainly changed through rhythmic descnsitization. The body, he said, holds the score. It carries a tune, sings a song and it is the interaction between the singer/ speaker and the listener that invites a new narrative or rendition. It was an eclectic approach and carefully planned sequencing of events - education, psychotherapy and algorithmic desensitization that enabled my student, Harumi, to hear and respond to her body's story.

Harumi was born in Japan. Her biological father claimed she was not his child, and refused to have his name placed on her birth certificate. One year later Harumi's brother, Yuta, was born. She loved him deeply. She remembered her mother being "very sad", "never smiling" but having "an old, serious face" whenever she looked at Harumi. She had no memory of her father's face. Harumi began school at the age of five without the cheering or expressed excitement as when Yuta began his education. She protected him from her parents' turmoil-filled relationship or what she called "chaos in the house." Until the age of five years six months, Harumi was repeatedly sexually abused by her father. Although she has no recollection of the sexual abuse, the following memory drives home any doubt of physical maltreatment.

Along with her brother, Harumi's father was giving her a bath. He filled an elephant toy bucket with burning hot water and continued to spray her with the trunk despite her shrinking away and cries for him to stop. To be freed, her father insisted she repeated the words, "I am a bad girl."

Harumi and her brother enjoyed playing between walls much like children would enjoy exploring the intricacies of a maze. During what she recalled as probably the height of her abuse, she remembered getting stuck between two walls and was perplexed about how to escape. Around the same time, she began feeling unsure and afraid at night and was confused about how to use the toilet without wetting her underwear. Harumi's most cherished game was assembling puzzles. Her getting stuck, feeling muddled and welcoming the challenge of solving puzzles, were gestures symbolic of the confusing, puzzling, and overwhelming experiences she tried to make sense of.

When Harumi was five years old her baby sister was born. Six months later, Harumi was given to an adoptive agency because of behavioural problems and a lack of finances. Interestingly enough, Harumi's father was a teacher at a private Junior high School, her mother was a house-wife and her grandmother lived next door to them. According to Harumi's adoptive, mother who was American but connected to the Japanese culture, Japanese people are very sensitive about adoption. Abortions are legal and made accessible to unwed mothers. Because family is so deeply respected and private in that culture, very few Japanese children are adopted. Furthermore, in an attempt to keep shameful behaviours from leaking out, adoption outside of Japan is almost taboo. Harumi's biological mother, relinquishing all parental authority, said she "would be grateful to have Harumi adopted by an American." Explaining to neighbours and teachers Harumi's new home was a Japanese orphanage would at least have drawn suspicion and at worst, would have exposed her biological parents' immoral and indecent behavioural problems. All ties would be broken between Harumi and her biological family. Without emotional farewell, Harumi's grandmother would give her away. It was only on insistence from Harumi's adoptive mother that some of her baby pictures were released.

Medical examination revealed extensive scarring of the internal and external genital areas. For one and a half years following the adoption, Harumi, in mourning the world's lovelessness had difficulties sleeping and was afraid of storms and thunder. In an attempt to escape from herself, her life and memories of persecution and rejection by her family, her body would suffer violent attacks of seizures at least two times a week. Her new-found mother noticed the attacks usually followed attempts to address Harumi's behavior. During one episode, her mother seizing the moment threw a glass of water in Harumi's face. The massive spasms vanished. Also at the age of seven, Harumi held her chest and said to her mother, "for the first time I feel love in the morning." Years later Harumi was asked where she believed her body's story was stored. Unhesitatingly she replied, "In my heart." Those who knew Harumi well came to experience her heart's story by studying her face.

It was amazing to witness the demeanor of a miserable, old hag, a witch-like face on a girl of 19 years. When Harumi, soldered to her past, was consumed with hatred toward herself and her father, feared that she resembled her biological mother and when her maturational deficits manifested in struggles to control herself and her world, she looked withered and beaten. Ill-disposed to handle her emotions, her shoulders slumped and her stare was cold and hard. It was as though she were plotting revenge, moving against herself and everyone else. Her teeth protruded from behind dried lips and her skin looked dull.

Harumi beat and bandaged her breasts to stunt their growth. She described her chest as becoming so tight it was as though her heart was being squeezed and pushed to the back. To the contrary, when she felt needy, inadequate and unable to function without the help of those she deemed important, and when she became obsessed with thoughts of abandonment, the face of 19-year-old Harumi regressed in temperament to that of a wide-eyed, 5-ycar-old, giggly, happy-go-lucky child. Unreflective and scattered in her thinking, her voice got syrupy and she became clingy and insatiable. During these periods she described her heart as feeling big and pushing forward. Harumi, at age 20, confronting her heart's struggles and the slings and arrows she received as an infant, slowly came to discover and to eventually flaunt her mysterious beauty.

Appreciation of her personal aspirations and her reparative strivings resulted in the gait and posture of a maiden. The sparkle in her sensitive, oriental eyes and the luster in her skin reflected her joyful, loving, generous heart; the heart she came to experience as light, centered and nurturing.

As a function of its metabolic genius, the body has the ability to preserve stories. Through rhythmic fine tuning and excellent timing the body is also able to catabolically destroy its toxic stories and anabolically construct healthy ones.

Conclusion

The values-based stories I shared addressed, on the one hand, how exclusion, alienation, intolerance and the aim to reduce humans to objects of hatred resulted in intense suffering. On the other hand, every story depicted human resiliency. Native rhythm, circadian rhythm and algorithm were present in every story. For example, Santos' sense of inner culture and her ability to stay grounded in the home within herself her native rhythm, enabled her to experience sisterhood among ethnic others, and fueled her passion to honour diversity. Circadian rhythm can be found in Harumi's story when in those defining moments, her receptivity to the sound of her self-loathe enabled her to make connections between her native rhythm story and her violent bodily abuse. Her here and now truths that resembled her dishonest, irresponsible and disrespectfulparents were likened to her hateful attitudes - her cardiac arrests. The idea of algorithm can be observed in my native rhythm story. At an unconscious level, a sequence of events - deeper understanding of Aristotle, Freud, Jung and tapping in to my artistic nature enabled me to purge my unevolved native rhythm story and create a new.

Relationships between professors and students are set in motion through stories and synonymous with story-telling is bestowal of gifts. The endowment of a professorship carries with it the obligation of exposing students to issues of tolerance, inclusion, acceptance and respect of self and others.

References:

Callahan, Roger. Tapping the Healer Within. New York: McGraw-Hill, (2001).

Das, Lama Surya. Awakening The Buddha Within. New York : Doubleday, 1997.

Estes, Clarissa Pinkola. Women Who Run With the Wolves. New York : BalantineBooks (1995).

Freud, Sigmund. Moses and Monotheism. New York : Random House, (1939).

Freud, Sigmund. Totem and Taboo. New York; W. W. Norton & Co., (1950)

Gay, Peter. Freud a Life for our Time. New York : W.W. Norton & Co. (1988).

Goldman, Jonathan. Healing Sounds : The Power of Harmonics. Rockport Massachusetts : Element Books, Inc. (1992).

Hay, Louise L. Heal Your Body A-Z. California : Hay house, Inc. (1998).

Harding, M. Ester. The of all Women. Boston Massachusetts : Shambhala Publications, (1970).

Howard, Gary R. We Can't Teach What We Don't Know. New York : Teachers College Press, (1999).

Jacobi, Jolande. The Psychology of C. G. Jung. New Haven : University Press. (1973).

Jagan, Cheddi. The West on Trial. London, England M. J. Ltd. (1966).

Jung, C. G. The Archetypes and the Collective Unconscious. Princeton, New Jersey : Princeton Univ. press (1998).

Lane, Robert C. Psychoanalytic Approaches to Supervision. New York : Brunner/Mazel, publishers, (1990).

Lavine, T. Z. From Socrates to Sartre : the Philosophic Quest. New York : Bantam Books, (1984)

Lear, Jonathan. Love and its Place in Nature. New Haven Connecticut : Yale Univ. press. (1998).

Macltyre, Alasdair. After Virtue. Indiana: University of Notre Dame Press, (1984).

Strtre, Jean-Paul. Extentialism and Human Emotions. Secaucus, New jersey : Citadel Press Inc. (1985).

Van der Kolk, Bessel. The Body Keeps The Score. Cambridge, Massachusetts : Harvard Press, (1994).

Chapter 16

Visual Instruction and Visual Aptitude

Richard E. Clark

This article intends frankly to discourage certain types of hypotheses which have proved futile in the past and offer a tentative direction for research in visual and verbal components of instruction to the special abilities or deficiencies attached to the media consumers. The discussion in the article provides understanding on issues like analysis of previous research questions dealing with visual treatments and abilities with particular reference to studies of various visual media, approach to visual aptitude and some tentative conclusions about a future direction for research which hypothesises between usual treatment and abilities.

Educators have long hoped that visual instruction would benefit students who do not learn easily. The limited verbal skills or motivation of some students are presumably overtaxed by the high verbal content of most instruction, causing them to perform below average. Researchers who have explored this issue (Cronbach & Snow, 1977) hope to learn enough to tailor the visual and verbal components of instruction to the special abilities or deficiencies of the student who fails to learn in the traditional (highly verbal) instructional setting. Much of the interest among educational practitioners in televised instruction (Salomon, 1978), visual learning (Leifer, 1976; Carrier & Clark, 1977), pictoral illustrations (Hagen, 1974), visual literacy (Levie, 1978) and mental imagery (Wicker, 1978) is based on similar expectations.

Although research questions about interactions between visual instruction and aptitudes are not new (Cf. Freeman, 1924); they were given strong support by Snow & Salomon (1968) in an article that has stimulated a good deal of research in the past decade. It should be noted however, that considerable confusion exists in this area of research as in other areas that deal with instruction and learning. The goal of this article is frankly to discourage certain types of hypotheses which have proved fruitless in the past and offer a tentative direction for research in the area. It will be recommended that researchers move away from research with media (pictures, television, print, movies, multimedia, and computer) and towards conceptualizing visual treatments in terms of the psychological functions they perform in learning. More specifically, the following discussion will provide : (1) An analysis of previous research questions dealing with visual treatments and abilities with particular reference to studies of various visual "media"; (2) A description of a line of research which follows Snow (1970) and Salomon's (1974) suggestion of a compensatory function for visual treatments; (3) An approach to visual aptitudes based on the work of Smith (1964), Cattell & Horn (1978) and Cronbach & Snow (1977), and (4) Some tentative conclusions about a future direction for research which hypothesizes interactions between visual treatments and abilities.

The Early Questions

One of the fruitless assumptions made by most studies which compared visual media is that there was one best media for teaching various concepts, subject matters or disciplines. A generally unrelated activity in the instructional development and curriculum areas was the attempt to develop "media selection schemes" (Clark, 1975; Clark & Snow, 1975). The strategy was similar to that employed in the wider area of instructional research, where the attempt was to find the best instructional technique for a given subject matter or task. Although many different types of questions have been asked in the past two decades, the following list will serve to represent the major efforts.

Is preference for visual instruction necessary for learning?

Both adults and children show a marked preference for visual information and are more likely to attend to visual than to verbal modes of instruction, (Stevenson et. al. 1971). This overwhelming preference for the visual has misled many researchers into asking whether students will automatically learn more from the visual than the verbal mode. In fact, a number of studies so designed have provided equivocal or ambiguous results (cf. Reviews by Bugelski, *et. al.* 1968; Christiansen & Stone, 1968; Smith, 1964).

With the benefit of hindsight, it is relatively easy now to see that preference for visual instruction is neither necessary nor a sufficient condition for learning. Similarly, attention to a treatment is a necessary condition for learning, but attention is difficult to measure (Clark, 1978). Also it is critical to note that attention is not sufficient for learning. There are, obviously, factors other than our attending to visual instruction that lead to variations in learning. Salomon (1978) has speculated that we should question, whether the compelling feature of visual instruction is that it allows for "shallow" information processing. It may simply be easier than verbal instruction because we are not forced to work very hard to "look" at something, whereas, reading demands greater effort. The drawback, of course, is that we might learn less, in many instances, from visual instruction. Willows (1978), for example, has reported that in some conditions, pictures interfere with learning. Researchers turned next to the identification of factors other than the "visualness" of an instructional treatment to explain differences in learning outcomes.

Is visual medium "A" more effective than visual medium "B"?

In the search for those crucial characteristics of visual instruction that enhance learning, we have learned much about the questions we should not ask any longer. We know, for example, that it is fruitless to ask whether medium A is better than medium B (Mielke, 1968; Salomon & Clark, 1977). If all factors, other than the medium of presentation, are controlled in an experiment, there would be no reason to expect differences between visual media. We would simply be comparing two devices for delivering the same piece of instruction.

Mielke (1968) argued that the results of such studies have tended to be more or less random. If studies, which found advantageous for medium "A" were placed on one end of a balance scale and those finding in favour of medium "B" on the other (with "no significant difference" studies placed in the middle at the fulcrum), the scale would balance and from a distance might even resemble a normal curve. Although studies which deal with this question continue to appear (most recently under the heading of multimedia versus single media presentations); it is time to put this particular issue to rest. As Schramm (1978) has a said : "Learning seems to be more affected by what is delivered than by the delivery system" (p 273).

Is instructional method X more effective on medium "A" or medium 'B'

We have also learned not to ask whether instructional technique X is more effective on medium A than medium B, such as presenting a programmed instruction text via print or television (Carpenter, 1955). We would not expect that presenting a written text via a visual medium would significantly change its impact. These delivery modes are not the crucial elements of visual instruction per se, although they may be useful variables to consider in the broader area of instructional techniques.

Research which contrasted different visual media that contained the same instructional technique was essentially a response to criticism of the medium A versus medium B studies. In this line of research an attempt was made to control the content of the presentation. However, when content was controlled, there were only minor differences in learning which generally turned out to lack instructional importance. For example, finding that the ability to review material when using a book is an advantage over the arbitrary rate with which a television presentation of the same textual material is offered to a student, may have more to do with the idiosyncratic needs of individual readers than with some essential instructional feature of the medium of presentation. Generally, it seems that when the content of instructional presentations is held constant, there are few educationally significant advantages among the various media.

Will a large scale instructional development project which employs visual media be more effective than "Traditional" instruction?

Salomon (1977) has noted that when ".. .an educational programme is propelled by a high powered technology, say television, (and it is)... compared with a failing programme or system - it makes an educational difference." Salomon goes on to claim that the difference is trivial since it would most certainly be due to the results of increased planning in the television treatment - factors such as sequencing, organization, clarity of information, pacing, and perhaps a Hawthorne effect. An example of this approach may be found in the Age Cohort Study, part of the first year research on Sesame Street (Ball & Bogatz, 1970). In that study, four-year old children were tested for achievement. One group watched the carefully produced television programme while the other did not. As would be expected, post-test score comparisons indicated that the television group achieved significantly greater gains in achievement scores than the control group.

Might certain disciplines or subject matters be taught more effectively through visual media?

Jamison, Suppes and Wells (1974) reviewed a large number of studies which had tested the effectiveness of various media to teach a great number of disciplines. These studies were based on the presumption that a single medium such as television might be more or less effective than, say, a computer for teaching mathematics or civics. Again, when the content of the medium was held constant, there were no significant findings of differences between media, even when visual and verbal mediums were contrasted for the same subject matter content.

The Transition Question

It is at this point that questions about visual instruction changed considerably the transition, to use the concept of Salomon & Clark (1977), was from research on media (comparison studies of different media) to research with media. The latter approach deals with those features of media which are critical for learning to take place for students who differ in relevant aptitudes. Research with visual media is based loosely on Olson's (1976) suggestion that all learning is of two kinds; information about the world (which Olson calls "knowledge") and information about how information is acquired ("skills"). While knowledge (content, subjects matter, disciplines) is obtained from a wide variety of experiences, skills, are acquired from a medium or "symbol system" in which they are encountered. The concept led Olson to claim that an important component of intelligence was "skill in a medium". What is proposed, in the terms of this article, is that some features of visual instruction (e.g. rotating objects, transforming visual figures, moving images within three-dimensional space) might be internalized by learners and stored for later use as skills in solving visual problems. It is the extent of the acquisition of these skills measured by the many pencil and paper tests of spatial and verbal ability (Olson 1976). This approach led to the most recent generation of questions about visual treatments.

Are there interactions between verbal and visual abilities and verbal and visual treatments?

After the failure of a number of lines of research which compared visual media, many researchers (e.g. Gange & Gropper, 1965) suggested that we ask whether visual medium A is better for high visual ability learners or verbal medium C is better for high verbal ability students. These studies tested the generic hypothesis that verbal aptitude predicted the ability to learn from visual treatments. The propounders of this approach made convincing arguments to the effect that previous research had been ambiguous, because visual treatments could not be expected to be superior, treatments for everyone. Highly visual students might profit mostly from high visual treatments, and conversely, highly verbal students would benefit from highly verbal instruction. Studies were designed to test different versions of this interaction hypothesis. Some of the more carefully conducted were the investigations by Allen, Filep & Cooney, (1967, Cf. a re-analysis of Cronbach & Snow, 1977). which dealt with the use of visual and verbal programming modes to teach crystallography to students, who differed in visual and verbal abilities; a study by Gange & Gropper (1965) on visual and verbal versions of seven lessons on mechanical advantage; by Snow (1965) in a study of various visual and verbal methods of teaching physics; by Crowley & Rudy (1969) in a study of videotaped versus printed script of a lesson in biology and by Koran (1969) who used the same treatment modes to teach higher order questions to teachers. All of these studies used some measure of verbal and visual ability, and hypothesized interactions with visual and verbal treatments. Yet, all of the studies yielded ambiguous results (Cronbach & Snow, 1977).

Although the difficulties encountered in these studies ranged from data analysis problems to a lack of specification of treatments, a common problem seemed to be a mismatch between the skills or strategies being reflected in aptitude measures and the actual skills required by the treatment and task. For example, Allison (1960), and Cronbach & Snow (1977) presented a number of concept attainment tasks to Navy recruits. Half of the tasks used verbal concepts and the solutions were semantic, the other half were spatial and the solutions dealt with the classification of shapes. Measures of verbal and spatial reasoning were taken and correlated with performance. Although the verbal measures correlated positively with verbal concept attainment, the visual aptitude scores did not. On closer inspection of the study it appears that the spatial tasks, although they utilized pictures and figures, simply did not call on the types of skills being measured by the spatial aptitude measures. What may have been occurring is that the figures which required identification by the subjects called on verbal skills in labeling rather than the visual transformation ability usually tapped by spatial measures. In one of the spatial task conditions, verbal ability actually correlated more strongly with outcome than the spatial measures.

At this point, a number of investigators became cynical about interaction hypotheses (e.g. Merrill, 1975), others recommended more complex interactions (Berliner & Cahen, 1973; Clark, 1975) or longitudinal studies (Cronbach & Snow, 1977), while others saw solutions in improved research designs (Salomon & Clark, 1977). While all of these alternatives are bound to bring added sense to the general area of visual instruction and visual abilities, the trend was to ask a more sophisticated question i.e. what cognitive activity is activated by visual treatments and how is that cognitive activity related to visual abilities and to learning? We have learned from past research that it is not sufficient to label a sequence of instruction "visual" simply because it requires vision to comprehend or because it provides pictures, figures, slides, television, video tapes or films. One current concern is with the mental operations that visual treatments are activating and/or providing for learners. In addition, we must be concerned with those aspects (characteristics, attributes) of visual treatments that are necessary and/or sufficient for activating or providing the cognitive strategies that are required for learning.

The next two sections contain a suggested approach to conceptualizing visual treatments and visual aptitudes.

Visual Treatments

The function of most instructional treatments is to facilitate learning by activating or providing cognitive processes that are necessary for learning. Visual treatments generally attempt to engage or provide visual/imaginal strategies for processing information. Research on visual instruction must be concerned with the psychological consequences of the treatment. It is no longer enough to present a sequence of visualated information to students who have varying levels of visual aptitudes and assume that those with high ability should learn more than those with low visual ability. When the treatment is planned we must be concerned with : (1) cognitive operations or processes that different elements of the treatment are designed to affect; (2) the expected extent (quality and quantity) of that effect; (3) the way in which the cognitive process determines what and how much is learned; and (4) the expected durability (lastingness) of the learning. Most important is that we find reliable and valid ways to measure all parts of the process whereby students encounter, assimilate, modify and utilize the strategies presented in the treatment.

Although there are very few examples of current studies which meet all of these suggested criteria, Snow (1970), Salomon (1970, 1972) and Goodman (1976) have provided what appears to be a productive way to hypothesize about the function of visual treatments.

Snow (1970) and Salomon (1972) have suggested that treatments be categorized in terms of their remedial, preferential or compensatory functions. Remedial treatments function to bring underachievers nearer norm levels, for example, reading, organizational strategies or study skills. When variance in learning outcomes can be attributed to task specific abilities then remediation is employed to improve the deficient capabilities of the learner.

Preferential treatments attempt to capitalise on the strengths of the learner. The major thrust of this type of treatment is to present a treatment which is shaped in the learner's preferred style or strongest aptitude mode of information processing. Alien (1975) has described one possible use of this type of treatment. In a review of visual treatment literature, he concluded that high general ability learners appear to persist longer and perform better, if a visual treatment is more information laden, more visually "rich", complex, and abstract. Low general ability learners however seem to perform better when treatments reflect shorter segments, and a more Spartan amount of visual information.

However, the most interesting model for the function of treatments for the purposes of this discussion is Salomon's (1972) description of compensatory treatments. Instead of improving the students' task, specific skills as in the remedial approach, or placing the amount and complexity of the treatment information in a preferred mode, the compensatory model stresses the concept of modeling strategies which the learner can use to learn a variety of tasks, including the one at hand.

Snow (1970) identifies modeling as a form of treatment which "compensates for each learner's deficiency by providing the mode of presentation that the learner cannot provide for himself" (p. 76). This model would emphasize a step by step depiction of a learning strategy with a great amount of repetition or redundance. Salomon (1970) assumes, this model would not necessarily benefit and may even be detrimental to high aptitude scorers. Compensating for the deficiency of a learner, by modeling, is similar to the approach which Goodman (1976) has called "exemplification". Presumably, the compensatory approach requires that we list the skills which learners need to possess to accomplish a task and ask which of the requisite skills are not available. Those skills which are available are merely "activated" or called forth in the instructions which accompany a treatment or are assumed to be implicitly understood by the learner (e.g. we seldom specifically ask students to use their knowledge of the English language in solving problems, the request is implicit in most problem solving situations). Those skills which are not available are provided for the learner in the compensatory treatment through explicit depiction or modeling. In this way we reason carefully about the skills necessary for learning, and compensate for deficiencies in skill acquisition by providing the skills that are necessary for learning. Most of the research, which has employed this approach successfully, reveals a thorough attempt to provide the requisite skill and to attempt to assess the psychological impact of the compensation. Studies by Gauverin (1967a, 1967b), Frandson & Holder (1969), Mayer (et. al., 1975), Koran, Snow & McDonald (1971), Salomon (1975), Carrier & Clark (1979) and Lucas & DiVesta (1978) serve as examples of various features of compensatory approach.

Gauverin (1967a, 1967b) first studied students' ability to manipulate letter tiles, when working on a series of six letter anagram problems; while in the second study no manipulation was permitted. His intent was to improve the performance of low spatial students. In both studies, the learners were divided into high and low spatial resoners. As hypothesized Gauverin found that the relationship between performance and ability was low under the "manipulation" while high under the "no manipulation" condition, *i.e.* low spatial students performed better when allowed to move the anagram letter tiles. Moving the tiles compensated for their low spatial ability.

Frandsen and Holder (1969) studied high and low spatial ability subjects in problem solving tasks of syllogisms, time-related-to-travel and logical deduction problems. Their results showed that under an activation strategy (instruction to solve the problems, no guidance) the high spatial ability students were superior to those with low ability. When a compensatory strategy was employed in which the treatment demonstrated how to use diagrams, symbolic maps and additional spatial devices for finding a solution to the problem, no difference was found between high and low spatial subjects. Mayer (et. al.,1975) demonstrates that the use of a compensatory visual model, to promote learning, provides the learner with a rich set of prior experiences by which new information may be organized.

Koran, Snow and McDonald (1971) studied individual differences in acquiring a teaching skill from compensatory written modeling, video-mediated modeling or an activation procedure. Results indicated that the video-modeling produced significantly higher performance skills than the written modeling, yet both modeling techniques outperformed the controls. Although this study strongly indicates the effectiveness of compensatory modeling as a teaching strategy, at no time were the subjects taught in an orderly fashion how to view the video-model or what they should pay attention to. The task chosen as a criterion measure was the teaching skill of "analytic questioning". If a spatial criterion measure had been used following the modeling technique which sequenced the learning of novel material from simple to complex, different or stronger interactions might possibly occur between modeling strategies and learner aptitudes. A method which would suggest a step process of modeling or guidance of a student, through learning tasks, would serve a compensatory function and would act to supplant, within the learner, a way or ways to process and apply the presented material. The converse would be to present the material, possibly with questions about the content, but without a step by step sequence of how to process the information. Learners who already possess the necessary strategies or have a preferred style, would hypothetically capitalize on the latter approach and thus do not need modeled strategies.

Salomon (1974) reported three studies in which he investigated the effectiveness of modeling certain visual transformations and the differential effect this modeling strategy might have, on learners with different aptitudes. The first study deals with the task of "zooming in" on relevant detail within a set of paintings. This process was thought to be analogous to the "analytic process of singling-out discrete components of a rich stimulus" (p. 501). Four experimental conditions were investigated. The first treatment consisted of three films. The first film depicts a wide shot of an entire painting; the next zooms in on a detail within the painting; then the third zooms out to show the whole painting. This process continues eighty times. The second treatment (a short-circuiting stimulus condition) consists of three series of slides corresponding to the material in the earlier films. The same procedure is employed, first the entire painting is shown, then a detail, then the entire painting, etc. Because the learner does not witness the "zooming in and out procedure" of experiment one, Salomon believes no transformation have been shown. In the third treatment, the activation condition, only pictures of the entire painting are shown to the subjects, thus providing neither modeling nor modeling with transformations. The results show that all three groups outperformed the control group on the criterion post-test measure of reported number of painting details. The modeling group did not differ significantly from the activation group, but both did better than the short-circuited or partial explicitness group. There was a negative relationship between prior ability to covertly perform the transformations and modeling technique designed to teach that skill, i.e. as ability increased, performance decreased.

The second study (Salomon, 1974) deals with the addition of verbalization to the original problem. The short-circuited group or procedure was dropped. The modeling group was divided in two halves. One half was asked to be prepared to read aloud the subscript shown on the slide. Then a student was called upon at random to perform this task. The other half was instructed to read the subscript silently. The activation group was also divided in two halves and the same procedure was employed. High verbal aptitudes benefited from the verbalization while less verbally able students did not. Students high in transformation ability benefited more from activation while those low in transformation skill preferred compensatory modeling.

In the third study (Salomon, 1974) the modeling strategy was a filming technique consisting of "laying out" solid objects and then folding them again. The modeling group was shown a film in which the object was gradually unfolded and refolded. The short-circuited group was shown slides first of the solid object and then the unfolded form. Five pairs were demonstrated. The post-test consisted to Thurstone's Surface Development Test used in the French Kit, French, et. al., (1963). Scores for the group were again significantly higher than the short-circuited group, yet both were superior to the control subjects.

Lucas and Divesta (1978) partially replicated Salomon's compensatory effect by manipulating the explicitness of instructions in a computer displayed three-dimensional tic-tac-toe game. In one condition (called "activation") subjects were given instructions on how to play the game and were allowed to practice. In a second condition they were given an elaborate spatial compensatory model which depicted strategies for representing and transforming the three-dimensional space in the game. A control group engaged in another game with the same computer system. Subject scores on the Space Relations Test (Form A) of the DAT were used to block them into high, medium and low spatial ability groups. All subjects played a series of tic-tac-toe games and their scores were obtained. Results of the study, in later trials, indicated that high and medium ability subjects had difficulty playing the games under the modeling condition and performed much better with the activation instructions. This finding illustrates the importance of employing compensatory treatments with low, rather than, with high ability learners. Presumably, high ability subjects already possess useful, although, idiosyncratic task relevant skills. Giving compensatory treatments to students who do not need them, probably, interferes with existing strategy deployment and results in a decrement in performance. The Lucas and DiVesta (1978) study also illustrates the necessity to look at the effects of strategy acquisition in compensatory treatments over a large number of trials. In the initial games the high ability students performed much better with the modeled treatment than with the activation condition. Lucas and DiVesta (1978) surmise that it is the high ability groups which learn the modeled treatment easiest (it is their preferred mode in many cases) and they initially do best with it. However, after a few trials their existing strategies are naturally brought into play and conflicting approaches result in decreasing success for the compensatory condition while the activation group increases in success. It is unfortunate that the study yielded ambiguous findings for the low ability group. Although data indicated that the trends were in the predicted direction (towards more success with modeling than activation or control) the task seemed to be difficult for lows and an extreme floor effect was found on outcome scores.

Carrier and Clark (1978) investigated the use of visual and verbal modeling in the solving of surface area problems in a geometry task for 4th and 5th grade students. They used the Primary Mental Abilities Battery (Thurstone and Thurstone, 1963) to assess both verbal ability and spatial ability. The task required the transformation of visual information about geometric figures, so they reasoned that the spatial relations subscale on the PMA battery would tap the relevant aptitude. The visual and verbal modeling treatments were contrasted with no modeling (activation) conditions. In the compensatory modeling treatments, students were encouraged to call on either their spatial or their verbal skills, in both, a preferential and compensatory fashion; in order to learn how to solve area and volume problemsin geometry. The results indicated a strong effect for modeling regardless of whether it occurred in the visual or verbal treatments and an interaction between aptitudes and the visual/verbal modes of the treatment. The interaction was weak, but as hypothesized, it indicated that the visual treatment was more beneficial for high spatial (and low verbal) subjects. However, contrary to hypotheses, the high verbal (low spatial) subjects tended to do equally well with the verbal and the spatial treatment. Upon closer examination it appeared that the modeling strategies simply were easier for the high verbal students to learn, than they were for the high spatial students (regardless of whether the modeling was verbal or visual).

It appears then that compensating for lacks in task-relevant skills by modeling the skill(s) under some conditions results in the formation of internalized strategies which can be used by learners to perform tasks. The compensatory function of treatments requires that an aptitude or trait, relevant to the learning, be identified. Where tasks are primarily visual or can be solved in the visual mode, spatial measures have been found to be appropriate. However, the specific visual attitude used must be chosen with considerable caution. Allison (1960) for example, found that a spatial measure did not predict performance in a task which required the identification of visual shapes. Most of the studies described in the preceding section utilized very different spatial ability measures. The most commonly chosen tests are from the French Kit (French, Ekstrom & Price, 1962), the Differential Aptitude test batteries Spatial Relations Test (Form A), (Benentt, Seashore, and Wesman, 1967), the Primary Mental Abilities Battery spatial subscales (Thurstone & Thurstone, 1963) and special measures such as those constructed by Salomon (1974). The fact that these measures represent many different facets and levels of spatial ability makes the results of the studies cited difficult to interpret and has occasionally resulted in equivocal findings within studies (e.g. Carrier & Clark, 1978; Lucas & DiVesta, 1978). What follows is a recommended approach to conceptualizing spatial abilities which may facilitate the successful prediction of results in studies which hypothesize interactions between spatial abilities and compensatory treatments.

Spatial Abilities

In many instances, previous studies of interactions between visual treatments and abilities provided instructional materials which were static and figural when aptitude and/or outcome tests measured the ability to mentally transform (e.g. rotate, twist, turn, and invert) visual 3-dimensional objects. This confusion of levels of visual ability and treatments must be cleared up first.

Although there are disagreements over the various components of visual or spatial ability (cf. Smith, 1964), there is some agreement that there are at least two subcomponents that should be taken into consideration in instruction. The first subcomponent has been called spatial relations and orientation (SRO) by French (1951). It is a composite to Thurstone's S_1, and S_3 spatial factors (Smith, 1964). Scores on SRO were thought by Michael, *et. al.* (1957) to be related to:

1. the ability to comprehend the nature of the arrangement of elements within a visual stimulus pattern;
2. the ability to remain unconfused by varying orientations in which a spatial pattern may be presented;
3. the ability to recognize the identity of an object when seen from different angles;
4. the ability to think about spatial relations in which the body orientation of the observer is an essential part of the problem;

The second subcomponent of visual ability has been called spatial visualization (Vz). It is similar to Thurstone's S_2 factor (Smith, 1964). The primary abilities represented by this factor are:

1. the mental manipulation of visual objects involving a specified sequence of movements within a complex stimulus pattern;
2. the mental rotation, turning, twisting or inverting one or more objects or parts of a configuration;
3. the recognition of a new position, location or changed appearance of objects that have moved;
4. the movement of images within 3-dimensional space and the manipulation of mental images.

Although correlations have been found between scores on tests that presume to tap SRO and Vz, researchers have argued strongly for their use as separate subcomponents of visual ability (Borich & Bauman, 1972; French *et. al.*, 1962; Carroll, 1976 and Berlyne, 1965). Although Guilford (1967) has argued for SRO and Vz. as completely separate abilities, Piaget and Inhelder (1962) have provided evidence that the ability to visualize a transformation (Vz) occurs later than the ability to recognize or recall a visual stimulus (SRO). They might imply that these two abilities are hierarchically related (with Vz a higher order skill than SRO), and that they appear developmentally at different times (SRO appears first, then Vz). This would imply that some of the ambiguous results of previous research have resulted from a failure to examine the separate scores on SRO and Vz. If an extremely high SRO score was present in enough learners to give the appearance of high "visual ability" and the treatment required transformations, it is doubtful if results would be interpretable. We also could expect that tasks which required spatial orientation and relations (SRO) skills should be able to be performed by more learners than tasks requiring transformations (Vz). We would also assume that all learners who could perform Vz oriented tasks could also be expected to perform SRO tasks, but that the reverse would not be the case.

Generally then, recent work on spatial abilities would lead us to suppose that performance on tasks which require the recognition of pattern arrangement, varying spatial pattern changes, object identification or observer body orientation would be correlated with SRO scores. Tasks which required the ability to mentally transform objects, to recognize a familiar object even if its position or orientation to the observer had changed and the transformation of the objects or images in three dimensional spaces would be performed best by those high in Vz ability. Presumably then, different visual treatments would engage each of these aptitudes, and this may account for some of the confusion in previous research.

An important cautionary note is in order at this point. Cronbach & Snow (1977) and Snow (in press) have noted that the aptitude measures which best predict interactions with variations in treatments are those associated with general ability (the g factor), particularly the approach and measures suggested by Cattell & Horn (1978). This view of human ability, unlike earlier work, places strong emphasis on subcomponents of general ability identified as crystallized ability (Gc) and fluid ability (Gf). Presumably, Gc scores reflect the notion that "over extended periods of life span development the many influences that promote... the intelligence of a culture work in loose harmony to produce the broad pattern of abilities" (p. 140). Gc thus reflects experience in acculturation including traditional schooling. It is loosely correlated with verbal measures. Fluidability (Gf) on the other hand reflects "... the many influences related to incidental learning and ... neurophysiological health" (p. 140) or a talent for making more or less successful responses to novel learning situations. Visual ability is some subcomponent of Gf in this system and as Cattel & Horn (1978) caution, it is difficult to distinguish in practice from measures of Gf. It would be most profitable to include predictor measures drawn from the Gd, Gf and Gv type tests.

Another ability which has promise of enhancing the adoption of instructional treatments for individual differences is some measure of the prior knowledge of learners. Since visual treatments and aptitudes in this suggested approach presume to affect the skills possessed by learners (Cf. Olson, 1976), it is clear that differential amounts of knowledge will affect outcomes. The best strategy here may be to try various combinations of Gc, Gf, Gv and prior knowledge in regression equations until the most accurate predictor of performance on a particular skill and task are available.

CONCLUSIONS & SUMMARY

During the past two decades, research on visual instruction and visual aptitudes has yielded a fruitful set of questions. Initially, researchers questioned the relationship between preference for visual materials and learning; the relative effectiveness of various types of visual media (e.g. pictures, television, movies, charts) for various subject matter areas and tasks; the impact of large scale development projects that utilized visual media and the interaction between visual and verbal modes of instruction for students who differed in visual and verbal aptitudes. It now appears that the questions which were asked were either based on invalid assumptions or produced ambiguous results.

A more productive approach has been the recent emphasis on studies which investigated the effects of compensatory model treatments (Salomon, 1972) which depict the critical features of learning "skills" (Olson, 1976). Presumably, subjects who are unable to perform a learning task because of skill deficiencies internalize the models and utilize them for successful learning. Evidence for the success of compensatory models in visual instruction has been provided in studies by Gauverin (1967a, 1967b) Frandsen & Holder (1969); Mayer (1975); Koran, Snow & McDonald (1971); Salomon. (1974); Carrier & Clark (1979) and Lucas & DiVesta (1978).

The compensatory approach to visual treatments presumes an interaction between the extent to which treatments functions to provide necessary skills and the existing aptitudes of subjects (Cronbach & Snow, 1977). Although there are a number of approaches to visual aptitude measurement the most promising for this type of research may be the scheme proposed by French (1951) and Smith (1964). Visual aptitudes are subdivided into two subcomponents which may be hierarchically and developmentally related. The first is spatial relations and orientation (SRO), which reflects skills in pattern recognition, spatial orientation and relationships; the second and higher order subcomponent is spatial visualization (Vz) which reflects skill in the cognitive manipulation of spatial objects including rotating, twisting, inverting and turning objects plus the ability to transform objects in three-dimensional space. Whatever the aptitude orientation chosen however, it is important that the aptitude which is measured reflects a skill which is necessary for the successful completion of the learning task. Many of the more common learning tasks or learning to attend to distinct cues in a complex stimulus (e.g. volume and area problems in geometry) seem to require visual skills which are reflected in SRO and Vz subcomponent scores.

References

ALLEN, W.H., FILEP, R.T. & CONNEY, S.M. (1967). Visual and audio presentation in machine programmed instruction. Unpublished report. Department of Cinema, University of Southern California. ERIC ED 016400.

BALL, S. & BOGATZ, G.A. (1970). *The First Year of Sesame Street : An Evaluation.* Princeton : Educational Testing Service.

BERLINER, D.C., & CAHEN, L. S. (1973). Trait-treatment interaction and learning.

In F.N. Kerlinger *(Ed.) Review of research in education.* Itasca, 111. : Fe Peacock. pp. 58-94.

BERLINE, D. E. (1965). *Structure and direction in thinking.* New York : McGraw Hill.

BORICH, G.D. & BAUMAN, P.M. (1972). Convergent and discriminant validation of the French and Guilford-Zimmerman spatial orientation and visualization factors. *Educational and psychological measurement,* 32(4), 1029-1033.

BUGELSK1, B.R., KIDD, D. & SEGMAN, J. (1968). Imagery as a mediator in one trial paired associate learning. *Journal of experimental psychology,* 76, 69-73.

CARROL, J.B. (1976). Psychometric tests as cognitive tasks: a new structure of intellect. In Resnick, L. B. (Ed.) *The nature of intelligence.* Hillsdale, N. J,: Lawrence-Erlbaum Association, Published.

CARPENTER, C. R. & GREEN HILL, L. P. (1955). An investigation of C.C.T.V. for teaching university courses. 1TV Research Project # 1. University Park, Pennsylvania: The Pennsylvania State University. ERIC ED 013540.

CARRIER, C.A. & CLARK, R.E. (1978). The effects of presentation modes, explicitness and student aptitudes on learning. *Educational Communication and Technology,* In press.

CATTELL, R.B. & HORN, J. L. (1978). A check on the theory of fluid and crystallized intelligence with description of new subtest designs. *Journal of Educational Measurement,* 15(3), pp. 139-164.

CRISTIANSEN, R. & STONE, D. (1968). Visual imagery and level of mediator abstractness in induced medication paradigms. *Perceptual and Motor Skills, 26,* pp. 775-779.

CROWLEV, R. J. & RUDY, D. J. (1977). Student aptitude and the instructional function of videotape. 1969 Unpublished report. Colgate University. In: Cronbach, L.J. & Snow, R.E. *Aptitudes and instructional methods.* New York; Irvinglon Press.

CRONBACH, LJ. & SNOW, R.E. (1977). *Aptitudes and Instructional Methods.* New York: Irvington Press.

CLARK, R. E. (1975). Constructing taxonomy of media attributes for research purposes. *A V Communication Review,* 23(2), pp. 197-215.

CLARK, R.E. (1978). The engaging features of broadcast television and their relationship to the engaging features of instruction and schooling. Symposium presentation for AERA convention, Toronto, Canada.

CLARK, R.E. & CARRIER, C. (1977). Current research on visual abilities and instructional methods. *Education Technology.*

CLARK, R.E. & SNOW R.E. (1975). Alternative designs for instructional technology research. *A V Communication Review.* 23(4), pp. 373-394. FRANDSEN, A.N. & HOLDER, J.R. (1969). Spatial visualisation in solving complex verbal problems. *Journal of Psychology,* 73, pp. 229-233.

FREEMAN, F.N. (Ed.) (1924). *Visual Education.* Chicago : University of Chicago Press.

FRENCH, J. W. (1951). The description of aptitude and achievement tests in terms of rotated factors. *Psychometric monograph,* No. 5.

FRENCH, J.W., EKSTROM, R.B. & PRICE, L.A. (1962). *Kit of reference tests for cognitive factors.* Princeton, N. J.: Educational testing Service.

GAGNE, R. M. & CROPPER, G. L. Individual differences in learning from visual and verbal presentations. Unpublished report, American Institutes for Research, Pittsburgh, PA. ERIC ED 010337.

GAVUR1N, E.I. Anagram solving and spatial aptitude. *Journal of Psychology,* 1665. 1967(0). 65-68.

GAVUR1N, E.I. The relationship of mental ability of anagram solving. *Journal of Psychology,* 1967(6), 66, 227-230.

GOODMAN, N.M. (1976). *The languages of art,* (2nd Edition). Indianapolis, Indiana; Hackett.

GUILFORD J. (I967). *The Nature of Human Intelligence.* New York; McGraw

Hill.

HAGEN, M.A. (1974). Picture perception : Toward theoretical model. *Psychological Bulletin, 81,* 471-479.

JAMISON, D., SUPPES, P. & WELLS, S. (1974). The Effectiveness of Deteriorative. Instructional Media : A Survey. *Review of Educational Research,* 44, pp. 1-68.

KORAN, M. L. (1969). The effect of individual differences on observational learning in the acquisition of a teaching skill. Unpublished doctoral dessertation. Stanford University. University Microfilms No. 69-17, 435.

KORAN, M. L. SNOW. R. & Mc DONALD, F. (1971). Teacher aptitude and Observational learning of a teaching skill. *Journal of Educational Psychology,* 62(3), 219-228.

LEIFER, A.D. (1976). Teaching with television and film in NL Gage (Ed.) *The Psychology of Teaching Methods,* N.S.S.E. Yearbook (Part I). Chicago: University of Chicago Press.

LEVIE, W.H. (1978). A prospectus for instructional research on visual literacy. *Educational Communication and Technology,* 26(1), 25-26.

LUCAS, P. & DIVESTA. F. (1978). Diagonality in Adults and the Interaction Between. Spatial Ability and Explicitness of Instruction. Paper read at the annual convention of the American Educational Research Association, Toronto, Canada.

MAYER, R., STIEHL, C.C. & GREENO, J.G. (1975). Acquisition of understanding and skill in relation to subjects' preparation and meaningfulness of instruction. *Journal of Educational Psychology,* 64(3), 331-350.

MERRILL, M.D. (1975). Learner Control : Beyond aptitude treatment interactions. *A V Communication Review,* 23(2), 216-222.

MICHAEL, W.B., GUILFORD, J.P., FRUCHTER, B. & ZIMMERMAN, W.S. (1957). The description of spatial visualization abilities. *Educational and Psychological Measurement,* 17, 185-199.

MIELKE, K.W. (1968). Asking the right ETV research questions. *Educational Broadcasting Review,* 2(6), 54-61.

OLSON, D.R. (1976). Towards a Theory of Instructional Means, *Educational Psychologist,* (2) pp. 14-35.

PIAGET, J. & 1NHELDER, B. (1971). *Mental imagery in the child.* New York: Basic Books;

SALOMON, G. (1970). What does it do to Johnny? A cognitive functionalistic view of research on media. In G. Salomon & R.E. Snow (Eds.) *Viewpoints*

(Bulletin of the School of Education, Indiana University), 46, 33-62.

SALOMON, G. (1972). Heuristic models for the generation of aptitude-treatment interaction hypotheses. *Review of Educational Research, 42(3)*, 327-343.

SALOMON, G. (1974). Internalization of filmic schematic operations in interaction with learner's aptitudes. *Journal of Educational Psychology*, 66(1), 40-48.

SALOMON, G. & CLARK, R.E. (1977). Re-examining the methodology of research on media and technology in education. *Review of Educational Research*, 47(1), 99-120.

SALOMON, G. (1978). On the future of media research, *Educational Communication and Technology Journal*, 26(1), pp 37-46.

SCHRAMM, W. (1977). *Big Media, Little Media.* Beverly Hills, California: Sage Publications.

SMITH, I.M. (1964). *Spatial ability : Its educational and social significance.* San Diego, California: Robert R. Knapp, Publisher.

SNOW, R.E. (1965). The importance of selected audience and film characteristics as determiners of the effectiveness of instructional films. Unpublished report, Purdue University, 1963. Also reported in: *Journal of Educational Psychology, 56, 315-326.*

SNOW, R.E. (1970). Research on media and attitude. In G. Salomon *&* R.E. Snow (Eds). *Commentaries on research in instructional media: an examination of conceptual schemes. Viewpoints* (Bulletin of the Indiana University School of Education), 46(5), 63-91.

SNOW, R.E. (1979). Aptitude processes, Snow, R.E., Federico, P.A. & Montagere, W., (Eds). *Aptitude, Learning and Instruction: Cognitive Process Analysis.* In preparation, Stanford University.

SNOW, R.E. & SALOMON, G. (1968). Aptitudes and instructional media. *AV Communication Review*, 16, 341-356. ERIC ED 023295.

STEVENSEN, H.W., HALE, G.A., KLEIN, R.E. & MILLER, L.K. (1971). Interrelations and correlates in children's learning and problem solving. *Monographs of the Society for Research in Child Development*, 33(7).

THURSTONE, L.L. & THURSTONE, T.G. (1963). *SRA Primary Mental Abilities Battery.* Chicago: Science Research Associates, Inc.

WICKER, F.W. (1978). Out Picture of Mental Imagery: Prospects for research. *Educational Communication and Technology.* 26(1), 15-24.

WILLOWS, D.M. (1978). Individual differences in distraction by pictures in a reading situation. *Journal of Educational Psychology.* 70(5), 837-847.

Chapter **17**

Telecommunication and the Political Development

Indu B. Singh and Michael Delli Carpini

Social changes of lasting import cannot take place when only one aspect of the social structure is involved. For telematics to seriously affect the democratic process, parallel changes must occur in the political, economic, social, and educational systems which complement and support this technical development. And yet change must begin somewhere, and telematics are clearly changing the way people are structuring their daily lives.

Much has been written about future developments in communication technology. From the perspective of an arm-chair thinker to actual practitioner, the future of telecommunications holds bright prospects. Recent innovations in electronics, computer sciences, and communication engineering and related fields have provided a springboard into practical "futuristic" telecommunication systems.

The present technological innovations will have significant impact on the development of telecommunication systems of the next decade. The key innovation technologies of the 80's include microelectronics, computer graphics, speech synthesis and voice recognition, database management and the on-going computer revolution. These and many other technologies provide the wherewithal to ensure effective, efficient telecommunications in the decades ahead.

The emerging telecommunication technologies and systems will have significant impact on the office of the future and the home communication environment. The future communication systems for the office of the future will provide the capability to transmit voice and data to any worldwide site or computer network. Business and corporate structures will be equipped with teleconferencing studios, abundant computer terminals, central communication control and roof-top earth stations. It has been predicted that by 1990 well over half the total communication dollars spent by business will continue to be in voice-based technology. Computers will play a larger role and there will be increased use of time sharing, interconnected data banks, distributive processing and electronic fund transfer.

The home communication environment will blur current distinctions between home and the workplace, time and distance, transportation and communication. In the area of education, we will see emphasis on continuing education via satellite television transmitted to homes. These and many other new services will be available for the office of the future and home of the future.

Future predictions about the impact of communication and information technologies on society are both fascinating and mind boggling. To an optimist, the sky is the only limit in deriving benefits from the "information society." To a pessimist, however, the unforeseen consequences of using these technologies are full of doubts and dangers. One thing seems crystal clear : to a very large extent, technologies are neutral. The positive and negative consequences are not linked to innovations but to applications.

This article examines development in the major communication technologies and systems for the rest of the century. It also explores the potential impactof these changes on the dynamics of mass politics in advanced industrial societies such as the United States.

Future Telecommunication Technologies

The communication revolution, according to Anthony Smith, has passed through three overlapping technological stages that have taken place during the last 150 years. The first of these was the Wire Age (1844-1900) , the second was the Wireless Age (1900-1970), and the third is the one we are now entering - the Integrated Grid Age, in which wire and wireless technologies are brought together to form the structure of the future global information utility.

In the Integrated Grid Age, one can examine two levels of technological developments. One deals with new technological innovations and the other with continued development in existing communication technologies. Although mans intellectual curiosity and desire to explore new technological frontier will continue to produce innovative forms of telecommunications, it is our belief that most future development will be the more extension of the present technologies. This does not indicate a defeatist attitude, but a reflection on the simple logic of technological development. In the past, the basis for most new technologies has always been the structural framework of the existing systems. The major progress in communication technologies and systems for the 1990's and beyond can be grouped into the following four categories : (1) development of high speed, efficient, user friendly and super intelligent computer systems; (2) development of advanced interactive communication systems; (3) development of advanced laser and optic fiber based communication systems; and (4) development, launching and establishment of geostationary space platforms.

The structural frameworks for these technologies exist now. Thus, the continued growth in technology will multiply our ability to create new communication systems to perform new functions.

Computers

The science fiction of the past two decades is fast becoming the reality of today. In the computer field, especially, the dynamic relations between man and the machine, is being intertwined in the most stimulating way. Computer technology and systems are experiencing exponential growth. Computer applications are becoming omnipresent in individual households and multinational corporations as well as in governmental and educational institutions. For example, at the end of 1982, computers were in about 4 million homes, less than five percent of all households and the estimates are that by the end of the century their numbers may run as high as 80 million. The sales of computers for the home will soar from 2.4 billion dollars in 1983 to more than 7 billion in 1988. In some ways, therefore, computers may define and determine society's future for decades to come.

Computers are the main engines of post-industrial society. The supercomputers are fast becoming the life-blood of the post-industrial, high technology oriented society. Since its inception, computer technology has passed through three phases. They represent the changes from vacuum tubes to transistors and from transistors to large scale integration (LSI) and very large scale integration (VLSI). Today's VLSI chips represent, in the views of many scientists, the physical limits of miniaturization.

Aside from the mechanical and circuitry changes of the past, modern computers are changing physically. The first electronic computers were large, freestanding information storage and calculating boxes without outside links. Subsequently, computers were developed with the additional capability to store programs and interconnect remote terminals. During the mid-sixties, computers became more sophisticated, being able to facilitate interactive networking at higher speeds and efficiency. Today's computers are equipped with extraordinary memory capacities which have the ability to perform independently of a main computer and can provide linkage to vast data networks. These networks store, process and retrieve huge amounts of information accurately and rapidly.

The 1990's decade can be classified as the era of supercomputers. As Lloyd Cowling notes, we are on the verge of delivering large scale computing power and world-class data bases into the hands of anyone who can carry a briefcase, or less. Current research on the Josephson Junction Computer (JJC) will change the face of large scale computation significantly. Josephson Junction Computer circuits use superconducting metals. At very low temperature these metals lose all resistance to electricity, enabling computation in trillionths of a second. At this speed, JJC will be a hundred times faster than existing computer circuits. This cryogenically cool mass of circuitry will outrun and out store anything we have seen thus far. Some have predicted the arrival of JJC by 1985; others have targeted 1990. We believe that JJC would become reality by the early 1990's.

Amidst the myriad of changes in computer technology, it seems obvious that the Computer industry has shown evolutionary rather than revolutionary changes. Designs of future computers will continue to be fast, efficient, compact, user friendly and intelligent. The 1980's launched the evolution of the so called personal computers. This has been made possible through developments in microelectronic circuitry, minicomputers and their terminals. These computers are capable of talking and listening. They can understand spoken instructions and provide feedback with synthetic speech. They are capable of storing voice messages in digital form. The 1990's will witness the evolution of more advanced forms of intelligent computers. The introduction of intelligent copiers by the Xerox Corporation signifies only the beginning of an era that is bound to expand our creative imagination and intelligence.

Cowling points out five major factors that will continue to influence the future development and use of computers. In his view : (1) corporate marketing strategies have been known to produce interesting products in the past. The marketing force often affects the product that the end user sees and feels; (2) home services are now, and will continue to be, a most rewarding end user field. The demands of the home will be the drive behind many innovative computer products and directions; (3) increased demand for business data bases will spur the need for new computer systems; (4) increased demand by the scientific and research community will enhance the requirement for large memory sizes in large scale computers with increased speed; and (5) the human factor - the need to overcome the feeling of being separated from others and from this world. This will increase the demand for advanced intelligent computer systems capable of providing human-like interaction. These are some of the driving forces behind future developments in the computer technology.

Interactive Communication Systems

The decade of the 1990's will see a continued advancement in the interactive telecommunication systems for homes and offices. These systems are called videotex. Videotex is a generic term for systems that transmit textual information for display on television screens. It involves two new technologies for carrying information over long distances - Videotex and Teletext systems.

A videotex system makes the connection via telephone or two-way cable lines and thus is interactive. This allows users to perform various transactions with the main computers. Teletext, on the other hand, is a one-way system that relies on signals broadcast over the air by television stations. Since it is a one-way system of communication, no transaction can be performed by its users.

We are currently witnessing the growth of many different types of videotex systems, some with proven capacity and other still in experimental stage. For example, videotex systems such as Prestel, Ceefax and Oracle of Britain; Antiope and Telematique of France; Captain of Japan; and Teledon of Canada are becoming known to many individuals and institutions around the world. Most of these systems have progressed beyond the experimental stage and have entered the commercial phase of their operations. The next decade will witness a rapid growth in such interactive systems for the purpose of providing news and information, home shopping and banking, electronic mail and other services.

In the United States, the videotex systems are in an infant stage. As a matter of fact, there are no commercial systems operating in the United States today, although many such systems are in the planning stages. It seems reasonable to predict that within a few years, we will see the commercial operation of videotex systems in the selected areas of the United States. And perhaps the beginning of the 1990's will see its market penetration rapidly increased.

In most cases, where videotex systems have been set up for residential use (in countries outside the United States), videotex services have fallen short of the anticipated level of success. Perhaps these early systems were not very user friendly; or perhaps the information was not all that gripping. In the United States, however, the popularity of computer games may be the ticket to promoting videotex as a home information and education system. In the corporate world, however, electronic message receptors, word processors and electronic mail are now saving time and inducing more critical thinking. Telecommunications and microelectronics will continue to set new standards of productivity for individuals as well as for businesses and industries.

Fiber Optics

Fiber optics, a new type of cable technology, promises to bring about radical changes in the wire communication system. The optical fiber or optical waveguides, as they are also called, is a hair-thin strand which transmits information in the form of light waves. The fiber itself is typically made of glass (silica).

According to Pastelis and Stubbs, today fiber optic systems are proving themselves. Improvements have been made in terms of long wavelength transmission, larger bandwidth, lower attenuation and greater repeater spacing. Now systems are available which transmit a longer wavelength (1300 mm), and feature a higher bandwidth (over 1 GHz) and lower loss (less than 1 dB/km). Also, greater distances between repeaters are possible, reaching up to the 25-km range.

There has been a variety of applications of fiber optics. Although in the beginning the technology was used primarily by the government and military, its commercial applications have recently bloomed. It is being used for instrumentation, in equipment boards, sensors, medical equipment and interferometric devices. Optical fibers are frequently being used in; local data communications for point-to-point links. In the future interactive communication systems including cable television and telephone companies may emerge as heavy users of this new technology.

The extraordinary signal capacity of optical waveguides has been the main reason for its wide application in communication systems. For example, a one-quarter-inch diameter containing two optical fibers can carry the same volume of communication as a three-inch-diameter communication cable containing 20,000 copper wires. This size advantage is further augmented by fiber optic characteristic of being lighter, smaller and more flexible than copper cable as well as having large bandwidth, and lower losses.

In the future, a number of new market segmentation will emerge for the use of fiber optics. We should see its widespread application in long-haul terrestrial routes, customer premises usage and local area networks, cable television and satellite business system networks. Overall, fiber optic transmission offers significant advantages over copper and microwave transmission of the full range of voice, video and data. As a result, the future applications of optical waveguides for telecommunication systems will increase by leaps and bound.

Geostationary Space Platforms

There have been two proposals for the creation of space platforms. They are called small space platforms and large space platforms. However, according to today's standards, even the small space platform will be larger than a typical-size satellite of 1,000 kilograms. The basic structure of this proposed platform will be made of light-weight metal. The small platforms will carry large antennas and a variety of equipment packed into separate modules. These modules would be designed to perform missions such as mobile telecommunications, meteorology, point-to-point telecommunications, space broadcasting, navigation, geological survey, etc.

Artists' concepts and drawings of geostationary space platforms have shown them bristling with many antennas of different types and sizes. Thus, the platform has been termed "Orbital Antenna Farms" (OAFs) by Edelson and Morgan.

The proposal to erect and establish large space platforms are even more far-reaching and intriguing than the concept for small space platforms. The large space platforms will weigh hundreds of thousands of kilograms and will span several square miles. Some envision these platforms to be as large as Manhattan Island. As Pelton writes, such space platforms would be manned and would serve a different range of functions. These missions might include energy concentration, generation and distribution, manufacturing in space and on-going research projects. Products manufactured in space would be those which require zero gravity, a near-perfect vacuum, extreme temperature differentials or a combination of these qualities. The large space platforms, thus, could give a new meaning to the existing concept of the space-industrial complex.

Both small as well as large space platforms are predicted for the 1990's. However, their realization would depend on finding solutions to problems in three major areas. According to Edelson, these areas are related to : (1) a number of problems involving interconnection, mutual support and prevention of interference between and among payloads (2) technical problems in erecting and assembling large structures in space, and combining the pay loads of several orbital; and (3) institutional concerns regarding the ownership and control of such structures.

The solution too many of these problems are being sought. NASA is currently developing an experimental space platform system in conjunction with the Fairchild Industry targeted to be launched at the end of this decade. The NASA experiment will serve as a model for operational geostationary platforms.

Although many critics argue that the space platforms are merely figments of imagination in the minds of technologically-inclined individuals, it is our belief that the birth and operation of space platforms is only a matter of time.

Overall, the future technological trends can be summarized in the following way :

1. An increased trend toward the technological development of telematics principles. This would create an increased emphasis on the convergence of telecommunications and computers;
2. Continued development of the small, portable, user friendly, super intelligent computer communication systems;
3. An increased emphasis on the integrated digital communication and data processing networks;
4. A changing emphasis from quantity - to quality-oriented communication systems;
5. Development of advanced satellite communication systems for the purpose of massive transfer of data and information;
6. Continued development of advanced, intelligent, interactive communication systems for the home communication environment;
7. High level integration of communication systems with everyday life. For example, an increased alliance between

telecommunications and transportation systems, and between telecommunications and the health care systems; and

8. A new partnership between education and telecommunications where home will become a center for education and training.

The Potential Impact of Telematics

Technological change has always had both planned and unintended impacts on the social and political systems in which it takes place. The printing press increased the availability of great works of literature, religion and philosophy, and led to a wider and faster dissemination of news about current events. It also became a central tool in first the maintenance of authoritarian forms of government and later the demise of many of those same systems. The mechanical innovations of the industrial era had profound effects on the distribution of economic and political power (both within and across nations) that went far beyond the anticipated impacts on the efficiency of production. The development of first radio and later television forever changed the way citizens of advanced industrial societies choose their political leaders and think more generally about politics in ways unimagined by the early developers and sponsors of these new entertainment media.

The recent technological development in communications described before are also having (and will continue to have) impacts on political and social structures and norms. Central to these changes are issues which revolve around the concept of information. More specifically, mass democratic politics can be affected by changes in the home information environment in three critical areas : control over or sources of the information provided by the new technologies; the content, quality, and extent of the information provided; and perhaps most importantly, access to the new technology itself. These three aspects which have been affected by the growth of telematics, in turn affect the democratic process in three ways: the formation of attitudes and opinions; the nature and locus of political deliberations; and the extent and type of political participation. The flow of this relation is presented schematically below.

In the remainder of this paper we explore the theoretical linkages among telematics, information and mass politics as outlined above, focusing on several critical junctures where the direction we move as a mass democratic society will depend upon the way telematics evolve in the next decade.

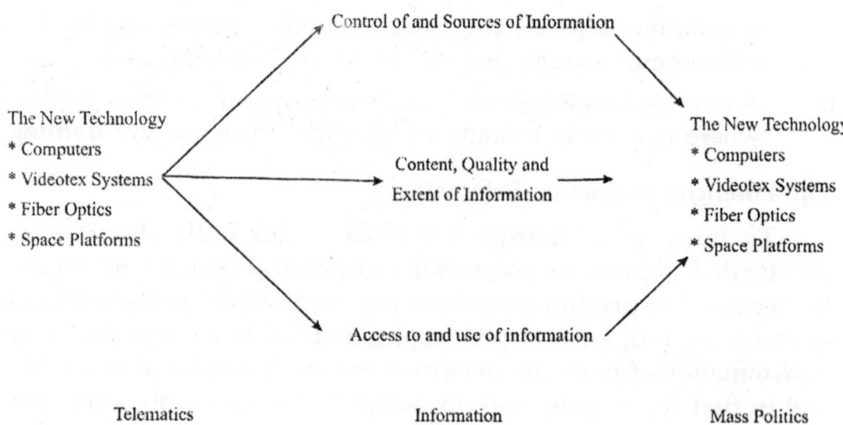

Figure 1 : A Conceptual Framework to Study Relations Among Telematics, Information and Mass Politics.

Information and the New Technology

Central to the democratic political process is the availability of information for citizens to use in evaluating the world in which they live and in acting upon those evaluations. At the heart of the recent advancements in telematics is the availability of more information, the availability of information in less time than ever before, and the availability of information directly into the homes of a large and growing sector of the population. While much of this information is largely entertainment oriented, some are not. The availability of 24 hour news from either cable stations, satellite news stations, or directly from wire services is one example. The electronic transmission of newspapers, journals, and books is another. Databank services such as COMPUSERVE provide instant access to thousands of different information sources on topics ranging from adult education to job openings to world news. As advanced societies move from an industrial base to a more mixed, service based economy, information becomes a critical currency which rivals capital and labour in its importance. The information explosion of which telematics is a critical part is both a cause and a result of this new balance of power.

Availability of information is not the only issue to consider, however. It is one of the ironies of today's society that information of all kinds is more available today than ever before, but because of this very fact, information is more difficult to store, synthesize, and use in the way normally available to the average citizen. The changing home information environment provides a geometric increase in the ability of citizens to both obtain and to store such information, and ultimately, to use it. These technological changes, while clearly offering the potential for a more informed body politic, are not unambiguously positive, however. The three issues of control, content, and access need to be examined in more detail.

Control

Of critical importance when anticipating the possible effects of the evolution of telematics is the diversity of information sources. Focusing solely on the user end of the process reveals a seemingly endless variety of hardware and software options, scores of cable and satellite stations, and a multitude of data bases and information services. One must be clear, however, on whether this represents greater and quicker availability and potential use of information which has, in recent times, always been relatively accessible (not an insignificant accomplishment in and of itself), or if it truly represents an expansion of various dimensions of public information itself : the type, quality, amount, and sources of information. This is centrally an issue of the control of information.

Consider, for example, the expansion of the news which has resulted from several related changes : the availability of 24 hour cable news stations such as CNN; the success of Independent Nightly News; the actual or planned expansion of PBS's McNeil/Lehrer show and the Network's evening news to an hour; the availability on some cable systems of direct linkage to the wire services; the development of "focused" news shows such as those available on the Christian Broadcasting Network; and the availability of major newspapers and magazines through video database services. On the surface it would appear that information about the world which is available to the average citizen has increased dramatically in the last few years. And yet what does this information explosion really mean? Does it mean more information? To an extent, since not all the various outlets duplicate the story selection, lengths, or the order in which they are presented. Yet the degree of variation is less striking than the overlap. The 24-hour news stations depend on repetitive 18 or 20 minute cycles, based on the same structure as the network news broadcasts, with a few new items (usually features) added each time. The expansion of the length of both local and national news has always meant more of an increase in canned, special segments than of hard news. And the video newspaper/magazine services simply ease access to already available sources - they do not add to them.

At the core of this issue is the question of the sources of information. The major wire services (AP, UPI, Reuters, NY Times) still dominate the information gathering and dissemination process as they have since the late 19th century, while the prestige publications (the New York Times and the Washington Post) and to a lesser extent the networks still dominate the news agenda process. The availability of actual wire services in the home eliminates one or two layers of the news distillation process, and also expands the direct availability of information but again, the source remains ultimately unchanged.

Perhaps most important in the actual expansion of information available are the database services which, in combination with home computers, allow detailed searches for information on specific topics. While, again, most of the information which is gathered in this way was available prior to the recent telematic developments, it can now be gathered and processed by the individual him or herself rather than by a single network, journalist, or editor. It allows the user to receive a more complete array of facts and opinions on a single topic from sources ranging from the encyclopedia Britannica to government publications. Equally revolutionary, it allows this access to take place when the individual citizen is in need or is interested. It is important to realize that even in this case, however, what is different is the ease and timing of the accessibility of information, not the sources. The structure of the control of information, in telematic societies has changed very little. Perhaps the one possible exception to this is the Congressional News Station (C-SPAN), which presents largely unmediated broadcasts of congressional hearings.

Content

A related but essentially distinct issue is that of the content of the information presented. In particular we are interested in the effects of telematics on the points of view available. Seldom is information truly neutral, especially information of social and political import. Either "by the choice of what information is made available or by the way it is presented agendas are shaped and options are foreclosed. Have recent developments in telematics expanded the points of view available to users? In most cases the answer is no, at least not yet. The expanded news programming has remained committed to what has been called the social responsibility theory of the press and has therefore, been reluctant to present extremely distinct viewpoints or take controversial, partisan, or ideological stands. The expansion of information outlets could potentially signal a return to a more libertarian approach to information, in which citizens learn by exposure to diverse opinions and arguments. To date, however, this has not occurred. The one partial exception to this is the Christian Broadcasting Network, which to some extent shapes its news agenda and form of presentation in a way consistent with its religious perspective.

This middle of the road approach has led some critics of the current information environment to argue that the media preserve the dominant social order rather than observe and report on it. To the extent that telematics serves to increase the dissemination of this information rather than to help expand it, it is subject to the same criticism. The information which is available through the database services does little to challenge this critique. Most information available through such services is largely factual documents, government reports, or mainstream publications. Little effort has been made to date to include information sources that are ideologically very far to the left or to the right of center.

Access

Most revolutions in communications have relatively quickly expanded to include the entire population of the societies involved. This is so because the costs of production decreased rapidly with further technological development and with increased demand and the related economies of scale. Also, the technology tended to require little technical skill on the part of the user and to involve few choices beyond the decision to "buy in" (one, for instance, simply acquired a telephone, radio, or television set). While the more recent telematics may follow a similar pattern (indeed there is evidence that at least concerning certain aspects of cost it is following such a pattern), certain qualities of the technology suggest alternative scenarios. Costs are much greater relative to income then with prior communication developments, and even with drops in cost will continue to be a relatively large expense for many. In addition, hardware varies in quality and capability in ways that are fundamentally different from the choice of, for example a colour vs. a black and white TV, or a dial vs. a touchtone phone. What one can process, at what speed, what information one can access, etc., are all still critically linked to the capabilities of the hardware purchased, and the decision of what to purchase requires both financial resqurces and knowledge of the options available. In addition, the separation of hardware and software as distinct aspects of the home information environment means that even with similar machines, access depends on knowledge of at least what software is available and a desire and ability to purchase it and use it. Such decisions are not "one shot deals" either, since software is upgraded and developed at a rapid pace, as are the number of and type of databases available. Finally, the ability to program oneself, while not a necessity, is a tremendous advantage in fully exploiting the advantages of telematics.

The current communications revolution, it would seem, has the potential for moving advanced industrial democracies in two very different directions. To the extent that access and use is equalized, one can imagine a systematic upgrading of the quality of democratic decision-making. To the extent that different sectors of society benefit from these developments at different rates however, the effect could easily be to accelerate rather than close the gap between the haves and the have knots between the information rich and the information poor.

Telematics, Information, and Mass Politics

Thus far we have focused exclusively on the linkages between telematics and information, arguing that information is critical to the democratic process, but leaving the specific ways in which both telematics and information translate into politics largely unstated. Let us now examine these linkages in somewhat more detail. Since the deliberation process and the formation of attitudes and opinions are closely linked, we will consider both of these political processes together.

Deliberations and the Formation of Political Attitudes and Opinions

Politics in a democracy is based fundamentally on the open exchange of ideas. How might the aspects of telematics discussed above effect such a deliberation process? Deliberations require information, and the extent to which more information, better information, or more accessible information is available should contribute to the deliberation process. Again, however, unequal access and the limits to the type and source of information are critical. Consider, for example, the issue of toxic waste disposal. Discussing this issue required technical, political and economic information. It also requires the summation of this information, along with personal values and more general political attitudes into a coherent opinion that is both internally consistent and consistent with other opinions which make up the individual's political outlook. A citizen, armed with factual information and exposed to a wide range of arguments concerning the problems, solutions and tradeoffs involved in dealing with toxic waste is a citizen who should be able to develop coherent opinions and discuss them intelligently with fellow citizens and, ultimately, with governmental decision makers. A less informed and involved citizen is at a clear disadvantage in such a process, and if such a citizen develops opinions at all they will tend to be less crystalized and consistent, and perhaps more superficial and subject to easy manipulation.

The deliberation process itself may be affected by the growth in telematics. Recent developments have continued trends toward moving the loci of work, education, and entertainment/social environments away from the workplace, school and Public Square and to the home. Such a movement shrinks the amount of time spent directly interacting with other people and replaces it with much more indirect interactions and reality tests. Deliberations become much more of a process of self-deliberation. The extent to which this is a negative development (beyond the argument that direct human interaction and contact is good in and of itself) for the political process depends centrally on the ability of telematics to present the citizen with diverse opinions, arguments, and bits of information, and on its ability to react to and interact with the individual in an iterative process that can supplement or supplant the traditional process of opinion formation, expression and reformulation. Interactive systems have some potential in this regard, but they are clearly the most complex and least developed aspects of telematics. In addition, unlike direct human interaction, where both (or all) parties involved are simultaneously learning and developing, interactive systems are usually designed with particular goals or outcomes in mind. They are, therefore, necessarily better at teaching particular skills, knowledge or preordained values and opinions than having opinions form out of the deliberation process itself. To the extent that telematics is a tradeoff for group and one to one interaction, the effects may be profound and costly to the democratic process.

It is possible to exaggerate the extent to which human interaction will be limited by developments in telematics. In fact, electronic town meetings and interactive systems which involve communication among users could conceivably add to the amount of deliberation possible. Even in this case, however, telematics can have indirect but equally important effects on the deliberation process. As information becomes more widely disseminated from centralized sources unbound by geographic or time constraints, the information and opinions which people bring to the marketplace of ideas will be more and more uniform. What was once an exchange relationship could easily become a reinforcement process in which attitudes and opinions developed in seeming independence are corroborated less because of the force of the logic, information and ethic they combine than because they are based on a common pool of fact and opinion disseminated by the telematic networks.

Political Participation

It is in the area of actual participation that telematics offers the greatest potential for change, partly because of the nature of the technology and partly because to date this is the area in which the least has been done. Telematics can impact on the level and type of participation in three ways. First, participation in politics is known to be strongly related to motivational forces such as a sense of civic duty, an interest in politics, and a sense of political efficacy. To the extent that greater exposure to political, social and economic issues and to the political process more generally can instill such attitudes and values, the level of participation should be increased.

Second, participation requires information if it is to be more than symbolic: citizens need to know what issues are important, what options are available, and where government officials, candidates, and parties stand on them. To the extent that telematics can provide quicker, more detailed, and easier accessed information on these topics, the quality of participation should be improved.

Finally, telematics offers the possibility that the modes of political participation available to citizens can be both expanded to include more "semi-direct" forms of involvement and easier, more accessible forms or involvement. Electronic town meetings, such as those experimented with in Columbus, Ohio, allow large numbers of people to listen to political discussion and actually vote on certain issues or express their opinion on them without ever leaving their living room. The potential for telematics to shrink problems of time and space and allow for a return to a more participatory form of democracy seems limitless. It is not hard to imagine the U.S. president addressing the entire nation on some issue of great import and then waiting, momentarily, to see if the nation supports or rejects the president's proposal. It is not hard to imagine the use of such interactive systems at the local, state, and even national level for voting on binding or non-binding referenda, or perhaps to even choose our political leaders. In short, the technology exists to radically change the way in which people participate, giving them a larger segment of the responsibility of governing themselves.

All of this speculation concerning political participation presupposes that the issues of control content and access are settled in ways beneficial to the society as a whole. If information sources do not expand, if the content of information remains culturally and ideologically constrained, and if access is allowed to continue to depend on personal skills and resources, then the picture is less optimistic. Greater participation by citizens who are unprepared to handle that responsibility is an image that has frightened political thinkers throughout history. Participation that is manipulated and influenced by the careful control of information is equally deplorable. And laissez-faire politics of the sort we have engaged in for two centuries would make a mockery of the term democracy, as access to the means of participation became more and more skewed. Again, the future depends critically on the specific ways in which telematics becomes woven into the social and political fabric of our nation. There are issues concerning telematics and participation that are more subtle than those discussed so far, however. Is participation in group deliberations when none of the group actually comes physically together the same as more traditional discussion and debate? Are the outcomes, the decisions of equal quality? Does the sense of community, of belongingness which is so critical as both a cause and an effect of political participation develop fromelectronic democracy? These are questions that must be answered as we move at an accelerating pace toward such new forms of politics.

Telmatics, Politics and Social Structures

The particular ways in which telematics will ultimately interact with politics depends as much social structures in which this interaction takes place as on any technical or political development. Central to this interaction is the role of education. The educational process can both affect and be affected by the growth in telematics. The use of computers at the earliest stages of education can increase the individual's competence and skill with such technology while at the same time speed the rate of learning in specific subject areas (consider, for example, the PLATO programs of Control Data). Children raised in an educational environment that includes computers will certainly be more comfortable and skilled in their use outside the classroom. Again the utility of this development for the political system as a whole depends upon how widespread this training is. Currently there is tremendous diversity in the extent to which computers (and personnel trained to use and teach with them) are available, and this diversity exists from the level of day care and kinder garden straight through college. In some cases computers and computer programs are simply unavailable, in others the quality of the equipment and/or the training of the staff is sorely inadequate. Such discrepancies, especially when reinforced from the earliest ages can lead to parallel discrepancies in both general educational development and more specifically in the ability to work with computer-based information in later walks of life.

The level and distribution of income is yet another environmental consideration which will critically shape and be shaped by developments in telematics. Indirectly it will affect the availability of telematics within the educational system, since money for both public and private schools is tied to the availability of funds from either the general population or specific benefactors. In addition, the purchase of home computers, cable, satellite hookups, etc., is all contingent on the disposable income to do so. This again interacts with the educational process, since a student who is able to extend his educational experience into the home is at a distinct advantage to one who is not in much the same way that having a piano at home facilitates the learning of that instrument. The extent to which telematics affords the opportunity to provide new "power brokers" in society, who these new power brokers are, and how large a segment of the population they represent depends in part on the degree to which telematics is able to disengage itself from the more traditional class and socio-economic status distinctions that already exist. If it is unable to do so then telematics could easily represent yet another rather large step toward a society made up of the technologically competent, the technologically superfluous, and the technologically obsolescent.

Conclusions

In this article we have tried to show the various changes which are taking place in communications technology, focusing on those changes which will directly affect the home information environment. We have then examined the ways in which these changes can impact on mass political attitudes and behaviours, suggesting that how positive a change telematics represents for the system as a whole depends on the control of, content of, and access of such information.

What, then, can be done? Do we simply stand by, as a society, and await the outcome of forces beyond our control? While all major social changes have an element of uncertainty or chance associated with them, it is possible to be more or less prepared for them. It is, of course, beyond the scope of this exploratory place to try to recommend specific steps to be taken, but we can point to the sectors Business and Labour (both within the telematics industry and in the larger community that directly and indirectly uses telematics) must become more aware that it has both a stake in and a responsibility for these developments and the directions they take. Their stake is the need for a technologically sophisticated and adaptable workforce; their responsibility is society's need for technologically sophisticated and adaptable citizens. Whether it is specifically in the development of socially beneficial software and databases, the creation of on the job or at the job education programs or greater support for the educational system, business and labour must be involved in the process of preparing for our move to a telematic society.

The academic community needs to devote increased attention to basic research in the areas outlined in this paper, specifying with greater precision the linkages among the various components. Such research should take an interdisciplinary approach, drawing on the perspectives and paradigms of history, political science, psychology and communications. In addition, the academic community must devise new and creative ways of integrating telematics, both as a subject to be taught and as a tool for teaching other subjects, into the traditional curricula and to do this at all level of the educational system. Education must involve the full development of the individual's potential as workers, citizens and human beings. A commitment to the integration of telematics into the educational process must go beyond some variant of vocational training to include its use as a more general educational tool and as a tool for the development of an informed and active citizenry. Our educational system must be capable of producing graduates who are comfortable with reading Shakespeare and scattergrams, who are capable of participating as well as programing.

Ultimately it is government which must oversee this transformation, for only government has the potential to consider these issues from the perspective of the public good and the power to implement the details of such a perspective. Government regulations need not mean overly bureaucratic and centralized control of the telematics industry, however. In fact, experience suggests that such a model is more responsive to the narrow demands of the industry than to the broader concerns of the public interest. Perhaps a more decentralized model based loosely on the citizen boards which help shape and regulate cable television policy in local communities might be more appropriate. In any event, unless citizens' interests are systematically included in the process of determining the direction of telematics, this new technology seems destined to be not a tool for the construction of a more open and participatory democracy, but yet another wedge between the theory and the reality of mass politics in advanced industrial societies.

The question then becomes do we attempt to nurture and channel these changes in positive and new ways, or allow them to flow down the well worn paths of past mistakes.

References

ANTHONY SMITH, "The Wire and the Wavelength - An Historical Study", in Cable : An Investigation of the Social and Political Implications of Cable Television. A Report published by the Standing Conference on Broadcasting, London, P.24, 1974.

U.S. News & World Report, July 25, 1983.

Lloyd Cowling, "Computers : The Future", in the Indu Singh (ed.) Telecommunications in the year 2000 : National and International Perspectives, Norwood Ablex Publishing, p.110, 19B3.

"Talking Terminals and Listening Computers overcame Toy Image," Info systems. pp. 52-56, Sept-1980.

Lloyd Cowing, op. cit., p.108

Anthony Pastelis and George Stubbs, "Fiber-Optic Technique Work Well". Telecommunications, Vol. 15 No.12, pp.16-17, De. 1981.

Burton Edelson, "Satellite Communications in the Age of Space Shuttle," in Indu Singh (ed) Telecommunications in the year 2000: National and international Perspective, Norwood: Ablex Publishing, p.92, 1983.

Joseph Pelton, Global Talk The Netherlands: Sijthhoff and Noordhooff p.66, 1981.

Button Edelson, op. cit., p.92.

Fred S. Siebert, Theodore Peterson, and William Schramm, Four Theories of the Press. Urbana, Illinois: University of Illinois Press, pp. 9-71, 1974.

Karl Marx's Capital: A Critique of Political Economy provides the classic explication of this relationship. New York: International Publishers, 1967 (From the English Edition of 1887).

See Scott Keeter and Cliff Zukin, Uniformed Choice: The Citizen in the Presidential Nomination System, New York: Praeger Publishers, 1983.

David L. Paletz and Rober M. Entman., Media Power Politics. New York: The Free Press, 1981.

Edward Jay Epstein., News From Nowhere. New York: Vintage Books, pp.87-91, 1973.

Jeremy Tuns tall., The Media are American. New York: Columbia University Press, pp.20-37, 1977.

Herbert J. Gans., Deciding What's News. New York: Vintage Books, pp. 91, 126, 180-81 & 266, 1980.

Todd Gitlin., The Whole World is Watching. Berkleys The University of California Press, pp.249-292, 1980.

Scott Burns., The Household Economy. Boston: Beacon Press, 1975.

Sidney Verba and Norman H. Nie., Participation in America. New York: Harper and Row Publishers, 1972.

Alvin Toffler., The Third Wave. Toronto : Bantam Books, p.427-431, 1980.

Water Dean Burham., "American Politics in the 1970's: Beyond Party?" in Jeff F ishel, Parties and Elections in an Anti-Party Age., pp.333-341.

Erwin Krasnow and Lawrence D. Longley., The Politics of Broadcast Regulation. New York: St.Maratin's Press, 1982.

Chapter **18**

A High-Tech Route to Development

A. Singhal and E. M. Rogers

Don't show me silicon and plastic. Tell me what it can do.

Satyan Pitroda, Advisor on Technology Missions to Indian Prime Minister Rajiv Gandhi, quoted in Dataquest (1988, p.71):

The history of scholarly work on development communication has passed through several eras : From an initial era after world War II until about 1970, when the mood was optimistic about the potential of mass communication for furthering development in Third World nations; a second era of the 1970s, in which pessimism replaced optimism: and the era of the 1980s, in which the applications of new communication technologies like computers, communications, and other microelectronics innovations have led to a guarded resurgence of positive thinking about the role of communication in development. During the 1980s, however, communication is still seen mainly as a *channel* through which development messages flow to some segment of the population (for example, to farmers, parents of school children, fertile-aged married couples, illiterates, and to Others).

An entirely different role for new communication technology in development is to think of it as an industry (such as high-tech microelectronics) which can provide jobs and create taxable income. In several Western industrialized countries, high-tech microelectronics and its applications in telecommunications and computers have come to represent a highly dynamic industry. For example, before its divestiture in 1982, the sales of American Telephone and Telegraph Company (AT&T) were greater than the individual gross national product (GNP) of 118 Third World countries (Dizard, 1982). In 1989, the sales of International Business Machines (IBM) were about $65 billion, higher than the GNPs of about 120 Third World countries in Asia Africa, and Latin America. Several Third World countries now realize the growing potential of new communication technologies as an "industry", and have instituted government policies to attempt a "high-tech" route to development.

The present paper discusses the validity of this conception of new communication technologies as a growing "industry" that contributes to development in Third World nations. A historical perspective on theories of communication in development is provided. Lessons learned from Western nations about the development of high-technology microelectronics industries are discussed, with implications for the role of high-tech industries in the development of Third World nations.

Background of Communication in Development

Development is the top priority of the national government in every Third World country. Development is "a widely participatory process of directed social change in a society, intended to bring about both social and material advancement (including greater equality, freedom, and other valued qualities) for the majority of the people through their gaining greater control over their environment" (Rogers, 1976). Development Communication is the uses of communication to further development (Singhal and Rogers 1989).

Influential scholars like Daniel Lerner and Wilbur Schramm became involved in development communication in the 1950s, soon after several Third World countries in Asia.

Africa and Latin America gained their independence. Lerner (1958) argued that the mass media could help "modernize" Third World individuals. Lerner's argument was consistent with such psychological explanations of Western industrialization as Weber's (1930) "Protestant ethic", McCielland's (1961) "achievement motivation", and Hagen's (1962) "innovative personality". Weber (1930) argued that a capitalistic industrial society flourished in Europe because the Protestant reformation movement ended the long-established authority of the church, encouraging rational scientific theory. McClelland (1961) argued that Western industrialized nations contained individuals with high levels of achievement motivation, the desire for excellence. Hagen (1962) argued that Western societies contained more innovative personalities.

Schramm (1964) argued that a Third World nation could facilitate development by expanding the mass media. The mass media were considered to be a "magic multiplier" of development during the 1950s and 1960s. A general optimism about development communication was widespread. This optimism was based on the so called dominant paradigm of development, which stressed economic growth through industrialization (Rogers, 1976). Walt Rostow, an influential American economist, emphasized GNP (gross national product) as a measure of a country's progress in development (Rostow, 1960).

The dominant paradigm fell into general disrepute in the 1970s, and criticisms of development communication appeared : Beltran (1975), Schramm and Lerner (1976), Rogers (1976), and Dervin (1960). Scholars realized that the role of mass communication in facilitating development was often only indirect and contributory, rather than direct and powerful (Hornik, 1980). Critical scholars like Nordenstreng and Schiller (1979), Hamelink (1980), and Dorfman and Mattelart (1975) asked such important questions *as* "Who owns and controls the communication systems?", "who gains and who loses?" Several Latin American scholars espoused a dependency theory, the idea that underdevelopment may be due to the dependence of Third World nations on Western industrialized countries, rather than due to resistances to development lying within a Third World nation (Frank, 1969). A questioning stance about the development communication field characterized the 1970s.

In the late 1970s, the dominant paradigm of development gave way to a new development paradigm, which was more pluralistic recognizing many pathways to development (Rogers, 1987). Further, the new paradigm was less Western in its cultural assumption. Key elements in the present day development communication paradigm include; (1) greater equality in the distribution of development investments, information, and the consequent socio-economic benefits, by focusing development activities on the weaker sections of society, (2) popular participation, knowledge sharing, and empowerment to facilitate self-development efforts by individuals, groups, and communities, (3) self-reliance and independence in development, emphasizing the potential of local resources, and (4) integration of traditional and modern communication systems (Rogers, 1974; Singhal and Rogers, 1989).

One cause of guarded optimism about the potential of development communication today is the social impacts of new communication technologies like satellite-based television broadcasting and long distance telephony, video cassette recorders, and applications of microelectronics in computers and telecommunications. Communication technology is the hardware equipment, organizational structures, and social values by which individuals collect, process, and exchange information with other individuals (Rogers, 1986).

It is not yet clear whether the development promise of the new communication technologies will be fulfilled. Many of these new communication technologies are at an early stage in Third World countries. Many past communication technology projects failed because these First World technologies did not fit Third World contexts (Mody, 1985) Too often in the past, the importance of the social, organizational, economic, and logistical context of a communication technology has not been realized, and thus the communication technology project fails. For example, the electrical power surges that often occur in many Third World settings play havoc with microcomputers (Singhal and Rogers, 1989).

Another cause of recent optimism about development communication is that communication technologies can become something more than just a delivery system (or channel) for development messages. They also represent, at least potentially, a new type of industry one that some enthusiastic observers claim can allow a Third World nation to leapfrog the industrial era, in order to become an information society.

The United States, Japan, and most Western European nations in recent decades have passed through an important transition in the makeup of their workforce, the basis of their economy, and in the very nature of their society (Singhal and Rogers, in press). These nations have become-information societies, countries in which a majority of the workforce is composed of information workers, who produce, process, or distribute information, and who produce information technology (Rogers, 1986). The U.S. became an information society in about 1955, when the number of information workers exceeded the number of industrial workers, and far surpassed the number of agricultural workers. The transition of the U.S. to an information society has been the focus of considerable scholarly research (Machlup, 1962, Bell, 1973; Forat, 1978; and Beniger, 1986).

Information is patterned matter-energy that affects the probabilities available to an individual making a decision (Rogers, 1986). Information behaves somewhat oddly as an economic resource in the sense that one can sell it (or give it away), and still have it. In an information society, information is the crucial ingredient, and it is increasingly of an earlier era. The computer is the most important tool in the information society, just as the steam engine was the basic technology in the industrial society (Singhal and Roger, 1989).

The high-technology microelectronics industry is one of the driving forces of an information society. A high-technology industry is one in which the basic technology underlying the industry changes very rapidly. A high-tech industry is also characterized by (1) a high proportion of highly skilled employees, many of whom are scientists and engineers, (2) a fast rate of growth, (3) a high ratio of R&D expenditures to sales (typically about 1 : 10), and (4) a worldwide, highly competitive market for its products (Rogers and Larsen, 1984).

Microelectronics, the part of the electronics industry centering on semiconductor chips and their applications (such as in computers and telecommunications), is usually considered the highest of high technology. The important technical advances in microelectronics in the past two decades have spurred the information revolution. The growing miniaturization, lower cost, and increased reliability of semiconductor chips translates directly into more ubiquitous computers, and eventually toward an information society. So here we see how advances in communication technologies contribute to the information society.

Lessons Learned from the West

In a review of literature on high-technology-based economic development, Segal Quince Wicksteed (1987) concludes that high-technology industries can substantially aid local economic development. Their review, based on several studies of high-technology areas such as Silicon Valley in Northern California, Route 128 in Boston, Research Triangle Park in North Carolina, and Cambridge in England concludes that high-technology industries (1) are high-growth industries, (2) have a tremendous potential to create taxable wealth, and (3) have a high job creation capacity (Segal Quince Wicksteed, 1987). High technology industries increase their employment faster than general manufacturing industries, and small high technology firms increase their employment several times faster than small manufacturing firms (Segal Quince Wicksteed, 1987).

While high-tech industries directly create several jobs, they also exercise a significant multiplier impact on jobs in other sectors for each high-tech job that is created, several service occupations usually follow. These support jobs include the employees of restaurants, travel agencies, and so forth (Rogers and Chen, 1988). These support jobs grow rapidly, eventually becoming more important than the high-technology sector of the economy (Segal Quince Wicksteed, 1987).

Experiences from Western industrialized nations show that the high-technology microelectronic industry forms these bases of a highly lucrative information industry. A major center for high-tech microelectronics in the U.S. is Silicon Valley in Northern California, which has been called 'the world capital of the information society' (Rogers and Larsen 1984). Silicon Valley is the world's largest and most widely known techno polis. A techno polis is a geographically-concentrated high-tech complex which is characterized by collaborative relationships between government, private firms, and research universities, and by the presence of venture capital and entrepreneurial spin off firms (Singhal and Rogers, 1989). One reason for the fame of Silicon Valley is the immense riches that have been created there by converting technological innovations in microelectronics into successful business ventures.

Cambridge, the prestigious university town in England, is the site of a burgeoning high-tech industry. Cambridge's 450 high-tech firms account for 15 percent of the total employment in the local area, and have a total sales output of about $3 billion (Segal Quince Wicksteed, 1985). While the techno polis on Silicon Valley and Cambridge represent high spontaneous high-technology development, several techno-polis were highly planned by national and/or local governments : For example, Research Triangle in North Carolina, Sophia Antipolis in Southern France, Tsukuba Science City in Japan, the Shenzhen techno polis (near Hong Kong), the Beijing techno polis. In China, and the Hsin-Chu technopolis in Taiwan (Rogers and Chen, 1988).

What factors influences the success of a techno polis? A heavy investment, for example, Tsukuba Science City in Japan represents an investment of $5 billion in public funds (Rogers and Chen, 1988). Long-term political and government commitment is necessary to create a planned techno polis. A techno polis must have its own source of cutting-edge technology, such as from a research university or from a government-funded research institute, so that technology can be transferred most easily to neighboring companies. The presence of venture capital helps create spin-off firms, which in turn creates new jobs. Venture capital is money invested in new or young high-technology companies with a high potential for growth. Nuchi (1984) advocates collaborative relationships, which he terms 'M-Form' social organization, as key to success in a techno polis. In an M-Forrn arrangement, the participating units cooperate to mutual advantage while remaining independent.

Why is there currently such high interest by governments, private companies, and scholars in the techno polis? To governments, a techno polis mainly represents the creation of jobs and the production of income (and hence, taxes). Governments also are motivated by a desire for their firms to remain competitive in the worldwide microelectronics industry that drives the information society (Rogers and Chen, 1987). To a private company, a techno polis is perceived as an unusual business opportunity and, for many firms, as a place where new competitors are likely to be spun-off. For social scientists, a techno polis represents an important contemporary type of social change that deserves careful study (Rogers and Chen, 1987).

Government officials enthused about jobs and wealth often are less aware of the negatives that usually come with microelectronics based techno polis. For example, Silicon Valley is characterized by extreme socio-economic inequality. The 250,000 microelectronics workers include not only white-collar engineers and managers, but also about 110,000 manual workers, mostly Third World women, who perform monotonous assembly-line tasks (Rogers and Larsen, 1984). Silicon Valley also has horrendous traffic problems, a level of air pollution that rivals Los Angeles' smog, and underground water contamination from leaking chemicals.

Further, high-technology growth is 'elitist' in that it directly benefits the best-educated members of the workforce. Because information is such a highly valued commodity in an information society, individuals who produce new information (scientists, R&D workers' and engineers) are treated as the super-elites of the information society (Bell, 1973). Further, high-technology growth deskills the skilled occupations. Deskilling is the process through which new communication technologies downgrade an occupation to a lower socio-economic status by replacing human skills with information handling equipment (Rogers, 1986).

Most Third World nations are still agricultural societies today. The majority of their workforce is employed in agriculture, and the next most numerous occupational category are industrial workers. So one might think that Third World countries are unlikely candidates to become information societies, (Regers and Chen, 1987). However, several countries in Latin America, Africa and Asia are pursuing national policies to move them rapidly to become information societies (for example, Brazil, Singapore, Korea, Taiwan, Hong Kong, and India). Do these countries need to go through the same stages that Western countries went through in order to become an information society? Or can they leapfrog the industrial society to directly become an information society in the future? Little scholarly attention (at least from a development communication perspective) has been given to Third World countries who pursue national policies encouraging an "industry" role for the new communication technologies.

In India, since Rajiv Gandhi became Prime Minister in 1984, the Indian government has promoted indigenous businesses in semiconductors, telecommunications, computers, and computer software. From 1984 to 1988 in India, the number of computers increased ten-fold, computer software exports increased five-fold, and the computer industry's revenues increased four-fold (Singhal and Rogers, 1989 (see Figure 1).

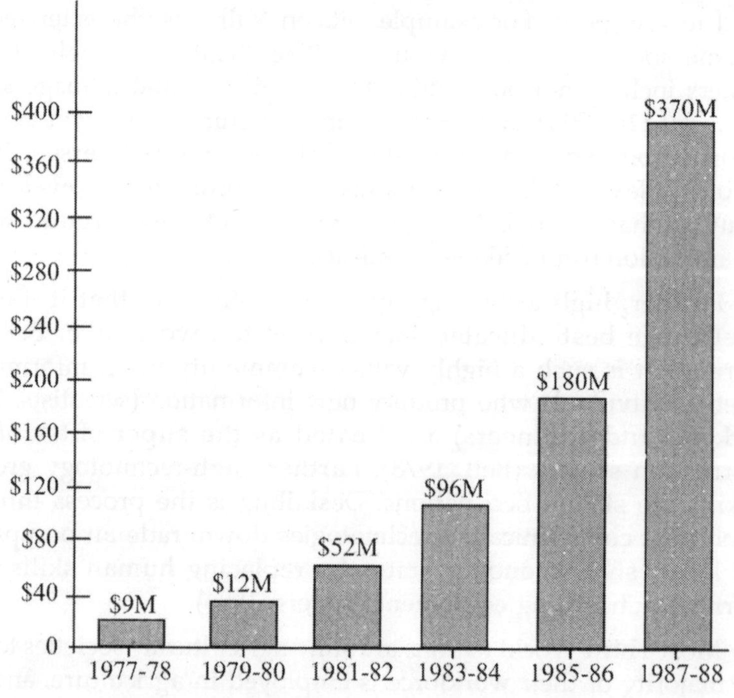

Fig. 1. Sales of Indian Computers

India is primarily an agricultural nation. In 1989, 70 percent of India's workforce are engaged in farming, although agriculture accounts for only 33 percent of India's gross national product (GNP). Some 20 percent of the Indian workforce is employed in industry, which accounts for about 25 percent of India's GNP. The remaining 10 percent of the Indian workforce are white-collar workers employed in service and information jobs. They account for 42 percent of India's GNP (Singhal and Rogers, 1989). India is far from becoming the type of information society represented by the U. S. or Japan, where a majority of the workforce is composed of information workers. But if one counts money instead of workers as the best index of progress toward becoming an information society, India is well on the way.

Governments of several Third World countries actively encourage the development of high-tech microelectronics industry. In 1980, the government of Taiwan launched the Hsin-Chu Industrial Park to fuel economic growth, and to create a Taiwanese showcase of future high-technology development. By 1987, about 60 companies were located in the Hsin-Chu Industrial Park with annual sales exceeding $750 million, and an annual growth rate of 66 percent (Rogers and Chen, 1988).

In China, the Shenzen Special Economic Zone is the largest of the four SEZs (special enterprise zones) set up in the 1980s by the national government to implement a new economic strategy of attracting foreign capital and technology. During 1980 to 1986, these four SEZs accrued $1.3 billion in overseas capital. Shenzen is a sharp contrast with the rest of China, economically, ideologically, and culturally (Rogers and Chen, 1988). Its average incomes are two to three times higher than for the rest of China.

The Korean government aggressively promotes high technology through such policies as providing tax incentives to local corporations (Singhai and Rogers, 1989). Encouraged by their government, Korean entrepreneurs-increasingly combines capital and technology to create new-high-tech ventures. Korea has a highly skilled, strongly motivated workforce who labour for relatively low pay. Korea is heading for the high tech route to development.

Implications for Third World Development

Previously in this paper, key elements in the present day development communication paradigm were discussed : (1) greater equality in the distribution of development investments, information, and the consequent socio-economic benefits, (2) popular participation, knowledge sharing, and empowerment to facilitate self-development efforts, (3) self-reliance and independence in development, emphasizing the potential of local resources, and (4) integration of traditional and modern communication systems (Rogers 1976). In the following section, the development related implications for Third World countries pursuing a high-tech-route to development are discussed within the framework of the present day development paradigm.

Socio Economic Benefits

The high tech route to development can result in substantial economic gains for a nation : (1) Wealth creation (and hence, taxes), (2) a higher GNP growth rate, (3) creation of new jobs, and (4) moving a nation toward becoming an information society. Also important are the social impacts of the high-tech microelectronics products (improved telephones, lower-priced computers, etc.), which can aid business development, increase productivity and efficiency, and improve individuals quality-of-life. However, on the basis of other nations experiences with high-tech microelectronics. Third World countries should expect that these benefits will come at a certain human cost.

Socio-Economic Inequality

Inequality is the overriding policy issue in contemporary development literature. Equity concerns dominate most communication policy issues, particularly concerning the new communication technologies impacts. Third World nations need to guard against the "elitist" nature of high-technology growth, which can promote an inequitable distribution of socio-economic benefits. For example, as stated previously, 10 percent of the Indian workforce who are information workers earn 42 percent of the GNP. Obviously these information workers are earning about four times as much as others in the workforce. These 10 percent elite information workers have political clout and economic power far beyond their numbers in the population. Those who govern Indian society must constantly keep in mind the 400 million Indians (50 percent of India's population) who live below the poverty line (Singhai and Rogers, 1989). Equity issues are also central to the applications of high-technology industries in computers, which have the potential to replace, displace, and deskill certain occupations.

Related to the issue of socio-economic inequality is the issue of equality in the distribution of development investment. High-tech microelectronics requires heavy capital investments, the cost of which may decrease spending for education, health, and other development problems. Given their scarce resources, Third World governments need to (1) carefully weigh potential costs and benefits involved in taking the high-tech route to development, and (2) its probability of success, given existing infrastructural, social, political, and economic conditions (O'Connor, 1985).

Dependency

Dependency is defined as one country's reliance on another for material, or non-material support. Since international competition and technology transfer are particularly unique characteristics of high-technology industries, Third World countries should consider dependency relationships and other external factors which might impede their national development : International terms of trade, the economic imperialism of multinational corporations, and the vulnerability and dependence of the Third World recipients of imported capital and technology (Singhai and Rogers, 1989).

Technological Suitability

"Big", imported communication technologies may not always serve the best interests of a Third World country (Schramm, 1977). "High-tech" and "low-tech" need tc-co-exist in Third World countries, and should supplement each other. High-technology in isolation is not a panacea for Third World problems. The Third World's challenge is to internalize, absorb, and manage technology to serve socio-economic development goals. For example, the Indian Center for the Development of Telematics (C-DOT) has indigenously developed state-of-the-art rural telephone exchanges which do not require air-conditioning, which tolerate electrical power surges, extreme temperature variations, high humidity, and a dusty environment. C-DOT's rural telephone exchanges are spectacular successes, aiding socio-economic development in rural India (Singhal and Rogers, 1989).

The use of chicken mesh satellite antenna in India's Satellite Instructional Television Experiment (SITE) represented another spectacular example of combining "high-tech" with "low-tech". For one year, during 1975-76, India's Space Application Center (SAC) used chicken mesh satellite antenna in the 2,400 Indian villages receiving direct television signals from NASA's Applications Technology

Satellite (ATS-6). Made of aluminum, SAC's three meter diameter chicken mesh antenna were good conductors of electricity, light in weight, easily portable, and cost only $100 (U.S.). Installation took only a few hours in each village (Singhal and Rogers, 1989).

Impacts on Culture

Culture consists of material and non-material aspects of a way of life, which are shared and transmitted among the members of a society (Rogers, Burdge, Korsching, and Donnermeyer 1988, p. 29). The high-tech "culture" may not fit well with the traditional values of several Third World countries. For example, cut throat competitiveness, a stress on material gain, and high divorce rates characterize U. S. techno polis like Silicon Valley, and clash with fundamental Indian cultural values.

Conversely, certain traditional values in Third World countries may be an anathema to high-tech growth. For example, in India, a high value on job security restricts the spread of entrepreneurial fever. Also, while collaborative relationships seem a natural part of Japanese culture, and are gradually being learned in the highly-competitive U.S. microelectronics industry, in India such collaboration is not very highly valued at present.

Production Centered Versus People Centered Approach

Third World governments who are planning to take a high-tech route to development should guard against the pitfalls of a production-centered approach, which has come under criticism from contemporary development scholars (Korten, 1984). *The* Production Centered development approach emphasizes economic growth through the production of goods and services. This approach stresses (1) industry over agriculture, (2) urban development over rural development, and (3) the utilization of capital resources over human resources. The needs of the production system take precedence over the needs of people. Needs of the rural poor are neglected, and the result is greater socio-economic inequality (Korten and Corner. 1984). In a people-centered approach to development, the needs of the people take precedence over the needs of the production system. Such an approach is consistent with the present day development paradigm. Central themes of people-centered development are (1) empowerment of the people, and (2) equality in the distribution of socio economic benefits. Economic development achieved via growth in high-tech microelectronics industries (and consequently growth in the information economy) primarily represents a Production-centered approach.

The village of Kukshi, located in a tribal district of Madhya Pradesh state in India, is home to a fast-developing re-chargeable flashlight industry. In 1982, Lakhan Pal Kumawat, a Kukshi farmer, developed a re-chargeable flashlight, specifically adapting it to the needs of fellow farmers. The local bank provided Kumawat with $4,000 (U.S.) to start his manufacturing operations. Today, about 30 small-scale flashlight units have sprung up in this backward village, providing employment to 1,000 villagers, and generating annual sales of $1.5 million (U.S.) The Kukshi case illustrates economic development via a people-centered approach.

Conclusions

In Third World countries, the design and production of new communication technologies is an industry that can help (1) create wealth (and hence, taxes), (2) provide jobs, (3) create a high rate of economic growth, (4) move a nation toward becoming an information society, (5) increase productivity and efficiency, and (6) improve individuals quality of life. However, these economic benefits may often come 31 a price : (1) Greater socio economic inequality, (2) decreased spending for other development programs, (3) greater dependency on Western countries for capital and technology, (4) adoption of technology which might be unsuitable for local conditions, (5) an erosion of traditional cultural values, and (6) an emphasis on production processes, instead of on people.

So how valid is the conception that new communication technologies as a growing "industry" can contribute to development in Third World nations? New communication technologies, as an industry, (1) may contribute to development in Third World nations in an important way (by creating wealth, new jobs, etc.), and, (2) at the same time, may produce certain undesirable effects, detrimental to a nation's development (like socio economic inequality, dependency, etc), Development problems are far too complex to justify a simple technology-driven solution. High-technology-based economic growth does not cover the full range of meanings associated with the term "development."

Third World countries should guard against an extreme position of technological determinism, the belief that technology is the main cause of social change in a society, (Rogers, 1986, p. 23). Other political, social, cultural, and economic factors along with technology help shape a society. Third World governments should realize that high technology will bring both positive and negative effects. Their challenge is to attempt social, technological, and political reform to optimize their positive effects, while .minimizing the negative effects.

Notes

1. This article draws upon Singhal and Rogers (1 989).
2. Arvind Singhal is a doctoral candidate and Everett M. Rogers is Walter H. Annenberg Professor of Communications both at the Annenberg School of Communications, University of Southern California.
3. Historically, scholars have adopted one of three meta-theoretical viewpoints about technology's role in society (Gendron, 1977), *Utopians,* comprising of technological determinists like Fuller (1981) and Weinberg (1981), believe that technology "is an unalloyed blessing for mankind" (Mesthene, 1981, p. 99). *Dystopians,* such as Ellul (1981) and McDermott, 1981), are convinced "technology is an unmitigated curse" (Mesthene, 1981, p. 99). The *neutral* school argues that technology is not a problem solver nor a problem-generator (Barnow, 1975).

   Mesthene (1981) claimed that the Utopian, dystopian, and neutral viewpoints are themselves "unhelpful" to policy-makers. Inspired by Pool's (1977) dual-effects analysis of the social impact of the telephone, Mesthene (1981) proposed that technology "has both positive and negative effects, and it usually has the two at the same time and in virtue of each other." The role of the social analyst, therefore, is to identify the contingencies that determine the positive and negative effects of technology. This *contingency* approach recognizes the desirable and undesirable impacts of a technology and proposes that these impacts are differentially determined by the context in which the technology is introduced (Contractor, Singhal, and Rogers, 1988).
4. This case illustration is based on Singh (1988).

References

Barnow, E. 1975. *The Tube of Plenty.* New York : Oxford University Press.

Beltran, S. 1975 "Research Ideologies in Conflict," *Journal of Communication* 25 : 187- 193.

Bell, Daniel, 1973. *The Coming of Post-Industrial Society.* New York : Basic Books.

Beniger, James R., *The Control Revolution* - Cambridge, MA : Harvard University Press.

Contractor, Noshir S., Singhal, Arvind, and Everett M. Rogers, 1988. "Metatheoretical Perspectives on Satellite Television and Development in India," *Journal of Broadcasting and Electronic Media,* 32 : 129-148.

Cornish, Edward (ed). 1982. *Communications Tomorrow:*

The Coming of the Information Society. Bethesda. MD : World Future Society.

Dizard Wilson P., 1982. *The Coming Information Age : An Overview of Technology, and Politics.* Second edition. New York : Longman.

Dorfman, Ariel, and Armand Mattelart, 1975. *How to Read Donald Duck* New York : International General.

Ellul, J , 1981. "The Technological Society," in A. H. Teich (ed.), *Technology and Man's Future.* Third Edition. New York. St. Martin's Press.

Fuller, B. R., 1981. "Utopia or Oblivion," in A. H. Tech (ed.), *Technology and Man s Future* - Third Edition. New York : St. Martin's Press.

Dataquest, 1988. "Interview with Satyan Pitroda." March.

Deirin Brenda, 1980. "Communication Gaps and Inequities : Moving Toward a Reconceptualization," *Progress in Communication Sciences,* 2. : 73-112.

Frank, A. G. 1969, *Latin America : Underdevelopment or Revolution.* Monthly Review Press.

Gendron, B., 1977. *Technology and the Human Condition.* New York : St. Martin's Press.

Hagen, Everett F., 1962. *On the Theory of Social Change* Chicago : Dorsey.

Hameiink, Gees, 1980. *The New International Information Order : Development and Obstacles.* Vienna : Vienna Institute of Development.

Hornik, Robert, 1980. "Communication as Complement in Development," *Journal of Communication,* Spring, pp. 10-24.

Korten, David C., 1984. "Rural Development Programming : The Learning Process Approach," in David C. Korten and Rudi Klaus (eds.), *People-Centered Development' Contributions Toward Theory and Planning Frameworks.* Hartford, CT : Kumarian Press.

Korten, David C., and George Garner, 1984. "Planning Frameworks for People-Centered Development," in David D. Korten and Rudi Klaus (eds.), *People Centered Development : Contributions Toward Theory arid Planning Frameworks* Hartford, CT : Kumarian Press.

Lerner, Daniel, 1958. *The Passing of Traditional Society : Modernizing the Middle East.* Glencoe, IL : Free Press.

Maclltup, Fritz., 1962. *The Production and Distribution and Knowledge in the United States* Princeton, N.J. : Princeton University Press.

McClleland, David C., **1951.** *The Achieving Society.* Princeton, NJ : Van Nostrand.

McDermott, J., 1 981. "Technology : The Opiate of Intellectuals," in A. H. Teich (ed.), *Technology and Man's Future.* Third Edition. New York : St. Martin's Press.

Mesthene, E. G., 1981. "The Role of Technology in Society," in A. H., Teich (ed.), *Technology and Man's Future.* Third Edition. New York; St. Martin's Press.

Mody, Bella, 1985. "First World Communication Technologies in Third World Contexts," in Everett M. Rogers and Francis Balle (eds.), *Media Revolution in North America and Western Europe,* Norwood, N J : Ablex.

Nordenstreng, **Kaarle,** and Herbert I. Schiller, (eds.). 1979. *National Sovereignty and International Communication.* New York : Ablex.

O'Connor, David C., "The Computer Industry in the Third World : Policy Options and Constraints," *World Development,* 13 : 311 332.

Ouchi, William G., 1984. *The-M-Form Society : How American Teamwork Can Recapture the Competitive Edge* Reading, MA : Addison-Wesley.

Porat, Marc Uri., 1978. "Global Implications **of** the Information Society," *Journal of Communication,* 28 : 70-79.

Pool, Ithiel de Sola., 1977. *The Social Impact of the Telephone* Cambridge, MA : MIT Press.

Rogers, Everett M., 1976. "Communication and Development : The Passing of the Dominant Paradigm," in Everett M. Rogers (ed.), *Communication and Development : Critical Perspectives* - Newbury Park, AC : Sage.

Rogers, Everett M., and Judith K. Larsen, 1984. *Silicon Valley Fever : Growth of High-Technology Culture.* New York : Basic Books.

Rogers, Everett M., 1986. *Communication Technology : The New Media in Society* - New York : Free Press.

Rogers, Everett M. 1987. "Communication and Development Today". Paper presented to Seminar on Communication and Change, East West Center, Honolulu, Hawaii, July.

Rogers, Everett M., and Y. A. Chen., **1987.** "Technology Transfer and the Technopolis." Paper presented at the Pan-Pacific Conference IV. Taipei, Taiwan, May.

Rogers, Everett M., and Y. A. Chen., 1988. Technology Transfer and the

Technopolis, in Mary Ann Von Glinow and Susan Albens Mohrman eds.), *Managing Complexity in High Technology Industries : Systems and People*-New York : Oxford.

Rogers, Everett M., Burdge, RabelJ., Korsching, Peter F. and Joseph F. Dennermeyer, 1988. *Social Change in Rural Societies.* Third Edition. Englewood Cliffs, NJ : Prentice-Hall.

Rostow, Walt W., 1960. *The Stages of Economic Growth* Cambridge, England : Cambridge University Press.

Schramm, Wilbur, 1964. *Mass Media and National Development* Stanford, CA : University Press.

Schramm, Wilbur, and Daniel Lerner (eds). 1976. *Communication and Change : The Last Ten Years and the Next.* Honolulu, HI : University Press of Hawaii.

Schramm, Wilbur, 1977. *Big Media, Little Media.* Newbury Park, CA : Sage. Schramm, Wilbur. 1988. *The Story of Human Communication.* New York : Academic Press.

Segal Quince Wicksteed. 1985. *The Cambridge Phenomenon* Cambridge, England : Segal Quince Wicksteed.

Segal Quince Wicksteed, 1987. *Universities, Enterprise and Local Development: An Exploration of Links* Cambridge, England : Segal Quince Wicksteed.

Singh, N. K., 1988. "Torch Bearers," *India Today,* October 31, p. 98.

Singhal, Arvind, and Everett M., 1989. *India's Information Revolution.* New Delhi : Sage.

Weber. Max., 1930. *The Protestant Ethicandthe Spirit of Capitalism.* New York : Charles Schriber's Sons.

Weinberg, A. M., 1981. "Can Technology Replace Social Engineering?," in A. H. Teich (ed.), *Technology and Man's Future.* Third Edition. New York : St. Martin's Press.

Chapter **19**

Academics Vs Communication Practitioners : Implications of their Paradigms on Development Support Communication

Srinivas R. Melkote

The development of rural areas around the world, particularly in the developing nations, has been an important area of study and practice for both academics and communication practitioners. While the concern for the suffering masses of people in the Third World has been genuine, the academics and practioners have differed in their assumptions and approaches to ameliorating the problems. The academics, because of the necessities of academic research, have sought to develop broad theoretical generalization on such aspects as : how people acquire and accept or reject information, ideas and beliefs; the role of media in the modernization process, etc. Invariably, the academics have attempted to provide blueprints for change and seldom concerned with how such blueprints could be really put into practice across differing contexts and peoples. In short, the academics have been reluctant to get their 'hands dirty'. The practitioners, on the other hand, have been more concerned with applying the theoretical ideas into specific problem areas. Consequently, they have been more sensitive to the immense problems of development across different cultures and different people in a particular culture. Thus, the practitioners main concern has been to solve practical problems rather than building theoretical generalizations, which any way, may not hold across differing societies. As Diaz Bordenave notes :

Thus, although theoreticians tend to state their views in absolute terms, there are no absolutely 'right' or 'wrong' theoretical models for rural development communication. A model that seems to work in one kind of development situation may not be suitable for another. Today, theoreticians know much more about communication in rural settings than they did ten or twenty years ago; tomorrow they will need to know even more (10, p. II).

This paper will attempt to review and compare the pursuit of development from the context of the academics and the communication practitioners. It will then discuss the implications and consequences of these two approaches on development communication strategies and outcomes.

Pursuit of Development by Academics Historical Review

The transition of European societies from a traditional and feudal structure toward more advanced social stages has been an important concern of social theorists from the classical to the modern age. Considerable literature is available on the forces and set of relationships that propelled Europe from the largely agricultural feudal order to the complex, highly differentiated, capitalist industrial society. In fact, as Portes comments, "the issue of European transformation has given rise to what is, without doubt, the core of classical and of most modern sociological theory" (24, p. 56).

With the emergence of the Third World, theories and concepts which recapitulated the unique West European and North American experience, generated in the last few decades, a host of new models of development. An important concern, among the 'developed' nations was how to develop the under developed societies just emancipated from centuries of colonial rule. The main trend in sociological and communication theories was extrapolation of concepts generated from the West European and North American transition to the study of national development in Third World nations. As Eisen-stadt notes :

These studies developed the initial paradigm, of 'modernization', linking together study of change with earlier systemic and behavioural approaches, ultimately articulating some of the basic issues of sociological theory. In this framework, traditional societies were perceived as basically very limited in their capacity to cope with problems or master their environment, while modern societies were seen as coping with a continuously wider range of their internal and external problems and environments (13, p. 31-32).

The dominant paradigm in the West vis-a-vis the Third World development was based on an application of the evolutionary concept of Darwin to social change. In other words, the process seemed relatively irreversible.Once the necessary conditions were established for take-off, a country took off, became modern and stayed modern.

Though the evolutionary perspective to socialchange floundered after its initial popularity (14), it gave rise to a number of theoretical approaches influential to our times : first, the bipolar theories based on a dichotomous conception of tradition as against modernity (8, 20, 29, 34), and second, the çoncept of social differentiation of society which underlined the inevitability of coping with strains at a given point through increased specialization, differentiation and interdependence.

Beside the evolutionary approach, the psychosocial perspective of development was also influential in the West. These psychological theories derived from Weberian views of 'Protestant Ethics,' considered certain attitudesand values in individualsas prerequisites to creating a modern socio-economic polity. The search for mental viruses (21), modern spirit (18), empathy (20), and status respect (17) in individuals, so that they internalize the process of modernization by not only being able to cope with change but expecting and demanding it, constituted the basic orientation of this school of thought.

Problems with the Approach of Academics

The major problem with the academic approach to communication and development in the Third World was its highly prescriptive nature. Areas or regions in the Third World countries were regarded as ends in themselves. In other words, it was presupposed that these societies were 'traditional' and therefore needed to be 'modernized.' It was felt that enough was known about these societies to make formative research unnecessary. Krippendorff notes, "It was very talk oriented and seemed to view its function more as providing blueprints for change than to dispassionately observe and analyze the ramifications of social change as it actually took place in Asia and Africa" (19, p. 73). However, as Cox (9) observes, to view these societies as arenas for action rather than as means or ends in themselves would have provided a more realistic perspective. This approach would have permitted understanding, analysis, and the assessment of objectives and possibilities before the launching of any large scale action. Cox (9) cautions that understanding must precede action if success is to be more than accidental. After all, time and resources human, financial, and material are relatively scarce in the Third World, and the urge to change these societies without adequate observation, understanding, and diagnosis of the dynamics of life in these cultures raise serious ethical and methodological questions.

Mass Media : Harbingers of Change

The main obstacles for the developing nations to attain the levels of 'modernization' achieved in the West were ascribed, to their 'peasant' culture and 'traditional' habits. The Neo-Weberian concepts of 'modernization' included the component of empiricism, rationalism, materialism, individualism, and universalism. The developing countries seemed to lack all these qualities in their cultures, rooted as they were in sacred-value orientation, authoritarianism and particularism. Therefore, underdevelopment in the minds of academics such as Lerner, Schramm, McClelland, Pye, and Rogers in the late 1950's through the early 1970's was perceived as a cultural problem. The Third World needed to abandon its traditional cultures and support institutions if it wanted to taste the goodies of development. This kind of theorizing provided a tailor made role for the modern media of communication. "Out of this equation between development and re-education arose the all important role given to the mass media by the communication approach. Nothing quite compared to the media of mass communication as instruments for the new education" (19, p. 34). The interpersonal networks of communication were considered to be supportive of 'traditional' structures and authority and hence were anti-development. Mass media, however, were free from such control. The issue whether mass media were instruments of exploitation, tools of a small coterie of elites was not considered important. "The ubiquity, standardization, and the omnipresence of the mass media, which were found to be oppressive in the West, were the very features which, recommended them in the historic mission to develop the Third World" (19, p. 74-75).

The mass media, therefore, were considered to be harbingers of major change in the 'traditional' Third World nations, Through 're-education' they were supposed to help the peasants in these nations cleanse themselves of their traditional values, attitudes and cultural habits so that they could be anointed on the altars of modernity.

From Source to Receiver : Paradigm of Academics

Much of the thinking and research on development communication by the academics implied the conversion effects of the media. The assumption was that the media had direct unmediated effect upon their receivers. Exposure to mass media was seen as having direct effect of instilling modern perceptions in the minds of the people of the developing countries. Consequently, the model that followed was based on a rather simple, mechanistic view of communication as a process of message transmission (10). Graphically, it could be represented thus :

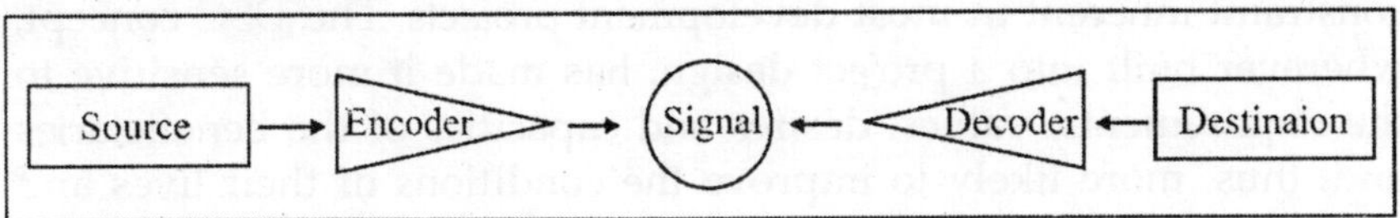

Other paradigms were also refined, especially after the Second World War, when research on persuasion and attitude change was transferred to areas such as agricultural extension, health education, etc., in the developing nations. The main thrust in the models was on communication effects such as acceptance of a new idea, habit, etc. Again, all of these models were basically one-way, top-down and linear. The preoccupation (especially in diffusion of innovations research) was with effects and not the relevance and accuracy of content. The dynamic process nature of communication was somehow obscured in these static and mechanistic models.

Pursuit of Development by Practitioners

Starting in the late 1960's development communication practitioners began to focus their attention on searching for factors which could presumably make projects more relevant to the needs of the people. They realized that many of the earlier projects had not given enough attention to the communication constraint. Their efforts, therefore, resulted in the conceptualization of communication as a dynamic support to development projects and activities. This gave rise to 'Development Support Communication,' a phrase originally coined by Erskine Childers, present director of information for the United Nations Development Program (UNDP). Explicating the DSC concept, Childers summarized that, it is a discipline in development planning and implementation in which more adequate account is taken of human behavioural factors in the design of development projects and their objectives. Then, on the basis of a behavioural analysis and the development of a feasible design, the requirements for technical human communication are built into that project as part of its own plan of operation and as a part of its budget (7, p. 87).

The Development Support Communication (DSC) analysis, especially in the UNDP, FAO, and UNICEF systems of the United Nations, has been an area of speciality attempting to put the spotlight on human communication factors which have tended to be ignored or omitted altogether in development project formulation and design. This speciality has thus sought to make itself that part of the practitioner's work attempting to overcome the communication constraint inherent in most development projects. The DSC concept, wherever built into a project design, has made it more sensitive to the requirements, values, desires and capacities of the beneficiaries and, thus, more likely to improve the conditions of their lives and livelihoods.

Thus, concepts such as the DSC are helping the practioners to be well ahead of academics. As Diaz Bordenave (10) notes, theory is lagging behind actual practice and technological competence since most theoretical models still reveal a transmission persuasion orientation.

The DSC approach, however, has its limitations. According to Ascroft and Brody :

The bottom line is that DSC personnel are no more than the mouth pieces of development, doing the bidding of development planners as best as they can, performing wherever and whenever possible a more sophisticated, technologically bound service than is within the capability of the average samaritan (2, p. 30).

Thus, the DSC practitioners while attempting to improve the efficiency and effectiveness of sources of development aid may not necessarily be promoting the interests of the recipients of that aid. The role of the DSC practitioner, therefore, will have to be expanded to facilitate the popular participation and action required for rural development. People should have greater access to communication media, not just as receivers but also as sources and actors.

Towards a Model for the Practitioner

The DSC idea fits well with the concept of open systems. According to systems theory, an open system is in constant interaction with, its environment. Through the processes of equifinalily and multifinality it regulates its behaviour in order to reach a specific goal or goals. Moreover, an open system is constantly taking m information from the environment in a purposive manner, internalizing the information to regulate its action, and then acting on the environment with the aim of transforming it. Thus, the system is acting on the environment and also being acted upon by that environment.

The systems model can be usefully adapted to explicate the DSC concept. First, this model gives a systemic view of the development process revealing the different interrelationships that exist between various factors. It thus helps communicators to look at the development problem from a much broader perspective. Especially, for the DSC personnel, the attention to objectives inherent in the systems approach will help them to assess carefully the communication implications involved and will also help for a careful analysis of problems right at the planning stage of a development project.

More recently, some researchers (3) have attempted to apply the ACB model of Westley and MacLean (37) to describe systematically the role of a development support communicator. The model can be depicted thus :

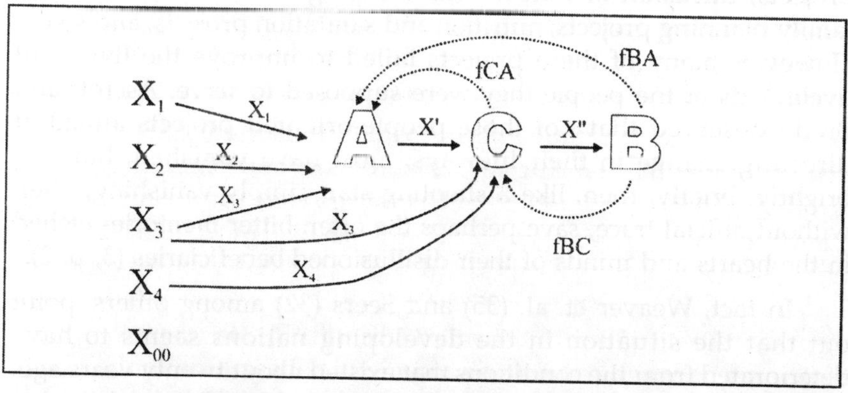

In this model, designed essentially to explain the role of mass media in society, the X's on the left are the different stimuli in the environment. 'A' is an advocacy role and she selects and transmits messages from these stimuli (X's) purposively to 'B', the receiver (or public). 'B' needs the information about his/her environment to help satisfy needs and solve security problems. The inclusion of the 'C' role in the model makes it very relevant to the proponents of the DSC concept. 'C' is in constant interaction with 'B' through, feedback loop 'fbc' and s/he selects and transmits non-purposively the information 'B' needs. Thus, 'C' extends B's environment and helps satisfy his/her security problems. The placing of 'C' between 'A' and

'B' makes the model useful in highlighting the importance of the DSC practitioner :

It is indeed very important to note that whereas As messages intended for B are always purposive, when passed on through C they cease to be purposive and become non-purposive. Thus, the effect of the addition of the 'C' role is to provide both 'A' and 'B' with more extended environment than is available to each on their own (3, p. 34).

If 'A' is thought to be the source of development and 'B' the recipients, then 'C' describes very articulately the role and function of a development support communicator.

A Research Stalemate

The theoretical ideas and assumptions of the academics generated an impressive array of development projects all over the Third World. Examples of projects undertaken with the main aim of attacking rural under-development fell into many categories : functional literacy projects, diffusion of innovations campaigns, radio farm forums, family planning projects, nutrition and sanitation projects, and so on. However, many of these projects failed to improve the lives and livelihoods of the people they were supposed to serve. Ascroft and Brody observed that : of those people-oriented projects aimed at directing change in their lifeways, most have vanished, burning brightly, briefly, then, like a shooting star, simply vanishing, often without official trace, save perhaps the often bitter memories etched in the hearts and minds of their disillusioned beneficiaries (3, p. 2).

In fact, Weaver et. al. (35) and Seers (32) among others, point out that the situation in the developing nations seems to have deteriorated from the conditions that existed about twenty years ago. Thus, there seems to be a stalemate in research efforts to bringing about a better quality of life among the poor majority in the developing nations. Today, many researchers argue **that** :

Development set-backs came more from inappropriate communication strategies employed by traditional extension systems. This shortcoming stems in turn from the failure of behavioural social scientists to involve themselves more actively in problem solving development roles. They have so far seemed content to stand by as objective observers, theorizing, criticizing, evaluating, but seldom actually participating. (1, p. 10).

Thus, the models of the academics have tended to be largely exogenous, top-down, and prescriptive. The major role of the academics has been at theorizing, evaluating, and criticizing theoretical strategies with little field participation in the solving of actual problems. The practitioners in the field have been unable to derive much use from the theoretical models in the solving of practical problems. For example, the diffusion of innovations research with its *post-hoc* preoccupation with adoption of innovations, made little attempt to identify, understand, and overcome bottlenecks (to adoption of innovations) the peasant faced. And, since there was little two-way participatory communication built into the Dominant Paradigm of Development (which greatly influenced the Diffusion of Innovations models), there were no channels by which development agencies could appreciate frustrations of farmers in their attempts at adopting innovations. Consequently, very often the innovations handed down were clearly inapplicable and sometimes even negative to the utilizer (11, 30). It is no wonder then that extension systems have been so narrowly focused on : a total technological replacement of all pre-existing traditional modes and practices, and a dyadic unidirectional model with authoritarian extension officers in super ordinate roles over peasants (1).

An important aspect in development research concerns the communication practitioner. Apart from brief flirtations with radio farm forums and functional literacy programs, academics have steered clear off from field experimentation. (4). So, much of diffusion research in the field has been done by practitioners not specifically trained for that purpose such as agronomists, nutritionists, family planning workers, etc. "Primarily trained to develop, test, and recommend adoption of needed life-improving innovations, they found themselves saddled with the additional task of planning and carrying out their own diffusion communication campaigns" (4, p. 4). Therefore, a majority of diffusion campaigns in the field have been conducted by 'non-communication' experts (5,23,26). This has been a serious anomaly. It is indeed paradoxical when one considers the fact that while an advertising communication expert would be needed to conduct a successful advertising campaign, or a public relations expert required to execute a PR campaign, such an analogy does not extend to a diffusion communication expert. No importance has been given to the training and emergence of a communication expert who has the theory based confidence and possesses necessary communication skills to handle communication field campaigns.

Considering all these, it comes as no surprise to discover a serious communication constraint in development projects (2, 22). Several biases that have contributed to this constraint have been identified and discussed extensively in diffusion of innovations research : pro-innovation bias, in-the-head variable bias (social-psychological characteristics of adopters), media effects bias, etc. (27). Ascroft and Melkote (2) have identified more biases in development communication strategies and campaigns : pro-literacy bias, source orientation bias, top-down message flow bias, etc. Since very little literature exists in diffusion research on the last three biases outlined above, they will be discussed in brief.

Pro-Literacy Bias

A majority of the rural people in the Third World are preliterate. Yet, there have been few effective strategies in diffusion research of communicating innovations to a preliterate audience (28). All strategies and innovations presuppose literacy and some level of formal education, which are by themselves, innovations in rural areas. Thus most development benefits have accrued to the large farmers and other elite groups since they possess the necessary prerequisites such as adequate literacy, education, previous knowledge of innovations, etc., for exploiting all the new information, methods and techniques (16, 33). The pro-literacy bias has acted as a major constraint to diffusion of information to preliterate audiences. It has prevented strategies of percolating information, knowledge and skills to an illiterate audience, which incidentally, forms the bulk of the population in rural areas.

These critical constraints, therefore, have led to easier information access by large farmers leading to relative successes and greater efficacy in adoption of productivity increasing innovations, whereas, lack of adequate and appropriate information to preliterate and subsistence peasants has led to relative failures.

Source-Orientation Bias

Little attention has been paid to locate any deficiencies or shortcomings of the source of an innovation, examine the relative merits and demerits of the innovation proposed and look into the quality of treatment accorded to the messages. The source has been considered faultless and blameless, the source-oriented message considered effective, and the innovation diffused considered beneficial to receivers There is even an explicit assumption that the source knows the kind of change desirable for the adopter. This can be seen in the way a change agent is identified : "A change agent is a professional who influences innovation decisions in a direction deemed desirable by a change agency" (27, p. 312). Thus, any anomaly in the diffusion or adoption process has been attributed to the recalcitrance of the receivers. This conceptual bias has masked the fact that weaknesses or shortcomings of the source and source-oriented messages may be responsible for some of the anomalies in the adoption behaviour of receivers.

Top-Down Message Flow Bias

The exogenous change orientation of diffusion research has acted as an effective constraint to an understanding of the flow of ideas as a multi-way process (15). Much of the diffusion research has implied a one-way flow of ideas, usually from the industrialized Western nations to much of the Third World, and within the individual countries, from the elites, such as scientists and government officials, to peasants. The lack of interest in a reverse flow of ideas and techniques from bottom to top, i.e. from the peasants to scientists, has resulted in many useful 'traditional' techniques being replaced by incompatible innovations introduced by external sources (6, 25, 36).

Conclusion

The DSC approach is appropriate to identify and overcome all biases in development communication strategies and thus overcome crucial communication constraints. Since it adopts a systemic approach, it constantly takes in information from the environment, internalizes the information to regulate its action and then acts upon the environment. Thus, in the DSC approach there is the possibility of acting upon the environment while being constantly acted upon by the environment. The bottom line is that in this approach, messages, and message strategies will be carefully evaluated in terms of fulfilling not only the objectives of the development project, but also in terms of their relevance to the unique needs, problems, and culture of the intended beneficiaries. Using DSC guidelines, Melkote (22) evaluated messages and message strategies of the World Bank's Training, and Visit (T and V) Agricultural Extension Project in India. The biases mentioned earlier in this paper, were found to exist in communication messages and strategies of the T and V project. Farmers were bombarded with messages using alien language, words, terms, and symbols. Consequently, the comprehension of messages was very low on almost all items, of knowledge diffused by the extension agents (22).

Several crucial areas in the design of communication' strategies are awaiting research and experimentation. The domain of message analysis (i.e. message design, construction, treatment, strategies, etc.) is still a seldom-researched area in development support communication. Crucial questions on message construction, design, treatment, etc., have been completely avoided by academic researchers probably because they cross the threshold of research and become more of a craft. However, messages construction and comprehension have important effects on the process of diffusion of information attempting to produce behaviour alterations among recipients. The DSC approach provides useful guidelines here keeping in mind the needs, capacities, problems, and culture of recipients. Last but not the least, this approach offers an excellent meeting place for the academic researchers and communication practitioners.

Notes

1,2. 'Purposive' as used in the model denotes a selfish interest on the part of 'A'. 'Non-purposive' denotes lack of self-interest on the part of 'C'.

References

Ascroft, J. : "A Curriculum for Development Support Communication". Unpublished document, 1985.

Ascroft, J. & S. Melkote : "An Ethical Perspective on the Generation of Development Communication Research Useful to Policy Makers and Practioners". Paper presented to the 33rd annual conference of the International Communication Association, Dallas, May, 1983.

Ascroff, J. & A. Brody : "The Role of Support Communication in Knowledge Utilization, Theory and Practice". Paper presented to the conference on Knowledge Utilization : Theory and Practice, East West Centre, Honolulu, 1982.

Ascroft, J. & G. Gleason : "Communication Support and Integrated Rural Development in Ghana". Paper presented to the annual conference of the International Communication Association, Acapulco, May, 1980.

Bhaskaram, K. : An Evaluative Study of the Training and Visit System in Andhra Pradesh". Paper presented to the workshop on Management of Transfer of Farm Technology under the Training and Visit System,

Hyderabad, India, 1983.

Bortel-Doku, E. : " A Fresh Look at the Traditional Small Scale Farmer. "*The Ghana Farmer* XVIII, 1978, pp. 4-6.

Childers, E. : "Taking Humans into Account. "*Media Asia* 3, 1976, pp. 87-90.

Cooley, C. : *Social organization*. Glencoe, Illinois : Free Press, 1962.

Cox, F. : "Introduction". In Cox et al. (Eds.) *Strategies of Community Organization*. Itascal, Illinois : F. E. Peacock Publishers, 1979.

Diaz Bordenave, J, : *Communication and Rural Development*. Paris : Unesco, 1977.

Diaz Bordenave, J. : "Communication of Agricultural Innovations in Latin America". In Rogers (Ed.) *Communication and Development : Critical Perspectives*. Beverly Hills : Sage Publications, 1976, pp. 43-62.

12. Durkheim, E. : *On the Division of Labour in Society*. New York : Macmillan, 1933.

Eisenstadt, S. N. : "The Changing Vision of Modernization and Development". In Schramm and Lerner (Eds.) *Communication and Change*. Honolulu : East-West Centre, 1976, pp. 31-44.

Eisenstadt, S. N. : "Social Change, Differentiation, and Evolution. "*American Sociological Review* 29, June 1964, pp. 375-86.

Fjes, F. : Communication and Development" Unpublished paper, 1976.

Gaur, R. A. : "Transfer of Technology : A Tool for Modernising Agriculture". *In Management of Transfer of Farm Technology*. Hyderabad, India : National Institute of Rural Development, 1981, pp. 103-105.

Hagen, E. : *On the Theory of Social Change*. Illinois : Dorsey, 1962.

Inkeles, A. "Making Men Modern : On the Causes and Consequences of Individual Change in Six Countries". *American Journal of Sociology* 75, September 1969, pp. 208-25.

Krippendorff, S. "The Communication Approach to Development : A Critical Review". In *Third World Mass Media : issues, Theory and Research*. William-sburg, Virginia : College of William and Mary, 1979, pp. 71-82.

Lerner, D. *The Passing of Traditional Society*. New York : Free Press, 1958.

McClelland, D. *The Achieving Society*. New York I Free Press, 1967.

Melkote, S. "The Biases in Extension Communication Revealing the Comprehension Gap". Unpublished PhD thesis, University of Iowa, November, 1984.

Nandapurkar, G. G. & R. R. Kulkarni. "Training and Visit System : Some Observation and Suggestion". Paper presented to the workshop on

Management of Transfer of Farm Technology under the T&V System, Hyderabad, India, 1983.

14. Portes, A. "On the Sociology of National Development: Theories and Issues". *American Journal of Sociology* 82, 1976, pp. 55-85.

25. Rahim, Syed A. "Diffusion Research - Past, Present and Future". In Schramm and Lerner (Eds.) *Communication and Change*. Honolulu. : University Press of Hawaii, 1976, pp. 223-25.

26. Reddy, S. V.; B. V. S. Rao & Pravenna. "Intensive Agricultural Extension System : A Critical Analysis". Paper presented to the workshop on Management of Transfer of Farm Technology Under the T&V System, Hyderabad, India, 1983.

27. Rogers, E. *Diffusion of Innovations*. Third Edition. New York : Free Press, 1983.

28. Rogers, E. M.; and Adhikarya. "Diffusion of Innovation : An Up-To-Date Review and Commentary". In Nimmo (Ed.) *Communication Year Book 3*. New Jersey : International Communication Association, 1979, pp. 67-81.

29. Rogers, E. *Modernisation Among Peasants*. New York : Holt, Rinehart, and Wins ton, 1969.

30. Roling, N.; J. Ascroft & F. Chege. "The Diffusion of Innovations and the Issue of Equity in Rural Development", In Rogers (Ed.) *Communication and Development : Critical Perspectives*. Beverly Hills : Sage Publications, 1976, pp. 63-78

31. Schramm, W. "End of an Old Paradigm?" In Schramm and Lerner (Eds.) *Communication and Change*. Honolulu : East-West Centre, 1976 pp. 45-8.

32. Seers, D. "The New Meaning of Development". *International Development Review* XIX, 1977. pp. 2-7.

33. Shingi, P. M.; and B. Mody. "The Communications Effects Gap". In Rogers (Ed.) *Communication and' Development : Critical Perspectives*. Beverly Hills : Sage, 1976, pp. 79-98.

34. Toennies, F. *Community and Society*. East Lansing : Michigan State University Press, 1957.

Weaver, J.; and K.. Jameson. *Economic Development. Competing Paradigms Competing Parables*. Washington, D. C. : Agency for International Development, 1978.

36. *Wall Street Journal*. "Sudan Farmers Find That New Methods Aien't Always Better". 1981, November 27, p. 1, Col. 4.

37. Westley, B. H.; M. MacLean. "A Conceptual Model for Communication Research". *Journalism Quarterly* 34, 1957, pp. 31-8.

Chapter **20**

Socio-Cultural Milieu of Peasants in the Third World: A Neglected Factor in Rural Development Strategies

Srinivas R. Melkote and Anantha S. Babbili

Two decades of planned development in many developing nations have little to show by way of improvement in living standards of their rural poor. In fact, the situation seems to have deteriorated since the 1960's. Poverty, income, inequality and unemployment are on the rise in the Third World nations. To use a familiar cliche : the rich are growing richer and the poor are getting poorer.

Authors have argued in favour of domestic policies in the Third World nations that concentrate on decentralized and bottom-up approach to rural development, keeping in mind the socio-cultural milieu of the people at the grassroots such as the peasants. They describe the social, political and cultural setting of people at the grassroots in the developing nations. In the process, the authors point out the misalignment between the present development models and socio-cultural milieu of the people at the grassroots. The authors then describe bottom-up approaches to development that could serve as alternatives to present top-down, linear models.

To begin with, this paper will examine important Western-oriented ideas and models that impacted-on the development planning process in many Third World nations. The paper will then discuss the social consequences of this planned approach on development in the Third World.

Throughout the first two development decades, development in the Third World was defined as "a type of social change in which new ideas are introduced into a social system in order to produce higher per capita incomes and levels of living through more modern production methods and improved social organization" (Rogers, 1969, p.18). This definition describes very well the economic planning and policy formulations in the newly formed Third World states since the 1960's right up until the early 1970's.

Concern with economic development is not a new phenomenon. Adam Smith in his The Wealth of Nations made 'economic development' his central thesis and this has continued to excite theorists and planners during most of the 200 years since he wrote his book. The conventional economic approach to development which started with Adam Smith was further developed·by Ricardo, Mill, Marshall, Schum-peter, Keynes, Arthur Lewis, Albert Hirschman and others. What are the basic goals of this dominant paradigm of Western economists vis-a-vis the 'good life' and 'good society'? A 'good life' is one where, one could maximize consumption of goods and services with minimum effort and work. The notion of a 'good society' follows directly from the idea of a 'good life'. A 'good society' is where people can get the most possible goods and services with the least cost and least effort (Weaver et. al., 1978). Most nations in Western Europe and North America came very close to representing the 'good society' and it was the promise of achieving this that development theorists tried to diffuse to the newly emerging states in the post-World War II period. The operational criterion for achieving the good society was simple : attaining the highest rate of growth of output (GNP) possible.

Rostow, Lipsett, Apter, Smelser among others were the main proponents for attaining the 'good society'. Rostow in Stages of Economic Growth (1960) identified universal stages of economic development through which every society will pass some day : the traditional society, the stage of establishing pre-conditions for take-off, the drive to maturity, and age of high mass consumption. In his Non-Communist Manifesto he proposed that the process of economic growth can usefully be regarded as centering on a relatively brief time interval of two or three decades, when the economy and the society of which it is a part transform themselves in such ways that economic growth is, subsequently, more-or-less automatic (Rostow 1956 : 25).

Rostow believed that there was a one-to-one relationship between economic and political development. So did Lipsett (1963), who visualized economic development as establishment of a western type democracy. And, the chances of establishing a stable democracy were related to the state of economic development.

Much of the theoretical and conceptual work in the sociology of development consisted of a wide assortment of bipolar theories which compressed the former evolutionary stages into ideal typical extremes. Here the general features of a developed economy were abstracted as an ideal type and then contrasted with equally ideal features of a poor economy and society (Durkheim, 1933; Cooley, 1962; Toennies, 1957; Lerner, 1958; Hoselitz, 1960; Parsons, 1964).

The Western theories on development had several conceptual problems. First, these theories implied that development was synonymous with Westernization. Second, much of the literature on development was a historical. In particular, both Lipsett (1963) and Rostow (1960) adopted a limited historical perspective. They analyzed the situation at a particular period in history (when Western nations were developed) and then did a cross national comparison of countries which were different stages of development. Aas Frank comments, "this entire approach to economic development and cultural change attributes a history to the developed countries but denies all history to the underdeveloped ones" (Frank, 1969, p. 40). Third, as much of the literature was written during the Cold War period, one cannot help noticing the ideological underpinnings. Many new states were coming into being as a result of the post - World War II decolonization and there was a vested interest in attracting them to the "liberal camp". As Pye (1971) notes, there was a bias for western type democracy.

Dominant Paradigm of Development :

The Dominant paradigm of Development, a product of the Western experience in development, has been very influential in development research and planning in the developing nations since the 1950's.

Rogers (1976, p. 49) distilled the following four main elements of this model :

1. Economic growth through industrialization and accompanying urbanization, approximately equal to passing through the Industrial Revolution. It assumed that development performance could be quantified in economic terms : GNP, per capita income, etc.
2. Capital-intensive, labour extensive technology mainly imported from more developed nations.

3. Centralized planning, mainly by economists and bankers, in order to guide and speed up the process of development.
4. The causes of underdevelopment lay mainly within the developing nation, rather than in their external relationships with other countries.

This model of development, essentially a West European and North American formula for modernization, has been very popular with the Third World policy makers. It seemed to provide a blueprint for speedy 'development' for all those developing nations stagnating in the backwaters of underdevelopment. As a paradigm, however, it was patently top down. It reflected powerful political and corporate interests. This approach did not consider the question of equality in the distribution of the benefits of development. Little attention was paid in this model to the fundamental contradictions of societies in the developing nations where economic and political powers are concentrated in the hands of a small elite and where large sectors of the population have no significant participation in the share of political and economic resources. Also, the dominant paradigm focused its attention on development constraints within a nation state. Invariably, the reasons for underdevelopment were thought to be the social psychological makeup of individuals or the 'traditional' culture of peasants in the Third World.

Social Consequences of Development :

The conceptual weakness of the Western-oriented models aside, the temptation of achieving the 'good society' was irresistible to policy makers and administrators in the developing nations. And the path to this was laid out in the dominant paradigm : economic growth through industrialization and accompanying urbanization, capital-intensive technologies, centralized and top-down planning etc. Many developing nations incorporated these ideas in their five-year and ten-year plans.

Over this whole period of the planned approach to development, there were a number of other trends which were much less positive. Weaver et.al. (1978) cite three vital areas that are of most interest : unemployment, income inequality and absolute poverty. In general experience, unemployment rose despite high growth rates. Also, inequality in income distribution increased substantially. In many Third World nations, small elite earned a substantial part of the national income. A growth in gross national product (GHP) did not lead to the 'trickle-down' of benefits to the poor. Another area of concern was poverty : the inability of persons to provide for their basic needs. In most Third World nations, it was common to find that as economic growth proceeded, the share of the bottom 60 percent of the people fell relatively and absolutely.

Peasant : The Neglected Role Player :

What do these developments imply to the administrators, policy makers and communication researchers? Clearly, the Western oriented development models are inappropriate for the Third World. In the Western models of development, the point of departure has been the structures and institutions at the Centre and not those in the periphery. These models and paradigms highlighted the supremacy of the nation state and its concomitant 'modern' structures and institutions over the periphery. The periphery, therefore, had to be replaced by the centralized set-up. Dichotomies persisted that showed the central state as 'modern' while the rural sector was 'traditional'. Therefore, institutions and structures, innovations, etc., needed to flow from outside to these traditional enclaves and somehow 'modernize' them.

The failure of the planned approach (Weaver et. al., 1978) suggests that the time has arrived for rethinking on the popular conception about the people in the periphery. Researchers and administrators can not afford to further indulge in intellectual totalitarianism by denying the existence, or refusing to acknowledge the political, social and economic structures and institutions of peasant cultures. This paper will argue in favour of policies in the Third World that concentrate on decentralized and bottom-up approach to rural development, keeping in mind the socio-cultural milieu of the people at the grassroots such as the peasants. The view that peasants should be active role players is undisputed at least on the basis of their huge numbers at the grassroots. As a social class, the peasants constitute a large percentage of the population in Asia and Latin America (Hyden, 1980). Yet, how much do development planners and public administrators know about these people? The Western theories which guided development planning did not talk much about peasants in the Third World, except the stereotypical view that, as a social class, they constitute the traditional sector which is necessarily in conflict with modernization. This view, as literature and research projects in development suggest, seems overly abstract and unreal (Gusfield, 1971). Therefore, if public administrators, development economists, or rural sociologists, are interested in a decentralized and bottom-up approach to rural development, there are many questions that first need answers : What are the political, economic and social attitudes and behaviour of peasants? What are the technical, social and moral arrangements the peasant makes in order to function efficiently in a difficult environment? What is the peasant's attitude to risk-aversion? Is the peasant really a traditionalist unwilling to change? Have the diffusion of innovations researchers denied or refused to see the logic and rationality in the peasants' outlook to risk-aversion?

Subsistence Ethic of Peasants :

Scholars such as Scott (1976), Popkin (1979), Migdal (1974), and Hyden (1980) agree on the precarious existence of peasants in the Third World : living dangerously close to subsistence level, Scott posits:

If the Great Depression left an indelible mark on the fears, values, and habits of a whole generation of Americans, can we imagine the impact of periodic food crisis on the fears, values and habits of rice farmers in monsoon Asia? (Scott, 1976, p.2)

Scott (1976) argues that the fear of food shortages in peasant societies has given rise to a 'subsistence ethic'. This ethic is the consequence of living so close to the margin and can be compared to the ethic of peasants in nineteenth century France, Russia, and Italy. He places the subsistence ethic at the centre of analysis of peasant politics. The peasant family's major problem, to put it starkly, is to produce sufficient rice to feed itself, buy a few basic necessities such as salt and clothes, and meet the ever increasing claims of outsiders such as landlords and tax collectors. Living dangerously close to the margin, where bad crop means starvation and malnutrition, the peasant has perfected ways of staying afloat. For example, the local tradition of seed varieties, planting methods, and timing was designed over centuries of trial and error to produce the most stable and reliable yield possible under the circumstances. These are the technical arrangements the peasant has evolved to iron out the 'ripples that might drown a man' (Scott 1976 : 3). Scott points out that many social arrangements serve the same objective. Patterns of reciprocity among the villagers, forced generosity, communal land, and work-sharing has helped to even out the inevitable dips in a family's resources which could otherwise have thrown them below subsistence.

Safety-First Principle :

Scott (1976) asserts that it is the 'safety-first' principle which lies behind a great many technical, social, and moral arrangements of the peasant agrarian order. It is perfectly reasonable, he says, that the peasant who each season courts hunger and all its consequences should hold a somewhat different opinion of risk-taking than the investor who is gambling 'off the top'. As Scott puts it "the peasant household has little scope for the profit maximization calculus of traditional neoclassical economics" (Scott, 1976, p.4). Both Popkin (1979) and Scott (1976) point out that in the choice of seeds and planting techniques, the peasant attempts to minimize the probability of a disaster rather than trying to maximize his return.

Through the subsistence ethic and safety-first principle, the peasant is desperately trying to earn a minimum income. It is common in developed nations for the state to assume the role of providing a minimum income to people through such measures as unemployment compensation, social security, social medicine, negative income tax, etc. One effect of these measures has been to make it more rational for Individual to engage in profit-maximizing behaviour. The people at the grassroots in the Third World do not obviously have this base in order to engage in innovative, achievement-oriented and profit maximizing behaviour.

Misalignment of Development Goals With Peasant Needs :

What does all this information on peasants imply to administrators, researchers and policy-makers? Is it still prudent to frame development policies in the Third World on the basis of Western-oriented models and experiences? Importantly, what should be the relationship of the 'developer' to the 'underdeveloped'? The discussion in the preceding sections clearly indicates that the peasant is not a passive person waiting in the periphery for some external agent to 'modernize' him. He seems to be a very active person with a rational outlook to the whole issue of risk-aversion and adoption of innovations. He seems to know what he needs and how to go about getting it.

Any meaningful approach to development would need to understand the problems, needs and expectations of peasants and others at the grassroots. How are development goals perceived from the bottom? What are the general terms under which the people in the periphery will be open to modernizing innovations? Development goals in most developing nations have traditionally been articulated from the top with the assumption that these are in the interests of the people at the grassroots. However, these development goals have been in sharp contrast to the constraints, expectations and goals of peasants and others at the grassroots. For example, while development goals from the top emphasized economic growth, rise in national income, greater agricultural productivity, political socialization, national Integration, etc., the goals and objectives from the bottom highlighted (among other things) social and health care services, expanded educational facilities, political participation, and eradication of social and economic inequities.

Clearly, there is a misalignment between the objectives of present development models and the socio-cultural milieu of the people at the grassroots.

There is an urgent need to investigate how local communities function in the development process. How best could they be involved in developmental decision making? What are the most feasible and efficient ways of tailoring developmental projects to the needs of local communities? These are the kinds of questions that need to be given increasing attention in the next few years. It is high time, therefore, that policies concentrated on decentralized, growth-with-equity, and bottom-up orientation to development.

Decentralized and Bottom-up Approach :

For development to become a reality, it must start where the problems and real needs exist, i.e. in the poverty-stricken rural areas, urban slums, etc. People living in such peripheries must perceive their 'real' needs and identify their 'real' problems. To a large extent, these people have not been able to do so due to a lack of participation and input in development strategies ostensibly set up to ameliorate their problems. What is necessary is for the peasants and others at the grassroots to develop a critical awareness of the realities of their situation (Friere 1970). It is through concretization, participation, and collective action that peasants can understand their 'real' needs, identify 'real' constraints, and plan to overcome problems. What is needed, therefore, as an alternative to the exogenous models using top-down approaches is a true self-reliant, people's movement, working from bottom to the top? There are several case studies the authors would like to describe wherein alternative approaches have been used with great success.

Sarvodaya Shramadana Movement :

This experiment which was started in Sri Lanka more than 25 years ago makes active use of traditional forms and philosophies of communication to meet user needs in the rural and other areas where the problems exist. Unlike the earlier development strategies that adopted Western models, this movement fosters a radical rethinking of the issues and problems of development drawing on traditional philosophies, local resources and skills to confront and ameliorate problems. Dissanayake outlines the philosophy of development of this movement : "It is based on self-reliance, self-development, grassroots development, popular participation, maintenance of ecological balance, and the importance of culture in social development" (Dissanayake 1984 : 46).

The Sarvodaya Movement is a self-help organization, currently with development programs in more than 2,500 villages all across Sri Lanka. Essentially, it is an approach where local culture is considered as a facilitator of development rather than as an obstacle to social change :

> Rather than blindly following the developmental scenarios that have been written in the West the architects of the Sri Lanka Sarvodaya Movement are engaged in a timely and arduous endeavour to formulate and put into practice a development strategy springing from the deepest currents of the culture that permeates society (Dissanayake 1984 : 39).

Kantowsky (1980) notes that the Shramadana camps foster spiritual development of both the individual and the group. The development of the individual personality is brought about by the emphasis on four Buddhist principles : kindness, compassion, sympathetic joy, and equanimity. Similarly, four Buddhist principles guide group behaviour : sharing, pleasant language, joint work and equality.

All of these principles are put into actual practice through Shramadana or labour donating camps. Dissanayake (1984) notes that in these camps important village needs such as laying of roads, construction of schools, community centers and rural housing schemes, repair and construction of irrigation channels, clearing land for agriculture, etc. are performed through the active voluntary participation of local people. The other activities of this self-help, bottom-up development approach include : training in agriculture related activities, development of local arts and village industries, training of village pre-school teachers and health workers (Ratnapala, n.d.).

Dissanayake compares the model of communication as used in the Sarvodaya Movement with the typical Western or Aristotelian model. It is schematically represented as follows (Dissanayake 1984 : 49) :

	Aristotelian Model	Buddhist Model
1.	Emphasis on communicator	Emphasis on receiver
2.	Influence a key notion	Understanding a key notion
3.	Focus on control	Focus on choice
4.	Emphasis on outward process	Emphasis on both outward and inward process
5.	Relationship between communicator and receiver asymmetrical	Relationship between communication and receiver symmetrical
6.	Stress on intellect	Stress on empathy

The Sarvodaya communication model depicted above comes out as an interactive, two-way process bringing out the dynamic nature of communication, whereas the Western model uses the static, top-down, one-way, transmission model of Berlo (1960). Thus, in every way, the Sarvodaya Self-Help Movement represents a refreshingly new approach to development whencompared to the centralized top-down techniques used in Sri Lanka and other developing countries. In short, as Ariyaratne (n.d.) aptly notes, this approach attempts to do no violence to the local culture or the spirit of the individual family or village.

Action for Cultural and Political :

Change : While the Sarvodaya Movement puts the spotlight on self-help, a question that begs an answer is : Self-help to achieve what? If people are locked into an oppressive and inequitable socio-economic and political structure, is self-help going to be used to achieve minimum gains while maintaining the status-quo in society or is it going to be employed for social transformation?

The Action for Cultural and Political Change (ACPC), an organization of Harijan agricultural labourers in Tamil Nadu, India, believes in social transformation. It is a popular organization run by the peasants in whom drama occupies a central part of a socio-economic and political transformation process involving concretization, organization, and struggle (Kidd 1984). For ACPC, the cause of poverty and underdevelopment is not the inadequacies and ignorance of the poor; it is the structural relationships which keep the poor powerless, subservient, and exploited (Kidd 1984 : 118). Thus, their 'backwardness' is more a symptom rather than the cause of their wretched state. Organizations such as the ACPC consider the conventional strategy to development as the 'Band-Aid' approach : a situation where the wounds of soldiers returning from a war are treated with band-aid and they are sent right back to fight the war. What the ACPC is attempting through their peoples' movement is to do something about the 'war' which is producing so many wounded soldiers. This organization does not see any merit in the conventional approach to development where the victims of an inequitable and unjust structure are given token benefits in the form of handouts (i.e. band-aids) to temporarily relieve their misery and then promptly sent back to accept a compliant role in the structure which produced their misery in the first place.

The ACPC organizing process consists of a number of stages : 1 getting into, accepted, and grounded in an area; 2. adult education and literacy classes; 3. leadership training and action committees; 4. cultural action programs and mass meetings; and 5. struggle and movement (Kidd 1984 : 107). Unlike the conventional development approach which is essentially an exercise in top-down propaganda and bureaucratic paternalism, controlled and run by development planners, the ACPC approach is a self-reliant, people's movement at the grassroots The main vehicle of concretization, organization and struggle is the drama. The plays grow out of the situations, experiences, and analysis of the actors who are themselves villagers. They aren't handed centrally produced, prepackaged scripts and told to per for. They create their own dramas out of their own collective analyses of their immediate situation and the deeper structures in which they are embedded. This is a genuine expression of the people. (Kidd 1984 : 117)

As stated earlier in this paper, the ACPC approach is a fine example of how people living in the periphery perceive their 'real' needs and problems through active participation in the development strategies. Though reality-based village plays, the peasants are encouraged to develop a critical awareness of the realities of their situation, identify 'real' constraints, and plan collectively to overcome problems. What we see here is a self-reliant, grassroots movement that truly understands its 'real' problems and is attempting to achieve its goals.

The ACPC is not the only village-based group doing such work. Several grassroots community groups in the Philippines such as **Kulturang Tabonon Sa Dabaw** of Davao and the **Kabalaka Mobile Theatre** also advocate dramatic concretization (Van Hoosen 1984), These groups are attempting to bring awareness amongst the villagers of some of the socio-political factors affecting their lives. Other notable examples of participatory grassroots movements include the Bhoomi Sena Movement in India, the Mother's Club in Korea, the Ujaama in Tanzania, and several community groups in Colombia, Mexico, Brazil and Chile.

Conclusions :

All of the alternative approaches to organizing the rural people and involving them in the design, planning, and execution of development strategies have enjoyed remarkable successes, albeit on a limited scale. These approaches certainly provide a promising alternative to the conventional, top-down strategies used until now. In the conventional, approach to development the dominant philosophy has been one of Blaming the Victim. The poor are poor the argument goes, because of their 'traditional' habits. The only way they can improve the quality of their lives is by adopting the modernizing innovations preferred by the government and donor agencies and, of course, giving up their 'backward' cultural ways. This approach is top-down and one-way, since the poor have nothing to contribute to the interaction because of their 'lack of knowledge' and 'backwardness'. If development is now defined as (Dissanayake 1984 : 42) 'the process of social change having the goal of improvement in the quality of life of all, or the majority, of the people without doing violence to the natural and cultural environment in which they exist', then the alternative approaches ensure that the people who are the objects of development efforts are in control of the process that would alter their life ways so dramatically.

Political Constraints :

The alternative approaches presented in this study illustrate effective ways in which needs and expectations are perceived and articulated at the bottom and then communicated from bottom to top. The important question is : how feasible or practical is it to stimulate such grassroots movements all over the countryside in the developing countries given the prevailing socio-economic and political structures in these nations? It is very likely that vested interests may feel threatened by the activities of grassroots development efforts to change the status quo. Examples from Brazil and Mexico where physical force was brought on grassroots development groups, attests to this fact (Shaull 1970, Van Hoosen, 1984).

We believe that the bottom-up approaches described in this study provide an effective framework of action for the people at the grassroots. They help them to perceive and articulate their needs and expectations and strive to achieve their objectives through collective action. Such grassroots movements will be close to the indigenous culture using it as a facilitator of development rather than work against it as in the top-down, exogenous models. Also, as Rahman (1981 : 43) notes, 'Participation is a process whose course cannot be determined from outside it is generated by the continuing praxis of the people, i.e. by a rhythm of collective action and reflection. This is what makes the process people's own as opposed to the people being mobilized, led or directed, by outside forces'. However, the paradox is that such efforts may cause an adverse reaction from vested interests whose power may be threatened, leading to legal, administrative and physical interventions to stop such 'subversive' movements. The consequence of this, we feel, is that instead of climbing one step up in the direction of development, the communities may be falling two steps down.

In the final, analysis, we believe that the sock-it-to-them kind of top-down approach is not going to produce long-term positive results for the majority of people especially those living in impoverished rural areas. Linkages between the top and the bottom that emphasize a give and take approach on both sides, a dialogue between the two parties, have the best chance of producing long-term benefits for everyone concerned.

References :

Ariyaratne, A.T., (n.d.). Collected Works (Vol.1). Co Iombo : Sarvodaya Research Centre, as cited in Dissanayake, W. 'A Buddhist Approach to Development,' Continuity & Change in Communication Systems. New Jersey : Ablex, 1984.

Berlo, D., The Process of Communication : An Introduction. San Francisco: Rinehart Press, 1960.

Cooley, C.H., Social Organization. Glencoe : Free Press. Development 1981:1, 1962.

Dissanayake, W., 'A Buddhist Approach to Development; A Sri Lankan

Endeavour. 'Continuity & Change in Communication Systems. Wang and Dissanayake (eds.). New Jersey: Ablex, pp.39-52, 1984.

Durkheim, E., On the Division of Labour in Society. New York : Macmillan, 1933.

Frank A.G., Latin America : Underdevelopment or Revolution. New York : Monthly Review Press, 1969.

Freire, P., Pedagogy of the Oppressed. New York : Continuum. 1970.

Gusfield, J.R., 'Tradition and Modernity : Misplaced Polarities in the Study of Social Change'. Political Development and Social Change. Finkle and Gable (eds.). New York : John Wiley, pp.15-26, 1971.

Hoselitz, B. F., Sociological Factors in Economic Development. Glencoe; Free Press, 1960.

Hyden, G., Beyond Ujamaa in Tanzania : Underdevelopment and an Uncaptured Peasantry. Berkeley : University of California Press, 1980.

Kantowsky, D., Sarvodaya : The Other Development. New Delhi : Vikas Publishing House, 1980.

Kidd, R., 'The Performing Arts and Development in India : Three Case Studies and a Comparative Analysis.' Continuity and Change in Communication Systems. Wang and Dissanayake (eds.). New Jersey : Ablex, pp.95-126, 1984.

Lerner, D. The Passing of Traditional Society. New York : Free Press. 1958.

Lipsett, J. S., 'Economic Development and Democracy.' Political Man. Anchor Publishers, pp.27-63. 1963.

Migdal, J. S., 'Peasants, Politics and Revolution : Pressures Towards Political and Social Change in the Third World.' New Jersey : Prince ton University Press. 1974.

Parsons, T.H., 'The Social Systems'. New York : Free Press. 1964.

Popkin, S.L., 'The Rational Peasant : The Political Economy of Rural Society in Vietnam'. Berkeley : University of California Press. 1979.

Pye, L.W., 'The Meaning of Political Development'. Political development and Social Change. Finkle et. al. (eds.). New York : John Wiley, pp.43-51, 1971.

Rahman, M.A., 'Reflections'. Development 1981 : 1, pp.43-51 1981.

Ratnapala, N., (n.d.) 'Study Service in Sarvodaya'. Colombo : Sarvodaya Research Centre, as cited in Dissanayake, op. cit., 1984.

Rogers, E.M., 'Modernization Among Peasants' : New York : Holt, Rinehar & Winston, 1969.

Rogens, E.M., 'The Passing of the Dominant Paradigm : Reflections on D if fusion Research'. Communication & Change. Schramm & Lerner (eds.)

East-West Center, pp.49-52. 1976.

Rostow, W.W., 'The Take-off into Self-Sustained Growth', The Economic Journal. LXVI, 261:25-48. 1956.

Rostow, W.W., 'The Stages of Economic Growth : A Non-Communist Manifesto'. Cambridge : Cambridge University Press. 1960.

Scott, J.C., 'The Moral Economy of the Peasant : Rebellion Subsistence in South East Asia'. New Haven : Yale University Press. 1976.

Shaull, R., in 'Pedagogy of the Oppressed' by Paulo Freire, Op. Cit., 1970.

Toennies, F., 'Community and Society' East Lansing : Michigan State University Press. 1957.

Weaver, J. H. & Kenneth Jameson., 'Economic Development : Competing Paradigms, Competing Parables.' Washington, D.C. : Development Studies Program, Agency for International Development. 1978.

Van Hoosen, D., 'The Barefoot Actors : Folk Drama and Development Communication in Asia'. Continuity and Change in Communication Systems. Wang and Dissanayake (eds.). New Jersey : Ablex, pp.127-137. 1984.

Chapter **21**

The Heterophily Gap between Benificiaries and Benefactors and the Question of Participatory Decision-Making

R. Agunga

The consensus among policymakers and scholars of Third World development is that if progress is to benefit small-scale farmers then they must participate in decision, making on projects and programmes designed to benefit them. Robert S. McNamara (1973), in his reign as president of the World Bank, had said that : "No programme will help small farmers if it is designed by those who have no knowledge of their problems and operated by those who have no interest in their future", After nearly two decades of new directions, development policymakers and scholars seem agreed that small-scale farmers are yet to be involved in the development process. The argument in this paper is that in the last 20 years, the environment for participatory decision making in rural development to occur has been favorable. What has prevented it from happening is the existence of a heterophily gap between development benefactors and beneficiaries which prevents increased interaction at the government/ farmers interface. This communication gap becomes apparent if we examine the common characteristics of the participants in development decision making.

In the context of Third World development, participatory decision making means the involvement of rural people in project or programme decision making. Lisk (1985) says that, with regard to rural development, participation includes-people involvement in decision-making processes, in implementing programmes, their sharing in the benefits of development programmes, and their involvement in efforts to evaluate such programmes. Participatory decision making, therefore, implies two or more parties engaged in dialogue. Dialogue, according to Webster's dictionary, is "talking together", an "interchange and discussion of ideas especially when open and frank, as in seeking mutual understanding or harmony" (Webster's New World Dictionary, 1989). Mutual understanding naturally implies some degree of commonalty or similarity between or among the participants engaged in the dialogue. The idea of participatory decision making, therefore, raises several questions. Who are the participants in the decision making process? What are their characteristics? Are they compatible with each other as co-equal decision-making participants? If not, how can two parties of dissimilar communication characteristics are brought into dialogue?

Identification of the primary participants in development decision making depends on the level of decisions being made, ranging from the macro level (i.e., government to government) to the micro level (i.e., extension agent to peasant farmer). The salience of aid from industrialized nations as the fuel with which to motivate and accelerate Third World development suggests a macro-level two-party nomenclature of (1) *aid-giving countries on* the one hand, and (2) *aid-receiving countries* on the other. Aid giving countries are mostly the industrialized countries of the Northern hemisphere, but may also include such pre-industrial countries as the Arab Emirates.

At this level, decisions are made by bureaucrats and by technical specialists. Although significant cultural differences frequently distinguish aid-givers from aid-receivers, these differences are seldom compounded by communication incompatibility, especially by language. Government officials use professional interpreters, when necessary, and the language of science use by technical experts to communicate with each other is designed to maximize mutual understanding. This paper does not focus, therefore, on this level of participatory decision making. Rather, it is concerned with the micro level end of the continuum, where the forces of socioeconomic and technical change come face-to-face with the people for whom the change is intended.

At the micro (or project) level, both foreign aid donors and the local bureaucrats and technologists join together in the same camp of those bringing development to the Third World masses. Let us call them the benefactors. The rural peasantry and urban poor whose well being they seek to improve are the beneficiaries. These benefactors and beneficiaries face the greatest likelihood of communication incompatibility and, therefore, the greatest danger of mutual misunderstanding. Such misunderstanding may doom the feasibility of useful participatory decision making between them (see Figure 1). The next section examines these participants more closely.

Benefactor System	**Beneficiary System**
Bilaterla Aid Agencies	Crop Growers
Multilateral Aid Agencies	Livestock Farmers
Government Ministries	Women's Groups
Non-governmental Agencies	Youth Associations

Interface

Limited interaction between benefactors and beneficiaries at interface

Figure 1 : Development Decision-Making Participants at the Project Level.

The term benefactor refers or someone who confers a benefit. In the development context this benefactor is the change agency and, by extension, the change agent who carries out the development programme. Rogers (1969) defines the change agent as "a professional who influences innovation decisions in a direction deemed desirable by a change agency". As shown in Figure 1, several kinds of agencies are constituted to influence the innovation decisions of their intended beneficiaries. Multilateral aid agencies, for example, include the World Bank and many of the United Nation's specialized agencies, such as the Food and Agriculture Organization (FAO) and the United Nations Children's Fund (UNICEF). Bilateral change agencies are exemplified by the United States Agency for international Development (USAID) and the United Kingdom Overseas Development Agency (ODA). In addition, all kinds of private voluntary agencies serving as change agencies (e.g., Oxfam; Save the Children's Fund, and a plethora of church related organizations). All of these agencies are external to Third World and the role they assume is often to make up for the shortfall of development capital, equipment, and expertise. This activity is commonly known as technical assistance.

Third World Governments also have several kinds of their own change agencies, such as ministries of health, agriculture, para-state boards and corporations (or specialized agencies), and a host of nongovernmental organizations. In theory, foreign technical assistance works in harmony with these local technical agencies to deliver services to their common beneficiary, the Third World poor. Each foreign technician is supposed to have a local counterpart, who is either a fellow expert or is a trainee serving an apprenticeship with the expert under the principle that the foreign technician is successful if he (or she) works himself or herself out of a job. Since foreign technicians and their local counterparts have different problems and motivations, the following sections discuss them separately.

The basic ingredient of development aid is, of course, new ideas in the form of techniques and technologies commonly available in the industrialized nations and held to be the foundation of their advancement. Under the rubric of technical assistance, experts in these new ideas are sent from the industrialized countries tothe developing societies to impart their Knowledge and skills. President Truman (1951) was the first to introduce this principle of technical transfer, in his Point Four Programme :

I believe that we should make available to peace-loving peoples the benefits of our store of technical knowledge in order to help them realize their aspirations for a better life. And, in co-operation with other nations, we should foster capital investment in areas needing development.

Our aim should be to help the free peoples of the world, through their own efforts, to produce more food, more clothing, more materials for housing, and more mechanical power to lighten their burdens. (See Daniels, 1951).

According to Lethem and Riley (1982) technical assistance has remained a strong component of such development-oriented agencies as the World Bank :

It has long been recognized that the transfer of financial resources for investment purposes is a necessary but not a sufficient condition for achieving development. Thus, the Bank has helped to finance several types of complementary activities (technical assistance) to ensure that projects are efficiently operated and, beyond that, to promote the development of human resources and the capacity for institutional and economic management in member countries.

A question immediately arises regarding the purposes of technical assistance. Indeed, the question is not whether technical assistance is necessary but whether it meets its fundamental objective, which is helping the poor improve their quality of life.

This issue has been studied for many decades. Most scholars of international development believe that technical assistance contributes little to improving the lot of the poorest of the poor. Scholars have differing opinions, however, about why this is so. One school says that technical assistance fails because it is actually intended to benefit givers rather than the receivers. The other school says that even when the motives for giving aid are altruistic, technical assistance still fails to achieve designated goals because aid givers understand very little about the aid receivers and their needs. These two schools of thought are examined next.

Technical Assistance as Self-Serving Neo-Colonialism

Ward (1970) claims that, in its present form, development benefits the givers, politically and economically, more than the receivers. Likewise, Black (1968) says that U.S. "exports have been expanding, partly, to be sure, as a result of foreign aid shipments commodities". Black also notes that the U.S. depends largely on the developing countries for commodities for which substitutes are nonexistent or unsatisfactory (e.g., chrome, manganese, nickel, tin, graphite, industrial diamonds, jute, sisal, .quinine, shellac, coffee, tea, cocoa, and palm oil).

Ficket (1966) points out a political benefit to technical assistance : by helping the poor countries to become strong, the West hopes to use them as a shield against the spread of Communism. Furthermore, the challenge of development, Ficket says, is the challenge to the Western World to impart its values and techniques to the developing societies, "It is obvious that the long-term survival of the Western system is going to be dependent in large measure upon its acceptability to the great majority of the world's people, who live in the so called underdeveloped areas".

Arnold (1979) claims that aid is now big business. Despite the amount of aid given (and it is often substantial), givers, seem largely uninterested in whether it indeed solves the development problems of the beneficiary nation. According to Arnold, every year a mass of statistics is published to demonstrate the quantity and kind of aid flowing to Third World countries. All too often, however, relevant questions about the use or efficacy of development aid are not raised because the donors have a vested interest in ensuring that these questions are not asked.

Griffiths (1984) further asserts that, in many cases, technical experts are sent to Third World countries not for altruism but to make a profit. Seligson (1984) has shown how the income gap between rich and poor countries has grown dramatically between 1950 to 1980, when massive infusions of technical assistance were supposed to achieve the reverse effect. The average per capita income of developing countries in 1950, Seligson points out, was $164 whereas that of the developed countries averaged $3,841. This yields an absolute income gap of $3,677. By 1980, incomes in the poor countries had risen to only $245, while those in the industrialized countries soared to $9,648. This set the absolute gap in 1980 at $9,403.

Ndeti (1976), a Kenyan sociologist, also questions the motives of technical assistance experts. Many of them, he says, simply want a vacation or to gain experience in a developing country. They see technical assistance as the means to these ends. Referring to the case of the British in Kenya, Ndeti says,

I honestly cannot believe that a British expert will give Kenya's rural transformation the knowledge that will ultimately result in excluding British goods from the Kenyan market. This type of situation-holds true for all aids from whatever quarter they come. It is the height of absurdity to think that foreign experts will provide information that runs contrary to their national interests. This indictment is very serious and raises the whole question of foreign experts.

Thus, according to Ndeti and others (e.g., Goulet and' Hudson, 1971; Hayter, 1971, 1982; Payer, 1984), rural development and the supply of technical experts are only a smokescreen for monitoring and ensuring continued-dependency. Developing countries, by accepting technical assistance, merely play into the hands of the industrialized nations. "Are foreign experts not the new vanguard of neo-colonialism?" Ndeti asks. Papanek (1969) observes this same problem but states it differently: "Is it realistic, first of all, to suggest that a foreigner can be professionally loyal to another government and, second, are there more than a handful of governments in the less developed world whose policies can command such; loyalty?".

Technical Assistance as Altruism

Many scholars do not accept the view that technical assistance primarily serves the interests of the giver at the expense of the receiver. Instead, they see technical assistance as a valiant struggle to achieve the goal of developmenta struggle that fails because of insurmountable constraints.

Cardinal among these constraints is the socio-cultural dimension, which reveals how vastly different literally worlds apart foreign technicians are from their intended beneficiaries. They are different, first of all, in terms of their education-based knowledge and skills, as well as their culturally differentiated language communication skills. A profile of these foreign experts shows them to be, in general, educated (the vast majority have at least a college degree), and to come from a monolingual environment (some arebilingual, but rarely multilingual). Their languages are invariable word-rich, technically complex, capable of accommodating the idiom of science and each readily translatable into the other.

Foreign technicians also most likely have a machine-intensive background and, in this "high-tech" age, are computer literate. They rely heavily on the written word as both their form of information storage and mode of communicating their intentions to others. They believe strongly in the scientific method and tend to give short shrift to anything that does not agree with this belief. They are used to a life of plenty and are not usually willing to give up the many perquisites of modernization that they have earned and enjoyed. They usually belong to a single, organized religion and have recreational pursuits that are equally organized. They stand in stark contrast to their intended beneficiaries. Rogers (1969) puts it this way :

What are the main characteristics of less developed nations that act to limit the transfer of Western social science research methods? These countries are typified by such greater variability in most socio-cultural dimensions : their vast range of languages, life styles, levels of living, attitudes, ethnic backgrounds, religions, and other human characteristics, is almost impossible to imagine.

Technical Assistance Local Counterparts

It has long been recognized that foreign technicians have trouble interacting with Third World commoners especially when the content of the interaction is highly technical. It was pertly to overcome this problem that the concept of local counterpart experts was introduced. Under this approach, foreign experts would make their intentions known to the counterparts, who would there translate them to the local people in the local dialect (Posz, Jun, & Storm, 1973).

Ascroft (1974) describes the relationship between technical expert and counterpart as analogous to the blades of a scissors. Foreign experts may have planning know-how and technical skills that local officials lack. Counterparts, the other blade of the scissors, may have the practical field experience, the operational know-how, the cultural compatibility, and the long-term continuity that the foreign experts do not have. Like the two blades of a scissors, neither foreign expert nor counterpart is useful without the cooperation of the other.

Unfortunately, the counterpart strategy has not worked well, mostly because the union of foreign expert and counterpart has been a rocky one from the start. Foreign experts, for instance, generally feel themselves saddled with incompetents for counterparts. Many of them claim that local counterparts are lazy, corrupt, and show little or no sympathy towards the poor (May, 1981). Local officials, on the other hand, often find their foreign colleagues reluctant to share strategic decision making with their counterparts, eschewing collaboration in favor of assuming full control of decision-making. For example, Ascroft (1974), in a study of a project in Zambia, found that the foreign expert often regards "local offerings as quaint, simplistic inconsequentialities". The local counterpart, he said, is often "reduced to the status of a bystander observing the antics of the technical experts, wondering no doubt when his international partner is going to learn the value of a two-bladed scissors". This lack of local participation, Ascroft adds, can be blamed in part on the habit of local officials to "suffer their frustration silently" as the result of their own "outdated customs" which deny them the right to discipline their house guests in a firm, no nonsense way. Ascroft calls this self-denial a "conspiracy of courtesy."

Finally, Ascroft blames part of the failure of local participation on the insensitivity of foreign experts to cultural nuances that they misinterpret as a lack in their counterpart's industry and commitment to the project. Nafziger (1979) also blames the ethnocentric attitudes of technical assistance experts, who tend "to appraise the developing countries from the perspective of U.S. standards, achievements, interests, and sympathies", As Nafziger adds, "There is a lack of sensitivity to explanations of the alien economy, polity, and culture in terms of its own goals, institutions, values, andcharacter". In short, the inability of both local and foreign experts to interact effectively with each other creates communication problems that must be resolved before they can work together to serve their intended beneficiaries effectively.

Local Technical Specialists

Foreign technical experts and their local counterparts constitute only a small part of the forces within a country that seek to improve the lives of the poor. Most of a country's development assistance work is in the hands of its own indigenous experts, many of whom have been trained at home and abroad to the same standards of technical competence as foreign technical experts, Rarely, however, are these trained specialists accorded the term "expert," which is usually reserved to describe foreigners (under the old saying that you cannot be an expert in your own home). These local technical specialists usually enjoy the advantage of cultural compatibility with their intended beneficiaries. Many were born and raised in the target environment and speak the appropriate languages.

Problems arise, however, because these experts level of education and standard of living, often equal to those of foreign technicians, creates an essential incompatibility with their roots. Having assimilated one or more languages of the industrialized nations, and perhaps a foreign religion and style of life, they have lost the ability to communicate effectively among their own people. Local technical specialists may have been raised in the target area but they may no longer understand the people. They still speak the local language, but now they mix so many foreign terms into it that they are no longer understood. Such technical jargon as "grams per gallon," "centimeters," "millimeters," "hectares," "acres," and "degrees Celsius or Fahrenheit" too often have no local equivalents in the languages and dialects of the people at the grass roots, yet they are very commonly uttered by extension agents in dialogue with small farmers.

Ascroft, Roling, Kariuki, and Chege (1973) as well as Holing (1974), observed that Kenyan extension workers were without the communication skills to be effective in their work. Says Rogers (1983):

> Change agents are often professionals with university degrees in technical fields. This professional training, and the social status that goes with it, usually means change agents are heterophilous from their typical clients, thus posing problems for effective communication about the innovations that they are promoting.

Heterophily, according to Rogers, is the degree to which potential communicants are different from each other culturally, educationally, in social status, and in other attributes. It means that the extension agent and the small farmer "simply do not talk the same language".

A related problem is these experts' belief in the scientific method and its value in bringing the resources of nature into the control of humankind. This contrasts dramatically with the values of most traditional Third World farmers. Scott (1972) and Toffler (1980) have found that local people are usually interested in achieving peaceful coexistence with nature whereas modern people seek to change or dominate nature.

The Development Beneficiaries

Prominent among the lowest level beneficiaries of Third World development aid are peasant farm families. Peasants, according to Shanin (1966) are the majority of humankind, yet inspite of their numbers they are the least understood. Rogers (1969) says that peasants "constitute at least three fourths of the population in most less developed countries". Their major characteristic is that theirs is a system of survival, which almost always includes some form of participation in agriculture (Roling & de Zeeuw, 1981). Rogers (1969) describes peasants as 'farmers oriented largely (but not necessarily entirely) to subsistence production'.

Relatively little is known about the response of peasants to the efforts of the well meaning benefactors who try to change them in the name of development. The first forms of organized aid aimed specifically at these Third World farmers began in the 1950s and 1960s. The results were discouraging in the extreme, leading to a suspicion that the local people's traditionalism rendered them resistant to change.

This notion of peasants intransigently turning their backs on life improving change gained particular currency with the publication of Rogers "subculture of peasantry" theory in 1969. Rogers (1969) characterized peasants as lacking in such "modern" attributes as innovativeness, need gratification, and empathy, while suffering from an excess of fatalism, familism and hostility towards government officials. Rogers was supported in this notion by other scholars, such as Inkeles and Smith (1974). Their book, *Making Men Modern,* held that the traditionalism of peasants inoculated them against the incursion of new ideas and practices.

These notions are disputed by subsequent scholars who stabled present forms more closely. Dscroft Roling, Kariuki, and Chege (1973) in a study of Kenyan farmers, show that what these local people lack to facilitate adoption of proffered innovations are not any of Rogers modern characteristics, but rather the provision of adequate information, knowledge, and skills is ways that create mutual understanding. Ascroft et al. (ibid) also point out a number of economy and logistic constraints in the environment that work against adoption of the innovations.

The study (ibid.) further shows that lack of adoption is not due to social or psychological variables that demand some powerful motivational push or potent incentive pull as a prerequisite for development to occur, as Rogers (1969) implies. According to Ascroft et. al. (ibid), traditional farmers may be surrounded by factors in their environment that mitigate strongly against adopting available innovations, even if they want to. These environmental factors may include lack of an equitable delivery system for cognitive inputs, lack of an equitable system for delivery of entrepreneurial inputs, lack of marketing and infrastructural facilities, lack of augmentative rural employment opportunities, lack of governmental policies favuoring small farmer development, and lack of small farmer participation, among others.

Roling and de Zeeuw (1981) also point out that what small farmers lack is not empathy but opportunity. They state that farmers the world over, including those without education in the Third World, usually know what they are doing. They know how to optimize their chances of survival given the means and technology available to them.

Scott (1972, 1976) may provide the most useful perspective on understanding peasants and their economies. Scott shows that peasants do not necessarily resist change. What prevents them from risking change, he maintains, is their precarious existence living dangerously close to, if not actually below the level of subsistence. Scott (1976) makes a poignant observation that 'If the Great Depression left an indelible mark on the fears, values, and habits of a whole generation of Americans, can we imagine the impact of periodic food crisis on the fears, values and habits of rice farmers in, say, monsoon Asia?'. The fear of food shortages, Scott argues, gives rise to a 'subsistence ethic', which overshadows the peasants outlook on life. Their perennial problem, to put it starkly, is to produce enough rice to feed the family, buy a few basic necessities, and meet ever increasing demands for taxes.

In essence, Scott (ibid) believes that Third World peasants, through centuries of living close to disaster, where one bad crop can mean starvation and malnutrition, have evolved a 'safety first' principle for staying alive. For example, local crop varieties, planting methods, and timing of farm operations have evolved over centuries of trial and error to produce the most stable and reliable yields under local circumstances. It is therefore perfectly reasonable that people who seasonally court hunger and all its consequences should hold a somewhat different opinion of risk-taking than their counterparts in an industrialized country, who are 'gambling off the top,' secure in the knowledge of crop insurance, unemployment compensation, and social security. In the choice of seeds, planning techniques, and the number of children they produce, Third World peasants attempt to minimize the probability of disaster than maximize their return on investment. Scott (ibid.) concludes that if the Third World poor seem conservative or seemingly resistant to change, this is based on a rational belief that radical changes in their techniques for earning are too risky to undertake without careful and premeditated each step of the way.

Ascraft (1974) found the same principle to hold true among Zambian farmers. Ascorft observed how news of a technology that was adopted and failed soon became common knowledge to whole regions of Zambia, making the peasants wary of adopting it. Knowledge of these failures reached them via the traditional communication systems which, like their seed varieties and planting techniques, were developed over millennia into a reliable information system.

Matovu (1984) asserts that traditional communication media have been largely overlooked by change agencies as tools for reaching peasants. The missionaries of change have never really understood let alone tapped into these traditional communication systems. They are dismissed as primitive and non-technological and thus irrelevant. As Ascarft, Brody, and Matovu (1982) point out. "How could they suspect that the message pursing from the taking drums or moving along the traditional grapevines were warning others to be wary of the risks involved in following Western technological advice?"

The Modern versus Traditional Technology Controversy

Change agencies need to consider that there may be considerably more method than madness in many traditional peasant techniques and technologies. The tendency for scientific men and women is to assume that peasant technical systems are without scientific value and must therefore be scrapped to make way for modernity. Such out-of-hand condemnation of traditional technology can backfire in unpredictable and often embarrassing ways. The following two cases illustrate this point.

Peasant farmers have generally been advised by agronomists to abandon 'intercropping,' the mixing of several crops in the same field at the same time, in favour of mono-cropping. Little empirical evidence exists to justify this change except that the change agents deem it desirable. Recent research suggests, however, that crop interaction, soil management, pest and disease control, and plant breeding under intercropping systems are likely to help achieve better yields (Food and Agriculture Organization, 1977). This suggests that traditional practices may have scientific validity, if researchers only take the trouble to study and understand them.

As another example, the November 27, 1982 issue of *The Wall Street Journal*, reported on its front page that Sudan farmers involved in the 'Gazira Scheme' in which the World Bank and several other multinational agencies have a stake, were finding the new methods less effective than their traditional farming techniques. Six years earlier, the expanse of farm land lying between the Blue and White Niles produced an average of five 'kentars' or bales of cotton per acre per year.

But that was before modern agricultural methods were introduced. Now, the yield is just two kentars per acre. The reasons for the decline are debated, but one thing is clear : while trying to get more out of the land, the Sudanese ended up by getting less, *(Wall Street Journal*, 1982).

For a poverty-ridden country like Sudan, these kinds of technological gambles can be ill afforded. As the article observed, Sudanese officials blamed Western experts for persuading them to put aside the traditional farm economy that had served them so well in the past.

What these two examples point up is the need for dialogue between the modern and the traditional technologists. The value of this kind of participatory planning is not so much to apprise peasants of the innovations that have been selected for their benefit, but to avoid embarrassing and potentially damaging strategies such as those just described. A critical question is what prevents this dialogue from taking place. Part of the answer may lay in the attitudes of technical assistants, both local and foreign, towards participating in agricultural dialogue on a co-equal basis a necessity for preventing both poor local involvement and inappropriate selection of technologies. Related to the issue of attitudes toward co-equal decision making is the problem of developing effective communication with the local participants.

Addressing Change Agency Attitudes

A major obstacle for many development change agents is that they have received little training in effective, intercultural communication or the social scientific background to go with it. When working in North America and Europe, the need for great acuity and sensitivity in the communicating with farmers is probably not as crucial. In the Third World context, where heterophily between beneficiary and benefactor is the order of the day, the skills required for effective communication become especially crucial.

Aside from problems inherent in the innovations themselves that may curtail local adoption, efforts at effective communication can be hampered by geographic inaccessibility, language differences, and differences in lifestyles, attitudes, ethnic backgrounds, religions, and other human characteristics almost impossible to imagine. The language barrier is a particularly serious problem. Africa, for example, has a population estimated at 344 million people who speak 800 languages and dialects. Even more extreme is the case of Papua New Guinea. It is estimated that in this island state, with an area encompassing 178,000 square miles and a population of only 2.5 million people, more than 700 languages are spoken (Wilcox, 1982).

The problem of rural people communicating with development officials is further exacerbated by the long distances that lie between where the farmers live and where the change agents are to be found. Many Third World villages are inaccessible to extension workers because of rugged terrain, dense forests, swamplands, and, or sand dunes. Anthropologists say that there are still places in Papua New Guinea, that are so inaccessible that these are "the last places where stone age man still resides".

Another obstacle to communication is the fact that an overwhelming majority of Third World citizens are preliterate, or their literacy is so marginal that they are unable to comprehend scientific material. When the word is removed from the list of usable communication skills, traditionally trained extension workers lose the tool on which they relay the most.

Coombs, Ahmed, and Kale (1976) emphasize the need for communication to "help local people see the relevance of the schemes to their own needs, to obtain local support to encourage people's participation, and to initiate and help develop local organisations". They note, however, that the communication expertise to do this work is usually lacking :

Too often in recent years, professional communicators have taken much too narrow a view-equating communications with the electronic mass media. They have treated communication components separately from other programme elements and have had too little regard for the characteristics of the clients being served and their social economic circumstances.

Requisites for Effective Communication with Subsistence Farmers

Ascorat and Brody (1982) identify a number of cautions for any change agency wishing to encourage peasants to modify their farming or other subsistence techniques. First, except when coercive methods are used, people will not change unless they want to change. Second, they will not want to change unless they come to know, in detail, of the changes required of them and why. Third, they will not learn about or accept these changes unless they are informed of them in an idiom they understand. Fourth, this information will not reach them *enmass* unless special, facilitating channels of communication are created, linking them with sources of this new information. Fifth, these channels will not be created unless the developers or benefactors acknowledge their shortcomings in mass communication across cultures. Sixth, this acknowledgement will not automatically make a difference unless the developers realize that Western style mass media established in the Third World link only a few rural elite to the sources of new ideas. If all of this is done and people still resist change, developers must learn first to look for the source of the problem within their own ranks, by examining their own policies, priorities, strategies, and biases, rather than immediately blaming the recipients of the aid.

Summary and Conclusion

In summary, there are presently two clearly distinguishable groups in development work-the beneficiaries and the benefactors. By virtue of the traditional training of the members of the benefactor system and the lack of education of the masses of members of the beneficiary system, there is very little communication flow between the two groups. The relationship is heterophilous. Extension educators, whose role it is to bridge this gap, lack the communication skills to engage in meaningful dialogue (Rogers. 1969, 1983; Morgan, 1984; Melkote, 1984; Ascroft 1985). What this suggests then, is the existence of a professional which the communication function yet unfilled.

In conclusion, the study proposes the practical involvement of trained communication professionals in the day to day operations of projects to play the communication functions. These functions will include promoting coordination and linkages between and among participating agencies; the training of extension workers in the preparation and use of communication teaching aids: assisting field workers in mobilizing, organizing, and training farmers' groups for participatory decision making.

Although the importance of communication in development has long been acknowledged (Lerner, 1958; Schramm, 1964). The tendency has been to think of communication largely in terms of media technology (Coombs et. al, 1976; Fuglesand and Chandler, 1986). Hardly any mention has been made of the need for trained communication professionals. One reason for this may be that there were very few such experts in the past. Today, however, scores of bachelorette, masters, and even doctorate degree holders are been produced who have been trained specifically in the use of communication in development (Ascroft, 1985; Ascroft et. al, 1987; Brody, 1984 Narula and Pearce, 1986). It is time for development promoters to seek out and utilize these professionals.

References

Ascroft, Joseph and Alan Brody, (1982). The Role of Communication Support in Knowledge Utilization : Theory and Practice. Paper presented at the East-West Center Conference on Knowledge Utilization. Honolulu, Hawaii.

Robert Agunga, Ab Gratama and Sipho Masilela, (1987). Communication in Support of Development : Lessons from Theory and Practice. Paper for Seminar on Communication and Change. The University of Hawaii and the 'East West Center, Honolulu, Hawaii.

(1974). A Conspiracy of Courtesy. *Development Digest.* Vol. XII, No. 3.

(1985). Report on the Workshop on Development Communication and the Workshop for Health Education held at The University of Punjab, Lahore, July, 1985 under the sponsorship of UNICEF, Pakistan. Iowa City.

Niels Roling, Joseph Kariuki, and Fred Chege, (1973). *Extension and the Forgotten Farmer.* Agricultural College of Wageningen, The Netherlands ; Bulletin No. 37. Wageningen : Afdelingen voor Societe Wetenschappeen aan Landbou-whogeschool Press.

Black, Lloyd, (1968). *The Strategy of Foreign Aid.* Princeton. N.J. : Van Nostrand.

Brody, Alan M., (1984). Development Support Communication Roles United Nations Programming : A Critical Review and Reconceptualization. Ph. D. Dissertation, University of Iowa. Iowa City.

Coombs, Philip, H.,(1980). *(ed.) Meeting Basic Needs of the Rural Poor*. The Integrated Community-Based Approach. New York : Pergamon Press.

FAO. (1977). A Return, To traditional Cropping. Ceres FAO Review on Agriculture and Development. July-August, Vol. 10, No. 4.

Fickett, Lewis Persley, (1966). *Problems of the Developing Nations.*

Fuglesand, Andreas and Dale Chandler. (1986). The Paradigm of Communication in Development. From Knowledge transfer to community participation. Lessons from Grameen Bank, Bangladesh. FAO Development Communication Study.

Goulet, Denis and Michael Hudson, (1971). *The Myth of Aid: The Hidden Agenda of the Development Reports.* Orbis Books.

Griffiths, Leuan L. L., (1984). *An Atlas of African Affairs.* New York : Methuen.

Hayter, Theresa, (1971). *Aid As Imperialism.* Penguin.

Hayter, Theresa, (1982). *The Creation of World Poverty'* London; Pluto Press in Association with Third World First.

Lerner, Daniel, (1958). *The Passing of Traditional Society. Modernizing the Middle East.* N. Y. : Free Press.

Lethem, Francis and Vincent Riley, (1982). The World Bank's Technical Assistance. A Review of the Bank's activities and the lessons of its experience. *Finance & Development.* December.

Lisk, Franklyn, (1985). The Role of Participation in Basic Needs-Oriented Development Planning. *Popular Participation in Planning for Basic Needs.* Franklyn Lisk (ed.). Longhead England: Blackmore Press,

Matovu, Jacob, (1984). In Search of Mass Communication Strategies of Facilitate National Unity in Uganda. Ph.D. Dissertation, University of Iowa.

May, Brian. (1981). *The Third World Calamity.* London: Routledge & Kegan Paul.

McNamara, Robert S. (1975). Annual Address by Robert S. McNamara, President of the World Bank and Its Affiliates, September 24-28,1973, In 1973 Annua Proceedings of the Board of Governors, IBRD. World Bank. Washington, D.C., February.

Melkote. Srinivas Rajagopal, (1984). The Biases in Extension Communication : Revealing the Comprehension Gap. Ph. D. Dissertation, The University of Iowa.

Morgan, E. Philip, (1984). Development Management and Management Development in Africa. *Rural Africans,* No. 18, Winter.

Nafziger, E. Wayne, A Critique of Development Economics in the U.S. *Development Theory.* Four Critical Studies. David Lehman (ed).

Narula Uma and W. Barnett Pearce. (1986). *Development as Communication.* A Perspective on India. Carbondale : Southern Illinois University Press.

Ndeti, K., (1971). Human Factor in Rural Transformation. *Strategies for Improving Rural Welfare.* M. E. Kempe and L. D. Smith (eds.). Institute of Development Studies, University of Nairobi.

Papanek Gustav F., (1970). The Economist as Policy Adviser. *Development Digest.* Vol. VIII, No. 2.

Posz, Gary S., Jong S. Jun, and William B. Storm, (1973). *Administrative Alternatives in Development Assistance.* Cambridge, Mass : Ballinger Publishing Company.

Rogers, Everett M., (1983). *Diffusion of Innovations.* 3rd Edition. New York ; The Free Press.

Rogers Everett M., with Lynne Svenning, (1969). *Modernization Among Peasants.* The Impact of Communication. New York : Holt, Rinehart and Wiston, Inc.

Roling. Neils and Henk de Zeeuw., (1984). Improving the-Quality of Rural Poverty Alleviation. International Agrarisch Centrum, The Netherlands

•Holing, Neils.. (1974). From Theory to Action. *Ceres.* March-June,

•Schramm, Wilbur., (1964). *Mass Media and National Development.* zStandford : Standford University Press,

•Scott, James C., (1972). The Erosion of the Patron-Client Bonds and Social Change in Southeast Asia. *Journal of Asian Studies.* No. 33, November.

•Scott. James C., (1976). *The Moral Economy of the Peasant : Rebellion and Subsistence in South East Asia.* New Haven : Yale University Press.

•Seligson, Mitchell A., (1984). Defining the Gap Between Rich and Poor. *The Gap Between Rich and Poor.* Seligson, Mitchell (ed.). Boulder, Colo. : Westview Piess.

Toffler, Alvin, (1980). *The Third Wave.* New York : Bantam, Books.

Truman, Harry, (1951). The Presidents Proposal. *The Point Four Program.* Walter Daniels (ed.). New York: H.W. Wilson.

Ward, Barbara, J. D. Runnalls, and Lenore D'Anjou, (1970). *The Widening Gap - Development in the 1970s.* Conference on International Economic Development, February 15-21, New York : Columbia University Press.

Wall Street Journal, (1982). *The Wall Street Journal*-P-1- Wilcox, Dennis, (1982). Radio : Its Nation Building Role in NewGuinea. *Gazette.*

Chapter **22**

Visual Communication : A Psychological Perspective

Aparna Chattopadhyay

People differ in the ways they characteristically operate in communication situations. In understanding what is happening in a communication situation, it is very helpful to know about some of these characteristic modes that people use. If you are part of the situation, it is helpful to know about some of your own personality characteristics and the reasons for them. Understanding others and understanding yourself won't guarantee successful communication, but it should certainly help.

You might have already observed one such characteristic : general persuasibility. Some of people are pushovers for every sales pitch that come along, and others are down right stubborn but there are other personality characteristics that are more interesting, and some of them are useful in understanding general persuasibility itself.

Those who are authoritarian or dogmatic, for example, communicate and respond to communication in ways quite different from 'those who are open-minded. People also communicate differently when they differ in self-esteem, security, aggressiveness, need for achievement, need, for affiliation, cognitive style, interpersonal trust, anomie, and need to influence (or "Machiavellianism"). Those are some of the personality differences that must be kept mind while studying communication.

There are many theories about why people differ in personality, about the ways learning modifies genetic differences. Hall and Lindzey describe twelve different theories of personality in their book by that name. If you are interested in the research findings which support the generalizations I am about make, many of those can be found in Personality and persuasibility, Chapter 7 of Persuasion : Speech and Behavioural Change, and Handbook of personality Theory and Research, especially Chapter 24.

Authoritarianism and Dogmatism

These two characteristics are discussed together because dogmatism appears to be a more general extension of authoritarianism. Both characteristics seem to be collections of symptoms, which makes them rather difficult to define. An authoritarian individual, as described by Adorno, is one who tends to use beliefs for the purpose of ego-defense and is especially likely to be influenced by what that individual considers to be an authority. Rokeach objected to the F-scale, the test of authoritarianism, on the grounds that it measured authoritarianism only among those on the extreme right of the political spectrum, whereas the symptoms it is supposed to measure may be equally characteristic of the left-wing radical. He defined open-mindedness, the opposite of dogmatism, as "the extent to which the person can receive, evaluate, and act on relevant information received from outside on its own intrinsic merits, unencumbered by irrelevant factors in the situation arising from within the person or from the outside. Obviously that should be interest to communicators. Suppose we consider, then, some of the symptoms of the dogmatic individual which are especially relevant to communication.

For one thing, dogmatic persons tend to maintain strong central beliefs which are highly resistant to change. These are generally religious, political, or philosophical values; beliefs about the nature of personal worth; and beliefs on the nature of authority. The dogmatic individual, for example, is expected to agree strongly with statement such : "It is better to be a dead hero than a live coward"; "The main thing in life is for a person to want to do something important"; and "Man on his own is a helpless and miserable creature." It is interesting that most studies have found that dogmatic persons are especially easy to persuade. This is probably because these studies have used messages designed to change what Rokeach has termed "peripheral" beliefs : the dogmatic person seems to be willing to sacrifice these peripheral beliefs in order to maintain those beliefs which are more central.

A second characteristic of dogmatic persons is that they tend to overlook the relevance of one of their beliefs to another, which allows them to maintain two contradictory beliefs simultaneously. Two test items with which the dogmatic person will agree are: "Even though freedom of speech for all groups is a worthwhile goal, it is unfortunately necessary to restrict the freedom of certain political group"; and 'The highest from of government is a democracy, and the highest form of democracy is a government run by those who are most intelligent." Rokeach has reported remarkable change when people are made aware of such inconsistencies in a subtle way, so that they do not feel the need to defend the contradictions.

A third characteristic of dogmatic persons is their tendency to organize the world into two primary categories: thing they accept and things they reject. This is evidenced by their tendency to agree with statements such as : "Of all the different philosophies which exist in this word, there is probably only one which is correct"; To compromise with our political opponents is dangerous because it usually leads to the betrayal of our own side"; and "There are two kind of people in this world - Those who are for the truth and those who are against the truth."

Further, it is especially difficult for the dogmatic listener to accept a person but reject something that person says, or vice versa. The dogmatic person relies heavily on the opinions of those that person considers to be authorities. Such a person agrees, for instance, that "It is often desirable to reserve judgment about what's going on until one has had a chance to hear the opinions of those one respect," and "In this complicated world of ours the only way we can know what's going on is to rely on leaders or experts who can be trusted." The dogmatic person would like to find someone who would tell him or her how to solve personal problems. On the other hand, the dogmatic listener is quick to reject someone who disagrees with him or her on an important issue; one test item with which such a person will be likely to agree is the statement, "My blood boils whenever a person stubbornly refuses to admit he's wrong."

A friend of mine worked one summer as one of a crew of four men charged with the job of building a road through a mountain pass in which the snow had only recently melted. With concerted and collective ingenuity the crew succeeded in miring the only available truck and the only available bulldozer in one of the many available mud holes. Trying to pull one with the other made both sink deeper. After the winch cable was fastened to a very sturdy tree, the bulldozer pulled itself onto solid ground using its own power and was then used to pull the truck free. This is mentioned not because a person talking to those who are highly dogmatic is in a very similar situation. If the person and the proposition are both somewhat suspect, the person coming out in favour of that proposition will only reaffirm the listeners' suspicion of both. Instead, one must find stable central beliefs of the listeners with he agrees and use those common beliefs to gain firmer terra for himself, once he is on solid ground, he is in a better position to advance the proposition. Dogmatic listeners will be especially likely to accept a person who agree with some of their strong central beliefs, and once they have accepted that person they will be especially likely to accept the arguments. Open-mind listeners may not respond as favorably to such treatment, but neither will they take offense as long as the approach is not obvious and insincere. As suggested elsewhere, a person is well advised, both ethically and pragmatically, to choose central beliefs with which he or she actually agrees. The listener is well advised to consider what effect his or her own dogmatism may have on response to communication.

Three words of caution seem appropriate. First, there is a tendency to think of the dogmatic person as less intelligent than the open-minded person. There is no research evidence that intelligence provides any guard against dogmatism. Second, it *is tempting to consider those who agree with us to be open-minded.* The reverse is more likely, since dogmatic individuals seem to be generally more easily persuaded. Finally, the fact that an individual take and extreme position on any one issue does not necessarily indicate that the person is dogmatic as defined here; he may simply believe that he right. An individual may be suspected of being dogmatic if that person takes extreme positions on *many* issues; changes many of his or her less consequential opinions often and especially in response to the influence of persons believed to be authorities; refuses to consider the possibility of being wrong in certain central beliefs; fails to see how his or her beliefs relate to one another; and rejects other persons to the extent that he or she rejects what they say.

The term "dogmatic" and "open minded" **carry immediate connotations with which we all identify.**

We generally identify positively with open-mindedness and negatively with dogmatism. Dogmatism still exists, however, and it probably is more a part of our own lives than we realize. Have you ever admired a particular professor to the point that he or she become an idol: or have you ever felt this way about a person of the opposite sex or a minister or a political leader. What dangers are inherent in such situation. How often have you supported a cause, only to find later that you changed your mind about it (or never changed your mind). Have your ever rejected a person simply on the basis of what he or she said. Have you ever unknowingly held conflicting view only to embarrassed when you discovered they were at odds with each.

Self-Esteem and Security

Research in this area seems to lead to the conclusion that the person low in self-esteem is especially easy to persuade. Of course, one will not often know which of the listeners are low in self-esteem. That information would not be very useful anyway; one could then predict which of the listeners would be most responsive, but could seldom base his appeal on such knowledge. However, it is sometimes used when one has limited time or resources and has a choice of directing appeals to different group. Politicians, for example, who have a limited number of speaking engagements, might consider which constituents are likely to be lowest in self-esteem and direct the majority of their appeals to those persons. If there are groups in the district who have been consistently exposed to experiences likely to reduce their self-esteem, the cunning politician may choose to appeal primarily to those. If, as I believe, awareness is the basis of ethical communication, such tactics should be exposed.

Sometimes people use such information to decide when to advance an appeal. They may wait until the group to be persuaded has some reason for being low in self-esteem. For example, national confidence probably reached a low point about the President Kennedy took office in 1961. Russia had put the first sputnik in orbit and followed that with a man in space. It was a propitious time for persuasion.

Sometimes people try to manipulate the esteem of others. Listeners may be reminded of facts and experiences which threaten their self-esteem. Obviously, the threat to esteem will probably come from a source other than the person speaking, and the speaker will probably assure the listeners that he or she does not hold them in low esteem; they may also be assured that the person has confidence in their ability to meet the threat and that he or she has a plan by which they may restore any esteem they have lost.

The self-esteem of listeners may also interact in strange ways with the esteem of the person they are listening to. A speaker will generally try to increase his or her own prestige and convince listeners that they have much in common. Sometimes this backfires. If the person speaking is high in prestige, those who see themselves as having a great deal in common with him or her will also be high in self-esteem and thus difficult to persuade. On the other hand, if the listeners are low in self-esteem, the communicator who is really sophisticated may try simultaneously to convince the listeners of their similarity without lowering his or her own esteem. That is a tricky proposition. It is in part the plight of a black college professor speaking to a group of young blacks in a ghetto high school, trying to convince them to continue their education. The job is to persuade his or her listeners that they have a great deal in common, and avoid any suggestion of being overly impressed with his or her own accomplishments.

Achievement and Affiliation

The person low in self-esteem is likely to have some needs that differ from those of people high in self-esteem. Two which appear at first most obvious are need for achievement and need for affiliation.

As with many personality characteristics, we need to distinguish between chronic and acute lack of self-esteem. It may be that the person who is chronically (consistently) lacking in self-esteem eventually abandons the *expectation* of achievement or affiliation and rationalizes, sublimates or somehow reduces his or her dependence on achievement and affiliation as sources of need satisfaction. If this is true, it may be that the person who is usually *high* in self-esteem, when that self-esteem is temporarily (acutely) threatened, will be the one who reacts by showing the greater need for achievement or affiliation.

Whether that is true or false, it is fairly certain that anyone whose self-esteem is *acutely* threatened will have a *temporarily* greater need for achievement and affiliation. Sometimes a person who perceives others to be suffering from an acute lack of self-esteem for one reason or another will try to persuade them by offering them opportunities for achievement and affiliation. For example, the person may present a plan by which the listeners can regain the confidence they have lost. Or, since people characteristically band together in the face of threat, the listeners may be encouraged to join together to overcome the threat to their confidence. If the listeners are aware of such tactics, they can react consciously in whatever way they believe appropriate.

Aggressiveness and Aggression

Similarly, it is possible to distinguish between chronic aggressiveness, adopted as a life-style, and acute, temporary instigation to aggression induced by circumstances with which one is presently confronted. These appear to be two very different concepts, and they seem to bear different relationships to communication.

Specifically, there is fairly good evidence that the person who is characteristically aggressive is very resistant to persuasion. On the other hand, while the evidence is not as extensive, it appears that those who are *temporarily* induced to aggression are probably *easier* to persuade. To further complicate the picture, acute aggressiveness apparently makes one more willing to accept plans involving punitive or harsh action.

Again it may be difficult to identify a person who is chronically aggressive without knowing that person for some time. However, it is sometimes possible to identify certain persons or groups having a history of aggressive behaviours or beliefs. People who know that their listeners are chronically aggressive sometimes try to state their proposed plans in aggressive terms. For example, consider a person who is trying to convince police that they should not deal harshly with a disturbance. Instead of trying to appeal to their sympathy, the person might point out that the leaders of the disturbance are deliberately trying to provoke harsh police action so they can charge police brutality and advance their cause. He or she might then argue that it is much more sadistic to feed out the rope and hang the protesters at a later date. Those who are trained in communication techniques will recognize such tactics and make their decisions on more rational grounds.

Acute aggressiveness, on the other hand, is generally induced by frustration or threat, and it is quite similar to acute lack of self-esteem. In fact, the same circumstances which induce acute lack of self-esteem may simultaneously induce acute and defensive aggressiveness. Individuals who have been frustrated or threatened are very vulnerable target for persuation, and this is especially true if a communicator is urging punitive action. Knowing this, some communicators may wait to advance their propositions until those they wish to persuade have been frustrated or threatened, or may remind them of frustrations or threats they have experienced or are likely to experience. It is to the listener's advantage to know his or her own emotional state and to know how someone else may be using those emotions.

Anxiety and Alertness

Alertness and attention are not synonymous. To say a person is alert implies that the person is ready to respond to external stimulation, but it does not necessarily imply that he or she is ready to pay attention to what someone else has to say. Again, anxiety is the term usually used to refer to a chronic personality characteristic, while alertness and attention are acute states induced by specific situations. That distinction is not perfectly clear, however, for we sometimes speak of acute temporary anxiety.

Alertness, arousal, and activation are used practically synonymously in the literature of psychology. When activation is distinguished from the other terms, it is used to refer more specifically to a state of the central nervous system. Thus psychologists usually speak of *cortical* activation and *autonomic* arousal. Alertness is a more general term, less technical, and thus more accepcable in non-technical discussions.

There appears to be a continuum of alertness or arousal ranging from sleep through what is generally termed alertness to states of high emotional arousal and outright panic. Further, increasing alertness seems to lead to better task performance up to a point, at which increased arousal begins to interfere with task performance. The point at which arousal begins to interfere with efficiency is very high on the continuum for simple physical tasks, but the interference comes much sooner for complex intellectual tasks such as listening to a complicated explanation or argument.

Obviously, alertness and arousal are closely related to states of aggressiveness and fear. This has been a real problem in research to determine how alertness and arousal are related to communication, since the states of alertness have been induced by introducing some situation in which the listener is threatened or frustrated or complimented. Then it has been difficult to determine whether the effect is produced by the general state of arousal or by the specific *type* of arousal. However, there are some tentative implications for communicators.

Certainly it is desirable that all the participants in a communication encounter be alert. *However, there are times when communication must be careful to avoid arousal levels so high* as to *interfere with the task of critical listening.* This is especially true if a discussion is complicated and if the communicators are chronically anxious or have been made acutely anxious by circumstances immediately preceding the communication event. At such times the use of intensely emotional language or highly emotional appeals is likely to interfere with understanding and critical evaluation of communication.

Distraction bears an interesting relationship to alertness. For example, speakers frequently have to deal with distracting stimuli which occur during speeches or which are recalled by listeners. If a person know that listeners are distracted by something which has occurred, the person may make reference to the event and then use it to capture or recapture the listener's attention. The same is true if the distracting stimulus occurs unexpectedly while one is talking. On the other hand, some researchers have reported that mildly "distracting" stimuli such as an irrelevant visual presentation may under some circumstances make listeners more susceptible to persuasion. There are competing explanations of this phenomenon and some question as to whether or not it actually occurs. It may be that the "distracting" stimuli, if pleasing, mask the effect of other more distracting stimuli, or listeners may deliberately concentrate more attentively on a message under certain distracting conditions. Some researchers have suggested that the "distraction" prevents listeners from building counter arguments as they listen to propaganda. One thing seems certain: even a mild, pleasant distraction results in less *retention* of material presented in a communication, even if it does not result in less opinion change, so whatever the effects of the distraction, it probably doesn't help the listener. Some persuasive messages, of course, are so bad that any distraction will be welcomed by the; if one is distracted, he or she may not react as negatively as with no distraction.

Dominance, Leadership, Need to Influence, Mach IV

The terms, "dominance," "leadership," and "need to influence" seem to be self-explanatory. Mach IV and Mach V are not. They are short for "Machiavellianism," and are two forms of a test designed to determine the extent to which an individual desires to manipulate others. These traits considered together not because they are identical but because people who are dominant, often leaders, high in need to influence other, or "Machiavellian" pose special problems in communication situations.

Such persons are easier to identify than some of those previously discussed. Person who are in positions of leadership are likely to be there because their need to influence is unusually strong, and the need to influence is likely to be unusually strong because they are in positions of leadership. Further, person as in professions which require close interpersonal relationships, such as psychiatry and sale, have been demonstrated to have especially high scores on Mach IV and V.

Finally, men on the average seem to be more Machiavellian than are women; as Christie and Geis put it, "Aside from Lucretia Borgia, most of Machiavelli's observations were of males (his title suggest this: *The Prince* rather than *The Princess;* The Discourses rather than *The Gossip*)." Forgive them, they were writing several years ago.

It is possible, of course, to fight Machiavelli with Machiavelli. If a person known that another member of a communication group is more interested in influencing than in being influence, he or she may use that knowledge by indicating that the other person has, indeed been influential. This may be done dwelling upon matters on which both agree or, in discussion, by demonstrating a willingness to change opinions in response to reasonable arguments. A subtle communicator may allow the other person to take credit of "discovering" the communicator's own arguments and propositions. Thus "inductive" organization leading form data to the formation of conclusions may be more effective with highly manipulative people than is an organization pattern in which the conclusions are stated and then supported. Manipulative people are especially susceptible to one form of the "bandwagon" device in which the proposed plan is presented as a new one which is likely to be accepted by others. This gives the manipulative person the "opportunity" to be among the early advocates of some thing which is likely to prove popular. If a communicator is subtle in the use of these methods, Machiavellians may depart congratulating themselves on the ease with which they convinced the communicator and eager to convince others in the same way. Whether such manipulation is justified even as a defense against manipulation is a decision you will have to make. Beware of such techniques if you are a "high Mach" yourself.

Cognitive Style

The characteristic of cognitive style seems primarily relevant to one's understanding of a communication.

Cognitive style has been measured by a variety of tests, and these tests are related to one another in a variety of ways. Basically, however, the theory behind such tests is that some individuals show a high degree of cognitive complexity, perceiving their environments in terms of many categories lying along many dimensions, and responding to strong need to impose order upon their perceptions of the external world. Other individuals have a low degree of cognitive complexity, maintain few perceptual categories along few dimensions, and are not driven by any obsession to order their perceptions. In short, people's images differ in complexity.

One of the symptoms of dogmatism is a simplistic cognitive structure in which the individual maintains little differentiation among belief and disbelief subsystems. The highly dogmatic person is one who, to some extent at least, tends to categorize other person and concepts as either "good" or "bad" and fails to perceive the relevance of his or her beliefs to one another, making it possible to maintain simultaneously contradictory beliefs. Thus cognitive style is one of the components of dogmatism.

Like dogmatism, simplistic cognitive style tends to be more characteristic of those who are less educated and those who are politically religiously, or philosophically extreme. These relationships are not high enough to be very dependable, however. Consequently, it is difficult to determine the cognitive style of another person.

Still, the characteristic does seem to be important in communication. Communicators high in cognitive complexity are more likely to impose organization upon disorganized communication, more capable, of drawing conclusions form data even when other communicators draw no conclusion, and more inclined to perceive relationships between new information and their personal beliefs. For those low in cognitive complexity, on the other hand, communicators will have more effect if they state clearly the intent and organization of their messages, rely heavily upon internal summaries and transitions, draw explicit conclusions and emphasize the implications of these conclusions for the subsequent behaviour of the listeners.

Interpersonal Trust, Anomie and Mach IV

Interpersonal trust" seems self-explanatory. It is another component of the conglomerate of symptoms Rokeach has termed" dogmatism." Christie and Geis have been puzzled by the fact that their measurement of desire to manipulate others bears no demonstrable relationship to measures of authoritarianism and dogmatism, but it is significantly related to "anomic disenchantment," a species of interpersonal distrust. They have concluded that there are two types of interpersonal distrust, one of which is characteristic of highly authoritarian and dogmatic persons, includes an apathetic response and moralistic judgment, and amounts to saying "People are no damn good, *but they should be*" The other, characteristic of the

Machiavellian individual, is neither apathetic nor moralistic, and it amounts to saying "People are no damn good, so why not take advantage of them?"

The distinction seems to be useful in communication. In either case a communicator may have the problem of convincing people who are characteristically distrustful thathe or she is a trustworthy person. The one type of interpersonal distrust, however, seems to be most characteristic of older persons and those of low socioeconomic status, in particular. Such people will probable be more responsive to a communicator who says, in effect, "People are not to be trusted, but one should be able to trust. I trust you and would like you to trust me because I have your interests at heart." The untrusting opportunist, on the other hand, may be more responsive to a communicator who says, "Granted that people in general cannot be trusted, but you can trust me in this specific case because my goals are the same as yours."

In both cases, however, as in most communication situations, the listener's responsiveness depends upon a communicator's ability to induce empathy and demonstrate that their interests, attitudes, and intentions are really similar.

Summary

People who differ in *personality* characteristic also differ in the ways they communicate and respond to communication. Being *sensitive* to one's own personality characteristic and those of others helps one understand communication better. Some of the personality characteristics which relate to communication are: dogmatism, self-esteem, need for security, needs for achievement, for affiliation, aggressiveness, activation level, dominance, leadership, need to influence, Machiavellianism, cognitive style, interpersonal trust, and anomie.

Suggestions for Developing Awareness

1. Consider how affiliations with groups, organizations and people (a) affect your image and (b) become manifestations of your image. As examples, you might use social and intellectual groups, schools, prominent individuals, leaders, wealthy persons, and poor persons. How do some affiliations hamper self-actualization and how do others enhance it? How do affiliations sometimes affect the reliability of your information gathering?

2. The ability to understand another person rests, in part, on how much information one has about the other person's background, interests, goals, etc. Generally, accuracy in predicting another person's behaviour increases as the amount of information you have about that person increases. The following exercise will illustrate this notion.
3. Find another person in class, preferably someone you don't know well. Each of you should select four things form you wallets that represent "frustration," "happiness," "security." and "freedom." Choose one item for each feeling. Now, exchange those items and discuss their meaning. Then try to predict, based on your information about each other, what gift the other person would most like to have. List the following things: a puppy, a stuffed teddy bear, a free dinner, theater tickets, and a book. Which gift would the other person most like to have? Are your predictions accurate?
4. If you try this exercise again later after an exposure to communication training, you'll be able to see how much you've learned about each other and how much each of you has changed.
5. Find a copy of Rokeach's dogmatism test (it appears in *The open and Closed Mind).* Take the test or, if possible, make copies so that the entire class can take it. Score yourself and compare your score to those of others. Do the items seem valid? Do you believe that "dogmatic" responses on some items are justified? Do others in the class agree with you? If not, why do they insist on being so dogmatic? Repeat your position to make it clearer to them.
6. Recall times when you have felt especially low in self-esteem or have been frustrated in some important way. Were you especially susceptible to persuation when you were low in self-esteem? Did you feel especially aggressive when you were frustrated, or did you become apathetic? You might want to share these incidents with the class.

Part II

Chapter **23**

Gandhian Challenge to the 21st Century

Kusum Singh

Mahatma Gandhi's Seven Deadly sins :

Wealth	without	work;
Pleasure	without	conscience;
Knowledge	without	character;
Commerce	without	morality;
Science	without	humanity;
Worship	without	sacrifice;
Politics	without	principle.

Tyrants (or would-be tyrants) nd use hyprocritical visions of public interests or the "common goods" etc. to help get, keep and use concentrated power. When forced to however, in order to increase their power they will even do some very good things for other people-even the weakest and poorest. To use one of Le Rouchefaucauld's *maxims*, (which helped send him to the Bastille) they will "do goods in order to do evil with impunity."

- *Bertram Gross*

I am very glad to be here and to have the opportunity of sharing my research of many years with you. I hope you will not deprive me of your comments and questions after my brief presentation.

My research, supported by an earlier sabbatical and a Fulbright fellowship in 1991 brought me to India where I interviewed people from all walks of life during a six month period. I stopped in Prague on my way to India to interview some members of the Charter 77 and other who participated in the "Velvet Revolution". I now have over 200 in-depth interviews that include testimonies from people in Moscow, South Korea, Australia, Japan, Brazil and the Netherlands that I visited leter over the years. Through extensive research I have been developing an integrative model based on the movements led by not only by Mohandas Gandhi (who practiced what Tolstoy and Thoreau preached) but also by many others. These include Vaclav Havel (the former Czechoslovakia), Martin Luther King Jr. (USA), Aung san Suu Kyi (Myanmar), Andre Sakharov (the former Soviet Union), Lech Walesa (Poland), Rigoberta Menchu (Guatemala), Medha Patkar (India) and many others not so well-known.

I have undertaken this work at a time when, in **Max Kampelman's words, humankind is "entering new worlds" in which a fundamental challenge is "what societies and peoples might do to resist evil." Resisting evil, I think, requires both high ideals, without which power is barbarous, and pragmatic power, without which ideals are futile.** For any serious resistance, I argue, person and groups need adequate information and full access to channels of communication. The dizzying pace of information technologies promises both but delivers neither. Rather, it burdens people with an overload of useful information sprinkled with an immeasurable amount of misinformation and disinformation. Besides, many of us filter our perceptions through a stubborn array of old world customs, ideologies and vested interests. Difficult adjustments are needed in our tacit assumptions, in the way we communicate with each other and in the duties we assume or evade.

One of the greatest miracles in human history is how Mohandas Gandhi and his followers succeeded in transmitting very dangerous messages to hundreds of millions of people who were mainly illiterate and had long been held in bondage by an extremely clever and powerful British imperialists. Something similar has been accomplished by such modern day Gandhians has Martin Luther King, Jr. Andre Sakharov, Vaclav Havel and Nelson Mandela. Then there are much larger number of unsung heroines and heroes who have sought communication with their people despite exclusion from the media of their own countries.

One of the most eminent Gandhi scholars in India, B. R. Nanda, has written many volumes to help rescue Gandhi from both deifiers and debunkers, In one of his books he portrays Gandhi's role in "major revolutions" against racialism, colonialism and violence. Unfortunately, each of these three evils are still alive and well. Racialism operates in many forms throughout the world Old-fashioned colonialism has been replaced by oppressive systems often labeled "neocolonialism," "indirect imperialism" or "the first stage of truly global capitalism." Under these circumstances we see Gandhian ideas as far more relevant now than ever in the past.

Mohandas Gandhi had a vital, although too often forgotten, concept of civilization. He envisioned a world that would be neither western nor Eastern. He criticized the failings of both. He saw the possibility that people of different cultures, religions and ethnic groupings would view each as kindred humans and peacefully share responsibility for helping the helpless and supporting the rights of every one. These values had been earlier voiced in the ancient texts of Judaism, Christianity, Islam and Ḥinduism and also in Buddhism, Taoism and Shintoism. **Gandhi brought the best of these together under the unifying and open ended concept of nonviolent activism.**

A reporter once asked Mohandas Gandhi : " Mr. Gandhi, what do you think of modern civilization ?"

Gandhi's answer; "It would be a good idea." The word "civilization" has long had many meanings. The first human tribes branded as "uncivilized" or "barbarian" **any people who looked different and spoke differently.** As powerful "civilizations" emerged, their ruling elites quoted holy texts selectively in proclaiming the divine right of rulers. Repressing humanitarian instincts, they acted on the axiom that "Might (or Male) Makes Right." They tyrannized women, poor people, colonies, foreigners, slaves and ethnic or religious minorities. They legitimated their right to do so by stigmatizing their victims as "inferiors." **By the nineteenth century, under the banner of "Civilizing Missions," the industrial nations behaved like true barbarians.** They fought each other savagely in establishing their rights to dominate - and "civilize" – people, cultures and economies in Asia, the Middle East, Latin America, Africa and Oceania. **"Might, Male, Money and Light-skinned Made Right."**

Let's listen to Mahatma Gandhi himself. In 1946, long before Marshall McLuhan advertised the worldas a "global village." Gandhi envisioned another kind of one world :

"In this structure composed of innumerable villages, there will be ever-widening, never ascending circles. Life will not be a pyramid with the apex sustained by the bottom. But it will be an oceanic circle whose centre will be the individual always ready to parish for the village, the latter ready to perish for the circle of villages, till at last the whole becomes one life composed of individuals, never aggressive in their arrogance but ever humble,

sharing the majesty of the oceanic circle of which they are integral units."

In this image of the "oceanic circle" Gandhi foresaw all of humankind as part of one world. It would be a world with people brought together by a common cognizance of the mutual interests of all individuals as members of one human race, the only race that now exists. That is what he meant by ever-widening.

But what did Gandhi mean by "never <u>ascending</u> circles?" Was it not an assault on the pyramids of power that dominate both Eastern and Western cultures? Was he suggesting an alternative to top-bottom superior-inferior male-female images that dominate our so-called "civilization?" Could his "ever-widening, never-ascending circles" help repress the almost instinctive arrogance with which people of one place, class, religion or gender regard themselves as superior to other people? On this circular globe, in the Gandhian view, all people are on much the same level, whether they live in the valleys or the mountains, in one storey shacks or in sky scrapers. Communication among people is both horizontal and circular as information among senders and receivers moves back and forth through the curved apace of the global planet called Earth.

Fortunately, throughout human history many brave **women** and men have tried to resist brabarism through the civilizing power indeed, **the sacred power of no violence.** Why do I place women first? Not because the female rebates against barbarism have been well recorded in history. I **place women first because over the centuries women have had no recourse other than practicing peaceful conflict resolution as an escape from oppression.** Thus it was with Aspasia (who wrote the speeches for Pericles of Athens), Sappho of Lesbos (the poet of peace), Mirabai (the Indian poet), Sister Juana Inez de La Cruz (the Mexican scientist, dramatist and poet), and Harriet Tubman (who led slaves to freedom through the "underground railway") we could mention hundreds of others. Then, of course, there are such male pre-Ghandhians as the Greek stoics, Thomas Jefferson, Henry David Thoreau and Leo Tolstoy.

My basic research has been to **identify leaders and social movements that have used nonviolence to work toward a more civilized in a world of increasing barbarism.** In doing so, I identify the difference between different kinds of nonviolence that may be nonviolent in their methods but not necessarily nonviolent in their goal. That distinction is the sacred nature of struggling against evil and oppression without hatred and ill will toward your adversaries. In Simone Weil's words,

> *The needs of a human being are sacred.*
> *Their satisfaction cannot be subordinated*
> *either to reasons of state, or to any consideration*
> *of money, nationality, race, or color, or to the moral*
> *or other value attributed to the human being in*
> *questions, or to any consideration whatsoever.*

A new paradigm for such research is slowly emerging. People in many countries have long been integrating nonviolence in its many forms (negotiation, democracy and protest) with female-male partnerships and support of basic human rights. Although scholars and researchers have tended to specialize on a few aspects of each. Activists have not had the time or resources to formulate a fully integrative approach.

There are many **powerful barriers to such integration. In an age of accelerating specialization, separatism has become a dominant message.** People have become separated from each other by gender, religion, race, nationality, occupation and political or sexual orientation. It is difficult in any one country for many people to learn about the combinations of progress and regress in their own country, let alone in other countries.

Above all, those concerned with nonviolence, feminism and human rights have understandably been reluctant to recognize the need for **the trained and credible physical force (local, national or transnational) without which violence cannot be prevented or controlled.** Untrained or uncontrolled, such force leads to the horrors of police brutality, civil war or war among nations and even to the legitimation of violence in the home, neighbourhood and workplace and the schools.

Sam Goldwin, one of the founders of Hollywood used to say, "For your information, let me ask you a question." So let me ask you four Gandhian questions.

My first question is :

1. "Can violence be curbed?"

We live in a cultural environment of violence-aided and often legitimated by the media. People suffer from TV barbarism. As the

number of available channels enlarges (even to include worthwile programs), TV watching displaces direct communication among people, provides escapist distraction from the real world and disseminates misinformation or disinformation alongwith imagery that has the effect of legitimating violence. This seems to permeate the entire post-cold war disorder. Bloody wars rage in many parts of the world. Civil war or terrorism grip Asia, Africa and the Middle East. Even in "civilized" countries rich in wealth and technology, people suffer the insecurity created by violent crime, domestic violence, murder, and terrorism. A cultural environment of glorified violence dominates our life in homes, neighborhoods, schools and workplaces. Many people justify extreme violence for various reasons like retribution, nationalism, religious extremism or ethnic superiority. Thomas Bailey Aldrich warns us, "Civilization is the lamb's skin in which barbarism masquerades."

What are the roots of such violence? According to media critics much of the violence in the U.S. is blamed on the huge amounts of violence on the air **waves. But what about poverty, prejudice, oppression and discrimination? Do they not provide the breeding ground for unscrupulous leaders who gain followers by appealing to the worst of all prejudices and emotions?** Indeed, is **democracy possible without nonviolence? Is** nonviolence possible without human rights?

And what about us? yes, you and me? Years ago, before liberation from the British empire, Mohandas Gandhi put it this way : "I have three adversaries the British, the Indian people and myself." In his experiments with truth, he too like all of us had to overcome instinctive reactions of anger, rage, frustration and revenge. That is not always easy to do.

2. So my second question to you would be :

"Is there a new global empire that needs to be demystified by communication researchers?

"Who will tell the people?" This is the question that William Greider asks in a book by that title. He answers it himself by documenting a new form of tyranny in the world.

I can summarize his words with one phrase : watch out for a new global empire. In 1975 many people called for a New International Economic Order. Then came the idea of a New International Communication Order. Well, a new world Order is, really emerging.

It is an economic order, a communication order, a political order and above all a frightening world order. It is **a form of empire that has no precedent in human history. It is** operated through clusters of private public and private organizations, constellations of clusters and larger complexes that have become the standard mode of structuring the planning and control activities of private and public banking, agribusiness, and mass communications."

The people in less developed countries are now more clearly the subordinate parts of a multi-polar system dominated by transnational corporations.

3. My next question naturally would be:

"Can non-violence under these circumstances ever really work?"

Most people today see nonviolence as a Utopian idea held by marginal people who do not understand the realities of politics and power.

Yet so-called "maginal people" have often won the power to make changes that had earlier seemed impossible. In the earlier years of this century, after years of being jeered at or jailed, non-violent female activists, with the support of some men. Won the right to vote in the United States. The civil rights movement helped improve race relations, thereby setting a powerful example for movements against war, environmental degradation, sexism, and homophobia. In India almost fifty years ago, the Gandhi led liberation movement achieved freedom from the British. In Asia and Africa, other colonies followed suit in most cases without bloodshed.

Later, the non-violent Solidarity Movement under the leadership of Lech Walesa peacefully displaced the Communist dictatorship in Poland. This was followed by the so-called "Velvet Revolution" in Czechoslovakia, the fall of the Berlin Wall, the peaceful dissolution of the Soviet Bloc and the disappearance of the Soviet Union itself. Non-violent activists displaced the Marcos regime in the Philippines. In Greece, Spain, Portugal, Argentina, Chile, and Brazil authoritarian regimes were displaced without war. And in Nicaragua and El Salvador, civil war came to an end.

More recently, without the civil war that was earlier regarded as inevitable, the African National Congress headed by Nelson Mandela displaced the apartheid regime headed by President de Clerk's National Party. With the help of United States Armed Forces, Jean-Bertrand Aristide (previously overthrown by the Haitian military) became president of a Haiti in which long traditions of violence are being slowly overcome.

I do not overlook the fact that in all of these countries; **victory has been accompanied by failure. The high spirit of former non-violent movements seems to have faded in the face of powerful assaults on economic, social and environmental protections.** The doors of opportunity for a better educated minority have not been open enough for the majority of the population. Most people are plagued by the barbarous insecurities of poverty, joblessness, bad jobs, and ever present threats of physical violence. With a decline in civility and civil discourse, new fascist movements have raised their ugly heads. It is the time for demagogues to exploit the anxieties of people for their own ends.

Under these circumstances, one has to go back to Gandhi for hope. As Vaclav Havel put it, "Hope is definitely not the same thing as optimism. It is not the conviction that something will turn out will, but the certainty that something makes sense, regardless of how it turns out."

Unfortunately, no one has yet attempted to bring together the many factors underlying these and diverse events of progress and regress. This is not for the lack of documentation. On each of these events there is a rich literature of detailed histories, journalistic reportage, biographies, autobiographies, memoirs and recorded speeches. There is even a treasure house of video tapes, well-organized documentaries and insightful films and novels. **What has been lacking, rather, has been a deep enough understanding of non-violence and the circumstances favourable to its use.**

My research indicates many **varieties of non-violence.** Usually the term "non-violence" is used in referring only to the many protest activities some legal and some in violation of law-used **by Mohandas Gandhi, Martin Luther King Jr., and non-violent activists around the world.**

There are other varieties of non-violence that are even more important. Think of what women have done from time immemorial to win some freedom from male domination : **talk, negotiate, bargain and when necessary manipulate.** Did not those efforts lead eventually to the constitutional processes of democracy litigation, elections, legislation, political parties, majority decision, and minority rights?

Integration was the central message of Mohandas Gandhi's experiments in struggling against violence. His "oceanic circle" connected women with men, means with ends, nonviolence with democracy, research with action, education with life, individuals with groups, nation with world and above all, **responsibilities with rights.**

4. So my last question is :

"What are the major obstacles to moving toward a more civilized, non-violent world?"

One of the miracles of our times has been the globe-spanning communication networks of transnational **capitalism** that has truly unified the world as one. But, by Gandhi's or **even our own standards, it is not a civilized world. The twentieth century world has been- and still is a cruelly barbarous place.**

Ralph Waldo Emerson pointed out,

As long as our civilization is essentially one of property, of fences, of exclusiveness, it will be mocked by delusions. Our riches will leave us sick; there will be bitterness in our laughter; and our wine will burn our mouth. Only that good profits, which we can taste with all doors open, and which serves all men.

My research identifies further some of the **myths that interfere with productive discussion of the subject. Let me spell** out these myths more specifically :

Charismacho myth. Indeed, nonelitist, democratic leaders were all "**Great** Communicators" who succeeded in communicating with thousands of millions of people across linguistic, religious and class barriers of all kinds. Was it their charisma that mesmerized the people to follow them or their macho leadership style? No, not if you understand charisma the way Max Weber defined the term. Efforts to deify them as superhuman persons with a Weberian type of charisma would have tended to undermine the power of their movements. They appeared more superordinary with a "common touch" who were able to build organizations and coalitions with many others in active leadership roles. Discouraging over-dependency on themselves, they have usually welcomed spontaneous activities initiated by others. They set personal examples of unusual courage.

"My life," said Gandhi, "is my message." While well educated (with the exception of some), they remained in touch with the less educated or illiterate. They were not "charismacho" leaders!

Invincibility of Mass Media myth. Many of these leaders became temporary "media stars." When denied access to major domestic media, they have often won attention from the foreign press and influential organizations abroad, including the United Nations or even a number of foreign governments. They successfully circumvented the so-called invincibility of mass media through other modes of communication : face-to-face relations and religious or political organizations. Top-top and middle level discussions- seem to have been most successful when more circular communication (including "bottom-sideways" and "bottom-up" communication) takes place.

The myth of Non-violent activism as passive resistance. The best known methods of "creative nonviolence" have been public demonstrations, strikes, memorial services, civil disobedience, and sanctions. These have been most effective when associated with the use of fact finding, negotiating, litigating, legislating, voting (when available) and the slow and difficult process of generating or improving the formal machinery of constitutional democracy. The major principle here is : that all people, no matter how selfish or brutal they may be, have something good in their souls, and that violence, even when justified as the "last resort," usually breeds more violence. This activism requires the training without which people would use violence instinctively in response to violence. Also these leaders recognized various needs for physical force. Personal security including that of nonviolent activists is not possible without protection by the very police power of the state that, if undisciplined, may undermine personal security. The threat or use of armed forces may also be necessary to regulate the production and distribution of deadly weapons, protect providers of humanitarian relief for starving people or, under special circumstances, to promote international peace keeping or peace making.

Myth of non-violent activism limited to technique and political strategy rather than commitment to a few high moral ideals.

Nations and civilizations, like individuals, do not live by abstract principles. Moral principles arise from experience; they are tested and revised in the face of new experience, and they must be embodied in experience and the collective memories that result.

In 1948 human rights advocates won approval, against tremendous obstacles, of the Universal Declaration of Human Rights by the United Nations General Assembly. The soaring words of this comprehensive resolution and its follow-up treaties have been historic forward steps-despite insufficient attention to duties in setting forth both high ideals and specific kinds of action needed for strengthening nonviolent democracy. Yet these ideals are flagrantly violated. The Universal Declaration's preamble urges that "every Individual and every organ of society.. shall strive by teaching and education to promote respect for these rights and freedoms. "No nation and no national or local education system has really done such striving. In a truly civilized world, there would be many more opportunities available for freely-chosen useful work. Instead, human dignity and self-respect are being threatened or destroyed by the worldwide scarcity of good jobs at decent wages. According to Benjamin Disraeli "Increased means and increased leisure are the two civilizers of man."

In Will Durant's words,

> *Civilization is a stream with banks, the stream is sometimes filled with blood from people killing, tealing, shouting and doing the things historians usually record, while on the banks, unnoticed, people build homes, make love, raise children, sing songs, write poetry and even whittle statues. The story of civilization is the story of what happened on the banks. Historians are pessimists because they ignore the banks for the river.*

When and if civilization comes alive in our time, we will see nonviolent activism for economic rights and duties. In Gandhi's words, rights to "living wages and jobs for all." We will be lobbying, litigating, voting and yes, demonstrating, against the economic and social conditions that promote drug addiction and violence in streets, schools and homes. What are we doing? What more might we do? George Orwell warns us that "To accept civilization as it is practically means accepting decay." Let's hope this conference might help the international community engage in the painfully slow tasks of structuring global communication that may lead to more civilized societies. I am not unduly optimistic. Are any of us? But are we hopeless? I hope not.

The truth, as I see it, is that there is a great struggle for secular democracy in India today as the Hindu nationalist BJP grow in power, those courageous men and women who are willing to engage them in a battle of ideas and moral values are rapidly becoming as important for us to know about, to understand and support as the dissident voices in the old Soviet Union used to be.

Several fundamental values which existed, and still exist, in Indian tradition must be identified and re-examined. Some critical values relate to looking at life in its totality and all its richness. The creative spirit exists in all human beings and does not always manifest itself spontaneously. The challenge is to bring out this quality and the strength and the dignity of the people that goes with it.

Gandhi once remarked that recorded history was a story of "every interruption of the even working of the force of love." This was a way of suggesting that the violent behaviour of slave masters, monarchs and warlords was only a small part of the entire - but mostly unrecorded - story of human life on the planet. To the extent that this is true, most people have been Gandhians without carrying that label, or even consciously dignifying their actions with Gandhian philosophy. In 1960, for example, African American students fought segregation by sitting down at a lunch counter in Nashville, Tennessee. Some of the people who marched in their defense were, like the Reverend James Lawson, students of Gandhian techniques. Daine Nash, the Fisk College student, who confronted the mayor of Nashville, was not. She was just doing what came naturally.

Despite skepticism and serious obstacles, I see nonviolence already happening in many homes, some schools, various neighbourhoods and, above all, through the formal machinery of constitutional democracy when it is available for use and used. A scattering of agencies-a few firmly established, others struggling for existence provide necessary training. Important efforts are being made by the Einstein Institute in Boston, the Centre for Negotiation, the U.S. Peace Institute in Washington D.C., the Martin Luther King, Jr. Institute in Atlanta, and Gandhi Institutes in India and elsewhere. New ground is being ploughed by some agencies of the United Nations and other transnational organizations in the defense of human rights, the provision of humanitarian aid and the care of refugees.

Today, among these transnational webs are the many dedicated people in scientific associations who reject the notion of charismatic solutions to democratic problems. Their members, like the mass communication researchers at international conferences, talk with each other at small and plenary sessions and debate critical or trivial questions. We meet each other in corridors and dining rooms. We come to know each other as human beings. We come to know that the world is also being unified by the human rights movement and by the newly strengthened United Nations. All these are ways of holding back new waves of barbarism and friendly or unfriendly fascism. They help nurture the "democratic global consciousness" through which we might just possibly attain peaceful coexistence and a more civilized world by the first decades of the 21st century. The moral vision that Gandhi envisioned can move the hearts and souls of people and would surely include :

- **the *strength*** that utterly rejects the tradition that Might (in the form of violence) Makes Right and is based rather on non-violent resistance to militarism and aggression;
- **the freedom** that consists of a fabric of rights and rules that, as Gandhi put it, "guarantee to our opponent the same freedom we claim for ourselves";
- the genuine ***individualism*** that can thrive healthily among persons who see themselves as part of a human community rather than lonely pecks of dust;
- the true ***patriotism*** that recognizes interdependent interests in a deeply divided world and, in Gandhi's words, "admits of no enmity or ill-will" towards people of other nations and is based upon open debate concerning national interests rather than acceptance of the narrow interests of self-serving leaders;
- ***sincerity*** reflecting deep conviction and selfless action towards the building of a solid democratic society.

I have examined these activities in the light of Gandhi's concept of "experimenting with truth" and his emphasis on duties and responsibilities as the foundation of rights. Through the spirit as well as the machinery of democracy, "Might (widely dispersed) might make right." Whether it does so or not depends on millions of people acting as leaders in tenacious action. For this to be done, **human rights would have to be backed up by female- male partnerships, better networking among now fragmented human rights groups, and by stronger private and public structures-local, national, regional and global of responsibility, action and resources.** International human rights law might thus become a living force as an ever-evolving global bill of human rights and responsibilities.

Idealistically, I join with those who view the 2000 as marking the first decade of a "Human Rights Century." The decline of present barbarisms could signal the emergence of **the first world civilization.** For this to happen, more persons, parties, enterprises, organizations and movements would truly move along the path of nonviolent campaigns against **real enemies : poverty, hunger, disease, environmental destructiveness, ignorance, prejudice, repression, nationalist extremism, dangerous and dehumanizing technologies, violence of all sorts and war itself.** More and more people could thus begin more fully in the words of the International Commission of Jurists-to "enjoy, exercise and utilize under the Rule of Law all his (her) human rights, economic, social, cultural, environmental, civil and political."

Kusum Singh, Professor of Communication, Saint Mary's College of California. Fulbright Research Scholar in India, Fall 1991. Author : "elites vs. Democratic Communication: Ronald Reagan and Mohandas Gandhi," Janet Wasko and Vincent Mosco, eds. *Democratic Communication in the Information Age,* Garamond Press, 1992; "Mass Communicators for Peace: Another Way, "in Elise Boulding, ed. *Peace, Culture and Society: Transnational Research and Dialogue.* Westview, 1991; "Adventure in Learning", (co-author), Marien and Jennings, eds. *What I Have Learned,* Greenwood Press, 1987; "Mass Line Communication : Liberation Movements in China and India" and "The MacBride Report: The Results and Response, "both in Gerbner and Siefert, eds.. *World Communications: A Handbook.* N.Y. : Longman Publishers, 1984; "Bottom-Sideways Communication Bottom-Up Democracy, "*Interaction,* Journal of the National Council of Developmental Communication, India, Fall 1983; "People against Charisma," *Communicator,* October 1981; "Gandhi and Mao as Mass Communicators," Journal of Communications, Summer 1979; "Elite Control and Challenge in Changing India," in George Gerbner, ed.. Mass *Media Policies in Changing Cultures,* Wiley Interscience, 1977.

Chapter **24**

The Global Village in a Digital Age

Hope Warren Shaw

"The Power of Digital" was the title of the 1998 conference of the National Association of Broadcasters attended by over 100,000 people from all over the world. The NAB is also the largest single exhibition of broadcast and multimedia equipment in the world. Slogans for companies and exhibits predominantly featured the word DIGITAL "Digital Mastery is Closer Than You Think-Tektronix; "Digital Reality" Sony, "DTV Powerhouse" Phillips; Globe cast the shortest route between you and your audience" -France Telecom to name just a few. Conference seminars included such titles as : Digital Video, Digital Studio Command and Control, Digital Sound Broadcasting Becomes A Global Reality, International Standards for Digital Terrestrial Television Broadcasting, Implementing Virtual Set Technology, Understanding the Future of Video Editing in a Digital Era., New Graphic Systems for the Digital Production Environment, and Where Are We Headed in the Digital Arena? And that is exactly the subject of (his discussion : What are some macro implications of these micro chips?

First, a simplistic definition of digital is the ability to reduce any component to either ones or zeros. Computer information exists in sets of ones and zeros. To translate video into these components allows us to have images of much higher quality than the current analog standard. Digital allows the conversion of an analog signal into binary form. In digital recordings, digits represent quantities, and digits in rapid sequence represent varying quantities. Digital video has the potential of making possible a much better broadcast system than the current NTSC used in the United Stales today.

Of course video on digital television will be digital but that already exists on CD ROMs, on computers, and on the Internet. One problem for video producers is putting digital video on computers for non-broadcast applications. Technical issues such as memory, file size, computer speed and access from a CD ROM or the Internet have to be considered. To deal with these issues the video has to be compressed. One way to visualize video compression is to conceive of video as a series of pictures in which some parts may change and some may remain the same. In a video of Ranthambbore Tiger Preserve one of may favorites the fortress around the park stays the same but the water on the lake moves. In an analog picture, the whole picture is painted in every frame whereas digital video remembers the part of the pictures that doesn't change and does not paint them on every frame but only changes the parts of the picture that do change such as the moving water in the lake. As compression quality has improved, digital video, even on medium priced digital cameras and editing systems, can approximate broadcast and even motion picture quality standards. And that is important, because no matter what the application, most people have seen television or motion pictures and have a psychological impression of what good quality looks like. Digital video images can be broken down so that parts of frames can be changed. They can be downloaded from the Internet. They can be manipulated and edited on a personal computer. You can take the fortress of Ranthambbore and put a tiger on one of the ramparts there some of the time and others not. Perhaps what is most important, materials can be created fairly quickly and easily that can be interactive.

One of most important things to consider in creating interactive multimedia is the way the target audience the people for whom it is designed work with it. There is associated with almost all video whether it is for training, for instruction, or for entertainment, a story. In a linear frame, you would lay out the story in a series of scenes with a beginning, middle, and end. In an interactive frame, you have to deal not only with the characters in the story but the users of the video or CD ROM and their roles vis a vis the characters. Do they have a part in the decisions the characters make? Is the story one that lends itself to interactivity? What is its style and does it involve sorting of information or working out of a complex series of events like a mystery? Then you have to consider what you can do to get the users to participate. Could the characters serve in this respect or is there some unique device to facilitate interactivity? Is there information that is necessary to the piece and how can you get the users to study it? Is there parallel action in the story with which the users can interact? Are there simulated exercises that come from the content of the story such as learning how to use a computer program or other device? Are there decisions upon which users could have an influence? Could users meet on a Web site outside the story? These are but a few of the questions involved in the creation of interactive multimedia materials. Many of these same questions are asked by video producers in creating linear training pieces but digital technology has taken us literally into a new dimension.

My own immediate concern and that of this paper is how to deal with this ever-evolving technology; how to understand its cultural implications; how to teach students to use it appropriately and, on a practical level, how to manage the financial aspects of changing software and computer platforms.

As San Diego City College and in our Communications Department, there is an equal emphasis between the academic and the vocational. Some of our students are graduates seeking to acquire vocational skills to change from one profession to another. In fact, current predictions show that in the United States, people may change careers three or four times in their lifetime. Student expectations are quite different from those of my parents or even my self. When I first started teaching, after having been a documentary television producer for the Time-Life station in San Diego, the emphasis was on teaching broadcasting. The professional organizations to which faculty belonged were broadcast and motion picture-oriented as were the careers in which students were interested. So, too, was the coursework at universities and graduate schools. The word "multimedia" was not part of our vocabulary.

When I first attended the National Association of the Broadcasters annual meeting, it was entirely broadcast oriented and attended by fewer than 20,000 people. In the past, the NAB and particularly the exhibits, have been a giant equipment fair, in which broadcast managers, engineers, and equipment vendors have gotten together to sell and spend millions of dollars on exclusive proprietary equipment. At the 1994 NAB and even more at the 1995 I found a striking difference in previous NAB's I had attended. This was illustrated in a session on graphics and editing. Seated beside each other, somewhat uneasily, were representatives of the British giant Quantel, manufacturer of high end proprietary editing system selling for three to four hundred thousand dollars; and Avid Corporation, whose high end product sells for around 80 thousand. This was but a portend of things to come as the show opened the next day to reveal not only the familiar major manufacturers but literally hundreds of "mom and pop" companies touting all kinds of production functions until now available only to broadcast stations and large production companies. All these products are made possible through digital technology. The personal computer and its developers had arrived and where literally in the middle of traditional broadcasters.

By the 1998 NAB, there had been an additional building allocated to "multimedia" and also higher and multimedia shown along with broadcast equipment. There was a curious visual mingling of broadcast types in suits and ties along with blue-jeaned young computer whizzes. Along with the word digital in almost every booth was the word virtual. One of the most interesting demonstrations I saw was how Silicon Graphics created the digital people on the decks of the Titanic. This technology included every detail of dress and gesture and looked extremely realistic. So, we now have virtual people as well as virtual environments. Although some of this technology is still only available to Hollywood and the major production companies, the "power of digital" has opened a new world of multimedia possibilities to the personal computer user. As Professor Clifford Christians points out in his paper, *A Theory of Normative Technology*, "technology proceeds out of our whole human experience and is directed by human commitments." Therefore, the commitment by developers to produce cameras, editing systems, and other equipment that would be affordable to small businesses and consumers has given some artistic and information control back to individuals. What is happening here is affecting broadcast and related industries on a daily basis. Purely technical quality, long the gatekeeper and the basis for the hegemony of networks and broadcast stations, is now within the reach of non-broadcast consumers. This has created more immediacy and flexibility, for example, the use of home video on news, documentaries, and entertainment programs.

Electronic Guru Marshall McLuhan coined the phrase "Global Village" in the 1960's. The term has never been more relevant than it is today. According to McLuhan :

"Societies have always been shaped more by the nature of the media by which men communicate than by the content of the communication... Electric technology fosters and encourages unification and involvement. It is impossible to understand social and cultural changes without knowledge of the workings of media."

The thrust of McLuhan's philosophy is that media and technology shape every aspect of our lives. But since his emphasis is on the medium itself rather than the content, the question of responsibility for those media remain vague. Whether "electric technology indeed fosters and encourages unification and involvement" or does exactly the reverse will be discussed a little later. I would first like to mention some other theories of technology and its effects on society. In "The Gutenberg Galaxy," McLuhan heads one of his chapters with the cryptic title, "Heidegger surfboards along on the electronic way as triumphantly as Descartes rode the mechanical wave." In his essay, "The Question Concerning Technology," Heidegger questions the essence of technology, saying that purely technical frames of thinking and discussion are not appropriate. He says that the essence of technology is ultimately a way of revealing the total human being. Heidegger suggests that :

"We shall never experience our relationship to the essence of technology so long as we merely conceive and push forward the technological, put up with it, or evade it. Everywhere we remain unfree and chained to technology, whether we passionately affirm or deny it. But we are delivered over to it in the worst possible way when we regard it as something neutral; for this conception of it, to which today we particularly like to do homage, makes us utterly blind to the essence of technology."

The digital world laid out before the National Association of Broadcasters in April is, in fact, determining the cultural landscape of the globe. Satellite communication has made it possible for a guerilla leader, Commandant Marcos in Chiapas, Mexico, to speak to a conference in New York and that same leader to communicate his ideas via the Internet. Ironically, the digital television technology bringing the World Cup to the farthest reaches of the planet has provided a distraction allowing the Mexican government to crack down on dissidents in remote Indian villages. But these government suppressions would not have been reported at all a few years ago. Now, the digital presence of the situation in these remote villages via satellite television, consumer cameras, and the Internet, has forced the world to pay attention.

In the Communications Department at San Diego City College, we have recently introduced an emphasis in Multimedia including an introductory course, digital audio, advanced multimedia design and production, and nonlinear editing all possible because of Digital Technology.

Nonlinear editing means that ideas and shots can be put in any order and can be easily rearranged in the same way that words can be rearranged in a word processor. Obviously, this has great significance for the editor. Frank Lloyd Wright said : "Conceive the piece in (he imagination, let it live there, gradually taking more definite form. When the thing lives for you, start to plan it with tools. Complete the harmonious adjustment of its parts." Being able to try things out, creating "what if scenarios with and without the client, without having to make creative compromises, is what nonlinear editing is really about. Digital nonlinear editing can now bring together a variety of media that previously could not coexist in one environment. It allows the editor and the producer to concentrate on the needs of the project. It permits the integration of many different kinds of media such as film, slides, video, graphics etc. into a completed program. It allows the producer to save time and money because a couple of key strokes can now rearrange scenes audio, video, and graphics without having to completely junk the original program.

If you looked up the word linear in a dictionary, you would find that it is an adjective that is relating to or consisting of or resembling a straight line. If you looked up the word nonlinear in a dictionary, you probably would not find it. Nonlinear refers to the concept that the physical nature of the medium and the technical process of manipulating it do not dictate the method by which the material is physically ordered.

When nonlinear editors first came out, they were almost exclusively used for off line editing that is to output a list that would then be taken into an on-line suite. However, with the advent of better compression systems, nonlinear systems are beginning to be used much more widely to create an on-line finished product an extremely important characteristic of the nonlinear editing system is the ability to create multiple versions of the same sequence. In linear editing copies could be made but it would mean a generation loss a degradation of the picture and the rebuilding of the sequence. With nonlinear editing multiple versions are a given. What used to be a nightmare scenario the client sitting there saying "I really don't like that scene, or that music etc." becomes a very palatable part of the editing process and the client-producer relationship. Clients can truly become a partner in the vision instead of the voice of doom. Just as word processing evolved, so digital media processors are also evolving. The evolution of these machines means that the hardware will no longer be the media processor but that increasingly sophisticated software will take over. Now it is technically possible to include in the post production process slow motion, repositioning shots, painting, audio manipulation, and digital video effects. Nonlinear actually emulates the way we think which is free form a kind of random access process. Randomaccess is a system that allows the editor to look for a particular section of material without having to proceed sequentially through material to reach that location.

Laotze :

"Thirty spokes are made one by holes in a hub,
By vacancies joining them for a wheel's use;
The use of clay in molding pitchers
Comes from the hollow of the absence;
Doors, windows, in a house,
Are used for their emptiness;
Thus we are helped by what is not,
To use what is"

McLuhan :

"Electric circuitry is Orientalizing the West. The contained, the distinct, the separate our Western legacy-are being replaced by the flowing, the unified, the fused."

When I was in India on sabbatical in 1992, I followed the progress of a family planning video being produced for UNESCO. I was extremely impressed by the quality of the pre-production and the great attention to detail that went into every aspect of the presentation. There was an enormous amount of pre-testing of the script as well as every detail of costuming and locale. It occurs to me now that process might have been made somewhat easier and faster if interactive simulations could have been created allowing for various nuances of language, scenarios, and outcomes. I was also struck by the possibility of doing field research with video given the portability of digital equipment. Of course, considerable research would have to be done to determine the validity of the approach for the target audience.

In the United States, digital cameras, editors, effects, and animation have just become cost effective to be used in universities and colleges. The possibilities are endless but there are also pitfalls. Christopher Lukas, in an article in *Television Quarterly* poses a key question; "Is there too much High Definition hype for hardware, technology for technology's sake?" He adds :

"What I don't like is technology which gels in the way of content, **or technology for technology's sake (the Insta Cam on the**

freeway those 11 P.M. local news remote "standups" when there is really nothing to see; the use of wipes to get from one banal scene to another; the sequence of fast cuts when the story line is what should be concentrated on.)".

This is an example of where the technology has established a *culture* of production. Sometimes the availability of special effects is too great. Often a simple dissolve or even a cut is far more effective than some MTV digital effect sequence. There is a real danger in the "garbage in, garbage out" syndrome.

John Pavlik, Director of the Center for New Media at Columbia University, also writes of the pros and cons of the new digital era. "Will digital TV merely means better pictures or will it lead to better content? ..... the fact is that much programming is still very marginal in quality. It is especially marginal for audiences that provide little market for advertising dollars, such as children." Sampling of American programming like Dallas and Dynasty and American cartoons can attest to "glitz" without content. But Pavlik goes on to show why the new mix of digital contributors might be a healthy addition to the television/multimedia scene.

> "There are three reasons why it might. First, the introduction of digital television and corresponding advances in compression technology means expanded channel capacity. This ...means programmers can introduce additional programming or new forms of programming....
>
> Digital television also means new creative opportunities for program producers. Interactive multimedia content is today seen only in any real manner on the Internet, but in a digital television environment, interactive video is a near term possibility...
>
> Finally, digital television means the introduction of new players in the programming mix. Computer hardware and software providers and consumer electronics companies represent significant entrants into the world of digital television and their impact on programming."

After these positive remarks, Pavlik ends his essay with the caveat that we may find ourselves "abandoned in a digital wasteland."

I think if we are not aware of the effects of the digital environment, we may indeed be ourselves manipulated by it. We may find ourselves in artificial backgrounds transitioning in and out by the spectacular effects, that digital video can make us feel are natural. Neil Postman in *Technopoly* believes that :

> "New technologies alter the structure of our interests : the things we think about. They alter the character of our symbols: the things we think *with.* And they alter the nature of community : the arena in which thoughts develop."

As long as we accept the fact that technologies are a part of our culture; as long as we realize that we do indeed think in a digital mode rather than a linear mode; and as long as we are prepared to deal with *digital* digitally, we will be able to shape the information highway not simply travel on it.

> "The digital gate has been opened and we have entered the path of multimedia and communications convergence; there is no turning back...Regardless of whether this volatile period in information and communication technology that we now embark upon is self-induced or part of our human nature, it is our choice either to go boldly into this next millennium or to be swept into oblivion."

We cannot put the genie back in the bottle- In all the magic of special effects and virtual reality we must not lose sight of the human factor, the people for whom we make the videos, the CD ROMS, the interactive simulations, and stories. In the *macro* culture of technology, we must not lose the *micro* cultures of villages and communities. As we pursue innovative compelling methods of communicating globally, we must not lose sight of the human, the individual, the one-on-one. As the information highway reaches across the globe, we must not[1] lose sight of the neighbourhood.

> "Like the 15th century Europeans eager to carve up the unknown, companies all over the globe are eager to colonize new virtual territory. Virtual reality interfaces promise to be the ships that will ultimately carry individuals along fiber optic rivers to the vast data seas of cyberspace."

As we go along those fiber optic rivers we must not make the mistakes of the European colonists. We must not subvert the culture of the villages along the way.

I once had the pleasure of sitting on a television panel with Americal Science Fiction Writer Ray Bradbury. He made the point about science fiction writing that we are only human and that the best of science fiction is the humanity that we bring to the characters regardless of their strange appearance or milieu. In Bradbury's "The Illustrated Man," there is a sequence in which a futuristic house has a Play Room that can create many different environments in which children could play and in which they could control the sequence of events.

Like so much of science fiction, that virtual reality is now our reality. The question is what to we do with it? The technology is now available to everyone. It is no longer the exclusive property of film and television studios. On the international super highway, there are as many opportunities for research in development communication as there are for corporate Internet takeovers. In a village in any country, primary research can be gathered with a digital camera, entered into a computer, and used for archival research and future training videos relating to national development.

The Power of Digital is limited only by our foresight and imagination. We must find the human face of this technology that shapes our cultures. We must search beyond the m/cro-chips of digital multimedia to find the *macro* meaning of a true global village.

Chapter **25**

The Peasants of Global Village; Feudal Age in the Information

Woong K. Park

In traditional times, communication was limited to physical dialogue (ex. face-to-face) among people in the same physical time and space. For centuries, people had little need for long distance communication. For some people in the medieval age, their entire life was spent within radius of no more twenty-five miles from the place of birth (Frederick, 1992).

However, development of communication and transportation technologies allowed people to cross the boundaries of time-and space limitations, and have connected the world together into a wider network. With the advent of telephony, fax machines, digital video disks, and computers, people can communicate with each other practically anywhere in the world and anytime of the day. As McLuhan predicted in the early 1960s, the notion of "global village" has come into being today. The concept of global village is that the electronic communication, technologies suchas telephone would help construct new economic and social geographies and new forms of spatial division and integration (McLuhan, 1964).

However, if global village exists in today's world, who are the villagers? Who are these people and where are they? Based on the discussion of McLuhan and Bagdikian, this study aims to examine and identify the villagers of the global village. Specifically, the study will look into secondary data of electronic communication media such as televisions, radios, video cassette recorders, telephones, and computer networks of various worlds. Thus, despite the possibility of somewhat inaccurate representation of data, it provides a starting point where one can grasp the (non) existence of global villagers.

During the past four decades, the world has undergone considerable transformations; this is especially evident in the areas of transportation and communications. Now, one has easy access to greater places in the world. Of course, this has only been possible because of the developments of new technologies (i.e. computers, modems, etc). Many scholars and others take the notion of global village for granted (see Choi, 1994; Jabbar, 1995).

However, in order to understand the notion of global village, we must understand the concept of "globalization" and "village". Albrow (1994) argued that globalization is the process whereby the population of the world is increasingly bonded into a single society. According to Webster's, village is defined as a settlement usually larger than a hamlet and smaller than a town (Mish et al., 1994, p.1317).

Simply put, the concept of global village is that the large world is being condensed into the scale of small village through the electronic communication technology such as television. As Choi (1994) argued, "rapid advances in telecommunication and computer technologies are making it possible to generate information which can be transmitted instantaneously around the world, resulting in the so-called 'global village'. (P.165) Frederic (1992) termed such notion as more of a "global civil society", and argued that :

Today neither community nor dialogue is restricted to a geographical place. Modem media have expanded our sense of place by reallocating space and time. In the past, personal relationships relied on meeting at cafe, signing a contract together, shaking hands, or interacting in the village square, (p.7).

Thus, it is argued that global village exists in the current world where villagers are able to communicate freely by the use of electronic telecommunication tools. The villagers are able to defy the traditional time-and-space boundary of village. McLuhan's notion was to capture the possibility of global communication and to suggest that the instantaneous reception of far-off sights and voices changed the content of society and culture. He thought that new worldwide communication would transform local life. Thus, global village is born.

To estimate the current whereabouts of global villagers in the world is not so simple as it may appear. The various data differ greatly while others appear not all the reliable. However, an attempt to put together - unfortunately, incomplete - pieces of the puzzle is still a worthwhile undertaking, if only to give us an idea of how important the global village is and of its role in the world.

The Global Villagers

No-one knows the exact world population. However, it is estimated that there are about 5.6 billion people in the world.

In 1990, according to the World Tourism Organization of the United Nations, 429 million people - almost 8 percent of the global population - traveled from one country to another country as tourists (Barnet & Cavanagh, 1995). In 1994 the figure was 531 million people almost 9.6 percent of the global population (World Tourism Organization, 1995). By looking at this figure alone, it can be speculated that about 10 percent of the world's population are global villagers who actually took the time and money to spend time outside their normal residency with or without the aid of electronic communication tools. However, who are these villagers and what are their characteristics?

The characteristics of global villagers are as follows : First, villagers can communicate to others that are outside the vicinity of their physical contact. Thus, a villager in Somalia can communicate to a villager in Norway without having to step out of one's house or town. Such communication can be done in various ways - for instance, a villager can send a hand written letter (considered a traditional communication), or a faster communication can be possible by faxes or electronic mails over the phone lines.

Second, villagers can communicate to others by the use of electronic communication tools. For example, by using mobile cellular phones, villagers can send and receive data practically anywhere in the world where cellular services are available.

Third, villager must share cultures that are within the scope of the village. It is argued that by the process of cultural homogenization around the world, many people will eventually share the same culture. In some vital respects, such as education, life-style, income levels, attitudes, etc., the people on and around Broadway Street, New York and Soweto in Johannesburg can have more common characteristics. These are rudimentary characteristics of global villagers.

Identifying Global Villagers

In order to identify the global villagers, there is need to have certain indicators as yardsticks of identification, and to assess the implication. (Traditional measures of development) Organizations like United Nations (UN) and (UNESCO) used figures such as income per capita, as a standard measure of living standard. Although data on income are valid and useful as economic indicators (such as the purchasing power of individual), these data will not be considered here, Instead, by looking at the availability of rudimentary communication tools, it is argued that we can capture the notion of global village within a certain framework.

To be a villager in the global village, a villager must have connection or access to various electronic communication media and this is vital to the concept of global villager notion. By looking at the availability of television receivers, radio receivers, video cassette recorders, telephones, and computers one can grasp the number of global village. This is the starting point.

Television

The number of television sets in the U.S. has increased about fourteen percent during the period from 1985 to 1994, whereas in China (a developing country), it has increased about three-hundred forty-six percent (International Broadcasting Audience Research, 1995). Yet the developing world's considerable increase was accompanied by rapid population growth. As a result per capita penetration increased at the same rate in most parts of the world.

The disparity in numbers of receivers between developed and developing countries is staggering (See table 1), For instance, Japan alone has higher numbers of television receivers than the rest of East Asia (excluding China). This figure is after including newly industrialized countries like Hong Kong, Singapore, South Korea, and Taiwan. According to 1990 figure, the number of receivers per 1,000 inhabitants was 485 in the developed countries, as compared to only 44 in the developing countries more than 12 to 1 ratio (Frederick, 1992).

Per capita penetration of television is also very uneven. While every two to three people in most developed countries have a television receiver, there are seven persons per set in Latin America, in Asia more than eighteen, and in Africa thirty-six. The disparity is acute in the least developed countries; in twenty-tow underdeveloped countries of the world, there are fewer than 10 television receivers per 1,000 people with eleven countries having only 3 or fewer. Countries like Cape Verde Republic and Eritrea had only one television set. In Africa, about two-third of the countries have fewer than 10 television receivers per 1,000 inhabitants (International Broadcasting Audience Research, 1995).

There are almost 4.3 billion people around the world without television receivers. By the availability of television receivers, only about 1.2 billion people are global villagers. Thus, only about 22% of the world population is in the boundary of global villagers.

Table 1 : Television Receivers Around the World, 1995.

	Population (Total in thousand)	No. of TV Receivers (Total in thousand)
Europe	5,594,000	1,242,000
Subtotal	500,000	253,400
FSU & S.W. Asia	437,000	131,600
Subtotal	437,000	131,600
ArabWorld (Inc. N. Africa)	249,000	38,000
Subtotal	249,000	38,000
Africa		
South Africa	41,000	4,500
Other countries	494,000	19,000
Subtotal	535,000	23,500
South Asia		
India	912,000	50,000
Other countries	282,000	12,800
Subtotal	1,194,000	62,800
Far East		
China	1,200,000	250,000
Japan	126,000	85,000
Other countries	575,000	67,400
Subtotal	1,901,000	402,400
Australia	28,000	10,600
Subtotal	28,000	10.600
North America		
U.S.	259,000	200,000
Canada	28,000	17,500
Caribean	12,000	2,600
Subtotal	299,000	220,100
Latin America	451,000	99,300
Subtotal	451,000	99,300

From "World Radio & Television Receivers 1995", by International Broadcasting Audience Research Library (1995), London.

Radio

According to Frederick (1992), the developing countries have made substantial progress towards increasing the availability of radio receivers, due to low cost of radio receivers and the large number of illiterate people in the developing countries.

About 39 percent of people in the world had access to radio receivers (See Table 2). North America and Europe alone had about 60 percent of the world's radio receivers. In Africa, the number of sets per 1000 is 1 (South Africa). According to UNESCO's very first development communication recommendation, every country should have atleast 50 radio receivers per 1,000 population (New York Times, 1962).

Table 2 : Radio Receivers Around the World, 1995.

	Population (Total in thousand)	No. of TV Receivers (Total in thousand)
World Figures	5,594,000	1,201,000
Europe	500,00	453,900
FSU & S.W. Asia	437,000	101,200
Subtotal	434,000	101,200
ArabWorld (Inc. N. Africa)	249,000	66,700
Subtotal	249,000	66,700
Africa		
South Africa	41,000	1,300
Other countries	494,000	67,600
Subtotal	535,000	80,600
South Asia		
India	912,000	90,000
Other countries	282,000	22,000
Subtotal	1,194,000	112,000
Far East		
China	1,200,000	300,000
Japan	126,000	120,000
Other counties	575,000	121,300
Subtotal	1,901,000	541,300
Australia	28,000	31,100
Subtotal	28,000	31,100
North America		
U.S.	259,000	577,000
Canada	28,000	43,000
Caribean	12,000	6,700
Subtotal	299,000	626,700
Latin America	451,000	187,900
Subtotal	451,000	187,900

From *"World Radio & Television Receivers 1995"*, *by International Broadcasting Audeince Research Library (1995), London: Internationel Broadcasting Audience Research Library, p.3.*

Twenty-five developing countries around the world still did not meet UNESCO's recommendations. Thus, the disparity between developed and developing countries is still huge.

Video

What will be the world be like without video cassette recorders (VCR)? VCR is, truly, the medium through which one can shift time-boundary. The pattern of VCR availability should closely follow the pattern of television receivers and this seems to be the case around the world. Only about 5 percent of the world's population has access to VCRs. About 5.3 billion people are not capable to viewing VCR's at their homes. (See Table 3). According to International Broadcasting Audience Research, Bermuda had the highest ratio of VCRs in the world - it had 43 VCR sets per 100 population, (However, this is due to the small number of inhabitants). Highly developed countries like U.S. and Japan had 31 and 24 VCR sets per 100 population, respectively, African (excluding Arab World) countries lag for behind in the availability of VCR sets.

Telephone

In 1989, the number of telephone lines in the world reached a record number of 405,848,805 (International Telecommunication Union, 1989). U.S. had the most telephone lines in the world. Again, among developing countries in Africa (excluding Arab countries) and Asia (excluding Japan, Hong Kong, South Korea, and Singapore), telephone coverage is very scarce. Although many developing countries are investing heavily in the telephone infrastructure (See Table 4), the disparity between have and have-nots is not narrowed (See Table 5).

Computers and the Internet

The Internet is the ultimate identifier of global village. The Internet is a hybrid form of electronic communication - it uses normal telephones lines and personal computers. Already people who are connected to the Internet can shop for records, cards, flowers, etc. At the touch of their fingertips on one's computer terminal and keyboard. Moreover, via Internet, people can listen to music (as well as radio), make telephone calls and look at movie clips (The Economist, 1960). Thus, the Internet has infinite possible application for global villagers. Also, there are new electronic devices, called "Web TV," where a simple cable - TV device-like module sits on the top of TV sets and one can access the Internet via ordinary TV sets. The Internet

Table 3: Video Cassette Recorders Around the World, 1995.

	Population (Toral in thousand)	No. of TV Receivers (Total in thousand)
World Figures	5,594,000	292,000
Europe	500,00	92,600
Subtotal	500,000	92,600
FSU & S.W. Asia	437,000	6,000
Subtotal	437,000s	6,000
Arbworld (Inc. N. Africa)	249,000	11,800
Subtotal	249,000	11,800
Africa		
South Africa	41,000	1,200
Other countries	494,000	3,500
Subtotal	535,000	4,700
South Asia		
India	912,000	5,000
Other countries	282,000	2,600
Subtotal	1,194,000	7,600
Far East		
China	1,200,000	10,000
Japan	126,000	30,000
Other countries	575,000	18,600
Subtotal	1,901,000	58,600
Australia	28,000	5,00
Subtotal	28,000	5,300
North America		
U.S.	259,000	80,000
Canada	28,000	7,000
Caribean	12,000	400
Subtotal	299,000	87,400
Latin America	451,000	18,100
Subtotal	451,000	18,100

From "World Radio & Television Receivers 1995", by International Broadcasting Audience Research Library (1995), London : International Broadcasting Audience Research Library, p.3.

is the buzzword in the 1990's and there is no doubt that the Internet could carry us to another playing field. According to the Network Wizards (1996a), a company that tries to keep track of Internet statistics, the Internet continues to double in size approximately every 12 to 15 months and there is no way to determine how many users are on the net. One can merely make speculations.

Table 4 : Top ten Phone Line Investment : 1993-2000

Country	Investment	Rank
Turkey	55.10	01
China	55.30	02
South Korea	43.70	03
Russia	23.30	04
India	13.70	05
Brazil	10.20	06
Mexico	09.40	07
Thailand	06.60	08
Malaysia	04.60	09
Poland	04.00	10

Note : *The table is* projected *spending for phone line in selected countries from 1993 until 2000. From "Third World leapfrog", by Engartiio, P., Stier, K., & Gatuzka, P. (1994, June 6). Business Week, p.49. Additional data from : "The race to wire* toe *world". (1994. May 23) U. S. News and Worig Report, p.18.*

Table 5 : Phone Lines Projections, by Country : 2000

Country	Phone Lines	Rank
Maritinique	89.09	01
Sweden	86.01	02
Denmark	74.83	03
Switzerland	73.84	04
South Korea	71.79	05
Canada	69.51	06
Finland	67.61	07
Luxembourg	66.80	08
U.S.	66.36	09
Hong Kong	65.84	10
Germany	65.65	11
Norway	65.04	12
France	63.56	13
French Guinea	62.25	14
Iceland	61.23	15
Malta	60.14	16
Netherland	59.08	17
Reunion	58.07	18
Bulgaria	57.52	19
Britain	57.13	20

Note : The *table is projection of the number of main phone lines per 100 people. Only main phone lines are considered. The forecast uses 1982-1991 population growth rate and assumes installation will meet targets.* **From** *"Call the future", (1993, June 7). Asia Week, p.10*

Again, the trends in the Internet closely follow the patterns of television, radio, VCRs, and telephone lines. The users are clustered around the developed countries (See Table 6 & 7). Due to the dynamic nature of the Internet, the figures always fluctuate but the gap between the developed and developing countries still persists.

Table 6 : The number of Internet by region.

Region	Number of Hosts ('000)	1994-95 increase (in %)
North America	3,372.60	100
Western Europe	1,039.20	88
Pacific	192.40	69
Asia	151.80	87
Eastern Europe	46.10	132
Central and South America	16.00	116
Middle East	13.80	98

From "The Internet. The accidental superhighway," by Anderson. C. (1995, January 7), The Economist, pp.5-10.13-20

Table 7 : Personal Computers in the U.S., Japan, and Europe.

Country	Number	Rank
U.S.	28.10	01
Europe	9.60	02
Japan	7.80	03

Note : *The table is presented in the number of personal computers per 100 people for the U.S., Europe, and Japan. From "The Comparative development of information age technologies in the United States, Japan, and Europe," (1994, January 15). Youth Report, p.5.*

A look at two countries, United States and Thailand, provides a brief comparison between developed and developing countries.

The Case of the U.S.

According to U.S. Central Intelligence Agency (The World Fact Book, 1995), U.S. had population of 263,814,032 people (July 1995 estimate). It had 126,000,000 telephones, 530,000,000 radio receivers. 193,000,000 television sets, 80,000.000 video cassette recorders and 432,727 Internet hosts (Network Wizard, 1996b).

About 48 percent, 200 percent, 73 percent, 0.1 percent of U.S. population had access to telephones, radios, televisions, VCRs and Internet, respectively. The available number of electronic communication tools is 452, 432,727 (According to International Telecommunication Union World Telecommunications Indicator

Database, the U.S. ranks number one in the world (See Table 8), in terms of multimedia indicators - which includes the number of telephone, TV, and personal computer density (per 100 inhabitants). The people in the U.S. have about two electronic communication tools each that will connect to the outside world. In 1993, 45,779,000 (17 percent of the population) American people actually took a trip abroad (World Tourism Organization, 1995). This can be translated that 83 percent of U.S. population stayed at home. The epitome of the global village can be found also in the case of Internet; only 0.1 percent of Americans are Internet users, thus 99.9 are left out.

A Case of Thailand

According to U.S. Central Intelligence Agency (The World Fact Book, 1995), Thailand had a population of 60,271,300 people (July 1995 estimate). It had 739,500 telephones, 11,000,000 radio receivers, 9,000,000 television sets, 1,500,000 video cassette recorders and 6,362 Internet hosts (Network Wizard, 1996b).

About 1.2 %, 18 %, 15 %, 0.02 %, 0.01 % of Thailand's population had access to telephones, radios, televisions, VCRs, and Internet, respectively. The available number of electronic communication tools is 22, 245,862, or an average of about 0.4 per population (According to International Telecommunication Union World Telecommunications Indicator Database, Thailand ranks number thirty-five in the world (See Table 8), in terms of multimedia indicators, which includes the number of telephone, TV, and personal computer density (per 100 inhabitants). Thailand has fewer than one electronic communication tools per person that will connect to the outside world.

In 1993 5,761,000 Thai people took a trip abroad, which is about 10% of the population (World Tourism Organization, 1995).

This can be translated that 90 percent of the Thai population stayed at home. The epitome of the global village can be found in the case of Internet; by looking at this figure alone (0.01 percent), we can grasp the notion of the global village and how the rest of the population (which converts to 99.99 percent) is left out.

Thus, when we look at the actual number of people who travel and who do not, there is still a staggering gap and this pattern also shows in the availability of communication tools. Although the cases between U.S. and Thailand look incredulous, they are indeed good examples of disparity of have and have-nots in terms of communication tools.

Table 8 : Multimedia Indicators.

Country	Tel. Density	TV Density	PC Density	Rank
U.S.	59.50	79.00	26.70	1.00
Denmark	60.40	55.00	19.30	2.00
Sweden	68.30	48.00	17.20	4.00
Australia	49.60	48.00	21.70	5.00
France	54.70	58.00	14.00	5.00
Switzerland	59.70	41.00	28.80	5.00
Netherlands	50.90	48.00	15.60	8.00
Germany	48.30	55.00	14.40	9.00
Japan	47.80	64.00	12.00	10.00
U.K.	48.90	45.00	10.70	12.00
Austria	46.50	48.00	10.70	12.00
Belgium	44.90	47.00	12.90	12.00
Singapore	47.30	38.00	15.30	12.00
Hong Kong	54.00	36.00	11.30	15.00
Italy	42.90	45.00	7.20	16.00
Spain	37.10	42.00	7.00	17.00
Korea	39.70	32.00	11.20	18.00
Taiwan	40.00	32.00	8.10	19.00
Hungary	17.00	42.00	3.40	20.00
Czech Rep.	20.90	39.00	3.60	21.00
Israel	39.40	30.00	9.40	21.00
Greece	47.80	22.00	2.90	23.00
Portugal	35.00	25.00	5.00	24.00
Argentina	14.10	38.00	1.70	25.00
Poland	13.10	30.00	2.20	26.00
Russia	16.20	38.00	1.00	26.00
Malaysia	14.60	23.00	3.30	28.00
Chile	11.00	23.00	3.10	29.00
Turkey	20.10	27.00	1.10	29.00
Mexico	9.20	20.00	2.20	31.00

Note : Main telephone lines, television lines sets and personal computers per 100 inhabitants, major countries, 1994, 1994. From "ITU World Telecommunication Indicator Database", (On-line),

The Non-Global Villagers

McLuhan's global village has not been realized in some parts of the world. First, there are more than 190 nation-states with several thousand of cultures and languages. Thus, there's no single McWorld (Barber, 1995) that can fit into a single country or village.

Second, the information infrastructure is vastly different in the countries around the world. Thus, a person in New York can log into a computer and connect to the world of Internet via local telephone and Internet providers. However, a person in Helsinki may not be able to do that even if one has a computer and Internet access, due to higher cost of making connection to the Internet.

Third, the developing countries do not have enough resources to put massive communication infrastructure in their countries; at least 80% of the world's population still lack most basic telecommunication (Panos, 1995). While the most advanced countries are awash in new communication technologies, the poorest nations have barely wet their feet. LeBlanc estimated (1981) that other countries will have to invest $50 billion a year over the next 30 years just to bring their communication systems up to present standards in the U.S.

The convergence of jet travel and electronic media have brought the world closer than ever before and currently, the technology can connect everyone in the globe. However, in cities around the world, disparities between neighbors have become sharper. Indeed, certain parts of cities located in continents far apart have nevertheless more in common with each other than they do with neighbors just a few blocks away in the same city. Is this what we call "global village"?

Also, though our world may be as instantly connectable and instantly covered through radio and TV, our world is immensely diverse and incredibly heterogeneous.

On the surface, it would appear that McLuhan's global village is with us, with all these communication technologies available. However, Me Luhan's notion of global village is based on an idealization of how technology can achieve its greatest feats, not the reality of the use of technology in the world.

Bagdikian (1989) said that few large corporate giants will control the most of the world's media and these "lords" of the global village will exert a great influence over people. It is indeed true that these giant multi-national corporations are having evermore power over people, due to technological changes. However, the changes and influence have not benefited everyone. Only a handful of these "lords" will benefit and rest of the "peasants" will suffer. It is an irony that history repeats itself in the age of so-called "information super-highway".

Table 9

Country	Tel. Density	TV Density	PC Density	Rank
Brazil	7.40	29.00	0.90	32.00
Venezuela	10.90	18.00	1.30	33.00
South Africa	9.50	10.00	2.20	34.00
Thailand	4.70	19.00	1.20	35.00
China	2.30	23.00	0.20	36.00
Philippines	1.70	12.00	0.30	37.00
Indonesia	1.30	9.00	0.30	38.00
India	1.10	5.00	0.10	39.00

REFERENCES

Anderson, C. (1995, January 7) The Internet. The accidental super highway. The Economist pp.5-10, 13-20.

Albrow, M. (1994). Globalization. In W. Outhwaite & T. Bottomore (Eds.), The Blackwell dictionary of Twentieth-Century social though (pp.248-249). Cambridge, MA; Black Publishers Inc.

Barber, B.R. (1995). Jihad vs. McWorld. New York:Ballantine Books.

Bamet, R.J., & Cavanagh. J. (1995). Global Dreams: Imerrial corporations and the new world order. New York: Touchstone Book.

Bagdikian. B. (1989, June 12) The Nation, p. 805. Calling the future. (1993, July 7). Asia Week, p.10

Choi, Y. (1994) The effect of social and physical distance on the global communication networks. Gazette, 54, 165-192

Engardio, P., Stier, K., & Galuzka, P. (1994, June 6). Third world leapfrog. Business Week, p.49

Frederick, H.H. (1992). Computer communication in cross-border coalition-building: North America NGO networking against NAFTA. Gazette, 50, 217-241.

International Broadcasting Audience Research (1995). World radio and television receivers. London: International Broadcasting Audience Research Library.

International Telecommunication Union (1989). Yearbook of Common Carrier Telecommunication Statistics, 16th ed. Geneva: International Telecommunication Union.

International Telecommunication Union (1994). Multimedia indicators (On-line). Available : http://html/fpd/technet/multi.htm.

Jabar, J. (1995) The global city, not the global village: The world as a unitary urban supermarket. 'Inter-media, 23(6), 44-45.

LeBlanc, R. (1981, February 1). Changes and opportunities in telecommunications. Telecommunications Industry Monthly, 116.

Network Wizards (1996a). Homepage (On-line). Available Internet: http://www.nw.com/zone/WWW/dist/bynum.html.

Network Wizards (1996b). Homepage. (On-line). Available Internet: http://www.nw.com/2one/WWW/notes.html.

Me Luhan, M. (1964). "Introduction." In HA Innis, The bias of communication. Toronto: University of Toronto Press.

Misch, F.C. et al. (Eds.). (1994) Meerriam Webster's Colegiate Dictionary (10th ed.) Springfield, MA ; Merriam-Webster Inc.

More news media urged by UNCESCO (April 10, 1962). New York Times, p.4.

Panes (1995). The Internet and the South: Superhighway or dirt-track? (On-line) Available http;//oneworld.org/panos_jnternet_ press.html#controls.

The comparative development of information age technologies in the United States, Japan, and Europe. (1994, January 15). Youth Record, p.5.

The economy of the Internet (1996, October 19). The Economics, p. 23-27.

The race to write the world (1994, May 23). U.S. News and World Report, p. 18.

United States Central Intelligence Agency (1995) The World Fact Book, 1995. (On-line), Available http://www.dci.gov/cia/ pubications/95fact/us. html.

World Tourism Organization (1995). Yearbook of Tourism Statistics, 1 (47), p.2.

Woong K. Park received a B. A. in Communication from University of Michigan, Ann Arbor (1991), and a M. A. in Advertising and Public Relations from Sogang University, Korea (1994). He has worked both in advertising and public relations agency.

Chapter **26**

Critical Premises Concerning Guidelines For Successful Information Campaigns in Developing Democracies

D. A. Newsom and B. J. Carrell

Information campaigns are one of the primary tools used in nation-building by the leadership of developing democracies. Some information campaigns are successful while others fall far below expectations. The purpose of this article is to examine the role of information campaign in developing democracies and to offer critical evaluations of guidelines that may improve their success.

A premise of this article is that in developing democracies information campaigns need not necessarily conform to Western models common in the early development literature as exemplified by such writers as Rostow (14). In fact, the structure of information campaigns is a political. But as they are used in developing democracies, information campaigns may take on political dimensions when they are ignores information campaigns that are purely propagandistic or which attempt to gloss over bureaucratic apathy or ineptness. Instead, the focus is on genuine information campaigns directed by a country's leadership to its populace to get action or results that are in the best interests of that populace, e.g., information about Acquired Immune Deficiency Syndrome (AIDS).

Even, so for information campaigns to be successful, they must :

1. Have clear objectives, distinct and measurable goals.
2. Convey messages that are both relevant and desirable by receivers.
3. Offer sufficient guidance on which receivers can take action.
4. Explain rewards or benefits as incentives to receivers to take proposed actions.

5. Reinforce of messages conveyed and action taken.

Premises of successful information campaigns :

These guidelines, found in such sources as Quera (11), Newsom, Scott and Turk (8) and Rice and Paisley (12), are some simple that some key premises in support of them are often overlooked or ignored, thus setting the stage for information campaign failures.

Influence of the individual :

The same message may be sent simultaneously to millions of people but interpretations and responses to it are always individual and limited by the personal circumstances of receivers. That simple fact alone often confounds planners as they assess the effectiveness of information campaigns. This was apparent, for example, at a family planning conference (7) in New Delhi when conferees seemed to focus more on messages than on examining the reasons behind individual responses to these messages.

A basic premise in democratic system is that individuals are capable of thinking for themselves and drawing their own conclusions. If women in Indian villages are not literate, they can't read the hoardings (posters) that proclaim family planning messages. If they have transistor radios, or access to them, messages then may get through only if they are in languages spoken in those villages. And television is available to them only if they live in villages that have electricity. Even when messages get through, women may reject the contents because experience has taught them that more children, not fewer, to work family plots and tend the animals offer some measure of future success. They also know that the death rate in high among new-born.

This simply means that even if the barriers to delivery of messages can be overcome some social engineering must preceded messages if the messages are expected to bring about positive changes in behaviour. It must be possible for these women, as well as their spouses, to perceive that it is to their benefit to have smaller families before they will practice family planning. Additionally, they must have reasonable access to medical care. Because responses to messages are always individual, social engineering must prepare audiences properly to receive messages. Otherwise, messages are likely to be ignored altogether or rejected because of people's inability or unwillingness to do as the messages suggest. Messages are never adequate substitutes for good social policies.

Because messages are received, interpreted and acted upon individually, the emotional environment in which personal decisions are made directly influences the success of information campaigns. This environment is made up of myths, religion, traditions and social mores. For example, information campaigns generally are destined to fail if they convey messages that contradict a person's religious beliefs. Even when people receive and understand messages as they are intended by senders, most people are unwilling emotionally to shed the influence of social mores, religious conventions, family traditions and the mutual support derived from peer groups.

Individual responses must also be looked at with respect to the Maslow's hierarchy of need (5). People who spend all or most of their time and energy at the lowest levels are not conditions for information campaigns. They are too busy with daily survival and safety. When these are met, people then permit themselves to consider information campaign messages that talk about acceptance, self-esteem and self-actualization needs. This means that governments must provide adequate services and programmes before they can expect the desired responses to the advocacy implicit in information campaigns.

Enlightenment process as persuasion :

Implicit in the democratic process is the idea that government should be responsive to the governed. Authoritarian vestiges sometimes remain, however, to confound true responsiveness. Historically, this is because it is easy for those in power to rationalize that they know best what is the proper response. The temptation is to use force to compel compliance. It is easy to forget that persuasion is the only really effective tool governors can use in a developing democracy to bring about planned change. Even then, planned change occurs not because it is on governors' agendas but rather because it is salient to the governed. Hence, the effectiveness of any information campaign is tied to the degree to which governors observe the know protocols of persuasion.

Presentation and delivery of messages is the first step in the persuasion process. It is there that planners are often be guiled by the massive number of people to whom messages may be sent. Because the numbers may be so large, this may lead planners to believe that each intended recipient of messages is a genuine candidate for the messages, just because individuals fall into a category described as a target audience does not necessarily mean these individuals can be converted by messages. Because of this greater care must go into defining and describing target audiences. Care also must be taken to identify unique linguistic and cultural behaviors which must be observed when constructing messages and selecting the means by which they are delivered. For example, a message in Hindi directed at many residents of Tamil Nadu, India, will not be understood. In fact, if the message is not in Tamil, resentment toward the message and is senders may occur. Because of literacy and technological problems, it is also a fact that many people can't be reached through traditional mass media. Hence, the use of non-traditional media, such as village or tribal leaders, may be a more effective way of presenting and delivering messages to many audiences.

Mere delivery of messages does not constitute communication. It is only when audiences pay attention that the communication cycle is actually completed. Because these messages, for attention, it is necessary to repeat messages often and in different ways so that the largest percentage of a target audience may be reached. Even if individuals attend message, they may no understand the content of the messages. They may lack enough education, background information or experience with what is being conveyed. Because individual receivers, not senders, supply the meaning to messages, it is especially critical that messages be simplified as much as possible. Even if messages are understood intellectually by receivers, messages may be rejected, and often are, if they present information that does no reinforce their predispositions. Messages must be congruent with audience and individual beliefs and values structures.

Otherwise, the level of acceptance is predictably low. That means that the persuasive intent to convert is lost on most of the audience. Waste of resources is high because messages that threaten strongly-held beliefs don't produce much yielding. Classic research (3) indicates that yielding to incongruent messages is infrequent and will occur only under very limited conditions.

Even when individuals yield to messages, they may not retain their new positions long enough to act on them. Therefore, every information campaign must be concerned with reinforcing attitudes and behaviours. Since these new ideas and behaviours represent change, they may be modified or cancelled by counter messages. Especially from family or peer group leaders. Hence, messages must be reinforced often and regularly.

Governors look to the number of converts to the proposed actions as the measure of the success of persuasive information campaigns. But 'counting tha house', so to speak, assumes that there is a direct cause and effect relationship between message and behaviour. It further assumes the absence of intervening variables. The fact is that well-designed information campaigns may be successful in leading some members of audiences to the retaining stage of the persuasion process but no information campaign can overcome physical and cultural barriers to action. Women may be convinced that using sterilized water to mix infant formulas is best but the effort may be too time-consuming and hard to gather enough firewood to boil water for a minimum of 10 minutes. Even when people know that condoms are the safest protection against AIDS, condoms may not be available to them, or at a price they can afford. Even it women want fewer children, their partners may deny the use of birth control because of cultural or religious norms.

Campaigns are often directed to people who are least able to respond to actions proposed in messages. This fact alone accounts for many of the failures of information campaigns, if members of audiences are not able to respond as suggested, they should be excluded from the focus of the campaign. Problems such as literacy, economics or technology must be solved before directing information campaigns to people.

Governments often stop campaigns altogether or change them so quickly that cumulative effects are often dimished. When people adopt new behaviours they may revert later to former behaviors, hence, even when purposes seem to be accomplished, information campaigns must be continued as reinforcement necessary to sustain new behaviours.

Although information campaigns may be successful when asking people to make small sacrifices, campaign planners must remember that the behaviour of individuals are driven more by personal than social goals. Therefore, ways must be found to meld these in ways so that individuals can fulfill social aims as they seek personal goals. This task often calls for a high level of creative thinking. Since personal goals are dominant, one of the keys to successful campaigns is to make rewards or incentives, real or perceived, a central part of every campaign. The governed respond to information campaigns for their own reasons, not those of the governors. More often than not, personal rewards or incentives are pivotal when people consider making changes is their behavior. These rewards and incentives must be significant enough to people before they will spend to energy or resources to adopt new behaviour or to take the risk involved implicit in adoption.

Information flow :

It is usually the case that messages in mass media contribute more to indirect than direct action. This is because of the crucial roles played by opinion leaders (4) and inter-personal communication processes (13, 1) in developing democracies. Membersof the audience look to village, tribal and other leaders for cues as to what should be proper responses to messages. Thus, the campaign structure must include appropriate emphasis on local leadership and reliance on interpersonal communication processes (10).

In effect, all campaign plans must take into account how to recruit and convert innovators and early adopters (13) to what is being advocated. Innovators and early adopters become disciples and proselytizers for those programmes. It is well documented, of course, that face-to-face spoken communication is more persuasive than mediated communication (6). In fact, the less developed is democracy the more attention must be given to this part of campaign planning.

Delivery systems :

The means by which messages are delivered to target audiences may contribute significantly to the success of campaigns. One key rule about selecting delivery systems is that the media chosen must be capable to reaching intended audiences India's family planning campaign relies heavily on public service announcements on radio, television and posters. Since 25 percent of the villages do not have electrical service, messages via television cannot reach a significant percentage of the target audience. This is further excerbated by the facts that only 1.6 percent of all households have television receivers and radios are present in only 17.2 percent of the households (2). And since the literacy rate remains at 25 percent, posters cannot be read by 75 percent of the Indian population.

Credibility is as improvement as reach. If messages are not believed, their physical delivery is a hollow accomplishment. Although messages in mass media may be entirely true, they carry less credibility with individuals than do messages delivered by local opinion leaders. If messages contradict first-hand experience or seem to violate social norms, their credibility may suffer enormously when delivered via mass media. And if messages are complex, they are more likely to be understood via personal media.

The media used must be capable of adequate conveying messages. Some messages can't be carried by some media. This is especially true of complex messages. Personal and print media must get priority use if message are complex or if the behavior advocated defies prior experience. If an idea must be demonstrated, then television and film should get priority use. If simple awareness is all that is sought, then radio and posters are generally the most efficient if the idea can be reduced to a simple message statement.

A mix of media should be used because learning is enhanced when the same message is delivered via multiple means. Rarely can the purposes of campaigns be realized with the use of just one media form. A mix of media, based on primary and secondary values, is almost always in order. Even when it is clear that one kind of medium can carry the weight of campaign, it is generally wise to include a mix of media.

Campaigns must be tightly structured and controlled but they must be flexible enough to actively use small group and interpersonal communication processes to the maximum degree possible.

Information systems in organization or societies are neither completely open nor are they totally closed. Most systems function somewhere between the two, but generally closer to one than the other. The direction toward open or closed is set by the leadership. If the leadership changes, this usually means a change in direction which affects the entire system, even an entire nation. A most dramatic example of this is the Chinese student demonstrations for democracy in Tienenamin Square following a visit by Soviet President and Communist Party Chief Mikail Gorbachev in 1989. In developing countries, the tendency is to be closed until a system is fairly stable, then the leadership may move toward more openness. However, events may cause a leadership crisis which can shut down and close the system again. In developing democracies this is detrimental to the free flow of information needed for full participation by the governed.

Information campaigns in developing democracies usually are conducted in an environment of considerable stress Vestiges of political or social authoritarianism remain to confound budding democratic processes. A primary source of stress is an unwillingness to accept the idea that all members of society have the ability to reason and to make choices. In developing democracies, leaders sometimes don't fully trust the masses to make good choices. Governors generally are more comfortable personally with a system that is more closed than open because they equate the lack of education with stupidity or incapacity.

Information in open systems moves freely between the governed and governors. Implicit in this exchange is the fact that governors are expected to respond to the demands of the governed. That is, governors are expected to find out what is important to the governed and help them achieve it. Thus, the role of information campaigns is mostly open systems is that of educating, advising and persuading the populace on how it can achieve what it wants. If enabling programmes and services are needed to support these goals, governors are expected to create and implement them.

Information campaigns may also play a key role in accountability of governors to the governed. It is enough the free flow of information that the populace can check on how well leaders are fulfilling their responsibilities. Of course, both the governors and the governed have access to the means of communication to tell their points of view.

Making individual choices is another aspect of openness. It is presumed that every citizen has the right to self-determination. Even when their choices appear to be bad individually or collectively, as they often will be, they nevertheless are treated as sacrosanct.

Information in closed systems is shared generally on a 'need to know basis'. There may be veritable flood of information flowing from the leadership to the populace, but often it is propagandistic in nature, designed to put the best possible face on governmental actions. Public accountability is usually limited only to what government wants others to know.

One defense of a closed system is that the general population is poorly educated and uninformed. Even if they would be willing participants, society's goals will be realized quicker and with less trauma if decisions are made by those presumed to be more capable. Another defense is that the nature of the work of government is so complicated that it is not only beyond the understanding of the general public but it is also quite sensitive. Risks are simply too great to entrust such comlex and sensitive information to people unprepared to handle it. The role of information campaigns in mostly closed systems is primarily doctrinaire, not dialectical.

Reactions to campaigns :

People react individually and collectively to information campaigns. The first and primary reaction is always individual. It is only when enough individuals react generally in the same way, even if for different reasons, that collective reactions are evident.

Personal reactions to messages are subject to many influences but some of most significant are personality traits, level of educational attainment, life experiences, family friends, and religion and extent period of economic and political development. Each of these factors also may influence collective reactions, but they are especially important to personal behavior.

Individual reactions always involve personal ego. Personal egos are threatened by messages that create cognitive dissonance, challenge personal, group or social conventions or that call for too much change in behavior. In such cases, processes of selective perception, selective retention, selective forgetting and rationalization are used to blank out messages or to change their original meanings to less threatening forms. It is only when messages are clearly salient and offer sufficient rewards or incentives that they may overcome these personal barriers to successful communication.

When people have the same basic reaction to messages, even if their reasons differ significantly, collective opinion begins to emerge in support of or opposition to the actions proposed in information campaigns. This is when planners can begin to gauge the overall effectiveness of their campaigns. It may be that only a small percentage of the target audience accepts what is proposed, but if it is an active group, the impression may be one of much greater success, especially if those who have not accepted the proposed action are no more than mildly opposed to it. A small but active group can create enough attention so that more messages in support of the proposed (9). But it is clear that reactions to information campaigns, individually and collectively, are based more on factor than messages themselves.

Message perception :

Many factors affect the way people perceive messages, but in information campaigns of sources of information and the usefulness of information are especially critical.

It the source truthful, credible and authoritative? If there is a history of untruthfulness,new messages from a source will be viewed naturally with greater skepticism. Trust in a source by an audience is an extraordinarily valuable asset when asking people to change their behaviors. Without trust the ability to bring about change is severely limited in any democratic system. Does the source know what it is talking about? Does the source have adequate expertise to advocate this action with authority? Expertise is recognizable even by non-experts. Even if the source is highly authoritative, planners must be careful not to project arrogance in their messages.

Messages must convey a sense of usefulness to receivers. Unless receivers can perceive some utility in messages, often a reward or incentive that improve their own well-being they are likely not to respond to messages even if they perceive them as truthful and authoritative. Messages must demonstrate and prove, to the extent possible, the kinds of personal fulfillments that can be achieved if proposed behaviors are adopted.

Majority-minority dynamics :

The intent behind every information campaign is to reach a point where the majority in an audience embraces or supports issues being advocated. As this applies especially to public policies, there is a tendency to rely on the concept of 'majority role' and freeze out dissenting points of view. The minority voices of dissenters may also be ignored. This may forment problems because policies in developing democracies are so fragile, especially during early periods of development. Exalting majority views also may be premature as tomorrow the minority view may be the majority. This is because changes in leadership or events may inspire new personal and collective attitudes. Change dynamics are the only constant in developing democracies.

What information campaigns can't do :

Information campaigns can't solve social, economic or political problems. They are never adequate substitutes for needed programmes and services. Information campaigns can only communicate ideas that may inspire shifts in attitudes which may lead to some changes in behavior, although the volume of change attribute to information is rarely significant.

In the absence of personal rewards and incentives, information campaigns, alone can't sustain new behaviors over long periods of time. Even reinforcement fails without personal realization of fulfillment that come from rewards and incentives.

Information campaigns are incapable of getting people to do things that are unrewarding to them personally. Even if do things of society share a common goal of nation building, individual behaviors continue to be driven more by personal than social goals.

Information campaigns are capable only of communicating with a limited percentage, sometimes a very small percentage, of the members of target audiences. It is not realistic to try to communicate with 100 percent of target audiences, even when membership is small and carefully defined. In developing democracies where literacy, technology and socio-economic problems represent formidable barriers, the percentage that can be reached may be dramatically small. And only a small number of those who are reached may be swayed by messages.

Conclusions :

Campaigns obviously are launched for some designated results. The fact that so many campaigns have such limited results. If any at all in terms of retained behavior, is largely due to asking people 1) to do something they can't do (easily, if at all) or 2) that makes no sense to them to do in terms of their self-interest. Additionally, the messages are often in media that either don't reach the intended audience or have no credibility with that audience. Many information campaigns seem to be launched for ago enhancement rather than behaviour since the principles for reaching audiences are often ignored. The other major reason for failure is simply that messages often don't reinforce that part of that audience's life.

References :

Gouran, J. S., (1969) 'Variables related to consensus in group discussions of questions of policy.' Speech Monographs 36(3), August, pp. 387-391.

'India' (1989). 1989 Britannica Book of the Year. Chicago : Encyclopaedia Britannica, Inc., pp. 620-621.

Klapper, J. T., (1960). The Effects of Mass Communication, New York, NY : Free Press.

Ibzarsfeld, P. F. et. al. (1994). The People's Choice. New York, NY : Duell, Sloan and Pearce.

Maslow, A. H., (1954). Motivation and Personality. New York, NY : Harper & Row.

McGuire, W. J., (1969). 'Nature of Attitudes and Attitude Change.' In Gardener Lindzey and Elliot Aronson (Eds.), Handbook of Social Psychology (p. 225). Reading, MA : Addison-Wesley.

National Workshop on Family. Planning Programme in India (December 6, 1988). New Delhi, India, Claridge Hotel.

Newsom, D. A., Scott, A. and Turk, J.V.S., (1989). This is PR : The Realities of Public Relations (4th ed.). Belmont, CA : Wadsworth.

Noelle-Neumann, E., (1984). Spiral of Silence, Chicago, IL : University of Chicago Press.

Pratt, C. B. and Manheim, J.B., (1988) 'Communication Research and Development Policy : Agenda Dynamics in an African Setting.' Journal of Communication 38(3), pp. 75-95.

Quera, L., (1973). Advertising Campaigns : Formulation and Tactics. Columbus OH : Grid.

Rice, R. E. and Paisley, W. E. (Eds.)., (1981). Public Communication Campaigns. Beverly Hills, CA : Sage.

Rogers, E. M., (1962). Diffusion of Innovations, New York, NY : Free Press of Glencoe.

Rostow, W. W., (1960). The Stage of Economic Growth : A Non-Communist Manifesto. Cambridge, MA : Cambridge University Press.

Chapter **27**

Why Development Communication Fails? Some Methodological Thoughts

J. S. Yadava

Increased communication facilities are necessary but not sufficient condition for development. Individual is important but 'social structure' of which he is a part is equally important lot his innovative behaviour and participation in development processes. The source, message, channel and destination (SMCD) the usual paradigm for research inquiry in communication are useful but need critical re-examination in context of development communication. The social structure, though taken note of, need greater attention in development communication studies. The concept **of Sadharanikaran** helps in better understanding of communication processes andalso in more effective development communication.

Any agreement on the concept or model of development is not easy but it would be worth while to view it as a process which facilitates and results in participation and advancement, both material and social, of the widest possible number in a given society. The contribution of communication in development process is well accepted though its precise nature, extent and mechanism still largely elude proper understanding. We all work under the faith, to the extent of becoming a fetish that increased communication will lead to development. And so efforts are made to expand the infrastructures for communications.

In India media facilities have expanded enormously over the years especially since independence. Technically the radio signals cover almost the whole country and television 'reaches' nearly 70 per cent of the area. There are about 20 thousand newspapers of these 1,334 are dailies. The combined circulation of Indian Press crossed the 50 million copies mark in 1980. But owing to mainly four mutually reinforcing factors the reach of mass media in India on the whole is rather limited, especially in rural areas, among women and slum dwellers in cities. These factors are (a) low literacy, about 36 per cent; (b) low purchasing power, 40 to 60 per cent are living below poverty line; (c) poor means of transportation for timely delivery of newspapers, or maintenance of radio-TV sets; and (d) lack of relevant information if purposive communication is the aim of media policy. As a consequence the mass media besides being urban based are largely the class media.

Nearly 93 per cent of the total sale of daily newspapers is in large towns with population of over one lakh which account for only 10 per cent of India's population. The radio is by far the most extensive network. Even here, on an average there are only 4.4 radio/transistor sets per 100 persons. Not only this, of nearly 30 million radio sets in the country 80 per cent are in the urban homes leaving only about 6 million sets with about 525 million rural population. The access to television to rural people is even more scarce despite the 70 per cent reach of TV signals.

Even, at the risk of over simplification. It may be stated that in contents the urban stamp is obvious to merit further dialation. Consumerism and individualism are the dominant value thrusts in our mass media fare. And entertainment is the major gratification the people seek from the radio, TV, and film media. Even newspapers beside giving hard news alongwith radio and TV news, generally serve entertainment functions as would be obvious from the popularity of glossy film and gossip magazines over the serious informative ones.

Be that as it may India has made substantial progress since Independence. The development and communication infrastructures have been enormously expanded. The result and communication and development processes have been strengthening and supporting each other taking the country forward. At the same time it is also a fact on which not many people would disagree, that both development and communication processes have not been equitable and participatory enough resulting in increase of disparities. It is a challenge to development communication scholars to study this question in detail and depth and suggest approaches and methods for more equitable and participatory communication and development.

II Rostowain economic theories and laissez-faire concept of market economy influenced development and communication thinking in most developing countries including India. "Trickle down effect" was central to this model of development. Daniel Lerner's Passing of the Traditional Society, emphasized the relationship between communication, urbanization and modernization which led to the thinking that the greater are the communication facilities the greater or even faster is modernization. Perhaps getting inspiration from such a thinking UNESCO in sixties laid down certain minimum norms for media structures and the countries not fulfilling the norms were considered as underdeveloped and deficient in communication, thereby emphasizing the need for expansion of media facilities.

"Diffusion of innovations" studies tried to investigate systematically the trickle down mechanism and processes. The research studies in development communication were strongly influenced by these two related strains of thoughts or theories. The questions of communication facilities and the processes of transferring information, ideas and/or innovations engaged the attention of many scholars concerned with development of traditional societies. In the first infrastructures for communication and in the second the receivers were main focus of research inquiries. The individual's traits in terms of his being 'cosmopolitan' vs. 'traditional' etc. were the leads for understanding his 'innovations' and adoption behaviour. These studies made valuable contribution to development communication theory research and practices.

However, in late sixties and seventies 'holes' became visible in such a development communication approach. Development on the ground was not fast enough and certainly not with social justice. In India despite spectacular achievements the development strategies resulted in increasing disparities making even green revolution look red for those who could not take advantage due to lack of adequate resources of land, money, and irrigation facilities etc. The Reserve Bank credit committee warned in 1968 of the impending rise in rural tensions that may lead to violence if some measure of help was not given to those who had been by past by the development strategies which made green revolution possible in-certain areas of the country.

People started questioning and looking for alternatives or improvements in strategies for development. Integrated Rural Development and other poverty alleviation programmes are result of this re-thinking about development concepts and models. As opposed to trickle down of benefits through usual development processes the new IRDP and poverty alleviation programmes are in a way 'bottom up' approach to development. Unfortunately there is no matching 'bottom up' approach in communication with the results a substantial proportion of resources for helping those who are below the poverty line are being misappropriated by others.

III The 'failure' of development communication to achieve desired/expected level of social change or 'development' is a complex phenomenon amenable to simple explanations. However, alongwith many other reasons two postulates can be identified which in combination appear responsible for failures of development communication.

1. Coherence of Communication & Social Structure

The 'social structure' has been a favourite subject of study for sociologists and anthropologists for decades. First with the industrial revolution in the West and later with post World War II developments in Third World countries the focus naturally has been on investigating 'social change'. Different theories have been propounded over the years to analyses and explain the phenomenon and the process of social change. It is not necessary to go into this subject here, nor I'm offering a critique of the theories of social change. For our purpose, the coherence of the social structure and interdependence of its elements, a recurring theme, of course with varying emphasized, in study of social change, merit some consideration.

To put in simple them the coherence of social structure suggest resistance to change and interdependence of its elements suggests change in one will lead to changes in other elements as well. By implication, as supported by imperical evidence also, 'modernization' of 'traditional' societies is not easy because of close interrelationship which exists between technical, economic, cultural, religious and other social aspects. Even so, social changes takes place, which can be viewed as the aggregate effect of individual adaptations to some 'new environment'. Communication plays significant both in individual adoption, their aggregation as well as the creation of new environment itself.

But communication as we all know well is the essence of a social system both in its continuityand change. As such communication like social structure show similar kind, if not of the same degree, of coherence and interdependence of its elements. However, communication certainly is more amenable to human manipulations than the social structure. And development communication can be viewed as all such communication decisionsactions and inputs which facilitate the process of social change in a desired and predetermined direction in specified areas or sections of a given society.

2. Relationship rather than elements of communication need emphasized for effective development communication.

The second postulate in way emerges from the first. Coherence of social structure and communication demands attention on relationship between elements of communication in such a way that the social structure remain 'visible' while analysing dimensions of communications. The concept of Sadharanikaran for communication promises to be a rewarding approach both in analysis and practice of communication especially in context of development communication. Being a new concept let me explain it in some detail.

Its usage can be traced to Natyashastra as it has been used for communication philosophy expounded in this treatise on Indian Natya (drama) and (Nritya), (dance). The Natyashastra is often referred to as the fifth Veda it simplified complex ideas and philosophies of the four Vedas for the benefit of the common people. Bharat Muni, who is credited with writing of Natyashastra codified the principles of human expression. Its precise date is not known. Some put it to 2nd Century B.C. while others attribute it to the 1st Century A.D. Some even put this as 4th Century A.D. Besides giving practical description of various aspects of dance and drama to the minute details, the document is rich about the basics of human communication.

Later this word was extensively used for explaining the aesthetic aspects of poetry in literary circles. Today the term Sadharanikaran is often employed to convey the idea of simplification. But Sadharanikaran as a concept for communication promises some new departures from the prevailing thinking which serves as helpful theoretical position leading to meaningful questions for research.

Tenents of Sadharanikaran

Let me first discuss briefly the main tenets of Sadharanikaran as a concept in communication. The following five tenets of Sadharanikaran are especially relevant for explaining and understanding of development communication in Indian Society.

a) Being Sahrldayas

The significant postulate in Sadharanikaran is that the parties involved in communication should be Sahridayas that is having common sympathetic heart.

It is perhaps a poetic expression used for being or having common orientation. **Sahridayas** is not coter-minious with psychological predispositions. It is much more than personality characteristics. It means conscious effort on the part of the 'communicator' to identify with the 'receiver' of his 'message'. Gandhiji identified himself with the masses, becoming **Sahridaya** with masses, thereby succeeded in having effective communication with them. He dressed and lived like a common man. Wearing the sary typical of the region/place on the part of Late Prime Minister Mrs. Indira Gandhi was an effort towards identification with the people of that region.

The postulate is that the greater the identification the greater is the success of communication. If so, one of the main questions for communication research is how far the parties involved in communication are **Sahridayas** or Identify with each other.

b) Sharing

Alongwith expectation of being **Sahridayas** the philosophy of Sadharanikaran emphasized sharing and not persuation different from the western concept of communication. The western concept of communication can be traced to the further elaborations of Aristotle's concept of rhetoric, the art of persuasive speech. Essentially, communication in western thought amount to Dialogue between 'equals' with a view to 'persuade'. But in Indian thought as reflected in the concept of **Sadharanikaran** communication is sharing among/ between 'un-equals' but Sahridayas with a view to not just persuade one or the other as such but to enjoy the very process of sharing. This makes **Sadharanikaran** more subtle and effective instrument of persuasion in Indian society.

c) Rasa Utpatti and Rasa Swadana

In context of Natya (drama) and **Nritya** (dance), the two intimately linked and closed developed art forms, which forms the substance of the **Natyashastra,** the question of participation of both performer and the audience is central to the philosophy of **Sadharani**karan. According to Natyashastra, the fountain of head of Indian arts, the performer is to arouse permanent moods in audience which will result in unleashing of attendant **Rasa,** reeling or aesthetic pleasure among the audience, thereby completing the process of communication. Thus, in the frame work of Sadharanikaran, participation through arousal and sharing of moods and feeling or aesthetic through arousal and sharing of moods is considered as absolutely essential element of communication. Without passing any value judgments as such, the success of entertainment industry speaks a great deal about the effectiveness of this approach to communication.

d) Simplification

Simplification is the essential dimension of Sadharanikaran concept of communication. In the communication process; the complex concepts and ideas are simplified by the speaker (source) with illustrations and idioms appropriate for the understanding of the listeners (receiver of messages). This approach makes communication a dynamic, flexible, practical and effective instrument of social relationship, persuasion and control.

The roving saints and **Suffies** popularized the complex religious and socio-cultural concepts and normsby following the simplification and illustration approach of **Sadharanikaran** throughout the Indian sub-continent, thereby contributing towards emergence of a Pan-Indian cultural mosaic. At the community level, 'Brahmins' adopted a similar role of propagating the religious ideas and cultural norms through simplification and examplification. The rich tradition of folk arts/media in India also served similar purpose of popularizing the culture through the length and breadth of the country.

As a matter of fact this dimension of **Sadharanikaran** seems to have become a common heritage of the Indian people as would be evident from the prolific usage of illustrations and idioms by rural folks to make a point in their day-to-day conversation.

e) Asymmetrical

Although the purpose of Sadharanikaran is to achieve commonness or oneness the process itself is an asymmetrical one. There is unequal sharing between communicator and receiver; there is a greater flow of communication from the former to the later. In Indian society, they are not equal the source is viewed as 'higher' and the receiver as 'lower'. The relationship is hierarchical and that of 'dominance' and 'subordination'. However, the source is held in high esteem by the receiver of communication, a relationship, idealized and romanticised in guru-shisya relationship. Although the source and the receiver are unequal but they are Sahridayas, which maken even unequal relationship/communicationsatisfying and pleasurable to both the parties involved, through Rasa Utpathi and Rasa Swadana.

The asymmetrical aspect of Sadharanikaran helped in earlier days in blossoming of Indian civilization through efficient communication and division of labour. This was however, later taken to the level of absurdity, resulting in a highly rigid and hierarchical society to the extent of making it a more or less closed system, and thereby contributing towards its stagnation and decay. Further the emergence and continuation of fuedal system is also perhaps the result of asymmetrical of communication in India.

These postulations are tentative and need further validation through research. However, on the face of it these postulations about Sadharanikaaran help provide better explanation and understanding of both communication and Indian society as such.

If the concept of Sadharanikaran provides better framework for understanding of communication processes than the kind of questions development communication scholars should ask are as follows :

a) To what extent those involved in communication are **Sahridayas** i.e. have common orientation and to what extent they identify with each others?

b) To what extent the communicator succeeds in arousing feelings and emotions among the audience pertinent to communication objectives?

c) To what extent the relationship between the communicator and the receiver is asymmetrical? Postulation is that asymmetry facilitates communication process. That is higher esteem of communicator, arising out of factors like knowledge, power, credibility etc. compared to that of receiver facilitates communication if the two (despite the unequal esteem and a symmetrical relationship) are **Sahridayas.**

d) To what extent the substance of communication has been simplified so as to make comprehension easy and appreciation a pleasure on the part of receiver?

These are merely illustrative of the kind of questions that can be asked in this new model of communication. One thing that should strike on looking at communication from the perspective of **Sadharanlkaran** is that the communicator and the receiver are in a kind of dyadic relationship in all five postulations and hence provides for more meaningful approach both for analysis and practical use of communication.

IV. In the above suggested scheme of things the focus of inquiry shifts to 'relationships' from elements involved in communication. Instead of inquiring into innovativeness of audiences or communication abilities of the source or the content analysis of the message, new approach and tools need to be developed to investigate the relationship between the parties involved in communication. For instance, it will be worthwhile to find out as to what extent the agriculture extension agent, family planning worker, village level worker or others responsible for development communication and tasks are **Sahrldayas** with the people they are supposed to motivate or change or 'develop'. We have scales for 'modernity' and adoption behaviour of individual recipients of innovations. Should the communication scholars not be innovative enough to develop scales and other tools to study this question of being **Sahridaya.**

Both research studies and practical experience on the ground amply demonstrate the significance of 'identification' of development agents with the target audiences in development processes. Beside lack of resources, the inadequate identification of development agents with the poorer sections of our society, who need development inputs maximum, has been an important contributory factor to their being largely left out from the developments efforts so far.

With this the related questions of approach to and philosophy of communication need to be investigated. Do the development agents 'share' or 'transmit' Do they enjoy the process of communication or just take it as a duty to be performed? On the answers will depend the effectivity of development communication.

In development communication the question of simplification and exemplification acquires great signification. As development involved communication of new ideas, concepts, techniques and methods do the development agents simplify these to the level of understanding of the target receivers and explain them with examples and in their language and idioms to make comprehension easy and meaningful? We need to develop methodology to study this aspect as well.

Finally, the development communication is an asymmetrical process. The development agents enjoy higher status and are on the giving end. Despite this do the receivers of development messages hold them in high esteem? Perhaps many a time not. The not very laudatory image of development agents and bureaucracy in general effects adversely the development communication processes.

Thus, the indigenous concept of **Sadharanikaran** for communication helps in better understanding of issues relating to development communication. It not only focuses on the relationship between elements involved in communication but also bring the social system more sharply into the picture. The suggested approach to research studies in development communication will yield better results and the **Sadharanikaran** approach to communication by development functionaries will help them to be more effective communicators leading to more participatory development in the country.

REFERENCES

Beltran, Louis, Alien Premises, Objects and Methods in Latin American Communication Research in Rogers, E.M. (E.D.), Communication and Development - Critical perspectives CA, Sage Publications, 1976.

Bordenave, J. Diaz. "Communication of Agriculture Innovations in Lation America" - The Need for new Models". In Rogers, E.M. (ED.), Communication and Development - Critical perspective CA, Sage Publications, 1976.

Dissanayke, Wimal. Development and Communication : Four Approaches, Media Asia, 8(4), 217-227, 1981.

Dubey, S. C. "Mass Media and National Development" in Schramm, W. and Lerner, D. (Eds.) Communication and Change in the developing Countries. Honolulu, University of Hawaii Press. 1976.

Galtung, Johan., Cultural action and conseientization. Harvaard Educational Review 40: 452-78, 1971.

Inkeles, Alex and D. H. Smith, Becoming Modern : Individual change in six developing countries. Cambridge : Mass : Harvard University Press, 1974.

Lerner, D., Passing of the Traditional Society : Modernizing the Middle East, New York, Fee Press, 1958.

McBride, Sean et. al, Many Voices, One World Communication and Society : Today and Tomorrow, UNESCO, Paris, 1980.

Rao, Y. V. Lakshmana. Communication and Development - A Study of two Indian villages, Minneapolis, University of Minnesota Press, 1966,

Rogers, E. M. and Shoemaker, F. F., Communication of Innovations : A Cross-cultural Approach. New York, Free Press. 1971.

Rogers Everett M. and Adhikarya Ronny, "Communication and inequitable development : Narrowing the socio-economic benefits gaps", Media Asia, Vol.5, No.1 PP.3-9, 1978.

Roy Prodipto, Agricultural innovations Among Indian Farmers, NIRD, Hyderabad, 1968.

Seminar Proceedings, "Rural Development and Communication Policies". Indian Institute of Mass Communication, New Delhi, 1980.

Singh K. N., et. al, Research in Extension Education for Accelerating Development Process, Indian Society of Extension Education, IARI, New Delhi, 1969.

Soedjatmoko, Development and Freedom, Ishizaka Memorial Lectures, Centre for Policy Research, New Delhi, 1980.

Yadava, J. S. "Communication in India : The Tenets of "Sadharanikaran" in Communication Theory from Eastern and Western Perspective ed. D. Lawrence Kincaid, Academic Press, New York, 1986.

Chapter **28**

Development Communication in Rural Areas : A Participation Approach

Cherif Jdiaye

Repeated failures of the majority of development project in the Third World in the sixties have had, however, a positive aspect : they have "sown the seeds of doubt" in the minds of a good number of people and institutions involved in the development process. Doubls which have generated new alternatives, new paradigms opposed to the "vertical" strategies prevalent until now, and which were considered as "bureaucratic", "centralized and centralizing".

The change of paradigm takes place on the basis of the role and place given to people in the development process particularly in rural areas. H. Sethi considers that "conventional modes of rural development, explicitly or implicitly, treat people as object of change and the relation between the development agent and the people often takes the form of a subject acting upon an object" and suggests re-thinking of the praxis (re-focusing praxis) of development

1. From 1979 this is what the World Conference on Agrarian Reforms and Rural Development has recommended, stressing the necessity for making the participation of people and in particular the mass of poor rural people a fundamental element in the drawing up and the execution of development projects.

Even since, there have been many controversies and differences about the concept and the practice of participation......

What is the Participation Concept?

If theoretically, in documents, there seems to be a consensus concerning the concept of "participation of people in development", serious divisions are seen in the implementation.

From this practice two schools of thought and praxis emerge which use the same terms to identify two totally different even opposed realities. The first school considers participation as a simple input which has to be added to the already tied-up project with the aim either of mobilizing the people for the execution phase or of identifying the needs to be taken into consideration as additional elements to get profits in order to quantitatively increase the commitment to support the project. However, in spite of the reference to the participation of people, this approach named "participatory" brings in its wake the same conventional defect of the "top-down" vision of development : participation is a "manageable input" which improves the prefixed project.

The second approach known as participation in development is seen as the opposite of the first in that it is a "bottom-up" strategy, or more precisely an interactive strategy involving partners equally responsible. Here participation is considered as a process by which people are capable of organizing themselves and, by means of their own organizations, of identifying their own needs of participating in the elaboration, the execution and the evaluation of development strategies. The canal aim therefore of participation is directly linked to the reinforcement of the capacity of the rural people to continually play a role in the process. Thus the distinctive qualitative factor of the paradigm of participation is found in the permanence and in the structural nature of people participation.

In this light the necessity of an appropriate methodology of communication as support, as the basis of the participation strategy, appears clearly. A system of communication which is open, flexible, dynamic.

Which Methodologies of Communication?

Perhaps it is necessary to state at the outset that this paper has as "basic raw material" the state of methodological research from reports of project activities in Asia, in Latin America and in Africa (whether they are conducted by government or non-government institutions) towards a perspective of participation.

The first observation made is that one rarely finds a systematic analysis of methodological practices of agents and projects; although the World Council of Churches in 1981, Crowley (1987) and UNIA (1988) can be cited for having attempted, on the basis of their respective practices in diverse contexts, for suggesting a certain number of guiding principles. These combined with others revealed by a careful reading of project files, enable us to glimpse atleast a methodological framework. One can discern, among others, certain major characteristics :

(1) The structural nature of the methodology of participation. It should not be ad hoc, reactive and subject to impromptu maneuverings.

(2) It must be in accordance with the pace and rhythm of the life of the people : a communicational rhythm which speeds up daily life, which seeks an immediate quantitative result must be avoided at all costs. "We are well aware that the formation of groups among marginalized people is extremely slow; it is important to go forward with patience and hope that people will discover what they need. This does not mean that we have to adopt a passive attitude, but more that it is important to determine the moment in which to hold a training course, put a proposal forward, allow something to develop, but always in the rhythm of the people" (CEBEMO, Columbia, 1966).

(3) The nature of the first contact, of the first step shows a great significance which can influence the process. The style of the first contact must already give the tone to the nature of participation of the strategy. A classic example illustrates very well the importance of the first step : one of an animator in the midst of a colony of fisherfolk in the North-East of Brazil.

Oakley reports; "She spent the first nine months merely living in Bomtempo observing the fisherwomen and being observed by them. One afternoon one of the fisherwomen stopped to speak to her after a day's fishing. This was repeated on successive days. The animator was then invited to fish with the women her baptism in the mud and that evening sat chatted with a group of fisherwomen outside one of the houses. In the animator's words, she had broken the barrier and begun to establish friendship and confidence which she felt would be indispensable for her future work. Gradually and five years later, these marginalized, isolated women, by means of their organization and those which they inspired, presented one of them to the presidency of the colony.

Without being a universal model this example illustrates very well what the importance of the first step could mean.

(4) Group approach also shows itself to be important. Pluridisciplinary, moble, united and often on the field groups.

These principles and key elements, without being exclusive, prove to be fundamental in the light of the qualitative contribution which they can make to the participation communication, whose practice reveals three major phases.

Phases of the Process of Participation Communication

The conventional modes of communication, whatever be the nuances, a "prefabricated" message is transmitted by the staff of the project (SENDER) to the people (receiver) : the aim here being to inform them, sometimes to get their opinions, but always so that they can be used at such a stage of the project. Even cases of feedback are without major consequences on the already defined profile of the project.

The participation approach, on the contrary, by nature of the process aiming at making people conscious and more responsible proves to be a methodology promoting a dynamic of knowledge creation for all the protagonists in question. Critical knowledge active knowledge.

The second phase is that of the diffusion of this knowledge in concentric circles by possible and available means.

This knowledge, once created, diffused, and assimilated will go into action : it is used.

These three interacting phases, while opening clear perspectives on the process, also increase the analytical Capacity and reinforce the commitment to action of the target people. Such a process of awareness of oneself by each one and of otherselves by everybody in the same united dynamic is one of the distinctive qualitative signs of participation communication. Paulo Freire says of this method that it "break rural people's mental isolation, to reverse the deeply embedded feelings of inability and inferiority, liberating from the moulds of deference and importance and providing basis for their active involvement in development".

Here the major mode shows up by which participation communication will take place : dialogue between the protagonists through informal discussions, regular meetings in small groups.

Dialogue, the dynamic basis of the whole process, is indispensable to develop and sustain a continuous participation. Discussions and meetings enable among other things.

- the development of the participation process.

- a forum for better involvement of people.
- a more acute consciousness of the problems which crop up and appropriate solutions.
- the base for a future more formal structure.
- the creation of solidarity, cohesion and unity in actions.

In such a context, the staff of the project (external agent) is a partner, a facilitator who must avoid thwarting the process by a too interventionist role or by being an agent of the "banking system". Here the partnership moves in an interaction :

In this dialogue, the qualitative factor is in the application of Virgillio's theorem : "Speak with me" : "don't speak to me".

Catherine Joseph, setting off from the fruitful dialogue between the Besant Nagar fisher community and the Bay of Bengal programme says : "Fisherwomen have an innate knowledge of their own status and problems. Their occupation with daily subsistence stems from poverty, hence they do not consider ways of improving their status. Given an opportunity they are capable of identifying and tackling many problems which normally they tend to accept as part of their way of life. If this target group is encouraged to think, discuss and take initiatives it would be a positive step in the process of development."

Taking the necessary time and determining in an adequate manner the size of the meeting groups, atleast in the initial phase, are also two requirements of the participation methodology. Just as the occupation of space where it responds to socio-cultural values.

Instruments and Means of Participation Communication

Besides meetings in small groups and public meetings, their mass version, practice reveals other means of participation promotion. Among others :

- theatre (here it concerns popular theatre that is, made by people being inspired by their own reality and their own projections)
- song (in certain cases song proves to be an efficient vector of rich dialogues on various themes and often improvised)
- workshops (including seminars, camps etc.)
- "Small group media" (video, tape recorder, slides)

Whatever be the instruments used, the importance is in their appropriateness to the context, their innovative and imaginative nature. Moreover, these instruments are not mutually exclusive, they can even be combined. However care must be taken to see that the staff of the project is prepared for the use of the chosen instrument(s).

Some Critical Aspects

The participation approach such as it is conceived here certainly has its supporters but it has many more adversaries. A whole range of literature has been brought out which considers it "Populist", "Utopian" and/or "purely emotional". Others, like the World Bank while recommending a greater involvement of people, finds that the participation approach is a time-wasting factor, this is to say time is valuable. Others simply think that rural people are "ignorant" and cannot contribute anything qualitative to such technical fields as for example, agronomy and related technologies. Finally a last group of critics, for politico administrative reasons do not wish to hear about the prospects of making the rural people responsible.

Certain critical observations are definitely relevant particularly those pertaining to the thorny question of time. However, one of the basic principles of the participation paradigm is to underline the long term aspect as opposed to the short term one. So the "value" of time will be measured not only according to quantitative parameters but essentially according to the coefficient of quality induced by and in the participation process.

For those who are still doubtful about the innovative potential of the rural people, not only does there exist today a corpus of the results of the convincing research available but also, most important, the practice of projects in keeping with the participation perspective is present to shake their certainties. To the last mentioned, opponents, the majority of whom are allegedly inspired by democratic values, the counter argument is to be looked for in the hollowness of their conviction in the Third World the largest number (the rural poor) is excluded from the development processes by the politico-administrative minority labeled however "democratic" :

Conclusion

As we have seen, what we have been able to call "methodology" of participation is rather a framework which synthesizes a large number of projects put into practice but placed in the perspective of people's participation particularly rural ones in the whole development process with the aim of continuously reinforcing their role. In this way new qualities are created which in structural interaction feed and propel the process. These new qualities which are to be looked for are :

- effectiveness
- efficiency
- expansion
- united continuity

These much sought after qualities make participation methodology "something" different from conventional methodology, "something" not so constrained by fixed time limits, less subject to an immediate quantitative impact, less inclined to tell the people what should be done and therefore less constrained by a pre-determined line of action. For that it demands a whole set of qualities, competence and resources often very rare, "participation is not a smooth, easy or painless development process" (ORAP – Zimbabwe).

In contrast to conventional methodology where rural people are "Limited mechanisms of mobilization", participation communication is circular or rather spiral, bringing all the protagonists (developments agents and rural people) on the same "wave - length" into constant, critical and creative interaction (dialogue). Flexibility, openness, critical acceptance of the other and dynamism are the intrinsic components of participation communication.

However the time that one mustgive the process to mature leads to an important question; can one really promote the participation process within the restrained and restrictive framework of a development project? Should one not effects a shift of perspective from projects to development programmes?

Notes

- H. Sethi : (Refocussing praxis, New Delhi, UNU-UNDP interaction programme 1987).
- DAKLEY : (projects with people, 1981, pp 237-238).
- Paulo Freire : (Education, the practice of freedom. London, writers and Readers publishing Co-operative,1947)
- See also : a) (The meaning of conscienticao : the goal of paulo Freira's pedagogy : Amherst University, Massachusetts,1976). (b) (For the critical of Freire's ideas & methodology : the pyramids of sactifice in political ethics and social change, New York, Basic Books, 1974).

- "When we had just met, Virgilio stood up and said : Lucas' I understand you want to know. You are a scientist and you want to know. But there is only one way to know what I know about cassava. Speak with me; don't speak to me like others did. Ask me about my life and I will tell you about cassava" in (Farmer First I. T. P. 1989 : Loak. Box : P. 61-Dominican Republic).
- 5. Catherine Joseph (in paper presented at the 5th Asian Fishworkers Consultation. Jan 24-31, 1992, HAT-YAT-THAILAND).
- In certain socio-cultural regions of west Africa, particularly among "sereres" of the central region of Senegal, it appeared, according to several surveys (1966-1983) that the circular scene conveys the principle of equality in speech in midst of an audience in a better way. However here too, age and gender have a role to play in the composition of the audience. Therefore the model of concentric circles, where "those who can speck unrestrainedly" will come together to pass on the topics of discussions talks to the whole family, has been thought of Relations between the different circles obviously go in both directions moving from the smaller group (circle) to the bigger and vice versa.
- Midgeley in OAKLEY & AL : 1986
- See : P. Richards : Indigenous Agricultural Revolution (Hutchinson, 1985).

D. Brookensha : Indigenous Knowledge Systems & development (Washington DC, University press of America 1980)

ORLAMDO FALS-BORDA : Knowledge and people's power (Mexico, Srglo XXI Editors, 1985)

See also projects files of :

a) Small farmers development programme (SFDP) in Nepal

b) Grameen Bank (Bangladesh)

c) Organization of Rural Association for progress (ORAP-Zimbabwe)

d) PIDER (MEXICO)

e) Six-S. Association (Burkina FASO, WEST-AFRECA)

Chapter **29**

A Two Way Transfer-Self-sufficient Technology

S. Cooney

The communication in future in not going to be one way. The rich rural craft heritage of third world countries could be the starting point. Gradually, the exchange of self-sufficient technology may bridge the exhisting gulf between the North and South. The communication facilities and channels needed to facilitate the two way exchange of technology largely do not exist at present-but they are emerging on the horizon.

The transfer of technology between the rich and third world countries will take place in future, not from the rich to the poor alone, as it has been in the past, but also from the poor to the rich. This will come about because technical innovation is changing the balance of the world's economy thereby creating pressures for a return to greater local self-sufficiency, in which the poor are not at such a great disadvantage compared with the rich. For this reason, I suggest that the session in which this paper is presented has been mistitled. Calling it "Technology Transfer to the Third World" implies that the rich countries are the ones who have the relevant technology, and the main problems the poor have to solve is to get hold of this technology, use it and become rich.

There are three main reasons for questioning this assumption. First, while the rich may in the past have had a substantial monopoly in relevant technology, there are strong forces today which will gradually reduce the dominance of the large scale, industrial technology which the West has developed over the past 200 years. Secondly, at the same time, the small scale, self-sufficient technology which has always been the heritage of the poor (and used to be that of the rich) will slowly regain its previous importance as a means of meeting human needs. Thirdly, the *Third World* is less backward in the area of self-sufficient craft skills than in the large scale industrial technologies. The poor will, therefore, have much to teach the rich in future; the problem will be to create a network to permit information to flow effectively in both directions.

Because we do not have reliable system models of the world economy in an age of technical change, the transfer of technology can have effects, which are not always foreseen, much less welcome. Two examples of technology transfer will illustrate how benefits for some can harm others. In 1980 summer, I bought a shirt in Canada which was made in Korea. It came from an industry which is rapidly driving the European and North-American shirt industries out of business, by providing those rich populations with inexpensive shirts and raising the prosperity of East-Asia. On the other hand, when Tanzania wants to buy a tractor, it costs her six tons of tea today, whereas ten years ago it only cost two tons. These and many Other examples show that the flow of innovations from one region to another can have ripples of effects throughout the world economy.

Coming as I do from a country which was a colony until 60years ago and working on several information projects in Africa and Asia over the last 15 years, I have some experience with the effects of technology transfer. I want to distinguish between the two main kinds of the problems. One arises in relation to creating attitudes and building mechanisms favourable to information transfer with the object of promoting development; the other relates to the after effects of such transfer, both on the countries concerned, and on others. In this paper I want to focus attention on the second type.

Economic Growth versus Appropriate Technology

There are, at present, two competing concepts of the role of technology in developments whichare canvassed among the agencies concerned with the *Third World.* The first is the economic growth model, based on the pattern set by the rich countries over the past 200 years. In their search for progress, most third world countries have looked to this pattern. It is still widely believed all over the world that worked for the rich can also work for the poor. It is believed as part of this theory, that technical innovation promotes economic growth, generates more jobs, increases wealth and additional government revenue which can be used in social and infrastructure development. The way for poor countries to achieve progress, according to this view, is to develop or adopt technology in such a way as to increase output and productivity. This is assumed to generate economic growth in the form of new and expanded markets, by which more and more people are employed and more money earned. Nehru once said "it is science and technology which has made other countries wealthy and prosperous and it is only through the growth of technology that India shall become a wealthy and prosperous nation". In a recent report to the European Community's FAST programme in technology forecasting, 1 named this the "Linear theory of innovation, growth and progress" because it assumes that each step normally follows the preceding one.

The second concept of development is usually termed "Appropriate Technology". It originated with the work of" E. F. Schumacher which has been, widely influential. It is sometimes also called "Alternative Technology". Now, many countries in Africa, Asia and elsewhere have institutes for the study of this kind of technology, set up with aid from abroad and full support from home governments. Appropriate technology has been defined as an approach in which technical innovation and adaptation are placed, subordinate to social and cultural integration. It is not defined in terms of the high level of technology used, but by the attention given to adapting the technology to the local environment. To achieve development from within a population requires working with people in their existing situation, with their existing resources and skills. This definition emphasises reliance on locally available *inputs* of skills and resources. It calls for a return to traditional skills, such as those of the blacksmith and for the use of machinery and implementations which can be locally serviced, in place of imported equipment which must be supported by imported parts and fuel.

This concept of appropriate technology, to my mind, only goes halfway. Less frequently cited is the achievement of full local self-sufficiency in essential products by producing locally what is locally needed for survival. In this way communities and nations can be made less dependent on imports from abroad and on the increasingly insecure possibility of exporting local produce to pay for such imports. In my formulation, appropriate technology includes not only the use of locally available inputs, but on changing the mix of *outputs* so that more are in the forms needed for local use. For instance, many third world countries are no longer self-sufficient in food, because they have concentrated on export cash crops such as tea, coffee and cotton. Instead of exporting these and importing food, they could ensure that their own needs were provided first, locally. I made a plea in 1981 for this concept to be applied in aid programmes from rich countries. In 1982 1 was interested to find a strong emphasis on self-sufficiency in the development policy of Tanzania.

Failure of the Linear Theory

The economic growth model is predicted on past experience and cannot be assumed to be inevitably the only way for solving the future problems of the developing world.

Indeed, if we examine the recent record of economic growth in the rich countries, the model seems not to have worked for some considerable time past. Over the past 12 years, unemployment has steadily risen in Europe and North-America. Very few authorities now anticipate much improvement before the 1980s. The panel of futurists, who replied to questions through the "Consensor" at the 1982 World Futures Assembly in Washington, believed that there would be another very severe depression before the year 2000. Business is increasingly squeezed by competitive pressure, much of which originates from the rapid multiplication of new products and processes, many of which compete with each other. Economic growth, it seems has always been the reward of those who could achieve a corner in some market, a monopoly which enabled them to make good profit margins or value added. Innovation provided such monopolies for the time being.

Contrary to accepted economic theory, the creation of economic growth by means of new technology has not been achieved by free-trade. John Eatwell, the Cambridge (England) economist, pointed out in a recent TV series that the doctrine of international free-trade was first enunciated in Britain, early in the nineteenth century, when this was in her interests precisely because she had a monopoly in most of the new technologies of those days. Now that Britain can no longer keep ahead of other nations in technology, free-trade sometimes acts against her economic growth and she falls steadily back in the world league. The same thing appears to have happened more recently to the United States, especially in relation to industries such as automobiles in which she is competing with Japan (The Consensor panel at the World Futures Assembly believed that Japan would also have a very large impact on computing before the year 2000). In general, when technology becomes more and more freely available to all who want to use it, including third world countries (as it is today), the monopolies in technology are undermined. These monopolies formerly enjoyed by the rich and were the foundation of their economic growth. When I was in Ireland, I believed that our economic problems stemmed partly from the lack of any substantial mineral resources, such as those on which Britain had based her industrial growth. However, in the 1960s and 1970s, some rich and easily worked deposits of lead and zinc were discovered in Ireland. These deposits were expected to play a major role in promoting economic growth. In fact, their contribution had been disappointing because the world market today did not allow us any monopoly. In contrast, Japan, which depended on people and technology, forged ahead during those years. Eatwell also points out in the same TV series that Japan was advised after World War-11 to go in for labour intensive development. This notwithstanding her preferred a capital and technology intensive option. And because of the fact that she did it better than others, she gained an unbeatable monopoly position.

It has frequently been argued that, even if this all is true, it is nevertheless time that the poor countries of the world enjoyed a share in the prosperity others have achieved. It is necessary, politically, that they should do so in order to maintain world stability. This could, it is claimed, be done by letting the poor take-over the middle level technology, while the rich move on up into more advanced technologies. In this way the both can maintain economic growth. The figures however do not show this to be a successful strategy.

Firstly, third world countries fail to gain as much as they hoped from moving in and taking over the markets in textiles and other middle level technologies from the rich. In order to compete successfully, they must undercut the rich in their home markets. They must do this while transporting their products from faraway. Therefore, they cannot make as much money, or achieve as much growth, as is expected. Free-trade with open technology in a world market means that the rich get poorer because they lose industries to the poor. But the poor do not get that rich at all.

Secondly, the high technology the rich are supposed to move into is more and more based on microelectronics, robots and other labour replacing technology, so that employment is less than that with older technology. The system of distributing purchasing power by paying labour for its work begins to break down, as it is happening today. Fewer people in the rich countries are able to afford the products of technology, whether these come from rich or poor countries. This restriction of purchasing power, or effective demand, is characteristic of the current world recession. When the products of technology are not bought, production must be cut back; there is more unemployment and the cycle continues to reinforce itself.

Thirdly, there is no reason why the highest technology should, in a free-trade world, remain the exclusive monopoly of rich countries. In fact, advanced microelectronics is now being made in South-East Asia. The western countries are faced with losing both the middle and advanced technologies with the passage of time, because their costs are higher than those of the Third World. At the same time, the poor countries will find it harder to sell their products to an increasingly impoverished and competition-ridden world.

As the examples of Poland and other communist countries illustrate they are not immune to these problems. They deal with them in a different way, to be sure, but they are not able to dispose of them any more than the West can.

The Third World countries cannot solve their problems by a return to exporting the primary products they have traditionally produced. Even though they have the right conditions for producing pineapples, coffee, tea and other tropical goods, and so enjoy a monopoly, it becomes still more difficult to achieve growth this way. As time passes, the terms of trade move against them (as in the case of the tractor for Tanzania, which now costs more tea than before). If these countries raise output to compensate, they add to the world's surplus. They are then either prevented by quotas from exporting the extra output, or the glut causes world prices to fall another vicious circle.

There is a grain of truth in the Luddite argument after all. So long as technology transfer did not undermine monopolies, economic growth could continue. But the situation has changed. To day, we are in a world which is approaching closer to "pure competition", in which profit margins are cut close to actual costs of production. These costs are under constant pressure downwards, squeezing out labour as new technology provides cheaper ways of doing the job. The cost of a robot in the US today has been estimated to be $4 an hour, against $16 for a man, in the automobile industry. What used to be considered "risk" has become "uncertainty". Pure competition plus new technology brings not only lower profits, but more unemployment. Because government revenues are derived from profits, incomes and sales, they too are falling, and with them the level of social services and infrastructure. The "swing to the right" in many countries appears to arise, not from a change in public philosophy, but from the fact that people with falling real incomes do not want to pay higher taxes. Governments faced with revenue shortages, therefore, must justify spending less on social programmes; they can do this by a right wing philosophy.

It is unlikely therefore, that technical innovation and technology transfer will promote economic growth in future. Now that the monopolies on which such growth was based are undermined, few can hope to make a lot of money out of new technology. Of course, there will still be fortunes made out of particular technologies, but at the aggregate level of whole industries and nations, this will no longer be possible.

The Self-Sufficiency Option :

Some third world countries, such as Tanzania, are seeking to achieve self-sufficiency, though not yet with much success. There is also a small, but growing, movement in this direction at the household and local levels in rich countries. I feel this is due to the trends analyzed above in both cases. And in these cases the type of technology required is different from that used in large scale industry.

Third world countries, could become self sufficient in food which they have ceased to be for long. The Green Revolution, whatever its critics may say of its social and economic effects, is contributing to make some of them self-sufficient. Its scientists are now seeking methods of raising food output without excessive reliance on imported machinery, oil, weedicides and pesticides. In this way the producers will become less dependent on imports they may not be able to pay for.

On the other hand, projects which arc aimed at raising money incomes by producing surpluses for export have lesser certain prospects, because they lead to interdependence and vulnerability to increasingly tough world market conditions. Third world countries, instead of producing coffee, sugar and cotton in ever larger quantities, might be better advised to ensure first that they have produced enough food to feed themselves for the next two years, to provide in the event of bad harvests. Land and labour, which are surplus after basic local food needs are met, should be used for cash crops to be exported. In this way, each family, community and country can be less dependent on the world economy.

The technology for doing this will also be largely a matter of local initiative-appropriateor alternative technology. There is no point, in an age of costly imported energy, in using tractors which must use oil, when something more traditional, but improved by research, will make the community more self-reliant. Anyone who has seen the craft skills of, for instance, the South-East Asians can have no doubt about their capacity to create such technology. What is needed is technology which can be made and serviced locally with available skills and it should be used to meet local needs in the first instance. Local inputs should be relied on to produce (for the most part) outputs for local use, as mentioned above. I have dwelt on this elsewhere in relation to the type of aid the rich ought to give the poor,

With the unemployment rising and the incomes of the employed failing to keep pace with inflation, a similar lesson is gradually being followed in the rich countries. More and more people are being compelled to provide for themselves things which they used to buy or else to allow a fall in their quality of life. There is less eating out and more home cooking. People are, rather slowly, being persuaded to insulate their homes better and so save heating costs. If people do not find someone to pay them for their time, they must use this time to meet the needs which their wages used to meet. There is no magical fountainhead of new jobs or of raised real incomes, so self-interest must compel increased self-sufficiencybased on new and inexpensive technology. I have discussed some of the self-sufficiency options for those rich countries in another paper.

Transfer to Self Sufficient Technology :

We may now reconsider the exchange of technology between the rich and the poor countries. As expected, if the present trend continues, more people will be using self-sufficient strategies in the rich as well as the poor countries. In both cases, they will be using this strategy because it is the only one which protects some of their living standards. In the poor countries, self-sufficient technology will be used to avoid dependence on imports such as oil and machinery which they cannot afford while still producing enough to feed themselves. In the rich countries, this technology will be used by a growing population of unemployed, under-employed (work sharing) and underpaid, to ensure that they can continue to enjoy, by their own efforts, atleast some of the amenities they used to pay for.

The self-sufficient technologies of the rich and the poor will show significant parallels. They will be small-scale, locally oriented, inexpensive and designed to be operated by generalists who have no special skills. They will be concentrated on the areas of primary importance such as energy conservation and solar energy, food production processing and cooking in the home, and others of that nature. Generally, they will be geared to help the consumer to be his own producer. They will be "do it yourself" technologies allowing some tasks which require the co-operation of the community as well as the individual and household. Many precedents for local cooperation will be found in traditional structures from the past. This type of technology will replace a good deal of the present large-scale, interdependent production of consumer items by local substitutes, produced with local materials by the person who is to use this product.

Because the self-sufficient technology, and the needs it meets, will be similar in rich and poor countries, there will be much of it which can be transferred from one region to the other. Of course, not every such technology will be useful all over (many of the major industrial technologies of today are similarly confined to particular locations). There will however be a large and growing range of locally developed self-sufficient technologies which have a global application and can be adapted to conditions in many places around the world. Since information can be moved more readily than goods or services, the information society can play a role in this transfer. Networks can be developed through which any person or community, wherever they may be, will be able to search the world for a solution to his local problem. Many of these networks will no doubt be computerised. People will thus be able to seek solutions to their local problems by exploring these information avenues spread all over the world.

Transfer in both directions :

It is at this stage that the potential for a two way exchange of technology between the rich and the poor will be created. Because the rich have concentrated for so long on specialization and international trade to meet their needs, crafts have decayed and have lost much of the self-sufficiency they used to have in the pre-industrial age. But the poor have not yet lost their traditional skills, and village crafts still provide for many of their needs. They form a large part of the skills which to the poor still have today. Unfortunately, there is contempt for such skills in third world administration does not recognise the importance of these skills this may be changing now (17).

In the case of the crafts on which self-sufficiency skills will be built up, therefore, there does not exist the same wide between the rich and the poor as there is in industrial technology. Besides, relatively little as yet been done in using science to improve these skills. While domestic equipment (such as washing machines) are widespread among the rich, they have succeeded less because they increased household self-sufficiency as because they could be mass produced and mass marketed.

In developing and exchanging self-sufficient technology, therefore, rich and poor start on a much more equal basis than in relation to high technology for industrial use and the world market. For this reason, the transfer of self-sufficient technology, as this technology grows in importance, will be mutual between the rich and the poor.

The communication facilities and channels needed to facilitate the two way exchange of technology largely do not exist at present but they are emerging on the horizon. Barriers of language, distance and cost still effectively isolate communities in different countries from interchanging information directly. However, the "Information Revolution" will soon provide the technology to overcome many of these barriers. This technology using computers and telecommunications is already being used in richer countries for "net working", where people indifferent locations can exchange information directly with each other. This can be done at a relatively low cost, and the costs are falling all the time. Work is also going forward (for instance in Japan and Europe) on automatic language translation, which will enable users at each end of the line to present questions and answers in their own language, to be translated for those at the other end.

The capital cost of a computer terminal for networking would today be comparable to that of a large TV set. This brings it within the means of many third world village communities. As prices continue to fall, these facilities will become still more accessible. To use the facility involves mainly the cost of the phone line. If this cost were borne by the users in rich countries; information could be exchanged mutually at little or no cost to the poor participants.

There will be a trend in future towards transporting information instead of goods. Today many countries must import goods, paying for them with exports in a world where the terms of trade continually move against the poor. Using the new information technology, third world communities will be able to exchange information with the rich about ways in which both can meet their own needs locally, without having to import. Both parties can expect good value from such exchanges, because it costs much less to move information than the goods.

REFERENCES

COONEY, S. "Ireland and the Third World's Food Problems" World Food Day, October 1981 (Seminar organised by Gorta).

SIMMONDA, W.H.C "Aggregating Technology into Economics" National Research Council, Canada, 1980.

COONEY, S. "Self Sufficiency - An Alternative Option for New Technology" in "Information Society - For Richer for Poorer". Proceedings CEC/ FAST Conference on "Transition to an Information Society", London, January, 1982.

COONEY, S. "The African Literature Service", An Foras Taluntais, Dublin, 1968.

ALLEN, T.J., PIEPMEIER, J. M. & COONEY. S. "The International Technological Gatekeeper" Technology Review, Cambridge, Mass. March 1971.

COONEY, S. & VON REVICZSKY, M. "Agricultural Documentation Centre for Kenya" (Preparatory Mission Report) Food and Agricultural Organisation of the UN. 1972.

ALLEN, T. J. & CONNEY, S. "Institutional roles in Technology Transfer: the Situation in one Small Country" R&D Management 4(l), 1973.

COONEY, S. & ALLEN, T.J. "The Technological Gatekeeper and Policies for National and International Transfer of Information" R&D Management (Manchester) 5(1), 1974.

COONEY, S. "Criteria for User Oriented, Cost-Effective Information Service : An Analysis of the Fast African Literature Services" Special Libraries, December 1974.

COONEY, S. (ed.) "Information Services for Agricultural Research" Proceedings of a Workshop on the Management of Agricultural Research. SEARCA, Philippines, February 1975.

COONEY, S. & VILORIA, R.V. "The Agricultural Information Bank for Asia (AIBA)" Fifth World Congress of IAALD, Mexico DF, April 1975.

COONEY, S. "A Literature Service for Tanzania" Mission Report to the Tanzanian and Irish Governments, 1978. "Innovation in Small and Medium Firms" OECD, Paris, 1982.

COONEY, S, "Productivity, Progress and Innovation". Report to the Forecasting and Assessment in Science and Technology Programme of the European Commission, 1982, (400 pages) (The FAST Programme).

SCHUMACHER, E.F. "Small is Beautiful", ABACUS 1974. The Arusha Appropriate Technology Project, 1981(Tanzania).

KIRKNAES, J. Village Skills Survey (Tanzania), 1981 (Unpublished), EUROSTAT Statistical Abstract, EEC, Brussels, (See also FAST report. Chapter 7).

EATWELL, J. "Whatever happened to Britain?" BBC TV, 1982.

Note: This paper was first delivered at the World Futures Assembly in Washington, D.C., U.S.A. in July 1982 during a session entitled 'Transferring Technology to the Third World.

Chapter **30**

The Transfer of Instructional Technology Between Nations

Richard E. Clark

In order to prevent, the misapplication of technologies to inappropriate problems it would be useful to encourage technology transfer specialists to think of their professional roles. The client for technology should look for a transfer specialist who is expert in variety of techniques and who has experience matching problems with appropriate solutions. Applying technologies to the full range of stages in the acquisition and transfer of knowledge can be enhanced with careful analysis of the instructional task during the technology selection and adoption process.

Technology results from the application of science and intuition to the search for solutions to problems. In its finished and successful forms, it is the most valued of commodities exchanged between nations. Without such exchanges going on constantly, each nation would need to rediscover solutions to all commonly experienced problems in education, energy production, transportation, agriculture and health care. And yet, there have been increasing doubts about the success rate of transfers of certain technologies between nations and cultures, notably those in education.

In the few instances where careful, long-term study of the effects of an educational technology transfers are carried out, important features of the transfer have been found to be apparently unsuccessful. Professor Thomas's description of the use of television in American Samoa (1980) and Scribner and Coles (1978) study of nonformal literacy instruction among the Liberian Vai are important recent examples. In both cases an instructional technology was transferred to another nation in a large scale (and well publicized) effort to solve educational problems. In neither case was the transfer successful in the way it was originally planned and often the very visible problems it brought may have far outweighed some of the more invisible but suspected benefits. Of course, these two instances may be unique or unrepresentative. Since few long term studies are conducted, it is difficult to determine what the usual long term success rate has been for educational technology transfers. However, since we have some well documented instances of less than ideal successes, a closer examination of the processes and issues which surround this activity seem warranted.

One of the main reasons for the confused state of our present knowledge is that all of our current definitions of instructional technology include audio visual media. While television and other media are useful technologies for delivering instruction, there is solid evidence that they contribute nothing of value to learning and achievement, (Schramm, 1977). Media are more accurately conceived of as delivery technologies. Instructional technology should be defined more broadly as programmes, methods and techniques for training. This definition, employed in this presentation, rests on the assumption that, like other technologies, those designed for instruction derive from research on achievement and our prior experience with instructional solutions to learning problems. As such, the issues and problems associated with their use tend to be similar whether they are being employed within or between nations. There may for example, be more differences in successful transfer methods between the east and west side populations of Los Angeles than between west side Los Angelenos and south side residents of Bogota. Despite this fact, there is a consistent reluctance for developing nations to accept instructional technologies produced in more developed nations. To the extent that this reluctance is based on the success rate of these transfers, this presentation is directed towards issues and problems which are thought to influence the success and failure of instructional programmes.

Out of many possible issues which might have been selected, five major problems will be chosen for discussion : (1) Inadequate understanding of the problem to be solved by the technology; (2) Conflicting theories which lead to conflicting employment of instructional methods; (3) Misunderstandings about the science which underlies the technology; (4) Vague definitions of technologies; and (5) Unfortunate matches between problem and technology.

After a review of the problems raised by each issue, an attempt will be made to suggest a solution or an approach which may lead to a solution. The present discussion is based on the assumption that technology transfers are vital to the continuing growth of all nations and that the issues connected with this activity are large and complex. The ideas offered here are intended to contribute to the ongoing dialogue.

I. The Problem to be Solved by Technology Transfer is Poorly Understood by the Transfer Specialist.

There are at least three types of misunderstanding : (1) The cultural or national bias of the developer or the adopter prevent shared understanding of a problem; (2) The attempt to solve one problem leads to other unanticipated problems; and (3) The problem which is addressed is not the problem which the adopter in fact thinks is being solved.

Cultural Bias

In the first instance, there are ample examples of disagreements between transfer teams and adopting nations over the problem to be addressed. The most dramatic recent examples concern the problem of world hunger. Technologies which are proposed to solve the problem range from high yield grain stock to population control devices. These two technologies reflect very different perspectives on the "problem." Less developed nations tend to see hunger as the root problem to be addressed by food supply technologies. More developed nations tend to view the problem as one of overpopulation and recommend birth control techniques.

In education this type of misunderstanding often occurs in the area of access to instruction. The fact that education is simply not available to a large percentage of the world's population is variously perceived : (1) as a problem in teacher training, and school construction; (2) as a local government attempt to control information and social reform; (3) as a problem with the delivery of instruction to rural people; (4) as an example of economic regression and (5) as a population control problem. The technologies available to address each of these views of the access to education differ greatly.

Unanticipated Problems

The second feature of this barrier is the often unrecognized interrelationships between problems. For example, an instructional technology which increases achievement and learning may at the same time, reduce adopters participation in the instructional programme. There is increasing evidence that some instructional techniques which are the most effective in promoting achievement are also the most disliked by learners. Conversely, instructional methods which are liked may lead to minimal learning. Since continued willingness to participate in an instructional experience depends to some extent on the satisfaction experienced by students, it seems possible that we might solve a short term achievement problem at the long term jeopardy of the entire instructional program (cf. Clark, 1980). This problem interrelationship difficultyis also evident when a technology is resented by some sub group in the adopting setting. Teachers, for example, sometimes resent the use of televised instructional programs and occasionally work to dilute their effect. This may, in turn, reduce the morale of the teachers who end up working less effectively than they were before instructional television was introduced (cf. Thomas, 1980). The end result may be a greater overall instructional problem than the one the technology was selected to address.

Transfer Goals

The third major difficulty here has been called the problem of the "referent generality" of outcomes, (Snow, 1977). Here the issue if the extent to which the problem which is addressed by the technology transfer actually represents the major long term difficulty which the adopter hopes to solve, that is whether the solution (referent) to the problem addressed will generalize to the future goals of the educational program.

The goals of most educational improvement programs in third world nations tend to be such things as improved quality of life for citizens, increased self esteem and ability to solvepersonal and family problems more effectively, and more knowledgeable participation in the political and cultural life of the nation. Since none of these problems can be addressed directly by reasonably manageable technologies, a decision always has to be made about which of the many short term problems which can be addressed will actually generalize to the solution of the larger and longer term goals for the educational system. The line of generalization from the solution of short term manageable problems to the addressing of larger and longer-term goals is not always clear or very direct. For example, Schribner & Cole (1978) studied the effects of literacy instruction on the Vai of Liberia and found that contrary to general expectations, teaching the Vai to read and write their language in a non-formal setting did not generalize to areas typically considered to be related to literacy, i.e. increases in verbal reasoning skills or the ability to deal with more abstract concepts. It led the researchers to claim that in this instance, the expectation that relatively short term instructional interventions would generalize to the solution of longer term goals was unwarranted. In many cases, cognitive skills are cultivated as a response to local cultural demands and needs and are not transferred to new skills or tasks unless the technology and local needs include provisions for the use of the skill in novel settings and problems.

Recommendation : The first two transfer barriers raised in this section, agreements over exactly what constitutes the problem being addressed avid potentially antagonistic interrelationships between problems and solutions can be alleviated somewhat by careful prior agreements and planning. The decision about what constitutes the problem is clearly the domain of the adopter. It is equally clear that much of the controversy over technology transfers will continue to be due to the predisposition of more developed nations to negotiate over which problems experienced by lesser developed nations will be addressed before technologies will be provided. The second difficulty can be reduced somewhat by a combination of careful prior analysis of the potential each technological solution has to become an unanticipated solution and careful monitoring of the adoption process through formative evaluation and revision cycles (Forman, 1980).

The referent generality problem, while difficult, may be the easiest to solve given the present state of the art. There are many techniques available which will promote the generalization of skills acquired by learners through instructional technologies to wider varieties of problems. Essentially, this is a technology of the transfer of training within a nation and it involves presenting more complex opportunities to practice skills which are learned in increasingly greater varieties of situations. The many instructional techniques which define the technology include the use of identical elements, role playing, naturally occurring reinforces, and simulations (these elements are discussed more completely in the next section).

II. Existing Theories of Transfer Appear to Conflict and Suggest very Different Strategies for Promoting Successful Instructional Technology Transfers

This problem tends to be present in two areas: (1) All transfer specialists work with implicit or unconscious theories about the factors which enhance or inhibit technology adoptions; and (2) The two major formal and explicit theories of transfer which are based on existing research tend to conflict.

Unconscious Theories

Although many transfer specialists would reject the idea that they are constantly working with theories as they plan a transfer, in fact, all strategies which to used facilitate adoption of some instructional program spring from implicit theories about why one method or technique will work better than another. It is seemingly impossible to make decisions, personal or professional, without assumptions about the consequences of the decision. In this sense then, we all use theories. Assumptions, models, paradigms and rationales are all ways to describe our attempts to create a simplified picture of reality and use it to predict future events which would result from present activities. This natural inclination of both transfer and adoption planners is a difficulty only when we fail to make explicit the implicit theories which guide our decision. It often happens that there are very different assumptions about the factors which will make a technology succeed on the two ends of the exchange, which lead to conflicts and working at cross-purposes. Occasionally, for example, transfer teams will tend to assume that more participatory and self paced instructional methods will work best in the rural areas of developing nations. These types of methods are based on the theory that learners in this setting must be enticed to learn and will learn best at their own speed. Learners in a rural, poor adoption setting, on the other hand, often expect more didactic methods to be employed and are usually highly motivated at the onset of instruction. On a more general level the difficulty is often one of different assumptions about what motivates the transfer process. Transfer specialists and managers often assume that adoption officials are motivated partially by personal gain. This tends to result in the use of more highly visible and expensive techniques being chosen for transfer. For their part, adoption officials sometimes assume that the transferring group is motivated by political gain and this tends to lead them to resist outside advice and influence over the adoption process and to be suspicious of expensive techniques. The fact that these two conflicting sets of assumptions about motives are occasionally correct assessments of the situation at hand does not necessarily reduce the problem. Neither theory has much to do with the factors which promote the success of the transfer for the ultimate client...the learner. They are essentially theories about the threatening side effects of the transfer process not theories about what leads to the desired learning.

Conflicting Theories

The second major difficulty with the use of theory is that there are two existing psychological theories of transfer which suggest very different techniques for instruction. (Royer, 1979). On the one hand there is a considerable amount of research which has resulted from so called "identical elements theory". This theory is based on the notion that the critical feature of all transfer situations is the extent to which the adoption situation is identical with the setting where the instruction was originally validated. This theory was originally designed to explain the transfer of learning for individuals and some stretch of the imagination (and the available data) is necessary to apply it to transfers of techniques between nations.

Identical Elements Techniques

Research on the identical elements theory has led to some very useful instructional techniques which can be built into instructional technologies such as :

- *Modeling* or simulating the behaviour to be learned for the adoption group such that all crucial features of the behaviour are depicted.
- *Role playing* experiences are required during learning will allow the learners to simulate or role play the learning so they demonstrate it's use to themselves and others many times during the adoption process, and receive correcting feedback.
- *Ensuring identical elements* in both the training and application situations. Here, the trainers work in the setting where the learning will be practiced or they attempt to simulate the crucial features of the application setting.
- *Overearning* the knowledge to be gained tends to ensure that it will be practiced after the training is completed. This involves instructional programmes that extend considerably beyond the point where the new behavior or knowledge has been minimally acquired.
- *Increasing variability of practice setting.* After the knowledge is acquired and overlearned. Its continued use is enhanced by incremental changes in the settings where it is practiced. This enhances the chances that skills will be used appropriately over a longer time and in a greater variety settings.
- *Learning maintenance techniques* such as : (1) the substitution of

naturally occurring rewards for practicing the new knowledge (*e.g.* substituting kind words from family and peers for trainer reinforcement); (2) change trainers by substituting peer trainers, and (3) provide for the gradual reduction of naturally occurring reinforcement from peers so that it is intermittent and not constant. This serves to gradually increase learner control over the practice of what is learned by teaching learners to notice and control their practice.

- *Inoculation after delay,* build into the technology transfer the requirement to follow up the results of the training after a reasonable period of time to reinstate and redirect the practice of the learned behavior (*i.e.* to inoculate against the disintegration).

The second theory of transfer deals less with instructional techniques than with a description of the internal, mental or cognitive events which take place when learning is transferred from the place where it is learned to the place where it will be practiced. Here again some imagination and speculation is necessary to apply this cognitive theory to exchanges between nations. Cognitive theory is in a more formative stage than the identical Clements theory and there is relatively less research evidence for its success. It is based loosely on information processing theory and it assumes that identical transfer theory tends to promote learning without understanding or comprehension (by rote learning and practice for example). It presumes that the more *understood* a concept is the more able a learner will be to use it in a variety of self-determined situations. All of the techniques for instruction which flow from cognitive theory have one crucial element : it is necessary for the instructional designer to determine which of the many existing skills and understandings that have been acquired by the learner are applicable to the "to be acquired" knowledge. The presumption is that learning and eventual comprehension of new knowledge will be enhanced if previously learned information is seen as analogous to the new technology. Most of the instructional techniques which will be developed in future in the cognitive approach will be an attempt to increase the "richness" of the connections between previous experience and new information to be learned.

One of the more obvious implications of cognitive theory is that the transfer specialist must have as complete a picture as possible of the existing skills and knowledge of the learner population. Rather than being concerned with the source of rewards or simulation techniques which concern the identical elements theorist, cognitivists would attempt to use familiar experiences as analogies which would activate previously learned information and relate it to new information. For example, if the adopting group were agriculturally oriented, the cognitivist might assume that analogies which draw upon knowledge of complex systems of planting, harvesting, weather, irrigation, cultivation and the actions of insects on plants. Health care instruction might be related to procedures for protection against the actions of destructive insects on crops. These analogies tend to be of two types: (1) *What is involved* analogies and *(2) How to accomplish* analogies. What is involved analogies relate the elements of a familiar event to those which are to be learned. We might for example, make an analogy between the crucial reasons for digging a well (to be learned) and the decisions which surround the building of a safe dwelling (already learned). How to accomplish analogies are attempts to relate new procedures to those already learned. For example, the steps in digging a safe well might be associated with the steps necessary to plant a new crop.

It should be obvious from the above description that in cognitive theory, the critical aspect of transfer does not revolve around identical elements in the learning environment. It is instead an attempt to promote the retrieval of previously learned skills and knowledge and encourage their application to new information. In this sense therefore, there is a real difference between the basic assumptions of the two theories and the exact instructional remedies they suggest. It is one of the basic rules of science that two theories about the same phenomenon cannot both be correct. An attempt to use either theory then is coloured with a good measure of uncertainty and the chance that the technology transfer will be less than successful.

Recommendation : In all transfer situations there should be a systematic attempt by both transfer specialists and adopters to make the implicit theories contained in the technology and the transfer plan more explicit. This kind of examination can only strengthen both the technology and the transfer. When they are made explicit, our unexamined assumptions can be checked for soundness. If we suspect that some aspect of our implicit theories is faulty, then we must re-examine that part of the instructional, technology which resulted from the flawed assumption. If we reconfirm our trust in the validity of a particular assumption we have still contributed to the success of the transfer. The assumptions underlyingour instructional techniques are essentially reasons why they should work in a certain way. If they are found not to work, we have the advantage of questioning our assumption and revising it rather than repeating the same error again.

In the case of the two conflicting formal theories, it is possible that the conflict might be solved by using each of the theories at different stages in the transfer process. The transfer techniques suggested by cognitive theory are probably more useful in the initial stages of transfer planning and identical elements techniques should be used in later stages. This resolution of the problem more accurately reflects the research available to buttress the theories and makes it unnecessary to either choose between two apparently conflicting approaches or to combine them in some way.

In the initial stages of a transfer it would be most beneficial if the transfer specialist makes a systematic accounting of the existing skills and knowledge of the adoption group which are relevant to the acquisition of the to-be-transferred technology. Initial instructional plans then will be able to draw upon readily available experience in the adopting group which will promote the understanding and comprehension of the "new" technology. Most often this will involve the use of familiar and analogous examples from the everyday experience of the adopting group which relate to the crucial features of the technology, Wittrock (1979) and Royer (1979) have given examples of other possible uses of the cognitive approach in, for example, the design of technologies, and the evaluation of their success.

Once a rationale and familiar experience for the technology to be adopted have been firmly established, the identical elements techniques can be employed to insure that the important features of the technology are practiced reinforced and extended to the everyday behaviour of the adopters. Here the emphass is less on the understanding of the technology in familiar terms than it is on the establishing of habits for using the technology in daily life over extended periods of time. Where the cognitive theory based techniques attempt to insure that the adopters *understand* why the technology is important, what elements are *crucial* and how it should be *practiced;* the identical elements techniques attempt to insure that the technology is practiced *appropriately, successfully* and *consistently* over time. The identical elements techniques also attempt to insure that the practice of the technology is firmly in the control of the adopting group which reduces the demand for continued outside support.

III. When the Science Underlying the Technology is Inadequately Developed or Poorly Understood by the Technologist

What was the obvious strength of the technology developed for, the exploration of space, a strong and well developed underlying science, is potentially the most important weakness for the successful transfer of instructional technology. Most of the available instructional methods are based on behavioural science models which are in their infancy and therefore vulnerable to misapplication and misunderstanding.

The External Validity of Technologies

The specific problem one finds here is the misapplication of instructional models and theories beyond a reasonable range of general izability. This is what the research design specialist calls a problem in the *external validity* of findings from educational research. Those factors which usually limit the findings of existing research are their extension to different people, settings, and times nearly all of which are present in any transfer of technology situation.

For example, most instructional theories and techniques are validated in controlled studies which utilize very specific and narrowly chosen populations; what might be called YAVIS individuals... Young, Affluent, Verbal, Independent and Schooled. Since technologies tend to flow from more developed to less developed nations, populations in recipient nations tend to be RUPAS peoples. .. Rural, Unschooled Poor, Agricultural and Social or family oriented.

Another problem is the attempt by technologists to generalize from research studies which have tested the effects of instructional techniques with subject populations for the standard thirty minutes of experimental time. One might reasonably question the extent to which an instructional method which was tested with YAVIS students for a brief time would generalize to a group of RUPAS people who are using the method over periods of months or years. There is ample evidence from recent research that people react very differently to instructional methods over time. Those methods which are successful at the onset of instruction very often loose their potency later on and conversely those which are unsuccessful at first often prove to be useful in later stages of instruction.

Recommendation : Any attempt to solve these problems must involve both our existing science and the development and transfer of technology process.

Aptitude - Treatment Interaction

Researchers need to be encouraged to utilize theories which take relevant learner differences into account as main factors in the development of theories of instruction. The current model which holds the most promise for eventually increasing the transferability or external validity of instructional methods is called the "aptitude-treatment interaction" (ATI) approach to instructional research (cf. Cronbach & Snow, 1977). This approach, which promises to bring more coherence generally to educational research, makes it possible to generate models of instruction based on their appropriateness for different people. It is based on the presumption that different people learn differently and that the task of both the researcher and the developer is to find the most ideal match between different instructional methods and recipient groups or individuals. An attempt is made to use more or less universal characteristics of people such as fluid or crystallized ability, spatial skills, locus of control, scalability, conformity, independence, anxiety, need for affiliation plus aptitudes which are much more specific and relate to the prior knowledge of crucial elements of tasks which must be learned. The ATI approach, if properly applied in research, will also increase the chances that we will know why something is expected to work in a specific setting with a specific group of learners.

Formative Evaluation

Even if the ATI approach is applied universally in the science which underlies most technology, it will still be necessary for the transfer technologist to accommodate unanticipated features of a transfer situation. There will always be unique factors which effect the success of our attempts to adapt instructional technologies. The most valuable "adjustment" technique for all instances is *formative development and evaluation.* In this approach, various methods are available for small but important field tests and revisions of key concepts in the instructional program with local groups. The results of such formative tests is information about their local success, and the discovery of factors which are unique to the transfer situation which influence the success of the technology. Technologists can then use this information to make *field revisions* of techniques. Formative evaluation and revision, when properly applied, reduces the need for developing nations to spend disproportionate amounts of development budgets in building a local educational science. If properly conducted, formative development and evaluation may, lead to clearer understandings about the unique factors which enhance and inhibit educational programs (cf. Cooley & Lohnes, 1977). Of course formative activities are costly in time and effort. However, they *must* be considered in the planning of transfers. Current evidence suggests that the alternative is visible failures and invisible successes.

The Training of Transfer Specialists

Finally, it is necessary for transfer technologists to receive much more adequate training in the science which underlies instructional methods and techniques. Clark (1978), in a survey of doctoral programs which train instructional technologists found evidence for inadequate training in those skills which would insure the most accurate use of research based knowledge in the development and transfer process. Most alarmingly however, was the report by those who manage major U.S. doctoral programs that future plans call for an even further move away from research elements in instructional technology graduate curriculums. While the need for instructional technologists to understand the art of development is increasingly recognized and valued, the experience of more established technology based professions such as engineering and medicine suggest that a strong background in the methods and results of research are 3 necessary condition for the successful development and transfer of valuable educational innovations.

IV. Instructional Techniques and Methods which Form the Basis of Programs which are Transferred are not Clearly Conceptualized or Described by Developers

One of the most dramatic examples of this difficulty in instructional technology is the undue emphasis which we place on the role of communication media in education.

Technology as Media

In fact, there is clear evidence that such media do not and should not be expected to enhance educational achievement. They are simply delivery devices which transmit instructional programs. Since they are based on a well developed technology however, their use has somehow been confused with the much more important issue of which many instructional programs and methods are most useful in promoting learning. This misunderstanding has resulted in the development of many so-called "media selection schemes" which presumably match learning needs with the media which will most adequately instruct. If the problem to be addressed is one of transmitting instruction to remote groups or very large populations when the number of teachers or geographical barriers prevent delivery then the newer communication media are worthy of consideration. Wilbur Schramm (1979) has recently noted however, that learning and achievement are more affected by the techniques we employ in the instructional act than by the medium which we choose to deliver instruction. He argues persuasively that if achievement differs as a result of student exposure to two different media (for example, televised instruction versus teacher presented instruction), one would have to conclude that the difference was due to factors other than the television or the teacher. In order to establish the superior capability of one medium over another it would be necessary to control all factors which might contribute to learning other than the medium. In other words, the teacher would have to teach on the television and simultaneously in a non-TV situation. In such a test, one would not expect differences in achievement between TV and non-TV groups which would be due to the medium. Therefore, the most important issue, seldom addressed by instructional technologists, is which of the many instructional methods which are available should be transmitted by an instructional medium. This decision rests less on the medium of instruction than on our current knowledge of instructional research and development.

One additional factor is important. Since transfers tend to occur from more wealthy and developed to less wealthy and developed nations, the cost of the instructional delivery device is often paramount. Developing nations often articulate a need for the cheapest and most effective instructional technology for accomplishing their educational goals. When relatively expensive communication media are offered as technologies which will enhance achievement (even by implication), the technologists have misrepresented their capabilities and paved the way for a potentially expensive transfer failure. The time and expense invested in installing sophisticated communication media is not matched by effort invested in the more important and demanding but less visible instructional method selection and development.

The Specification of Technologies

Admittedly, the precision with which instructional techniques and methods are described is inadequate at best. The most popular current methods such as individualized instruction, learner controlled methods, mastery learning approaches, Keller plan, open classroom methods, audio-tutorial laboratory, multi-image instruction, computer assisted instruction and so on are so carelessly and generally described for those who wish to employ them that many are no more than slogans. The difficulty with this state of affairs is that when these approaches are transferred between nations, the specific instructional acts they designate are interpreted very differently by different adopters. The end result does not reflect the actual set of operations for which we have research evidence of success or developmental experience. The typical failure which results is often therefore interpreted as a deficiency in the original concept rather than a simple lack of adequate specification of the crucial elements of the technique.

There are of course many instructional techniques which have been specified with some precision. Among them are reinforcement and token instructional systems, programmed instruction, and a number of instructional support techniques such as cueing, underlining, advance organizers, note taking, modeling, role playing, inserted questions, and the use of imagery. Even here however the developer has a difficulty. It appears from recent research that nearly all of these better specified methods of enhancing achievement work with some learners and not with others (Snow, 1979). The current direction is to demand that instructional techniques not only be clearly specified by their crucial features but by a clear description of the function they serve in learning for different types of students.

Recommendation : There have been a number of recent suggestions for overcoming this problem, the most promising may be those described by Rhetts *(1974)* and Salomon (1979).

Instructional Development

Rhetts suggests a problem solving strategy which is similar to many of the current development strategies but which places much more emphasis on the specification and study of the instructional methods employed in an educational program. His procedure is detailed and is worth studying.

Functional Analysis

Salomon (1979) provides what he calls a functional approach to the conceptualizing of instructional methods. He describes three models for thinking about the function of all methods and gives a number of examples of each. His first model, remedial functions, is based on the large body of research and practice best represented in the work of Gagne, Merrill, Ausubel and to some extent Skinner. His compensatory model is based in the current work being conducted by cognitive psychologists and developmental psychologists and his preferential model, the most speculative is loosely based on the current work on human abilities and the field of special education.

V. There is an Inadequate Match between the problem and the Technology

There are two major types of barriers that are related to this problem :

1. The predisposition which most technology transfer specialists have to choose a specific technology and search for a problem which it will fit; and
2. the tendency we have to apply technologies inappropriately to features of problems which they were not designed to address.

Technology Advocacy

Whenever we become advocates for a particular technology the temptation is great to apply it inappropriately to a problem. This situation leads to unsuccessful transfer for obvious reasons. The most common example of this difficulty is the attempt to "sell" the use of electronic media to increase learning or to advocate an instructional technique such as programmed instruction regardless of the problem which needs attention. Although electronic media by themselves do not address learning difficulties, programmed instruction and other instructional methods have been found to be more appropriate for some learners and learning content than for others (recent research seems to indicate that programmed instruction works better with lower ability students and/or when highly novel content is being taught). Since nearly all technologies are limited in their scope in the problems which they can help to alleviate limitations on their use is a potential problem in all transfer situations.

Process Analysis

The inappropriate matching of problems and solutions also

extends to misunderstandings about the specific point in the learning process to which an instructional technique is directed and most effective. Relatively complex learning contains many different stages and process for each individual learner. Instructional strategies which characterize a given technology tend to be effective in promoting one or two stages in cognitive processing during learning but not at others. Although educational psychologists differ in the number and characterization of learning stages, most would agree that all learners must be engaged by the task *(i.e.* motivated, attending), that they must be able to decode the instructional materials and understand what is expected of them (the so called "demand characteristics" of the instruction); that they must be able to call forth experience and remembered skills which are required to manipulate or solve the task appropriately; and that the way they solved the task must be stored in such a way that it can be retrieved from memory and applied (generalized) to a similar task at a different time and place (retrieval and transfer of learning). Instructional technologies which would assist in the earlier stages of learning, say programmed instruction, would not necessarily be helpful in storage, retrieval or transfer. Similarly, the use of memory devices such as visual imagery which aid storage and retrieval would probably not enhance the solution stage. Transfer techniques such as identical elements and role playing may aid engagement and understanding the task as well as transferring what is learned to new situations but not assist in storage or manipulation of the task problem. The key point here is that any one of these technologies can be inappropriately applied at the wrong stage in the learning process. They can also be appropriately applied and learning or transfer can *fail to occur* because they were directed towards only one or two stages and not assisted through the entire range of the learning process stages.

Recommendation : In order to prevent the misapplication of technologies to inappropriate problems it would be useful to encourage technology transfer specialists to think of their professional roles as *brokers* of technology rather than as advocates for specific technologies.

Technology Brokering

The client for technology should look for a transfer specialist who is expert in a variety of techniques and who has experience in matching problems with appropriate solutions. In addition it would be useful for the developers of instructional technologies to describe them in such a way that the problems which they were designed and those for which they are contraindicated to solveare clearly described along with their other more obvious features.

Analysis Tasks

Applying technologies to the full range of stages in the acquisition and transfer of knowledge can be enhanced with careful analysis of the instructional task during the technology selection and adoption process. Various task analysis schemes have been proposed by Rhetts (1974). Most instructional technology development models contain a procedure which enhances the chance that instruction will nurture all aspects of the learning process. It only remains for the available models to be used in development and adoption.

References

CLARK, R. E. "Do students enjoy the instructional method from which they learn the least?" Paper presented at the Annual Convention of the American Educational Research Association in Boston Massachusetts. April, 1980.

CLARK, R. E. "Doctoral research training in educational technology" Educational Communication and Technology Journal, 26(2), 1978.

COOLEY, W. & LOHNES, P. "Evaluation research in education : Theory, principles and practice, : Irvington Publishers Inc, N.Y., 1976.

CRONBACH, L. J. & SHOW, R. E. "Aptitudes and instructional method," Irvington Publishers Inc, N.Y., 1977.

FORMAM, D. "The range of evaluation techniques for educational television and radio internationally," Paper presented at the XXIV Annual Conference of the Comparative and International Education Society in Vancouver BC Canada, March 22, 1980.

RHETTS, J. E. "Task Learner and treatment variables in instructional design," Journal of Educational Psychology, 66(3), PP. 339-471, 1974.

ROYER, J. M. "Theories of the transfer of learning," Educational Psychologist, Vol. 14, PP. 53-69, 1979.

SALOMON, G. "Interaction of Media, Cognition and Learning," Jossey-Bass Publishers, San Francisco, 1979.

SCHRAMM, W. "Big Media, Little Media." Sage Publications, Beverly Hills, California, 1977.

SCHRIBNER, S. & COLE, M. "Literacy without schooling; Testing for intellectual effects," Vai literacy project, Working paper 2 Rockefeller University, N.Y., 1978.

Chapter 31

New Communication Technology and the Developing World : A Theoretical Framework for Policy-Making

A. S Babilli

During the last two hundred years, most of the Western nations France, Britain, Germany and the U.S.A. particularly went through a metamorphosis that included three distinct revolutions : the first witnessed their claim to political independence and sovereignty, the second involved an industrial revolution which not only altered their economic landscape but also transformed intimate communities into mass societies, and the third we refer to as the "communications revolution." Most of the developing world, on the other hand, is being compelled by the world order an order driven by servitude to the status quo of nations based on power to leap from the first revolution to the third clearly bypassing the second. A forced metamorphosis poses drastic implications for the developing world.

Consequently, this paper attempts to develop a theoretical framework that explicates the consequences of a "communications revolution" in the developing nations. The paper contends that the technological inventions developed in the West are adapted to the conservative tradition of monopolies in communication which carry consequent disturbances to public opinion and to political organization in the Third World. It is imperative that communication scholars strive to develop theoretical formulations with a view to influence pertinent and democratic policies within a particular national character.

To understand the significance of any technological advance and its impact on a society social scientists must formulate a theoretical and philosophical framework. Theoretical perspectives, however, nascent in view of fast changing technology, are paramount to sound policy-making and to devising approaches to study the decisive impact of technology on politics, society and culture.

Policy, more than ever, must be based on sound theory (Morse 1976). It must be predicated upon an understanding of historical changes in, and consequences of, new communication technology. In this broadest sense then, this paper represents the beginning of a theoretical exegesis heretofore lacking in the study of communication technology at the macro level. Yet it is no more than a perspective. The policy implications that flow from it are both obvious and significant. The paper will begin with an overview of issues that accompany new communication technology, pursuingspecifically the direct broadcast satellite (DBS as an example); this will be followed by an attempt to formulate a theoretical framework for policy-making.

DBS Package

Direct broadcasting by satellite has attracted increasing attention over the past two decades. In large part, this stems from the "great promise" telecasting through space holds for all societies in the world as a revolutionary, low-cost, versatile means of voice, video, data and news transmission. Accompanying such an optimistic view, however is a growing concern with the ways by which technical performance will be maximized in a fashion that may jeopardize development and general welfare of the global mass (Irwin 1971).

Correctly anticipating the potential uses of **satellite** communications advanced countries particularly the United States began to diffuse its scope on an international basis. Various regulatory bodies were organized to oversee issues concerning business, ownership, programming and international cooperation. The U. S. played a vital role in incorporating the INTELSAT (International Commercial Satellite Communications Consortium) *by* virtue of the Communications Satellite Act and that UN's brainchild, the COMSAT (Communications Satellite Corporation). By the mid 1960s, UNESCO found itself active in the media supported educational activities between the member countries. They continued into the 1980s with several important projects in India, parts of Africa, Arab states and Latin America, Specifically, the utilization of satellites for development purposes, usually through dissemination of educational, scientific and cultural information, has become a UNESCO priority. This began explicitly with the 1972 General Conference of UNESCO which adopted the Declaration of Guiding Principles on the use of satellite broadcasting for the free flow of information, the spread of education and greater cultural exchange (UN 1973, Knight 1976).

Before embarking on major projects of telecasting for the less developing nations or less developed countries (LDCs) in the likeness of India's SITE program the UN has been studying the problems apt to be associated with the DBS use political, economic, social and cultural. Among them are control and ownership of the LDC satellite broadcasting systems, public versus private financing, supply of home receivers to designated communities or institutions for specific purposes, social policy, software construction, support infrastructure, etc. Receivers, for example, for both economic and political reasons, are likely to be simple installations, capable of receiving only the channels, perhaps the only channel, on which the state broadcasting entity operates (Masani, et.al. 1973). In such social conditions, access to the system and control over the programming will be firmly in the hands of political party in power or the state the most likely owner and subject to schedule of perceived priorities. Economically (and technically) on the other hand, satellite broadcasting will drain some of the scarcest resources of many developing nations. With a few developing nations as exceptions, the engineering and technical skills needed for design and operation of such systems are likely to be desperately in short supply. Simpler community receiving installations may be manufactured or assembled in countries which possess an industrial infrastructure. The satellite hardware itself, however, as well as launch services and complicated transmitting stations will have to be procured abroad at the expense of precious foreign exchange. Requirements for power supply, maintenance work, and operating personnel in remote, inaccessable parts of the country may mean big budgets. Moreover, many countries cannot support indigenous software production; consequently, even the software may have to be purchased abroad. The DBS software will carry with it all the sensitive problems of translation and adaptation of educational materials from one language and cultural setting to another. The dilemma extends to programming that depicts, even unconsciously, affluent western life-styles which stimulate demand of import of consumer goods.

Despite the success of SITE in India, educational telecasting through satellites in the developing world has not yet lived up to its predictions. UNESCO spent several years exploring the feasibility of regional satellites for the Arab states, south America and the sub-

Sahara (Knight 1976; UNESCO 1976). The planning, execution and evaluation of the SITE project in India, nevertheless, provided clues for the potential innovative technique offered to the developing nations.

However, predictions of DBS effectiveness in developing nations should probably be viewed with caution. The issues that are problematic suggest that satellite broadcasting is by no means the panacea for the educational and cultural needs of all LDCs. Costs, benefits and national priorities besides the program content with inherent political and social value overtones, have to be carefully evaluated along with the overall development plans (Masani et.al. : 1973).

To arrive at formidable policies, a deeper understanding of the ultimate impact of DBS upon the developing world especially on the kaleidoscopic cultures of the developing regions is paramount. Most of the issues for policy making stem from some indication of trends, possibilities and implications of satellite communications in the light of more clearly defined aspects. The UN Committee on the Peaceful Uses of Outer Space (Signitzer 1976) has had several working groups for formulating policies. The debate on DBS at the UN and UNESCO highlight concerns over two indisputable trends in the international flow of information : one-way traffic from the big exporting countries to the rest of the world, and dominance of entertainment material in the flow. (Katz 1973), among many others, feels highly dissatisfied over these trends. Katz also blames the importing countries for harboring misconceived expectations of this new medium via satellites. The nature of news broadcasting via satellites is another implication which has impact on developing nations. Katz calls it the "problem of homogenized news." By that term, he means that programming news content is done by people who want others to see the world from their point of view.

Implications surrounding the effects of DBS are more pronounced in the political arena. At the heart of the debate is the realization that communication is power and that control over the mechanisms and the content of a nation's communication system enables vested interests (private or public), to control important aspects of a society's decision-making apparatus as well as the cultural and political symbols that bind a particular society together (Katz 1976).

Against this broad background of national security and a concern over the cultural integrity of states, the prospect of the DBS has raised dramatic and complex political questions for policy makers. These questions extend to problems of dependency, cultural identity, national integration, political and sociological propaganda individual privacy, human rights, national sovereignty, the decision-making process and social policy.

How, then, can these issues be understood by communication scholars? What framework can be applied to analyze these issues and formulate workable policies? Let us explore a theoretical framework which can explain and help us understand the issues in direct broadcasting by satellite.

A Framework for Analysis

In the early 1950s, a Canadian economist-historian, Harold Adams Innis, (1950, 1951, 1952) observed : 'The future of the West depends on the cultural tenacity of Europe and the extent to which it will refuse to accept dictation from a foreign policy developed in relation to the demands of individuals in North America (1952),' While juxtaposing this observation with what he perceived as the American dominance in communications, Innis (1952) warned? 'Whatever hope of continued autonomy Canada may have in the future must depend on her success in withstanding American influence and in assisting the development of a third bloc (the present Third World) designed to withstand the pressure of the United States and Russia.'

Innis, in *the Changing Concepts of Time* (1952) voiced the fear that European countries stood directly exposed to American influence and to the threat that 'cumulative advantage of size and technological progress of the U.S.' may enforce uniformity and standardization with disastrous implications for the artistic culture of Europe and for Western civilization.

Innis, unfortunately, does not include Eastern civilizations and the ancient cultures of developing nations in his analysis. His work is indeed limited by time and place; most of his analysis was done in the late 1940's and early 1950's. And, being a Canadian nationalist, his studies focused primarily on Canadian relations with her mother country, Great Britain, and her immediate neighbor, the U.S. His death precluded analysis of television as a full-fledged medium of communication. Irrevocably, major developments in communication technology as witnessed through application of satellites for mass communication purposes and their inherent problems eluded his critical analysis.

Ironically yet, after nearly a half century, most of Innis' observations are now increasingly pertinent to the study of satellite communication technology. Strikingly, the basic concepts of Innis, and his propositions provide useful notions in examining the concerns relating to DBS. In the characteristic absence of an orderly, systematic argument in the works of Innis, the author will rely on his foremost interpreter, James Carey (1967, 1970, 1975).

Innis, like Jacques Ellul (1964) and Herbert Schiller (1969, 1873, 1976), is by no means a technocrat a positivist eliciting pragmatic adoration of technology as the neutral tool by which we can create the great and global society of the future in which all mankind's needs supposedly will be met by the inexhaustable resources of technology. But Innis, like Ellul, is by no means, a pessimist either; he sees though cautiously, major advantages in progressive communication technology despite its inherent special bias. In fact, through his *Empire and Communications* (1950) and *The Bias of Communication* (1951), Innis presents one of the most scholarly though not too systematic arguments that communication technology should be thought of as an Integral component of any social structure without which society itself would be quite different in a large number of ways. He presents these insights : that in the context of ancient Egyptian culture, writing on stone tablets permitted the maintenance of social structures in interior areas, but that with the invention of lightweight transportable papyrus, the maintenance of an integrated society extending along the Nile valley was possible. He also argues that Romans could not have long maintained a large land empire without both written communication and a road system. The development and maintenance in both Canada and the U.S. of a democratic political system that spans the entire continent depended in part on the road and the telegraph.

But the communication technology used by the elite in a society for imposition of control through space was what Innis concerned himself with at length. If the deterministic Marshal McLuhan, the disciple of Innis, implied technology is the means of fostering a global village, Innis would not have. Nevertheless, both Innis and his disciple converge on a common view, that of a social historian : the effects of changes in communication technology may be more significant than the transmission of the message itself. Within that context, it can be seen that the power elite of the world whether the U.S. or the Soviet Union can increase their span of control through the medium of communication satellites.

Innis felt that American media in the modern times-news-papers and the radio were inherently space-binding. The flow of American culture into Canada through the U.S. media while justified under U.S. doctrines such as freedom of information and of trade, was, however, radically problematic for Canadian culture. Similar argument is now seen in Schiller (1969), who thinks that the American postwar imperial thrust is unblushingly evident in contemporary satellite communication.

Tunstall (1976), similarly, thinks that American mass media output and dominance has spread to such an extent that no country in the world has escaped its commercial influences. More contemporary criticism is also offered by Carey (1979) on a broader level of Western influence though space-biased communication. Terming it 'the politics of popular culture.' The points to the dominance of the British through the BBC in the Republic of Ireland.

Technological Distortion : Time and Space

To put these observations in context, let us examine concepts of time and space in Innis' works. Along with social control, Innis divided communication into two major categories. *Time-binding* media the manuscript and human speech, favored closely-knit communities nurtured human speech and traditional authority. *Space-binding* media such as print and electricity, were connected with expansion and control over territory and favored the establishment of commercialism, empire and eventually, technocracy (Carey 1970). Innis only hoped that the American and Western bias for special communication would take into consideration other operating, traditional media in world civilizations. Innis (1951) wrote :

We can do little more than urge that we must be continually alert to the implications of this (space) bias and perhaps hope that consideration of the implications of other media to various civilizations may enable us to see more clearly the bias of our own.

He saw that an extended use of space binding communication and its pervasive influence will eventually create a civilization in which life and flexibility will become exceedingly difficult to maintain. He believed that the impact of a new medium like the DBS in this instance will lead to the emergence of a new civilization at the **cost of** the older richer one.

The satellite technology, specifically the DBS dominated by the West provides the modern example of space binding communication in Western civilization. This capacity of technology. Innis pointed out (to borrow from Carey), enhanced the capacity of imperial powers to brine their allies and colonies within the orbit of their control. No amount of rhetoric could varnish or reverse the pattern of technological control. But Innis held the last hope; "Only the workmanship of politics and scholarship, the consistent attempt to maintain another counter culture, offered any viability." (Carey 1970). As Carey observes, the demise of a culture can only be prevented by a deliberate curtailment of the influence of techniques and those institutions in which techniques were encased by a strenuous cultivation of the realms of art, ethics and politics. These observations of Innis can be identified with the present Third World attempts to resist unrestricted application of the DBS to safeguard their cultures which are invariably based on oral tradition. The debate at UNESCO can be seen in relation to what Innis calls workmanship of politics and-scholarship and consistent attempts to maintain countercultures as represented by those in developing countries. These debates to check the dominance of space-biased communication technology was exactly what Innis hoped for. To quote Carey (1970) again : The bias of technology can only be controlled by politics, by curtailing the expansionist tendencies of technological societies and by creating avenues of democratic discussion and participation beyond the control of modern technology.

However, Innis was pessimistic about the outcome of such debates to control communication technology. Historical events tended to support his pessimism; the development of radio end television, the influence of American influence on spectrum allocation and the expansion of communication industries in British colonies resulted in imperialism. In *The Changing Concepts of Time.* Innis charged that vast monopolies of communication occupying' entrenched positions involved a continuous, systematic, ruthless destruction of elements of permanence essential to cultural activity. Despite its shortcomings, the UNESCO debate can be seen as rooted in Innis belief belief that the speed and distance of electronic communication enlarges the possible scale of social organization and thereby greatly enhancing the possibility of centralization and imperialism in culture and politics.

The centralization has already occurred in satellite communication and thereby extending Innis' notion of monopoly of knowledge vested in political control of space orbit. It is evident through the establishment of the Soviet Union's own satellite communication system (Maddox 1972), America's COMSAT and COMSAT dominated Intelsat Control over not only hardware but also software is evident.

Monopoly of Knowledge

Knowledge is power. It can be monopolized like any other commodity. According to Carey et al. (1973), there are two interpretations of this notion. In the first, monopoly of knowledge is seen as simply a means of control of factual information; it allows groups to control such information and deny access to the uninitiated and disconnected. In the second, monopoly of knowledge has a more stringent sense of meaning; it refers to the control of entire systems of thought that determine what is factual, what the standards are for assessing the truth, and that, in turn, defines what can be accounted for as knowledge (Carey 1973).

The majority views held in the UNESCO debate are built on the second notion of monopoly of knowledge. The Third world nations argue that nations which control satellite technology do also monopolize their approved, sanctioned, official mode of thought, and even the definition of what constitutes the truth. Indeed this thought is persistent in Innis' theme : "Each medium has its significance for the type of monopoly of knowledge which will be built and which will destroy the conditions suited to creative thought and be displaced by a new medium with its peculiar type of monopoly of knowledge (1952). In *The Bias of Communication,* he notes : "We are compelled to recognize the significance of mechanized knowledge as a source of power and its subjection to the demands of force through the instrument of the state." Even worse, Innis observes pertinently, history has shown that technological innovations which result in monopoly of knowledge tend to create "consequent disturbances to public opinion and to political organization."

Schiller equates the monopoly of knowledge to what he calls the mechanics of cultural levelling. The content of programming in the DBS and television relays is seen by Schiller as embedded in the notion of monopoly of knowledge. As a consequence, Schiller (1969) observes, in this controlling instrument "content is all that matters, for what it broadcasts may determine, in large measure, the cultural outlook and the social direction of the new nations for generations." Studies by varis and Norderstreng, for example, point to the existence of such monopoly of knowledge (Varis 1973, 1974).

Communication technology has widespread physical, structural and cultural implications for a given society. Using evidence from Egypt, Mesopotamia, Greece, Rome and medieval Europe, Innis documented that the replacement of clay tablets, papyrus, and later the introduction of paper and printing created new forms of monopolies over knowledge and power. True to his observation, the result of applied satellite communications has created control of information by few at the center, while making it more difficult for feedback from people.

Structurally, this electronic revolution has led to the decentralization of communication activities and centralization of authority (Robinson 1975). Evidently, in international satellite broadcasting, as Schiller (1969, 1976) and Kat; (1973) argue, this revolution has narrowed down the cultural variety and will result in a homogenized conception of reality among different societies. These political and business powers controlling the satellite communications, Innis would have agreed, have acquired the social power to produce official versions of the world-fashioned from values stressing market perishability and political control rather than human needs As Carey (1967) suggested, these sort of developments which intensify special-bias, increasingly "transcend all national boundaries, thereby weakening nationalism or at least tending to undercut the parochial limitations of national identification. Further, such media are a potent force in generating a more universal, world wide culture which is urban, secular, and in Innis' terms, 'unstable.'

Also, against the background of national security and a concern over the cultural integrity of states, the prospect of the DBS has raised dramatic and complex questions regarding national sovereignty. These questions dealing with the political issues involved in the "governance" of broadcasting from space are clouded by the contradictory goals and principles set forth in the UN Charter and other international proclamations. The UN Charter, for example, is premised on "non-interference in matters within the domestic jurisdiction of any state. "Yet Article 19 of the Universal Declaration of Human Rights proclaims that" everyone has the right to seek, receive, and impart information and ideas through the media regardless of the frontiers." UNESCO's guiding principles on the use of satellite broadcasting, on the other hand, suggesting that states should take into account the principles of freedom of information to "reach or promote prior agreements concerning DBS to the population of countries other than the country of origin of the transmission." Unfortunately what is formally proposed in one international proclamation is often disregarded or contradicted in another.

Behind the principles expressed by the UN, and beyond this echo of the nations' expression of concerns regarding the implications of the DBS use, lies an essential value : the sovereign rights of states in matters pertaining to communication across national frontiers. Nations, especially those that are former colonies, treasure strict sovereignty .and regard the use of DBS as forfeiture of independence; they perceive themselves to be in an economically and politically dependent position vis-a-vis the relatively few countries which have centralized the technological prowess and possess financial capability to build, launch and maintain such satellites. A growing number of Third World countries also fear that their countries' dependency on foreign broadcast technology and programming will be intensified if they fail to take steps to protect native time-biased communications. A failure to stem the tide of imported technology will only reinforce the "nightmarish realization" that basic decisions regarding communication policies will continue to be "forfeited to outsiders." (Mattleart 1970). The senses of vulnerability perpetuated by unequal access to satellite technology is the most common political theme uniting the advocates of a strong nationalistic approach to the development, control and application of the DBS.

Technology and Informational Sovereignty

This view can draw inspiration from Innis, whose nationalistic stance implied protection of national sovereignty. Describing the power of space-biased communication relating to sovereignty, Innis (1951) says :

Political boundaries related to the demands of the printing industry disappeared with the new instrument of communication (radio). The spoken language provided a new base for the exploitation of nationalism and a far more effective device for appealing to larger members Illiteracy was no longer a barrier.

Also, imported software of the DBS penetrates the host cultures in ways that are yet unpredictable. Thus problems of national sovereignty stemming from dependency, cultural identity, national integration, political and sociological propaganda, individual privacy, human rights, the decision making process, and social policy are now among the matters of broad international concern. Because of these factors, which are parallel to Innis' concerns, international and national political systems are going through a profound change. Mowlana (1977) notes that the influence of communications satellite technology is transforming the traditional views about sovereignty which has always been expressed in special terms, to a new kind of perception which may be called inform national sovereignty.

Paradoxically, enormous improvements in communication technology, Innis (1952) says, have made understanding difficult. He argues that new innovations create political instability; this view can be stretched to the DBS impact on Third World political systems. He equates politics with economics; that the influence of communication (advertising) on the new media and, the "lumpy character of new developments in communications ..... accentuates instability if not insanity."

Dependency theorists and nationalistic media planners argue that satellite communications will only exacerate the existing inequalities in international communication flows, thereby weakening even further the position of the Third World countries (Ploman 1971). Moreover, cultural penetration through satellite telecasting takes place in various ways at different stages of media development in poor nations. At the early stage, direct foreign-ownership is usual. As the history of early television broadcasting in Latin America, the Middle East, Africa and Asia shows, British, American or French companies often directly owned or assisted the establishment of a television system. North American transnational companies can be seen jointly as another moving force in world commercial television as evident in Latin America.

Moreover, at a later stage of media development in the Third World, national production often begins to imitate the dominant Anglo-American way of production. The results are usually worse : national versions of globally marketed materials. As soon as the governments become aware of their dependency on American and other foreign material, they try to establish quotas and take other artificial measures to control 'unwanted foreign influence.'

These economic complications of communication technology are also pointed out by economist Galbraith (1974) : Modern mass communications, especially the movies and television insure that populace at large will see the most lavish comparisoning on the bodies. In South America, in the Middle East, and in Socialist India .... most people are still ..... poor and unable to afford the goods which advertise wealth.

Implications for Culture

Innis also attempts to show that sudden extensions of communication are reflected in cultural disturbances. McLuhan (Innis 1950), in his introduction to Innis' *Empire and Communications,* notes that Innis saw media, old and new, not as mere vortices but as 'living vortices of power creating hidden environments that act abrasively and destructively on older forms of culture.' This statement almost sounds like the echo of the present Third World argument concerning the effects of DBS on their older forms of culture. In his chapter on 'Industrialism and Cultural Values.' Innis implies that constant changes in technology particularly as they effect communication are deterrent to determination of cultural values of a society. 'The cultural values of an industrialized society,' Innis (1951) pertinently pointed out, are not the cultural values of other societies/ Traditional societies rooted in time as opposed to space, when imposed upon by new communication technology, offer a weakening possibility to predict impact of unrestricted DBS telecasting. Innis' observations in this regard can provide a valuable basis for the Third World charge of cultural imperialism.

In direct relation to the cultural segment of the UNESCO debate, Innis (1952) had warned Canada back in the 1950s :

We can only survive by taking persistent action at strategic points against American imperialism in all its attractive guises .... (by) attempting constructive efforts to explore the cultural possibilities of various media of communication and to develop them along lines free from commercialism...

Let us examine these observations of Innis relating to culture in the context of the DBS issues. Even though international satellite broadcasting is often perceived as one of 'cultural interchange,' the reality points to commercial interests. A majority of television programming in international distribution is produced for audiences in the United States, Canada, Australia, Japan and the United Kingdom and emphasizes entertainment. Hence it is not at all surprising to see that at the heart of the debate, there is an evident realization that communication is power and that control over the mechanisms, the content of the DBS communication system enables vested interests (private or public) to control not only important aspects of a society's decision-making apparatus, but also the cultural symbols that bind a particular society together (UNESCO 1976).

These implications arising out of cross-national telecasting have also aroused contemporary scholarly interest. Writing in defense of the 'viability, integrity and values of traditional cultures,' Carey (1979) attacks the larger states and the dominant class in those states who control the international means of production in communications. He fears that the "older forms of imperialism and colonialism have been supplanted at advance stages of capitalism by cultural imperialism : the use of the...... mass media to erode traditional cultures and to incorporate smaller nations into world-wide commercial markets." The importance of cultural survival is thus highlighted (Carey, 1979).

Culture is not merely a structure, a thing; it is a creative process of growth and achievement : a daily struggle that involves a selective recovery of the past in order to endow the present with meaning, significance, purpose and, above all, hope. To keep this process alive and vital among all the members of a community, to promote clear and sustained vision is the great challenge. And the real question is this does television kill this creative process? The answer at the moment, yes, for it conspires to send it elsewhere. The real issue is the distortion of life by the means of communication : the abstraction of culture from time and place. Television everywhere largely consists of messages from nowhere, aimed at no one and presenting nothing. And yet we gull ourselves into believing we are being entertained and informed...

Satellite communication can be viewed as a step towards more powerful and all persuasive mass media whose content binds individuals to a technocratic order this will lead to more intense propaganda and to cultural hegemony by that countries with technological and economic leadership.

Language dialectic dilemma is also one of the main defenses put forward by the developing nations in the **DBS debate.** As most international broadcasting also **carries** with it the inevitable involvement of transmission of foreign languages, this factor may have implications for the evolution if not the extinction of that vernacular languages. If DBS inherently fosters global communications, in what language will it broadcast? Dialectics form one of the methods through which Innis traces processes in communication technology developments. Consequently, Innis was conscious of the implications of new forms of communication technology on language and the vernacular. Links between the technological innovations in media and exploitation of foreign languages is clearly seen in Inms' observations in history (1952);

The British Empire has been exposed to the effects of increasing monopolization based to an important extent on language under the influence of mechanization of the printed word as in the case of the French in Canada, the Dutch in South Africa, the languages of India and Pakistan, and the attempt to revive the Irish language in Eire.

Innis' analysis of dialectics and the media can be related to the Third World viewpoints on DBS impact on languages. Innis (1950) observes that large-scale mechanization of knowledge is characterized by active creation of monopolies in language which not only prevents understanding of other cultures, but also hastens appeals to force. He relates commercialism to the creation of monopolies in language and maintains that monopoly of knowledge itself in one way resulted through struggle between language groups (1950). He observes that development of new media through, printing, by emphasizing on vernacular, created divisions between states based on language and resulted in the growth of intense regionalism. The press not only "accentuated the importance of the English language in relation to other languages, it has also created divisions between classes within English-speaking countries (1950)."

In essence, what Innis is attempting to explain is that new communication technology has been the cause for division and instability in the world. Innis (1950), observes :

The printing industry had been characterized by decentralization and regionalism such as had marked the division of the western world in nationalism and the division and instability incidental to regions-within nations. The radio appealed to vast areas and favored centralization and bureaucracy. A simple individual could appeal at one time to vast numbers of people speaking the same language and indirectly, though with less effect, through interpreters, to numbers speaking other languages. Division was drawn along new lines based on language but within language units centralization and coherence became conspicuous. Stability between language units became more evident and instability between language units more dangerous.

On a much wider scale, what Innis saw in printing and[1] radio is true with regard to DBS. Contrary to the optimistic prediction that "with the advent of the direct broadcast satellite, the influence on children and adults of one-language (English) TV programs and advertising will be enormous," many small and developing countries prefer to think they are faced with the glaring prospect of linguistic homogenization. One researcher points out : "If the world is thought of as a speech community, then English and a handful of other languages are competing for leadership in the same way as dialects within a nation [Ploman 1971 : 21]." The UNESCO debate has witnessed attribution of that pressure on international languages to global broadcasting; others have speculated on worse terms whether world television will lead to the extinction of minority languages (UNESCO 1977). Unfortunately, for most of the Third World countries, linguistic diversity sparks of major problems for their language planning and policies. Those countries which are multi-lingual face two major problems : first, the propagation of a national language within the country, and second, the role of language of wider communication. The problem, then, would be meeting these challenges on a national scale that focuses on either language form or language use or both.

Conclusion

This paper has attempted to juxtapose Innis' explanatory power with the current issues in communication technology at the macro level. When one studies Innis, it appears as though the DBS and cross-national television debates have roots in Innis' work. Canada, of all nations, introduced and applied in 1970 the notion of the right to communicate within her society and did much to popularize it among the Third World countries. The emerging concept of the right to communicate (Harms 1977, Richstad 1977) with its moral and ethical foundations, exhorts equal access to, and participation of the disadvantaged in communication processes. Frenchman Jean D'Arcy coined the phrase, the "right to communicate." Canada implemented it through statutes. And in the international realm, the notion of a right to communicate has given hope for the Third World to rectify the present imbalance in information flow whether it be through news agencies or the DBS.

The explanatory power of Innis may offer new insight into the problems associated with internal or external telecommunications. Communication scholars in the West and in the Third World must explore such a framework in search for sound and workable communication policies that are sensitive to the cultural needs of the developing nations.

References

Carey, James W., (1967). "Harold Adams Innis and Marshall McLuhan." in *The Antioch Review,* 28 : 1.

and John J. Quirk, (1970). "The Mythos of Electronic Revolution," Part I and Part II in the *American Scholar,* 39 : 3.

And John J. Quick, (1973). "The History of the Future," in *Communication Technology and Social Policy : Understanding the New Cultural Revolution,* George Gerbner, *et. al.*

(1975). "Canadian Communication Theory : Extensions and Interpretations of Harold Innis," in Gertrude J. Robinson, *et.al. Studies in Canadian Communications,* Montreal : McGill University.

(1979). "The Politics of Popular Culture : A Case Study," in *Journal of Communication Inquiry.* 4 : 2. Ellul, Jacques, (1964). *The Technological Society.* New York : Vintage Books.

Galbraith, John Kenneth, (1974). *The Affluent Society.* New York : New American Library Books.

Harms, L. S. and Jim Richstad, eds. (1977). *Evolving Perspectives on the Right to Communicate.* Honolulu : East-West Center.

Innis, Harold A., (1950). *Empire and Communication.* London : Oxford University Press.

(1951), *The Bias of Communication.* Toronto : University of Toronto Press.

(1952). *Changing Concepts of Time.* Toronto : University of Toronto Press.

Irwin, Marley R., **(1971).** *The Telecommunications Industry : Integration Vs. Competition.* New York : Praeger Publishers.

Journal of Space Law. (1973). 1 : 2.

Katz, Elihu, (1973). "Television as a Horseless Carriage," in George Gerbner *et. al. Communications Technology and Social Policy : Understanding the New "Cultural Revolution."* New York : John Wiiey & Sons.

(1976), In "Cultural Broadcasting," UNESCO Report on Mass Communication, No. 77.

Knight, Robert P., (1976). "UNESCO's Role in World Communication," in **Heinz Fisher, John Merrill,** eds. *International and Intercultural Communication,* New York : Hastings House.

Maddox, Brenda, (1972). *Beyond Babel : New Dimensions in Communication.* Boston : Beacon Press.

Masani, Ito and Abram Chayes, (1973). *Satellite Broadcasting.* London : Oxford University Press.

Mattelart, Aimand, (1970). "Aresion en el Espacio." Santiago, Chile; UN

Working Paper 2315.

McLuhan, Marshall, (1950). In Foreward for Innis' *Empire and Communications*. Toronto : University of Toronto Press.

Morse, Edward C, (1976). *Modernization and the Transformation of International Relations*. New York : The Free Press.

Mowlana, Hamid, (1977). "Political and Social Implications of Communication Satellites Applications in Developed and Developing Countries," Brent Ruben, ed. *Communication Yearbook*, ICA Publications. New York : Transaction Books.

Ploman, Edward, (1971). "Uses of Satellites for Educational and National Development," UNESCO paper 23-72.

Richstad, Jim, L. S. Harms and Kathleen Kie, eds., *Right to Communicate : Collected Papers*. Honolulu : Social Sciences and Linguistics Institute, University of Hawai.

Robinson, Gertrude Joch, ed., (1975). "The Politics of Information and Culture During Canada's October Crisis." *in Studies in Canadian Communications*. Montreal : McGill University Press.

Schiller, Herbert I., (1969). *Mass Communication and American Empire*. Boston : Beacon Press.

(1973). *The Mind Managers*. Boston : Beacon Press.

(1976). *Communication and Cultural Domination*, New York : M. E. Sharpe, Inc.

Signitzer, Benno, (1976). *Regulation of Direct Broadcasting from Satellites : the U N. Involvement*. New York : Praeger Publishers.

Tunstall, Jeremy, (1976). *The Media Are American Anglo-American Media in the World*. New York : Columbia University Press.

U.N. Doc. (1973). A/AC 105/109/corr. 1.

UNESCO Reports on Mass Communication No. 70. (1976), *Planning for Satellite Broadcasting : The Indian instructional Television Experiment*.

(1977). "Cultural Broadcasting," Report No, 77.

Varis, Tapio, (1973). *International Inventory of Television Programme Structure and the Flow of Programmes Between Nations*. Finland : University of Tampere.

and Kaarle Nordenstreng. (1974). *Television Traffic - A One-way Street*? Paris : UNESCO.

Chapter **32**

Emerging Trends in New Communication and Information Technology for Development in Asia

John A. Lent

The introduction and expansion of new communication and information technology in Asia now threatens every aspect of the continent's existence. Everything from the type of clothing individuals wear, the food they eat, and the way they use their leisure time, to the manner in which their national governments trade, negotiate, and integrate themselves into the international economy have felt the impact.

Yet, the hoopla emanating from the technocrats and other hucksters of new information technology clouds over any serious scrutiny of social characteristics and consequences of information. Computers and other such gadgetry have taken on a protected status - they are almost sacrosanct. They are beyond criticism, and those who dare question the merits of new information technology risk being labeled traditionalist, old-fashioned, anti-technology, or a "Luddite," a term one professor used in condemning non-users of such technology.

Everyone, especially those in poor countries, is told that to be modern, they must join the technology 'race'. More than a decade ago, critic Herb Schiller (1983:20) described this pressure situation. He said it goes something like this : "Unless the race is joined, you will be left behind. Entrance into the modern age, it is said, demands participation in the race."

However, Schiller cautioned that, "entering the race requires careful consideration. There are races to be run and races to be won. There are also consumer society, with information technology leading the way, is what people aspire to?"

In the past decade and a half, most Asian governments and mass media have been smitten by the lure of this new information technology. Japan pioneered in this area, computerizing, news rooms in the early 1960s, and earlier, launching newspaper facsimile networks. By the 1980s, Singapore set as its goal to become the information technology center of all of Asia, and in Malaysia, Japan, India, Singapore, and other countries, larger media operations transformed themselves into integrated information industries, hooking into business, industrial, military, and government networks. Correspondingly, a number of Asian countries launched their own satellites, among them Indonesia, India, China, as well as others.

The amount of information forthcoming from such technology has given some of these governments and especially corporations within them the edge in negotiations, trading, and sales. In some isolated cases, the technology has been used to enhance people's education, health care, nutrition, and cultural well-being. But these are rather rare consequences, since the chief beneficiaries have been, and will remain to be, the military, national governmental bureaucracies, and large transnational corporations (mainly form the West and especially from the United States).

In the hands of these entities, the new information technology has engendered patterns such as widening gap between information have and the many information have-nots and commodification of information, stripping it of all social relationships and oftentimes putting it out of the reach of the general public. Schiller (1983:19) showed how research reports have been priced at astronomical levels, how formerly-called "free television" had been converted to "pay television" through substantially - expensive cable subscriptions and "pay-per-view" fees, now public library services have been put on a paying basis, and how corporate sponsorship has invaded film, theater, museums, news, the arts, sports, and general education.

From the outset, the military has been involved in the design, funding, and development of most information systems, including radar, computers, satellites, lasers, and cable television. The United States military has been the major actor (and user for wartime purposes) in this regard, but those of Asian countries have also been involved in the setting up and use of advanced communications systems. For example, when Indonesia's Palapa was launched in the mid 1970s, the military and government were reported to be the major beneficiaries, using it for centralization of power.

As early as 1945, Norbert Wiener (quoted in Roszak 1986 : 178), the father of cybernetics, feared that new technology would be exploited by the military to wage war and misused by industry to deskill and disemploy workers. Wiener did not see these as separate maladies, but part of the same structure.

Computerization offers important advantages to government bureaucracies in processing masses of data. But, it also is extremely effective for social control (read social engineering in Singapore) purposes, capable of amassing and storing voluminous amounts of data on individuals' work habits, tax payments, credit ratings, consumer behaviour, dissident actions, and movement, at the same time increasing the authority of the technocrats.

Computer expert Theodore Roszak (1986:187) believes that by helping bureaucracies, new information technology actually hinders democratic growth, acting as a surveillance machine that invades people's privacy, and an election tool that sells candidates in tiny bytes of information in much the same way that commodities are sold. Biasing elections in favour of those who can spend the most for the best polling services, Roszak added, leads to a dismal style of politics that grows more and more obsessively focused on imagery, sloganeering, rhetorical legerdemain, and the huckstering skills of the market place. Such a political situation, abetted by new information technology, has already permeated a number of Asian countries.

Roszak (1986:209) is adamant that the benefits the citizenry receives from advanced communication system will never outweigh the costs, stating, the violation of privacy is the loss of freedom. The degradation of electoral politics is the loss of democracy. The creation of the computerized war machine is a direct threat to the survival of our species. It would be some comfort to concludethat these liabilities will result from the abuse of computer power. But these are the goals long since selected by those who invented information technology, who have guided it and financed it at every point along the way in its development. The computer is their machine; its mystique is their validation.

Although Roszak, Schiller, and others were calling attention to ill effects of information development in the United States, the impact has been worldwide, and perhaps especially in Asia, deemed the locale of major economic development in the 21st century. It has been suggested that if the nineteenth century was Europe's (especially Great Britain's) and the twentieth was the United States', then the twenty-first will be Asia's. Schiller (1983:20) made a point of emphasizing why peoples of other parts of the world cannot sit back and say "it's an American problem, why should we be concerned?" He said :

The United States is the hub of a still incredibly powerful international economic-military-cultural system. The impulses coursing through America do not remain confined within that national territory, encompassed by two oceans. They are driven outwards by very powerful forces, not least of which are the super, transnational companies, of which two-thirds (in 1982) are American owned.

When dealing with new information technology, most governments are unresponsive to social development needs. Sussman and Lent (1991:21) said : Ambitious for increased integration with First World capital, state technocrats in the Third World have paid little attention to the sometimes disfiguring effects and economic distortions of the introduction of advanced communication commodities from abroad. Poorly paid and hazardous assembly - stage computer chip production by mostly female labor in Malaysia and the Philippines; the eye-straining monotony of data entry and transmission work, often farmed out to South Asian and Caribbean workers' the surreptitious supervisory monitoring, facilitated by digital equipment standards, of harassed long-distance telephone operations;.... all expend the energy and attention of Third World citizens in forms that subsidize the life-styles of peoples far beyond their shores. Relatively few Third World workers are sharing in the 'revolution' in electronic science and technology, even as their exploited labour puts shirts on the backs and computers on the laps of subsidized Western consumers.

Governments throughout Asia have followed the lead set by the Reagan administration in the United States (and that of Thatcher in England) to deregulate and privatize transportation and communication. In the area of broadcasting, the rationale was that deregulation would ushser in multi-channel competition, but instead, it has meant unrestrained cross-media ownership and vertical monopolies that stifle competition and do away with the concept of broadcasting as a public service. Within days in the summer of 1995, in the U.S. Disney acquired ABC/Capital Cities; the nuclear company Westing-house bid on CBS network, and the large newspaper group Gannett took over Multimedia, a broadcast and television syndication firm. Already, the other television networks were in conglomerate hands - NBC owned by General Electric, an atomic relative of Westing-house, and Fox by media vulture Rupert Murdoch. All of this was sanctioned by the United States 104th Congress.

Murdoch, a powerful force in Asia now with his ownership of Star TV, has been a strong advocate for the relaxation and even elimination of public service and public interest obligations of broadcasters, claiming that more options will be the result. But as Melody (1995:67) wrote:; "Greater choice does not necessarily mean greater diversity, adequate service to either majority or minority interests, easier access to audiences for alternative programming, or a more informed public opinion." Obviously, one victim of the government's relinquishing its public service regulations is the concept of accountability. As conglomerates become immense and align with governments through electoral favors, they wield tremendous power, resulting in less government surveillance, less competition, and less demand for accountability from the public, the latter comatosed by government hype.

The third major benefactor of the new media and information technology is the transnational corporation. To ensure quick and efficient business transactions, these companies demand the most up-to-date information infrastructures. This is usually a prerequisite before they will set up an overseas operation. In this paper, the discussion will focus on the transnational media firms-giants such as Murdoch, Berlusconi, Time Warner, Paramount, Bertelsmann, etc.

Communications has become such an important commodity worldwide that these corporations and a few others have expanded much energy to be among the top five or six they believe will control the media of the world by century's end. Their work will be made easier as the trend continues towards national concentrations of ownership to a few groups. In recent years, Asian newspaper groups have become affiliated with transnational corporations such as Dow Jones and Murdoch's News Corporation, and combines of government, local business, and regional conglomerates have purchased parts of broadcast stations and newspapers in at least Hong Kong, Taiwan. South Korea, Indonesia the Philippines, Malaysia, Singapore, and India. Group ownership and cross media ownership are prevalent in many instances.

As possibilities of lucrative media acquisitions in the West dwindle, and as some Asian publishing and broadcasting outlets become more economically profitable, these outside conglomerates, in their lust for more and more properties, will increasingly look to

Asia and other parts of the Third World. Murdoch has already owned the South Chine Morning Post of Hong Kong, as well as Star TV; Dow Jones has had parts of media properties in Hong Kong and Singapore, and other media moguls, such as Maxwell, Sumitomo, Viacom, and others, have bid on newspaper, cable, television, and other media properties.

Asian mass media are increasingly found in some national conglomerates which connect to transnational corporations. For example, Samsung Group, owners at times of South Korean newspaper, magazine, and broadcast interests, controls a portion of Corning Corporation in the U.S. Other Asian conglomerates have had huge investments in communication are Indonesia's Bimantara and Malaysia's Fleet Group. The country's fastest growing corporation, Bimantara started Rajawali Citra Televisi Idonesia in 1988. With assets of more than US$1 billion, Bimantara has equity in 90 companies and close ties with the ruling establishment. By bringing commercial television to Indonesia, the company realized part of its five-year plan to become a major force in telecommunications, already owing Elektrinod Nusantara, Settel Technologies, and CakraNusa.

Fleet Group, part of which was sold in recent years, reached astronomical proportion, not only in mass communications, but in many other business and industrial pursuits. It controlled the commercial television channel, at least five major dailies, three Sunday newspaper, many magazines, book publishing firms, a food products company, and a major hotel chain. Fleet was made up of, or closely tied to, over 150 companies. Following a management by four executives of the New Straits Times Press (part of Fleet) in 1993, the NSTP and the commercial channel, TVS are now under the Malaysian Resources Crop. Bhd.

When Singapore's press was completely restructured in 1982-83, three dailies closed and three large press firms merged into Singapore Press Holdings. The new company encompassed not just the press, but the entire economic scene, becoming the number one Singaporean, publicly-quoted, industrial company. Singapore Press Holdings has ownership connections with many foreign companies, is involved in cross-media, ownership, and has joint operations and firms involved in non-communications fields.

As early as 1980, one of those companies merged into Singapore Press Holding had engulfed whole or parts of properties in England, Tahiti, New Caledonia, Singapore, New Zealand, Australia, Thailand,

Brunei, Malaysia, Hong Kong, and the U.S., with plans to acquire others in Indonesia, Sri Lanka, and the Philippines. Some of the foreign connections were with significant media concerns, including the Hachette Group, France's contribution to the world of media titans.

Cross-media and non-media ownership prevail among other newspapers of the region. Hong Kong newspaper groups owned other media and a travel agency; South Korean dailies were tied to travel agency, construction, automobile, and explosives business; Indian newspaper groups are owned by conglomerates in some instances, and Japanese media are characterized by concentration and monopolization. The five largest Japanese dailies account for sixty percent of Japan's hundreds of regional editions daily, and have many media and business holdings. For example Asahi publishes at least one English-language daily, three weeklies, three monthlies, and ten yearbooks, and maintains interests in sixty other enterprises, including radio-television stations, travel agencies, and real estate firms.

The close alliances between media barons and business, industrial, political, and military powers present serious limits upon the expression of diverse viewpoints and to the watchdog function some Asian presses have performed; instead, they lend continued support to power brokers and the status quo. There are other possible impacts of such concentrations of media ownership, of which three are singled out here.

1. The farcical bigness of mass media in parts of Asia and worldwide represents a serious threat to the democratic ideal of an open market place where all ideas can contend. These conglomerates are interlocked through ownership and directorship to huge banking and other business and industrial complexes, as well as the government and military, some important stories affecting vested interests are either slighted or completely ignored. In the West, one must wonder how much critical comment about military spending, defense weaponry, or even war finds a place in mass media owned by France's Hachette, which is merged with Matra, the country's largest defense manufacturer, or on NBC network in the U.S., owned by General Electric, which does more business with the military than any other company. Big business journalism, like big business generally strokes big government, and vice versa.
2. With the weakening of unions by big corporations and governments and the devil-may-care attitude of many media

moguls, disparities in the mass communications work place have been augmented, and the rank and file media workers find themselves in extremely tenuous situations. They do not know when the media mogul will whimsically close down their place of work, or sell or trade it; they watch helplessly as their chief executive officers gather large chunks of revenue to themselves. Those with talent even find the entry points to the creative world have been shut for all except the chosen ones of the moguls. No doubt, as the big corporations move towards maximizing profits, they close themselves to outsiders among journalists, authors, musicians, and other creative people. They buy every possible means of producing and distributing their products and prefer to use their own, rather than independently created material.

1. Homogenization, commercialization, and trivialization of media products have been the end results in many cases where conglomerates were in control. One would get very little argument that journalism and media entertainment have sunk to new lows where all parties involved look to ratings, sales, and possible other uses for media products within the corporation. Calling the result of all this the development of an idiot culture, Carl Bernstein (1992) of Watergate reporting fame recently wrote:

For more than fifteen years we have been moving away from real journalism towards the creation of a sleazoid infor-tainment culture in which the lines between Opray and Phil and Gerardo and Diane and even Ted, between the New York Post and Newsday, are too often indistinguishable. In this new culture of journalistic titillation, we teach our readers and our viewers, we pander to them... We are in the process of creating, in sum, what deserves to be called the idiot culture.... For the first time in our history, the weird and the supid and the coarse are becoming our cultural norm, even our cultural ideal.

Bernstein cited numerous media conglomerates that support and enthusiastically promote trash journalism:

And the great information conglomerates of this country (U.S.) are now in the trash business. We all know pornography when we see it, and of course, it has right to exist. But we do not all have to be porn publishers; and there is hardly a major media company in America that has not dipped into the social and political equivalent of the porn business in the last fifteen years.

Trash journalism has been promoted by broadcast networks and their affiliates since at least the early 1970s, in the guise of "happy talk" news, featuring facile faces, plastered pompadours, trivialized and tasteless talk, and what Max Lerner once termed the "daily disaster diet."

And recalling Schiller's earlier quote, this idiot culture phenomenon is not isolated to the U.S.; with the new information technology, such as satellites and cable television, it becomes available, and in some cases, fashionable almost anywhere.

More specifically, a look at what is being, or is likely to be, exported to Asia is in order. In news, both the program format and presentation style of United States networks and CNN, carried via satellite, have left their impacts upon Asian news casting. The format and style was described by Schecter (1994:5), who said:

Most of the networks clone each other's looks, homogenizing all differences away... We all know the news format: highly paid news stars 'anchoring' broadcasts that look the same, sound the same, and feel the same... The stories are short and self-contained, rarely linked to each other, and rarely illuminating larger systemic problems.

The content usually reflected western (United States) imaged and assumptions, often downplaying those of Asia. I recall while traveling through South and Southeast Asia during the summer of 1993, seeing over and over, in one country after another, the CNN pictures of the devastating flood that swept the Midwest of the United States. At the same time, as damaging or worse floods covered Bangladesh but were not covered. The result, as Thusse (1994:11) wrote, is that Asian countries must "live the western version of world events." He said :

When western strategic or economic interests come into play, the supposedly neutral and liberal media lose all pretension to objectivity and tend to follow the standard western line. This can mean slanting reporting in terms of tone, emphasis and space given to any story.

The most pervasive United States news casting operation in Asia has been CNN, the 24 hour daily network that reaches 30 million households in 130 countries of the world (see Hong 1994). Although CNN World Report has been credited with providing a balanced presentation of development and non-development issues, with most stories produced by journalists/news organizations from the participating countries (Fowler and Shafer 1994), the network has also come in for its share of criticism. Chief complaints have dealt with its broadcasting in English when only 20 percent of Asians understand that language, and its crisis orientation, repetitiveness, and concentration on United States news. As a result, many overseas viewers, especially in India, have drifted away, usually to BBC World Service. Some Asian news organizations, among them NHK in Japan and the Hong Kong-based Business News Network, are planning to expand news programming to the rest of Asia and the world.

Of more serious concern has been the impact these outside newscasters have had upon local political situations. In India during 1992-93, a debate was waged on whether foreign coverage and transmission via satellite television (CNN and BBC) helped spread violence by fanning communal passions. Official statements in India never specify religious groups involved in riots, whereas CNN and BBC did. Former Singaporean head of government, Lee Kuan Yew, in a South China Morning Post article of October 17, 1990, also blamed the Tiananmen Square massacre upon foreign news coverage. He said the Chinese students' experience of international television reporting of peaceful Philippine and South Korea demonstrations had given them the mistaken notion that protests would bring about rapid political change in China.

Much entertainment programming that makes up the American "Idiot culture" has found its way into Asian households. United States game shows are imported outright or cloned to meet local language or cultural situations. In the Philippines, for example, Michael Jackson look-alike contests have been televised, as was a "grandmother Madonna" competition, with elderly women dressed like the rock star and dancing on television to "Like a Virgin" (Kalter 1987:28). In 1995, Grundy Worldwide was aiming to put United States-style game shows in Chine, India, Japan, Hong Kong, Indonesia, Malaysia, and the Philippines. In each case, locally-made versions of such United States shows as "Family Feud" or "Wheel of Fortune" will be co-produced with Asian broadcasters or production companies.

Also transplanted or imitated are United States gossip and talk programmes, neither noted for accuracy or good taste. One writer, critiquing the American penchant for gossip, said, "Instead of finding meaning and moment in their own lives, Americans were encouraged (by gossip) to live in daydreams about the lifestyles of the rich and famous, to emphasize the material over the spiritual" (McDowell 1990:51).

United States-style TV talk shows have become popular in Japan, Singapore, the Philippines, and India, among other countries. Such shows of the United States often dispense advice on sex and other controversial subjects, and show people struggling or boasting about bizzare sexual and emotional tribulations. Talk show recently came under attack in a paper read before the American Psychological Association. The authors, Jeanne Heaton and Non Wilson, wrote :

Talk TV shows distort psychological problems both in terms of symptoms and frequency, provide short sound bit advise that misrepresents responsible treatment and exploits existing stereotypes about gender, race and sexual orientation... They use the most bizarre or fringe events as examples of what people might experience in their relationship, family or job. These situations are portrayed to seem like they happen every day and to everyone... 'Marriage with a 14 year old' or 'Men who sleep with their parents' sisters' are not common occurrences. This misconstrues reality for a lot of people since the audiences are reaching 15 million per show daily... (These shows) have become the leading source of information about mental issues (TV Talk Shows a Danger to Mental Health 1995).

Other denunciations of exported United States popular culture have blamed it for giving message damaging to young psyches and dangerous to society. Among some such messages are : violence is a justifiable means to resolve conflict; women are sex objects to be abused/used by men; to hate certain groups of people is all right; children's disrespect of elders is "cute," and so on. Discussing United States audiovisual messages in relationship to his homeland of Mexico, Sanchez-Ruiz (1994:76) stated that the economics of culture and communication cannot be left adrift to the 'invisible hand' of market forces... We are not talking about mere commodities as such, but also of assertion of cultural meanings, of a sense of identity, of affirmations of pride in collective being, of scales of values, social ethics and aesthetics, among other things... It is not a matter of closing borders. It is a matter of having some control of who enters your home to tell your children what the world is, what they should value, who they are and should be.

Sanchez-Ruiz's warning is particularly timely as global programming increasingly becomes the trend, at the expense of loss of programming of distinctive national characteristics-Coupled with global programming, of course, is a unified global mass marketing via television. Without international regulations, advertising already dominates television and other media crossing national borders and is quickly putting entire societies up for sale (athletes who look like walking billboards, museums and sporting arenas corporately sponsored, etc.). As Melody (1995:68) predicted :

> 21st century mass media television is taking on the characteristics of a global electronic billboard. It is likely to be 'entertaining,' offering a plethora of choice within the narrow framework of intensified commercialization, to the primary benefit of global mass markets. The public interest objectives of true diversity of programming in response to the diversity of interests and cultures in society, and of maintaining an informed public, are not likely to be high priorities on this system.

Asian countries have begun to introduce interactive television, opening up a whole new range of commercialization where shoppers will choose, order, and pay for products "unseen," while sitting in their living rooms. Further, telecommunications industries of Asia are mapping out plans for as many as 500 channels, which will turn homes into electronic shopping malls and target viewers as consumers, not citizens. Such television phenomena, along with the burgeoning of malls throughout Asia, will also mean the doom of street vendors and open marketplaces, noted for their variety, personal-ness, honesty, and freshness.

The 1990s has seen Asia inundated with new satellite and cable television systems, almost all of which were designed to import the aforementioned foreign programming. Western media programmers were keen to tap into the huge potential market of Asia, where by late 1994, there were more than 400 million television receivers, 230 million of them in China. The Chinese total is more than that of the United States, and the prediction is that by 1999, five percent of the set owners in China will subscribe to foreign television services. India, with 42 to 47 million television households, Indonesia, with 190 million people, and Japan are other countries foreign programmers have their eyes on.

Satellite television was virtually unknown in most parts of Asia until late 1990, when Hong Kong bilionaire Li Ka Shing launched Star TV. The potential lucrativeness of this system was apparent to Murdoch, who soon after purchased 64 percent of Star TV. He bought the rest of the shares in 1993. MTV Asia started up in September 1991 as a joint venture between Star TV and the United States media giant Viacom. Within two years, 38 Asian countries adopted MTV Asia. Other satellite television operations have been planned, including the 24-hour Mandarin news service aimed at the Chinese in Southeast Asia and backed by one Hong Kong daily paper, Ming Pao.

Some Asian governments, in efforts to protect their national broadcasting system and to counter the onslaught of foreign culture that satellite television brings, responded by barring satellite reception, but by 1995, most saw the futility of such, legislation and tried other methods. Among these are establishing more national television channels; providing a greater variety of quality local programming; launching their own satellites to offset those dominated by West and to broadcast to neighbouring countries, and doing co-productions with outside agencies.

The abundance of outside television brought in by Star TV, MTV, CNN, and others has had mixed blessings. In India, MTV Asia has been credited with lifting the music industry to new heights overnight. One company, Magnasound India, invested US$5,000 to produce a video of a Hindi rap tune that was boosted on MTV Asia; the result was 500,000 albums sold. Similarly, the Beijing based Tang Dynasty heavy-metal band was rescued from obscurity when MTV Asia aired its video; one-half million of the group's records were subsequently sold. The Chinese recording industry is relatively new, consisting of about 200 state-owned recording/distributing companies competing with one another (Riggs 1992).

In summary, at least the following trends are discernible in Asian communications and information:

1. An almost fanatical growth and expansion of new information technologies in all sectors of society;
2. a leassening of government regulation and the privatization of national media systems;
3. the swallowing up of many mass media by national conglomerates with the likelihood that they, in turn, will be bought by transnational;
4. the augmentation of close alliances between media barons and business, industrial, political, and military powers, with impacts on freedom of expression and the media workplace;
5. the continued homogenization, commercialization, and trivialization of media products, both those brought in from

the outside and local clones, and

6. the inundation of the continent with satellite and cable system.

REFERENCES

Bernstein, Carl. 1992. The Idiot Culture. The New Republic. Fit for Fat. 1995, June 19. Asia week, p. 17

Fowler, Gilbert L. And Richard Shafer. 1994. An Analysis of the CNN World Report Gatekeepers : A 1994 Survey. Southwest Education Council for Journalism, Flagstaff, Arizona, Oct.

Ghose, Bhaskar. 1993. Impact of Transnational Broadcasting on Regional Society and Culture. Asian Mass Communication Bulletin. Nov.-Dec., pp. 10-11.

Hong, Junhao. 1994. CNN Sets Its Sights on the Asian Market. Media Development. 4, pp. 10-13.

Kalter, Joanmarie. 1987. Guess Right-Or We'll Throw You to the Bess . TV Guide. July 11, pp. 26-28.

McDowell, Jeanne and Naushad S. S. Mehta. 1990. Pssst.. Did you Hear About? Time. March 5, pp. 46-51.

Melody, William H. 1995. Information Infrastructure of the Future: An Overview. Media Asia. 20-2, pp. 63-70.

Riggs, Peter. 1992. Country Report: China. Popular Music and Society. Winter, pp. 11-36.

Roszak, Theodore, 1986. The Cult of Information. New York: Pantheon.

Sanchez-Ruiz, Enrique E. 1994. The Mexican Audiovisual Space and the North American Free Trade Agreement (NAFTA). Media Information Australia. Feb., pp.70-77.

Sechecter, Denny. 1994. Human Rights and Network Wrongs. Towards Freedom. Dec., pp. 5-7, 35.

Schiller, Herbert I. 1983. The Communication Revolution: Who Benefits ? Media Development. 3, pp. 18-20.

Sussman, Gerald and John A. Lent. 1991. Transnational Communication: Wiring the Third World- Newbury Park: Sage.

Thussu, Daya Kishan. 1994. Subverting the South. Toward Freedom. Dec., pp. 11-12.

TV Talk Shows a Danger to Mental Health. 1995, Aug., 16. UPI dispatch, Korea Herald.

Chapter **33**

Asian Skies New Technology and Media Imperialism

Daiwon Myun

Standing on the threshold of the 21st century, the world has made phenomenal strides in the communication field. In some Asian countries, people have experienced product leapfrogging, such as in China, which has gone from no phones directly to cellular and from no typewriters straight to computers. The broadcasting media are no exception in Asia. Having purchased a TV set, many Asian people have at least 20 channels of both nationwide network television and cable or satellite broadcasting, confusing which is the right one and where it comes from.

New technology has brought both the subordination to the West culture and the ability to overcome the foreign media pressure in Asian countries. As we have seen drastic changes of the culture including clothing and food in China, media imperialism is becoming a major issue in some Asian countries due to the modern communication technology. This phenomenon seems to be transient or permanent according to their economical status or political background. The study will show different effect of modern broadcasting market or technology in several countries and regions.

The Advent of Cable & DBS Era

Now announcements are being made promising 500 channels. The production dimension, however, is conveniently ignored when new services are announced. More channels does not necessarily mean more different programmes. Historically it has mean more repeats.

In this respect, cable and DBS (Direct Broadcasting by Satellite) might be a very limitedservice, because distribution of more channels similar in content to those carried by existing cable channels, without the benefit of local station's programming, will face severe marketing difficulties. On the contrary, the value of network programmes has been enhanced by the increased investment and association with the programmes of local affiliates to compete with that of cable and DBS. This is exactly what we have seen in Korean broadcasting industries.

In general, cable broadcasting industries all over the world have to face three types of problems. First of all, most of the audience have had the opportunity to enjoy various programmes on free network television. They are not willing to pay for viewing cable channels even though they already have subscribed. Second, competition among cable companies means less advertising income, more production costs. Third, cable companies thought that they would be successful selling advertising time on a different basis, citing the potential rather than actual audience. There is usually a very large difference between the two. This confusion seems to come from the term "subscriber". For example, there are more than fifty-eight million subscribers on CNN, which is part of nearly every basic cable package. But when Nielsen measures the actual CNN rating, it finds that only a tiny fraction of these people watch. The prime-time rating of CNN is only 0.6.

In Korean broadcasting industries, competition between network television and cable leads to wasting money or resources and increasing deficits. In 1995, 20 cable programme providers and 53 cable station operators suffered from a $375 million deficit. One programme provider lost $1.25 million every month last year. News and movie channels such as YTN (news), Catch one and DCN (movie), and HBS (entertainment) experienced worse situations, losing at least $2.5 million a month each. Some critics expect that cable broadcasting industries in Korea can make a profit in five years. *(Dong-A Daily News, January 16, 1995)* There are at least two difficult problems in making this expectation happen in 2000.

Never before have Korean viewers been fascinated by the outstanding and very competitive programmes in over-the-air Korean networks. Some Korean-produced programmes such as Sand Watch were big hits with an over 60 percent rating in 1995. How Korean cable TV programmes can compete with these network programmes is the first problem to resolve. The second difficulty is how Korean cable TV operators can deal with the programme shortage and quality. They have to cope with the gap between small subscribers and big production costs. Han (1991) suggested limiting the cable channels to a maximum of 10, including the over the air networks; he thought this should be for at least few years, until the software production infrastructure is gradually improved and becomes relatively competitive.

In Taiwan, many subscribers don't pay, limiting the amount of money available to purchase programming. "No one is making money," says Lin Chong-Liang, General Manager of Pan Asia-Pacific Satellite Media Communications. "Only the big groups operating channels can afford it" (Keenan, 1995).

In India, the number of cable operators is starting to shrink as they're bought up by more powerful players like the Hinduja Group, RPG Enterprises, and Zee TV, of which Star TV owns 49.9 percent. Some are being undercut by the giants, which may charge as little as $1 per month for service compared with the $3 smaller distributors charge. It is predicted that within a few years, India's cable TV business will resemble its telecoms sector, with one or two operators per state.

Media Imperialism Revisited

Media imperialism can be defined as the imbalance and inequality in flow of mass media materials between developed and developing countries, and its subsequent effect on the developing country's society and culture. (Frederic, 1993) For example, in the Caribbean region, 79 to 88 percent of all television programmes are imported. United States entertainment programmes constitute the largest portion. At an estimated cost of $400 per hour purchased, Caribbean television stations spend more than $11 million annually on importing these foreign programmes. (Wert, M. & Stevenson, R. 1988) This dumping by syndicators can undermine local, indigenous television in the region.

The reason Third World countries import Western programmes are complex. We should also take into account other media channels in those countries, availability of home videocassette recorders, lack of production space and equipment, lack of engineering and artistic talent, lack of funds for local production, and pro-Western orientation of programmers due to their academic training in the West.

In their 1974 study, Nordenstreng and Varis reported two trends in the international flow of television programmes : a one-way flow from the big exporting countries to the rest of the world, and domination of this flow by entertainment programming. These two aspects constituted a tendency towards concentration, they stated. (Nordenstreng & Varis, 1974)

This is not just a Third World problem. Even rich nations, such as Canada, have had reason to complain about media imperialism. The influence of American programming on Canadian broadcasting is reflected in the following statistics. Of the 52,000 hours of English - language television programming available to the average Canadian each year, barely 370 hours are Canadian drama including sitcoms, movies, and miniseries. Only 28 percent of all programming available on English-language television is Canadian; of the total time Canadians spend watching English-language television, less than one-third is devoted to Canadian programming. (Mukasa, 1987)

In a follow-up study, ten years after the first in 1973, Van's found one thing held constant : In most countries, one-third or more of total TV Programming is imported. But a considerable effort has been made to change this dominance. There was a notable increase in regional programme exchange, particularly in the Arab states, Latin America, and Caribbean. About one-third of imported programmes in the Arab world originated in the region itself. (Varis, 1984) And Caribbean nations were trying to set up a regional television production centre called Carib Vision (Frederick, 1988).

The satellite telecommunication networks have also made possible the growth of regional television news and programme exchanges in Asian countries. The *Asia-Pacific Television News Exchange (ASLAVISION),* with headquarters in Kuala Lumpur, was started in April 1984, when a group of Asian countries, including India, Pakistan, Bangladesh. Sri Lanka, Brunei, Malaysia, and Indonesia, began exchanging television news weekly through INTELSAT satellites over the Indian and Pacific oceans. Following a principle of "free to offer, free to receiver or to refuse," ASIAVISION is now exchanging news daily. A second ASIAVISION network was also created linking China, South Korea, Japan, Hong Kong, Australia, and New Zealand.

What is it about local media products that have successfully emerged as alternatives in Asian, Arab, and Latin American countries? Some developing countries now have enough funds and personnel to make good - quality programmes to be able to appeal to the historical and cultural backgrounds of local peoples, and Asian people prefer these newly-made local programmes to old Hollywood movies. In Taiwan, viewers seem unimpressed by the programmes available on their cable services; Korean broadcasting also shows the same phenomenon. The top-ten broadcasting programmes are all Korean-produced ones.

Localization of Multinational Cable TV and DBS

There are two reasons for adopting a localization strategy to survive in Asian skies. First of all, there are dozens of different languages and cultures in Asia. English-language shows and movies have a definite limitation in that English is only spoken by a small elite. Second, it is predicted that the TV wars will soon begin to produce causalities. The field will become even more Towded with the launch of the AsiaSat-2 satellite. Digital compression will allow regional satellite network Star TV to air local language programming region-wide, placing greater demands on limited TV viewing time.

Multinational satellite TV broadcasters are now emphasizing local-language programming geared to individual countries. "Localization is the way to go in Asia," says MTV International's Tom Hunter. He admits that two and a half years of free-to-air, pan-Asian broadcasting was a good way to introduce his station to the region. But now "we are trying to set up commercial operations where profits are a key criterion," Hunder says "For that, the way to go is Mandarin, Hindi, perhaps later Thai or Indonesian & other Asian languages" *(Asia week, 19 Oct' 1994).*

The challenge is also to provide quality programming. With broadcasters tailoring channels to specific audiences, satellite television is becoming more attractive to advertisers. Companies recognized that their marketing power can increase if they beam in Hindi in India and Mandarin in China.

Star TV is preparing local-language broadcasts, expecting to launch a Japanese channel by April 1 and preparing 15 local languages, pay -TV channels for Indonesia. Star TV hopes that locally targeted channels will brighten its appeal to advertisers. After it was sold to Murdoch, Star TV updated its old programming and tailored it to specific audiences. Thus, in April 1994, Star TV launched Star Movies, a pay channel featuring predominantly Mandarin Films, and later four more targeted pay -TV stations.

TNT & Cartoon Network, owned by CNN founder Ted Turner, are aimed at attracting children and Hollywood fans, combining animation during the day with MGM movies from Turner's vast library at night. The channel has the cartoons, "Tom and Jerry" talking in Thai in Bangkok, Mandarin in Taipei and English in Hong Kong.

Like its STAR competitor, Prime Sports, U.S. sports channel ESPN is adapting to the region. Sports tastes vary in Asia. Indians want to watch cricket, Chinese, volleyball, and Koreans, baseball and soccer. Sumu (Japanese traditional wrestling) and Japanese based-ball games are top priorities in Japanese sports programmes. ESPN cannot just dish out American baseball or Canadian ice hockey programmes to Asians; more local sports coverage and planned dubbing in local languages are necessary.

HBO Asia also is looking increasingly at both subtitling or dubbing movies in local language and directly showing local films. Almost all movies that go into Thailand and Taiwan already carry Thai or Chinese subtitles CNN already has specialized local-language segments in Japan, and BBC World Service Television provides Arabic newscasts for West Asia and four hours of simultaneous local translation for its Japanese beam.

After a two and one half year partnership with STAR, MTV decided to bail out, leaving the Hong Kong based satellite network to devise its own all music channel. STAR music video channel (V) decided to target individual markets in the Asia region. The reason is that only 5 percent of the Asian audience is comfortable with English - language programming. On (V), Mandarin ballads mix the Indian Bhangra and Tagalog love songs follow Japanese pop *(Asia week, October 19, 1994)*. (V)'s northern beam disseminates eight hours of Mandarin programming, while its southern one transmits the same amount in Hindi. (V)'s local - focused strategy has been evaluated more successful in Asia than MTV was in its first two and half years. Despite its earlier reluctance MTV is going local too. Once digital compression is available, both (V) and MTV Asia will be able to beam shows in other languages such as Thai and Bahasa Indonesia.

Reverse Flow of Communication

In some other parts of the world, a noticeable change has also been to more local and regional production. Straubhaar (1984) found the decline of U.S. programming on Brazilian television; Rogers and Antola (1985) noted that television programmes exported to Latin America (the majority from the U.S.) decreased between 1972 and 1982 and that indigenous and regionally produced tele-movies, essentially soap operas, took the place of foreign programmes. Further, there seems to be the beginning of a reverse flow; Latin American tele-movies are increasing inpopularity in U.S. cities where Spanish-language television stations now attract large audiences (Stilson, 1986).

Brazil's tele-movies are as successful in the global market as most American television shows. However, as Oliveria (1992) has complained, this does not mean those tele-movies are serving Brazilian cultural needs or exporting Brazilian values. The shows are situated in a fantasy of developed affluence and feature stars whose looks tend towards a European model. Although domestic productions may be popular with local audiences, domestic producers are not necessarily in dialogue with their national culture.

Things are little different in Asia. Hong Kong director Wong Kar-wai has created a film, *Chungking Express* that completely visualizes the nature of Hong Kong. He tried to express the conflict, anxiety, or instability of a young generation in Hong Kong right before its historic 1997 change. The film suggests the possibility of freedom, opportunities, or release within the trapping of potential problems (Tsui, 1995). This most ethnocentric film is now a big box office hit in Asian countries. For instance, it is a very popular movie in the Korean video rental market ranking seventh of 20 best movies (Dong-A Daily News, Januarys, 1996).

The following case study of Korean film and broadcasting industries shows another instance in reverse flow of communication. The openness of the Korean film and broadcasting market and the advent of the cable broadcasting era have made Korean-produced programmes more competitive. Most of the top 20 television programmes have been locally produced since 1992. Among the 20 best films in the Korean movie rental market, four each from Hong Kong and Korea are included. This is quite a surprising change. Two years ago, the Korean video rental market was dominated by Hollywood films (90 percent of the total). Only five Korean-produced films were included in the top 100 video rental list in 1993 (Man & Won, 1995). This ethnocentric or local-focused phenomenon also has been found in the Korean pop music industry. The glory of American pop music in the sixties and seventies in Korea had already disappeared in late eighties. Several radio programmes of American pop music has been faded out or replaced by Korean pop music ones.

There are three desirable conditions for local TV programme production in Korea. First, the Korean conglomerates' (chaebols) participation makes possible enough funds to produce good programmes. Second, new Korean broadcasting systems (cable and commercial networks) require more programmes. Third, Korean producer's capability for excellent programming is now highly evaluated as competitive (Robinson & Man, 1992). South Korean films already are winning foreign festival awards for their high quality.

The advent of a satellite broadcasting era in Asia, with the launch of many local satellite including Korea Sat I *(August* 5, *1995)* and Korea Sat II *(January 13, 1996)* promise immediate Asian distribution for Korean video and film products. Korean products also have broken into the North American market. U.S. Korans spend an estimated $50 million ($250 per Korean household) for Korean videotape rentals per year *(The Chung-Agn llbo. May 6, 1991).*

In 1995, South Korea started cable broadcasting with 20 programme providers and 53 stations operators. Nine out of 30 major *chaebols* became key investors of programme provider (PP) in the Cable TV industry in 1994. They run 11 channels. SamSung which has owned and lost private broadcasting network TBC, fulfilled a long cherished desire by owning two cable channels - movie (Catch One) and documentary (Channel Q), Hyundai, well known auto maker, also owns an entertainment (Hyundai Broadcasting) and a music channel (Korean Music Broadcasting). DaeWoo, automobile maker and electronics manufacturer, was approved to have movie channel (DeaWoo Cinema Network), having already involved itself in the movie industry, especially home video market Lucky-Goldstar seized the last ticket to the Cable TV industry with a home shopping channel (Korean Home Shopping).

Korean *chaebols* are also rushing to fill South Korea's video programming gap. Domestic filmmakers are linking with the large conglomerates. DaeWoo has signed a long term contract with HyunJin Film. SunKyung, which lost its chance to own a Cable TV channel, decided to get into the programme production and distribution business by allying with Mido Films. Lucky-Goldstar has signed with TaeHung Film and HyunDai with HapDong Film. Some *chaebols* are entering directly into the production business, SamSung, which owns a home video production company, submitted the documents necessary for the approval of the Ministry of Information to enter the domestic film market; it and other companies are hurrying to set up or expand programming studios- On April 29, 1995, Cheil-Chedang, former Samsung's subsidiary and still run by Samsung's family, made a contract with Hollywood leading power figures, Jeffrey Katzenberg, David Geffen and Steven Spielberg, to create 'Dreamworks', a software production company, investing $300 million for an 11 percent stake (Fabrikant, 1997).

Conclusion

There are two major perspectives opposing cultural or media imperialism. One is the liberal perspective, which acknowledges the existing imbalance, but argues that market power will redress it. Technology will allow nations to remedy the imbalance and inequity. Another is the centrist perspective- Centrists think that even though the United States has advantages in contemporary global world, developing countries can mobilize all internal power to offset foreign media pressures.

Media imperialism still has the power, explain some case studies. However, it must be modified to fit some local phenomena happening in some countries of Asia such as South Korea, Latin America, and the Middle East. The centrist's perspective is more reasonable and powerful in explaining local case studies in these area than is media imperialism, which is a grand theory or paradigm to encompass all developing countries- Generalizing a theory, media imperialism, is less efficient and more dangerous in local studies in these days because many developing countries are no longer passive recipients.

Modern technology gives nations not only subordination to Western countries but also ability to overcome the subordination. Some nations are still in subordination status by the technology, but others, using the technology, started or are starting to mobilize their internal strength to make their own voices. In 21st century, the study of media imperialism emphasizing the contradiction between the West and the Third World in 70s and 80s, will be and must be localized to understand what is happening in a country.

REFERENCES

A TV extravaganza : In the no-dish State, a 30-Channel Menu on Cable. (October 20, 1995). *Asia week.* p.32

Battle for Asia's Teens: Broadcasters Are Trying to Cater to Local Tastes. (October 19, 19, 1994). *Asia week* p.53

Boyd, D. (1988). Third World Pirating of U.S. Films and Television Programmes from Satellites. Journal of Broadcasting & Electronic Media, 32(4), pp.149-161.

Dinosaurs without Gold Eggs: 375 million dollar deficit in cable industry in 1995. (January 16, 1996). *Dong-A Daily News* p.9.

Frederick, Howard H. (1988). Caribbean media workers combat information colonialism. *Extral,* January/February, p.14.

Fabrikant, G. (January 20, 1997). Great Expectations For a Fledgling Studio, *The New York Times*, D 1, 9.

Frederick, Howard H. (1993). *Global Communication and Internal Relations-* California: Wadsworth, p.135.

Man, G. {Winter. 1991). The flow of US telecommunications policy and the cable TV development plant in Korea: Lessons from US experience of competitions. *Journal of Broadcasting Research*, pp. 171-211.

Man, G. & Won, W. (1995). Building a telecommunication infrastructure for the 21st century public policy implications of the cable television development plant in Korea. *Asian Journal of Communication 5* (1), pp.1-15.

Jankowaski, G. & Fuchs, D. (1995). *Television today and tomorrow: It won't be what you think*. New York: Oxford University Press.

Keenan, F. (October 5, 1995) Visual Clutter: Asia may get more TV-viewing choices than it wants. Far *Eastern Economic Review*, p. 86.

Mukasa, S. (1987). The *Epistemology of the New World Informatio/ Communication Order*. Montreal: McGill University, Centre for Developing Area Studies, p.7.

Nordenstreng, K & Varis, T (1974). Television traffic- A one way street ? *Reports and Papers on Mass Communication*, No.70, Paris: UNESCO.

Over $50 million spend on video. (May 6, 1991). *The Chung -Angtlbo. p.1.*

Oliveira, O. (1992). Brazilian soaps outshine Hollywood: Is cultural imperialism fanding out? In K. Nordenstreng & H. Schiller (eds.), *Beyond National Sovereignty: International communication in the 1990s.* Norwood, UJ: Ablex Publishing. P. 128.

Robinson, D. & Man, G. (1992). US and ROK cable system and advanced communication networks. *Asian Journal of Communication*. 2 (2), pp.105-133.

Rogers, E. & Antola, L. (1985). Telenovelas : A Latin American success story. Journal of Communication. 35 (4), pp. 24-35.

Stilson, J. (1986). Latin beat heats on new TV outlets: Attracts lollars & viewers in top markets. *Variety*, pp.31,124.

Straubhaar, J. (1984). Brazilian television : The decline of American influence. *Communication Research. 11, pp. 221-240.*

Twenty best movies in Korean video rental market. (January 6, 1996), *Dong-A Daily News, p.9)*

Tsui, C. (Wilnter, 1995)- Subjective culture and history : The ethnographic cinema of Wong Kar-wai. *Asian Cinema*, 7 (2), pp.93-124.

Van's, T. (Winter, 1984). International flow of television programmes. *Journal of Communication*. Pp. 143-152.

Wasser, F. (December, 1995) Is Hollywood America? The transnationalization of the American film industry. *Critical Studies in Mass Communication, pp. 423-437.*

Wert, M. & Stevenson, R. (Spring, 1988). Global television flow to Latin American countries. *Journalism Quarterly,* 65. pp.182-185

Chapter **34**

Organizational Constraints in Indian Television System

V. N. Bhat

Television in India, to sum up, depends greatly on extraneous factors like political and economic considerations while deciding internal issues. Such a dependence on external forces has important implications for the internal structure of the organization, for the commitment of material and non-material organizational resources to outside interests is likely to affect various programmes and departments as well as the very core of the television system.

Modern organizations base their functioning on two decisively important questions : What are the choices available and what consequences flow when a particular choice is made? What are the factors that affect in significant ways the choice of a particular alternative or strategy? Related to these two transactional problems is the third issue of the consideration of the choice of wrong alternatives, inadequate techniques, or cloudy and culpable means-ends relationship. In this paper a modest attempt has been made to look in to the Indian television as a system of constraints by raising these questions and by attempting to answer them in terms of consequences while keeping in mind the fact that physical and social environments themselves impose numerous restraints upon any human organization.

Functional definitions of the terms to be used in the course of this paper are in order at this juncture. First term, which cries for a comprehensive definition not withstanding voluminous literature, is Organization itself, The term 'organization' in this paper is used to express *a system (structure) of consciously created coordinated, and interrelated personal activities or forces directed toward a final goal and influenced by the environment.* The second term *constraint implies conditions that are to be satisfied at least to a certain degree if an organization is to be considered effective.* These conditions are "operationally defined by a set of referents or standards." Failure to meet such constraining conditions represents a state of ineffectiveness of the functions of the organization.

The concept *structure* has been well explained by Perrow (1979) who says : "The object or 'raw material' (in an organization) may be living being, human or otherwise, a symbol, or an inanimate object. In the course of changing this material in an organizational setting, the individual must interact with others. The form that this interaction takes we will call the structure of the organization." Thus, organizational structure involves relationships that permit control and co-ordination of the tasks performed by an organization.

The approach adopted here to study television system in India as a system of constraints can be termed as an integrated structural process approach. These structural variables constitute division of labour, authority structure, span of control, patterning of command, and direction. The study, however, also looks into various processes of goal defining, interpersonal and small group interactions, resocioalization and anticipatory socialization of the members, and related processes. In other words, through an implicit way, the conception of organization developed by the 'contingency theory of organization' (Lawrence and Lorsh 1965) has been used.

In a situation of limited resources and numerous needs and the resulting constraints, relatively better efficiency can be achieved by effectively dealing with the constraining variables to increase each unit of input/output.

The idea of Indian national television on an experimental basis was debated at the 1956 General Conference of UNESCO held in New Delhi. As a result an agreement was signed between All India Radio and UNESCO in 1959 and a pilot project was set up to study the use of television as a means of education, rural upliftment and community development. The UNESCO and AIR agreement conceived this attempt as an experiment for assessing the value of different types of educational television programmes suitable for group viewing in rural and urban communities in such a way that the result of such an assessment may be useful not only for India but also for other Asian countries in the use of television for community education. The experiment was initiated in Delhi by the Central Government with a lot of back-up, and tele-clubs were formed to reinforce television messages. An evaluation study established significant benefit to the audiences; specially in the field of information gain.

School broadcasting for Delhi area (primary schools) was started in 1961 in collaboration with the Ford Foundation and 250 schools in the Central Province of Delhi were covered for in school viewing to begin with. By the end of 1965 this benefit was extended to as many as 36,000 students of science and 96,000 of general English. On August 15, 1965 a regular television service began in New Delhi. In October 1972, a new station was opened at Bombay and the following year, a relay Centre at Pune.

Television Centers at Srinagar and Amritsar (1973), and Calcutta, Madras and Lucknow (1975) soon followed. By 1976, Indian television catered to a population of 45 million spread over an area of over 75,000 sq. km.

Satellite Instructional Television Experiment (SITE), the result of a contractual arrangement to use ATS 6 satellite for one year, marks an important landmark in the history of Indian television. SITE (August 1975-July 1976) was the result of the programmed effort of NASA, Indian Space Research Organization (for hardware technology) and Ministry of Information and Broadcasting (for software technology and maintenance) in utilizing community television sets to beam educational programmes. Under this programme, community sets had been provided in about 2400 villages in six states. SITE was a unique enterprise which had hardly been tried with success elsewhere, and thus, attracted international attention. Of the six states which were selected for SITE, all of them have terrestrial transmission centers already working by now. Commercials on 'Doordarshan' ('views from afar' official name of the Indian television) introduced on 1 January 1976, are now being telecast for about 6 minutes of every television hour chunk from seven centers and four relay stations. Advertisements are booked as movies as well as static spots.

Television coverage is now available to about 83.2 million population (figures are for the year 1980-81) spread over an area of 2,00,000 sq. km. Doordarshan covers 56.1% of the area and 15.2% of the population of India (figures are for the year 1981-82). This possible television audience covers 35.5 million urban and 48.16 million rural viewers. The organizational set up of Indian television system is being presented below which gives an idea of this growing system and linkages. Indian television network forms a part of the Ministry of Information and Broadcasting, and is wholly owned and controlled by the Central Government.

India has had 56 years of radio; only five years less than the BBC; and 14 years of effective television, but with an almost negligible effect on Indian national life. On the cultural side, the effect has been somewhat marginal. It has been lamentably small on the administration and management of this huge populace. To be brief, Indian electronic media disappointingly fell short of expectations. What could have been the problems? Limited resources to satiate unlimited needs are obvious, but little can be done to change them radically. Instead, the concern should be placed in the sphere of those constraints which are accessible to manipulation and can be adjusted relatively easily. A good understanding of these constraints should help the policy makers as well as the programmers not only to develop a realistic picture of Indian television, but should also assist them in suggesting possible changes in the organizational structure as well as its functions toward a better efficiency.

Environment

Television in India limits its own function and operations because of its placement in the constraining physical and social environment. Nature falls short of doing its best for the working of TV as can be seen in the case of the rural TV-SITE Continuity television. In this instance, all but one of the so called 'grassroots' rural television centers are situated in the hearts of metropolises, away from the villages, making regular contacts between the source and the receivers almost impossible. Unmanageable terrain, "jeepable" all weather roads, supply of electricity, and so on add to this infrequent interaction between the rural TV and its target audiences.

Constraints in terms of social environmental constituents are even more powerful. As crisply pointed out by the Macbride Commission members (1979), communication "comprises first and foremost, a cultural dimension and constitutes a social phenomenon governing the relations between individuals and human groups. The scale and complexity of communication problems far transcend those relating to how the mass media are used." Some of the important domains of the social environment which have a heavy influence on the working of Indian television are related to the economic system, the system of political control and cultural domination, socialization process of the personnel, and, dominant and subordinate coalitions within the organization itself.

A. Television and the Economic System : Discussion on the production-distributionsystem in India as being unequal has become a bromide issue and it is a common place understanding that there exists a huge recess between the well-off and the weaker sections of the community. As Beteille (1966) aptly pointed out in his celebrated study on 'caste, class and power', ability to influence the destiny of the nation lies in the hands of a small minority constituting the upper class interests. Mass media in general and the electronic media in particular, project this feature of the Indian society powerfully.

It requires little effort to demonstrate the obvious upper class character of the mass media in India. Under the garb of "functional autonomy", radio and television, for instance, go out of their way to concoct and brew the images of the ruling class and men in power; few journals in the country come out openly against the ongoing oppression of the weaker sections of the population. Time and again the upper class bias of the television is revealed in no uncertain terms. To cite an example from the working of the rural SITE Continuity TV, the so called 'educational' television is currently engaged in telecasting full-length movies and the film music under the caption "entertainment", and there is considerable evidence to argue that this is mainly a gesture in response to the manifest pressures brought by the Indian film industry on the television management. Interference of and predilection for the upper class in the urban television centers is glaringly perceptible in the nature of TV shows and the content of the programmes. Suffice it to say at this juncture that the advertisements on the urban television control the nature of programmes in a major way.

B. Television and Socio-political and Cultural Dominance : Closely related to the issue of TV and the economic system is the question of sociopoliticaland cultural dominance-dependence relation which is a direct outcome of the former crux. 'Doordarshan' is controlled by the Ministry of Information and Broadcasting, and is wholly planned and financed (except for the advertisement revenue of the commercial television) by the Central Government. Hence, television has little freedom to air issues and ideas which may contradict the policies of the political elite. The governing elite, which basically is made up of the high class and high caste individuals, looks at the industrialized West as an ideal model for 'modernization' and is Western oriented in its broader values and life style. There undoubtedly exists a gap between the "progressive" outlook of the elite and the 'traditional' outlook of the masses and owing to this gap in perception of the world, often the TV looks at its audience as if it were a conglomerate of low level of intelligence and of even lower taste. Programmes are produced in isolation from the social realities of this 'passive' mass leading to a new form of cultural dominance dependence relationship.

National and local political powers constrain the freedom of the television to manipulate the network in another way. Hierarchy in the television structure is a direct consequence of the constraints imposed by the government authorities explicitly or implicitly, and serves as a good example of "constraints from above".

C. Socialization of the Personnel : The process of socialization also acts as a basic environmental constraint in the functioning of Indian television. One of the key problems that Indian television encounters is how to socialize members into its organizational culture, particularly when the new administrative creative members bring with them the values that contradict the standard operating situations. For instance, television producers often enter into the production units highly motivated, idealistic, and committed to the cause of meaningful mass communication. Idealism is quickly lost, and the new recruits learn during their resocialization process that zealousness in the performance of their duties only causes trouble in one way or another. Thus, there seems to be an important on going course of operation in the creation of the organizational staff constituting desocialization of earlier norms and resocialization in institutional values. At the stage of selection itself, the personnel are quite carefully recruited so as not to disturb the existing organizational standards. Specially in the case of administrators and the technical staff, many come from 'second class' training background. Their social origin is generally modest. During their studies they were only taught a 'technical language' and were hardly educated in larger issues of social importance. Institutions that train the television staff (technical and non-technical). Film and Television Institute of India, Pune and Mandi House, New Delhi, hardly focus upon real live concerns pertaining to present day Indian society, and the training largely consists of imparting information within the boundaries of four walls. This choice of mode of selection, training, and socialization process guarantees a lack of political appreciation by the staff of the organization's important decisions on planning and implementation policies, and effectively constrains the efficient functioning of television.

D. Dominant and Submissive Coalitions : Cyert and March's (1963) conception of organizational goals being determined in a process of bargaining and settlement among dominant coalitions, or Mills' (1965) concept of different elites or interest groups negotiating to their advantage in organizational goal setting are very true in the case of Indian television. Television network as a formal organization is divided between different interest groups such as engineers, administrators, the production staff, and the utilization and maintenance personnel, each group trying to promote its own interests.

Some coalitions are obviously more powerful than others due to their importance, indispensability, or accessibility to power and control. These dominant coalitions bargain for their interests and in this process impose conditions of restrictions on organizational decision making and action. Submissive coalitions like the production staff and the contract artists do not wield authority and therefore can hardly participate in goal setting and directional functions of television organization.

One constraint which has a determinate influence over the functioning of Indian television is the factor of intra-organizational conflict between different coalitions. Conflict within the television network, as it is in any similar organization, is mostly over the internal allocation of resources : who shall make the rules governing resources, what will be the content of those rules, who shall enforce them, and how. This conflict is directly related to increase in salaries, creation of new posts, promotion, and such issues which affect the life chances of members in the organizational hierarchy. The divergent motives of members towards these incentives are the major sources of intra-organizational conflicts. They are also sometimes over the problem of the administrators' desire for autonomy and participation as pointed out by Dahrendorf (1959).

A second source of conflict which restrains the functioning of Indian television is the ambiguity of rules and procedures in complex situations. Different groups interpret the same rules and objectives differently and interpersonal differences based on languages, region, ethnicity, sex and age impede the efforts to achieve a common understanding of the given situation. Mutual task dependence, task related dependence of one group on another relationship of performance to rewards, role dissatisfaction, etc; also contribute to intra-organizational conflicts and enhance the constraining influences on the working of television.

In a nut shell, environment plays a critical role in deciding on the choices that are available to the organization. The constraints that are placed on Indian television by the environment become an integral part of the organizational structureand frequently work even without being noticed by the participants themselves. Environment is a powerful constraint on the organization because it is the reality. As Matheson (1978) points out, "Sound implementation will depend for a large part on sound planning, providing always that the plan is in accordance with the practical environmental realities."

Technology

Organizational forms and functions are limited by the nature of the existing technology. Low quality of programming in the case of Indian television system is, to a considerable extent, due to low quality film and film processing, non-availability of high quality videotapes, bureaucratization of the selection procedure of the technical manpower, and the resulting vacancies and ad-hoc appointments, short supply of sophisticated equipments for film and VTR, editing, vision mixing, and so on. Indian television was black-and-white till recently, and at present there is a limited colour TV coverage in most-centers.

The larger and equally important question in this regard is the application of technology in formal organizational settings. Experts on classical management theory have striven to put forward principles which would apply to organization, and technology, from this point of view, was treated as ethically neutral and non-participatory in character. It should be realized, however, that technology in itself is neutral but its application within a set of socio-economic circumstances is not neutral. Even with the given level of meagre technological sophistication, television in India would have performed better but for the constraining socio-economic and political environment within which technology is supposed to act.

Technology and organizational structure are strongly related. This proposition goes to the Marxian thoughts on the technological basis of the divorce of capital from labour, and later to Durkheim on the very notion of the division of labour. In the case of Indian television, a low level of technology works as a mighty constraint to work satisfaction, efficiency, organizational performance, and overall weakening of the system as a whole.

Task

Proper identification and definition of the task of the personnel has been one of the major constraints in the working of Indian television. A basic problem with this set up is that its role and boundary are not clearly stated, leaving ample scope for every one's own interpretation as pointed out earlier. Responsibility, which makes the strength of the educational television meaningful, seems to have got lost in the huge hierarchy of bureaucratic set up. There is an atmosphere of insecurity, indifference, and opportunism in the television set-up which appears to grow out of a lack of any clearly and precisely defined role and responsibility.

There seem to be three primary variables in this respect, they are: pressure from lack of time; pressure from insufficient authority; policy as a constraint in performing the expected task.

A. Pressure from Lack of Time : Television personnel, specially those who are engaged in production related activities at the level of TV centers, are hard pressed for time. Majority of television centers employ technical and creative staff much too small to carry out the required task and to discharge the necessary responsibility satisfactorily. In some TV stations the producers are expected to fill-in almost 100 minutes transmission chunk every week. Thus, in this rat race, these people are on a constant pressure to deliver the 'messages' in time with little regard for its quality or content. On account of this short supply of qualified manpower and the resulting on rush for meeting the deadline, time constraint becomes primary and accomplishment and quality considerations are pushed into the background.

B. Pressure from Insufficient Authority : Another major constraint related to the task accomplishment is the lack of power and authority to the staff often resulting in despair and passive indignation. The bureaucratic regulations in television, coupled with basic insecurity among the lower ranks owing to the present 'contract' system, render the telecasting activity virtually frail and feeble. In this context, the working conditions of the television producer who has little authority and control over studies, has virtually no power to obtain beyond certain footage of film while shooting, or even to use editor or floor manager of his choice. In such a predicament the producer is a dummy in the hands of his administrative 'boss'.

C. Pressure due to Organizational Policy : It has already been recorded that the organizational policy decided by the negotiating coalitions constrains television performance. Any zeal and zest shown by the lower ranking television personnel may cause him the wrath from the top circles. This organizational policy which prohibits the employee from exhibiting any creativity and initiative, unwittingly (unintentionally though) leads to an inefficient system.

Control system

Victorious coalitions in the conflict for control of the organization generally use their power for repressing or disregarding demands of the submissive coalitions to get a bigger share of the 'limited pie' in a process similar to the "iron law of oligarchy". In the case of Indian television, the administrators and the top engineering staff having direct 'connection' with the political elite and the Directorate, use the power to control and direct the organization according to their whims and fancies. In one sense, this ruling coalition forms an intra-organizational political complex in an effort to control the goals, and performance of the organization.

The 'political system' of an organization like the Indian television which determines the vital economic choice is generally very centralized. The domination it exerts on the organizational structure does not necessarily require formal rules and bureaucratic regulations : for instance, in the television set up, routine staff inspections and daily meetings with the hierarchical superiors considerably influence the work itself and lead to greater elite control over the working system.

Hierarchical structure of Indian television network with its characteristic vertical and one-way communication system from the Ministry to the "grassroots", and the authoritarian order are direct forms of control system as a constraint to the actual functioning of the television set up. Control in several areas of decision making is over-riding. The legislative decisions regarding expansion of departments, implementing new schemes, special allocations and so on are generally kept as secrets and the concerned staff is told about these matters only at the last minute and is thus often caught unaware of the needed preparations, resulting in temporary and short-term planning for long-term goals.

Size

Blau (1970) Records : "By generating differentiation, large size indirectly raises administrative overhead (...) and if its influence on differentiation were unrestrained, large organizations might well have disproportionately large administrative machineries, in accordance with the bureaucratic stereotype". This analysis fits well in the case of Indian television to a large extent resulting in functional constraints. Large size of the television network increases complexity in the task structure, adds potency to the 'initiation' and 'negation' powers of the ruling coalitions, makes the submissive coalitions passive and uninterested in organizational affairs and thus increases inefficiency. Television, with its large size and organizational structure, is increasingly becoming bureaucratically rigid and static. This encourages both an inflexible, hierarchical, non-participatory control system. This greater structural complexity also intensifies problems of communication and coordination and makes new demands on the time of both the planners and the programmers to run the organization at a reasonably efficient level.

Size of the organization not only augments the power of dominant coalitions, but also increases impersonel and 'bureaucratic' relationship within and between different categories of workers. For example, whenever the employees have grievances in a large organization like Indian television, these grievances are routinely "processed" by the administrative staff, thereby effectively closing off these people from the top of the managerial and administrative hierarchy.

The constraining variables are closely related to one another in the larger environment of Indian television system. They have been dealt here separately only for the sake of analytical clarity and convenience. It is also not the contention that these are the only constraining variables that should be looked into for an adequate understanding of the working enigma of Indian television network.

Udy (1959) found out that the ownership, structure, recruitment practices, and the general reward and incentive systems in organizations are clear reflections of the economic and social environment of the society in which the organization exists.

The problems with television in India, as they are with other mass media of communication, are one of bridging the gap between the "have nets" in terms of information within the society. While pondering over world information order and international media dependency, it seems to be equally important to think of intra-societal information inequalities. With improved infrastructure for global communication, it is conceivable to bridge the present gap between the international elites. The "best people" in Delhi, Bombay, and

Madras may be at the same international information level as people in London, New York, or Tokyo but how about the 630 million others who cannot afford a paper, cannot buy a radio, cannot think of watching television? The so called 'community television' should strive to live up to its name and that can be possibly achieved only when these constraints are effectively dealt with.

Television in India, to sum up, depends greatly on extraneous factors like political and economic considerations while deciding internal issues. Such a dependence on external forces has important implications for the internal structure of the organization, for the commitment of material and nonmaterial organizational resources to outside interests is likely to affect various programmes and departments as well as the very core of the television system. Problems of short supply and coordination, as pointed out earlier, are likely to become particularly acute under these circumstances. The most effective way to overcome these problems is by heightening the frequency of internal communication and participation by the grassroots employees. Effectively dealing with the environmental constraints. The unequal economic system, political and cultural dominance of the elite sections resulting in high class dominant coalitions, training systems which encourage passive and non-human attitude towards the needy sections of the population, rigid, hierarchical system of control and communication, cloudy definitions of roles and responsibilities and so on seems to be the only way for Indian television to fulfill its basic obligations to the people in terms of meaningful communication order.

The structure of social communication in a country reflects the structure and bias of that society. And television is no exception to this general principle. When it is argued that the television as a formal organizational set up has to deal efficiently with some of the (hopefully, all the) constraints it is made to face with, it is not intended that the constraints of this powerful medium should be examined in terms of other social realities. An attempt to deal with these constraints of television, which are in most cases internal to the organization itself, must simultaneously deal with other larger issues which are vital constraint in people's day-to-day living.

REFERENCES

PERROW, C., "Technology in A. Etzioni and E.W. Lehman (eds) A Sociological Reader on Complex Organizations," Hoit, Rinehart and

Winston, New York, 1960.

LAWRENCE, P.R. AND LORSCH, J.W., "Organizations and Environment: Managing Differentiation and Integration", Graduate School of Business Administration, Boston, 1967.

The Background to the Meeting of the Macbride Commission in Delhi-Vidura, Press Institute of India, Vol. 16, No. 2, April 1979.

BETEILLE, A., "Caste, Class and Power", Oxford University Press, Delhi, 1966.

REFER SCOTT, W. RICHARD, "Organizations : Rational, Natural, and Open Systems." Ibid, for an interesting discussion on the topic of the dominant coalition concept and the arguments of Cyert and March (1963: 27-32).

MILLS, C.W. "Power Elite", Oxford University Press, New York, 1956.

DAHRENDORF, R., "Class and Class Conflict in Industrial Society", Stanford University Press, Calif., 1959.

MATHESON, R., "People Development in Developing Countries", John Wiley and Sons, New York, 1978.

MICHELS, R., "Political Parties : A Sociological Study of the Oligarchical Tendencies of Modern Democracy," Hearst's International Library, Co., New York, 1911/1915.

BLAU, P.M., "A Formal Theory of Differentiation in Organization", American Sociological Review, 35, 2, April 1970.

UDY, S.H., "Organizations of Work", HRAF Press, New Haven, Conn., 1959.

Chapter **35**

Measuring Communication Effectiveness

Anoop S. Sandhu

An attempt has been made to review the current approaches to operationalise the communication effectiveness and categories them into two broad approaches namely, effectiveness of communication encounter and effectiveness of communication performance. Keeping in view the limitations of these approach another approaches to measure communication effectiveness in terms of comprehension, perceived validity and utility on the part of the receiver have been suggested. This approach will not only include establishing commonness in terms of meaning but also in terms of perceived consonance, credibility congruity and relevance.

Man is a queer animal. He not only communicates, but thinks about communication. Wherever we think about this process, our major concern is to increase its effectiveness. In order to increase effectiveness it is necessary that we understand what effectiveness is and then be able to manipulate various factors in order to achieve effectiveness.

According to dictionary meaning, the word effective means (1) having an effect; producing a result, or (2) producing a definite or desired result, efficient. Accordingly we may say that communication effectiveness means a communication encounter that has produced desired results. However, when we try to operationalize the desired results, we feel greatly handicapped.

Two Approaches

Many attempts have been made to operationalize communication effectiveness. Some efforts to measure communication effectiveness have also been made. A review of such efforts show that these efforts can be categorised into two broad approaches namely :

(1) Effectiveness of Communication Encounter.

(2) Effectiveness of Communication Performance.

(1) Effectiveness of Communication Encounter

Effectiveness of a communication encounter refers to achieving commonness between the communicator and receiver in terms of comprehension. For achieving proper comprehension, it is necessary that the receiver received the symbols communicated along with their meaning and in the context in which the originator of the message wanted.

If the effectiveness of a communication encounter is to be viewed in terms of comprehension then, a question may arise about the extent to which the communicator was conscious and clear of the intended message, and the extent to which he happens to be successful in achieving that purpose by resorting to a particular communication act. Thayer (1968) also says in this regard that "the originator's interpretation of the message is not necessarily anymore "accurate" than the receiver's. The orginator may or may not have clearly in mind what he intends to accomplish. Indeed, the originator may have more reason to "distort" the meaning of his message than the receiver has.

A similar question may also be raised about the receiver's point of view. When a receiver engages himself in a communication act, he is likely to comprehend the message according to his communication capabilities, and in the context in which he happens to receive the communication at any point of time.

Since communication is a dydatic process, effectiveness in terms of comprehension can also be thought of in terms of both the communicator and the receiver's point of view collectively. Further, if they happen to engage themselves in the communication on the behalf of an organization, then objectives of the organization will also may have to be kept in view in counting effectiveness.

Of course, it is clear that cost factor is not as yet, being considered for determining the effectiveness. It is also being presumed that there are different qualities of messages which are also likely to affect the extent to which a communication encounter may be successful. Similarly, situational factors over which the communicator or the receivers have no control are also likely to interfere with what is likely to be accomplished in a communication encounter.

(2) Effectiveness of Communication Performance :

Under this approach can be broadly included all those attempts which measure the success of any communication in terms of subsequent behaviour on the part of the receiver. In this approach, as Shannon and Weaver (1949) point out, the problems of influence or effectiveness are concerned with the success with which the meaning conveyed to the receiver leads to the desired conduct on his part.

$$A \longrightarrow B = X$$

Desired conduct or desired results on the part of the receiver have been operationalized differently by different authors. Its may be appropriate to mention some of these attempts, before making a critical analysis of such attempts.

Berlo (1960) identified four factors namely attitudes, knowledge, communication skills, and position in a social-cultural situation, that govern effectiveness of a communicator and a receiver. Similarly, Thayer (1968) has mentioned factors like communicator receiver relationship, purpose or intentions, roles and rules, and the organization on behalf of which they engage themselves, which affects fidelity of a communication situation.

Thayer (1968) also states the following two variables that govern the success of a communication event :

(i) Value or utility of the resources each participant brings to the encounter : intellectual, attitudinal, communicational etc. for the other participant.

(ii) The relative commitment to the cooperative enterprise each participant makes, with no specific-objectives vis-a-vis the other.

Jha (1968) while measuring the effectiveness of a communication situation operationalized it in terms of attitude changed, knowledge gained and knowledge retained over a period of time and practices adopted, on the part of the receivers.

If we view these approaches critically, the following points may emerge :

(i) Variables like attitude formed, knowledge gained, or practices, adopted can be termed as effects of communication and not effectiveness of the communication, situation.

(ii) Certain personal characteristics that the communicator or the receiver may possess are likely to effects their communication behaviour, but cannot be said to be the result of a communicator's act.

(iii) Situational or contextual factors under which the source and the receiver operate are also likely to affect the way each will interpret the message. Hence, they are important but still they remain extraneous variables.

(iv) There are differences in the characteristics of message and communication performances which do contribute to the effectiveness or the ineffectiveness of the encounter. But these qualities liesin the positive or negative contribution they make to the overall success of the encounter Thayer, (1968).

(v) As we know subsequent behaviour not only depends upon comprehension of the message but also upon the receiver's ability and willingness to do what the communicator wanted him to do. It is quite possible and probable for a receiver to have a complete understanding of the message, but still not to do what the communicator wanted him do.

From the above considerations, one may tentatively conclude as Thayer (1968) does, that communication effectiveness in terms of subsequent behaviour on the part of the receiver may require the establishment and manipulation of conditions prior to, during and subsequent to, any act of communication. Hence, measuring communication effectiveness from the effects it has produced on the part of the receiver may meet limitations of this kind.

Effectiveness in Terms of Comprehension, Validity and Utility

To overcome some of the limitations and difficulties it may be appropriate to operationalize communication effectiveness in terms of comprehension, perceived validity and utility that a communication encounter may establish on the part of the receiver. Thayer (1968) has also discussed these qualities in the context of message treatment.

Communication
Effectiveness = Comprehension (clearness)
+ Validity (Consonance, credibility and congruity)
+ Utility (relevance)

Comprehension : It refers to the extent the message conveyed by the source has been clear to the receiver.

Validity : A message may be comprehended by the receiver properly, and yet it may not effect him as intended by the communicator. Those of us, who are involved in the study of human communication know fully well that if the source's message is not perceived as valid by the receiver, then its impact on him may be reduced or contrary to the expected one. As Thayer (1968) argues, a message must be comprehended before it can be validated. But its comprehensibility does not ensure its validity for a particular receiver in a particular situation.

There are three process of validation that may have to be taken into account. First is consonance validation. What is perceived as legitimate may also be perceived as consonant. Second may be credibility validation. It refers to the trustworthiness that a receiver may accord to a source at any particular time. Research studies indicate ample evidence to suggest the perceived credibility of the source influence his success as communicator.

Third is congruity validation. A receiver will usually test a particular message against, what Thayer (1968) calls, "our psychological system and his intentions and expectations."

Utility : Utility may be defined as a measure of the perceived relevance and consumeability of a message from the receiver's point of view at any given time. The more relevant a message is to our needs, purposes and resources, the more utility it is likely to have.

Communication effectiveness then can be measured in terms of comprehension, validity and utility on the part of the receiver. Subsequent behaviour of the receiver may still be considered independent of the communication effectiveness.

SUMMARY

An attempt has been made to review the current approaches to operationalize the communication effectiveness and categorise them into two broad approaches namely, effectiveness of communication encounter and effectiveness of communication performance.

Keeping in view the limitations of these approaches, another approach to measure communication effectiveness in terms of comprehension, perceived validity and utility on the part of the receiver have been suggested. This approach will not only include establishing commonness in terms of meaning but also in terms of perceived consonance, credibility, congruity and relevance.

References

1. Berlo, David K. *The Process of Communication* New York : Holt, Rinehart & Winston, Inc., 1960
2. B, S. Bhatia & A. S. Sandhu, Determinants of Job Effectiveness on the

Village Level Workers, *Indian Journal of Extension Education*, Vol. KI, No. 1-2, January-June, 1975, pp. 62-64.

3. R. Parshad & A. S. Sandhu, A Study on communication Effectiveness on the Village Level Workers in the context of High Yielding Varieties Programme, *Journal of Research*, Vol. XII, No. 2, pp, 173-179, June, 1975.
4. R. Parshad and A. S. Sandhu, Knowledge Test for village Level Workers in the Context of High Yielding Varieties Programme-Structuring and its standardization, *Indian Journal of Psychology*, Vol. 49, No. 3, pp, 213-219, 1974.
5. Jha, P. N., A Critical Analysis of factors associated with Communication Fidelity with respect to High Yielding Varieties Programme. New Delhi : I.A.R.I., Ph.D. Thesis, 1968.
6. Shannon, C.E & Weaver. W. *The Mathematical Theory of Communication*, Urbana : Uni. of Illinois Press, 1949.
7. Thayer, Lee, *Communication, and Communication Systems*, Homewood, Illinois : Richard D. Irwin, Inc., 1968.

Chapter **36**

Theory Building in Communication Research

Dipak De and V. K. Dubey

Theory is the foundation of accumulative knowledge. According to Popper (1965), "Theories are nets cast to catch what we call, the 'World' to rationalise, to explain and to master it. We endeavor to make the mesh ever finer and finer".

Since long the researchers in Communication Science have been gathering facts and speculating communication behaviour, the time has come now to undertake research interest in the Methodological issues of theory construction. This is only because the status of a scientific discipline is measured, how much mathematical logic it issues and how much of a generalisable theoretical framework it embodies. The use of mathematical logic has been growing, but the status of generalising principles and more specifically theories, lag far behind. This emphasises the need for theoretical systematisation of isolated facts.

Modes of theory construction differs in various respects and there may be several contenders in the field. No path is right or wrong and the superiority of one over another is a matter of opinion. Mode of theory construction has been characterized as positivism which has been following three characteristics : (1) theories should be stated formally; (2) theorise should be testable and (3) prediction power should be the primary criterion for assessing theories (Gibbs 1972).

What is theory

The tern theory is often used quite loosely in social science writings but it is capable of a fairly precise definition. There is a general agreement that a theory is set of propositions or theoretical statements (Zetterberg, 1965; Galtung, 1967).

According to John et. al. (1954), theory has been defined as a generalization of a high order which, in some sense, explains observed phenomena. Theory in this sense is bound by a real world of a essentially Kickable facts; but has the logical character of being general. This implies that it is possible to assert rules by which one can judge a statement about the world to be false.

According to Wieks (1972) "A theory is that form of prepositional knowledge, which postulates the existence of determinate relations between or set of variables, in terms of which - empirically ascertainable regularities - can be explained".

This definition emphasises that the aim of theory is to explain a wide variety of empirical phenomena in a succinct a way as possible, the explanation itself being dependent on testable statements about empirical data.

According to Hafe (1972) "A theory should contain not only concepts and statements but definition-both theoretical and operational and linkages, again both theoretical and operational. The concepts and definitions should be ordered into primitive and derived and the statements and linkages should be ordered into premises and equations".

Biesanz and Biesanz (1973) wrote in his book 'Introduction to Sociology' facts are verified statements about phenomena; concepts enable as to perceive or think about phenomena and communicate about them; and propositions are statements of relationship among concepts and/or facts. None of them, however, explains the phenomena and their relationship. What do the facts mean? Answering this question explaining or making sense of facts is the main function of theory.

A theory is a set of systematically related propositions. It may be empirically grounded, interpretative or philosophical.

In empirically grounded theory, theory enquiry and empirical fact ate interwoven in a texture of operation with theory guiding inquiry, inquiry seeking and isolating facts, and facts affecting theory. The fruitfulness of their interplay is the means by which an empirical science develops (Blumer 1969)

Interpretative theory may or may not be firmly related to empirical findings typically it is less rigorous but it is also legitimate and important, for its aim is not to from scientific propositions but to outline and define life situations so that people may have a clearer understanding of their world, its possibilities of development and directions along which it may move. In every society, particularly a changing society, there is a need for meaningful clarification of basic social values, social institutions, models of living, and social relations. This need can not be met by empirical science, even though some help may be gained from analysis made by empirical science.

Social philosophy is not generally considered social science theory, for it typically deals not with what is but with ideas of what ought to be of the ideal society.

According to Kerlinger (1978) "A theory is a set of interrelated constructs (concepts), definitions, and propositions that present a systematic view of phenomena by specifying relations among variable, with the purpose of explaining and predicting the phenomena". This definition implies :

1. a theory is a set of propositions consisting of defined and interrelated constructs;
2. a theory sets out the interrelations among a set of variables (constructs), and in so doing, presents a systematic view of the phenomena described by the variables.
3. a theory explains phenomena. It does so by specifying to what variables and how they are related, thus, enabling the researcher to predict from certain variables to certain other variables.

From the above discussions it can be inferred that a theory is a set of prepositional statements witch present the distinct picture of the phenomena by determining relations among variables with the purpose of explaining and predicting the phenomena.

Different categories of scientific theories and differences between them.

SCIENTIFIC THEORIES CAN BE CATEGORISED AS :

1. Law like theories
2. Causal theories
3. Axiomatic theories (Stinchcombe, 1968)

Such a classification is based on the premises that the logic of scientific inquiry can be either inductive or deductive. By induction, one means the process of scientific interference which moves specifically observed situation to generalized level of abstract propositions or principle. The process of deduction implies a movement in the enquiry from abstract principle or proposition to the situation to be observed or explained. However, if a system of theoretical proposition already exist the best approach is to use the process of deductive inquiry. Whenever it is felt that no such system of deduction is possible, an investigator has no choice but to use the system of induction.

It is somewhat a difficult task to assert that there are law like theories in social sciences even those situations or social phenomena which might conform to such a situation one likely to be observed with less frequency. In order for a theory to be a law like theory, it must be supported by directly observable empirical findings. Thus, if one were to find a law like theory, it must be supported at the operational level and concepts which can be used in such a theory must be concrete or totally operational. A low like theory must be totally deterministic i.e., the cause and effect sequence must be fully established with no exception Since total causality is an impossible task to achieve in social sciences, one would have to rely upon the other two forms of theories causal and axiomatic approaches

An axiomatic theory usually consists of definitions of concepts to be used both at abstract and operational level, statement ranging from axioms to proposition with a built-in logical system to derive them. In axiomatic theory construction each concept has to have empirical reference or operational definition. Each of the concepts must be having existence as wall as relational terms. The relational statement should be made ranging from the most abstract forms to most concrete forms. The propositions derived should constitute hierarchies which range from the most abstract propositions to most concrete proposition. Axioms have to be selected on the basis of their abstrations, logical interrelatedness, demonstrated causal sequence and self evidence.

The difference between causal and axiomatic forms exists at the level of axiomatisation. In the causal approach, it is assumed that each of the propositions are of equal importance, rather than any implicit or explicit.

The advantages of axiomatic format over the other theoretical models are as follows :

1. The concepts and postulates offer the most parsimonious summary research findings;
2. The axiomatized theory has the highest plausibility per amount of empirical data;
3. Through use of an axiomatic model strategic model strategic research problems may be located;

4. The source of failure of a hypothesis is to meet empirical test may be efficiently discerned, and
5. The axiomatic model permits clear distinction between propositions which are definitions and those which are hypothesis. Considering these points in mind, axiomatic theory building technique has been chosen for building up a theory in Communication Research.

AXIOMATIC THEORY

According to the father of axiomatic theory, Zetterberg (1965), axiomatic theory is also known as deductive theory. In Scharag's (1967) view, deductive theory as : Assumptions
Concepts Observed evidence

Winton (1974) defined axiomatic theory consists of logically related, logically consistent and logically complete definitions and propositions about human group life.

According to Singh (1975), an axiomatic theory consists of definitions of concepts to be used at abstract and operational levels, statement concerning the existence of concepts, a relational statement ranging from axioms to propositions, with a built-in logical system to derive them.

For this study purpose, axiomatic theory consists of a prepositional statements ranging from postulates to hypotheses with a built-in logical system for explaining and predicting Communication behaviour of an individual.

TERMS USED IN THEORY BUILDING

Axiom

According to Znaniecki (1934), a deductive science needs a set of fundamental truths axioms which would serve to test all the others but would not need to be tested themselves. These axioms would be derived either from experience or from some other source. If they were derived from experience, they would be inductive, and therefore from the point of view of the deductive standard should be again tested by deduction from other truths. If not, they would be inapplicable to empirical reality unless that reality were first made to fit them, which means that it would not be empirical reality, independent of mind but the work of mind.

According to Greenberg (1959) axioms are tested as premises from which the theorems are deduced.

According to Cohen and Nagel (1964), axioms are simply assumptions or hypotheses used for the purpose of systematizing and sometimes discovering the theorems they imply.

In Pooper's (1965) view, axioms may be regarded either (i) as conventions, or they may be regarded (ii) as empirical or scientific hypotheses. He has further pointed out that we may distinguish, within a theoretical system, statements on the highest levels of universality are the axioms statement on the lower levels can be deduced from them.

As per Galtung (1967), higher order assumptions or premises are called axiom.

According to Blalock (1969), axioms are propositions that are assumed to be true.

Postulate

Northrop (1967), the postulates of a deductively formulated theory are those propositions which are assumed in the theory in question as logically unprovable and which are sufficient lo enable one to prove, i.e., to logically deduce, the theorems.

According to Feigl (1956), postulate in modern science, especially in mathematics, assumption means which serves as a premise for deduction.

As defined by Zetterberg (1965), the postulates are chosen from the list of original propositions (inventories or metrics) so that all other theorems are capable of derivation from other postulates.

Proposition

According to Zetterberg (1965), propositions relate variables to each other.

In Dubin's (1969) view, proposition specifies the relationship of variable to each other.

Kckhardt and Erman (1977), a proposition consists of a stated relationship between concepts

Theorem

According to Northrop (1947), the theorems of a deductively formulated theory are all the empirical proposition in the theory other than the postulates.

As stated by B la lock (1969), theorems ... are derived by reasoning or deduced from the axioms.

Hypothesis

According to Zetterberg (1965), propositions for which more evidence is needed are called hypothesis.

In Dubin's (1969) view, the general form of a hypothesis is a conditional prediction bout the relationship between two or more things, followed by a figurative question mark.

According to Smelser (1969), hypothesis is a statement of the logically and presumably temporarily prior conditions under which dependent variables may be expected to vary in Certain ways.

Statements designated as hypotheses (Taken from Gibbs - 1972 : (87-89)

H1. "The more a member conforms to the norms of a formal organization, the greater the likelihood that he will be promoted" (Zetterberg, 1965 : 104).

H2. "The greater the socio-economic rewards derived from present American life in comparison with past cuban ones, the greater the integration" (Portes, 1969 : 511).

H3. "A positive association was anticipated between an individual's most-valued source of information and the time at which he adopted the study innovations" (Becker, 1970 : 272).

H4. "Only those actually sharing in societal power need develop consistence societal values" (Mann, 1970 : 435).

H5. "French Policies in the Caribbean were perceived, even by enlightened Antillean leaders, as having achieved a more-or-less satisfactory spread of equality, and this satisfaction, in turn, largely 'cut the ground from under' the desire for political independence there" (Murch, 1968 : 554).

H6. "The size of the nuclear family is an inverse function of position in the stratification system" (Smelser, 1969 : 14).

H7. "Work would have the most consequence for community status in the Costa Rican village, next most consequence in the Michigan village and would be least important in the Guatemalan village" (Faunce and Smucker 1966: 393).

H8. "The larger a city, the larger the range of its occupational composition" (Schwirian and Prehn, 1962 : 815).

Statements designated as propositions

Pr1. "The second derived proposition is that the larger an organization is, the larger the average size of its structural components of all kinds" (Blau, 1970 ; 207).

Pr2. "Hypergamy is more common that hypogamy in the United States" (Rubin, 1968 : 752).

Pr3. "By combining either propositions 5 and 2, or 3 and 4 we get the sixth proposition : 6. The actual responses of others toward the individual will affect the behaviour of the individual" (Kinch, 1963 : 482).

Pr4. "Values in the rest of the United States are opposed to the practice of segregation by the white south" (Heer, 1959 : 593).

Pr5. "The more a member conforms to the norms of his group, the more favourable valuations does he receive from his group" (Zetterberg, 1965 : 107).

Statements Designated as Theorems

Th1. "The primary structure of the human personality as a system of action is organised about the internalization of systems of social objects which originated as the role-units of the successive series of social systems in which the individual has come to be integrated in the course of his life history" (Parsons, 1955 : 54).

Th2. "The suicide rate of a population varies inversely with the degree of status integration in that population" (Gibbs, 1969 : 522).

Th3. "People generally what to do what they are supposed to do, and this is what the society needs to have done in order to continue" (Goode. 1960 : 485).

Statements Designated As Postulates

P1. "The greater the number of associates per members, the greater the division of labour" (Zetterberg, 1965 : 161).

P2. "There is a universal set of basic human needs which have attributes of their own which are not determined by the social structure culture patterns, or socialization processes" (Etzioni, 1968 : 871).

P3. "Although the development of the Puritan-Parliamentarin culture accelerated the emergence of industrial society in England, this cultural development depends on cumulative reactive interaction between its adherents and the adherents of the simultaneously developing Anglican-Royalist culture" (Israel, 1966 : 597).

P4. "The individual's self-concept functions to direct his behaviour (Kinch, 1963 : 482).

Statements Designated As Axioms

A1. "An increase in the number of associates per member will produce an increase in the division of labor" (Blalock, 1969:19).

A2. "The probability of remaining in any state of nature increases as a strict monotone function of duration of Prior residence in that state" (McGinnis, 1968 : 16).

A3. "Every social pattern continues to manifest itself in constantly recurring social action at an unaltered rate unless some social force modifies the rate or pattern of such action" (Catton, 1966 : 235).

A4. "Individuals who face common role obligations can generally fulfill them" (Godde, 1960 : 484-485).

From the above definitions and statements of the axiom, postulate, proposition, theorem and hypotheses, it can be inferred that the axiom and postulate are the same term used by different authors interchangeably. Because axiom and postulate are uneducable from any other theorems or hypo-these. It is also clear that there is a general lack of consensus about the uses of variety of labels (axiom, postulate, propositions, theorem and hypotheses) among the authors and there are no established conventions. Moreover, the authors use labels to identify types of empirical assertions and definitions. It suggests divergence rather than consensus in the use of labels. Needless to say, it would be desirable for the researcher to employ the same nomenclature in typifying empirical assertions, only the axioms appear to be consistent in from that is none of them were derived. In the present paper, postulate and hypotheses have been used for theory construction.

Steps involved in theory building.

Before discussing steps to be followed in theory construction it would be worth-while to look back to the historical prospective of the theory.

Grand theories are conglomerations of ideas so vague that no evidence would lead to their rejection. Foremost defect is in terminology. Grand sociological theories are not defined to promise empirical applicability. Consequently the test results of grand theories are debatable because the theories are vague and there is no effective consensus as to what would be an appropriate test or what would constitute negative evidence. In typical grand theories, testable and untestable statements are not distinguished rather it is lead to the audience not only to decide what component statements are testable but also to formulate a test procedure. In any case if only some of the component statements are testable, systematic evidence cannot be brought to bear on the theory as a whole without formally deriving testable assertions, and that mode of formulation in alien to the works of Grand theorists, they preferred the discursive exposition of a theory in which empirical assertions are interspersed with conceptional analysis, arguments and digressions over hundreds of page. Discursive mode makes it difficult to identify the central assertions of a theory and no less difficult to determine which statements are derived and how they are derived. Consequently apart from the problem of empirical applicability the asmorphous character of typical grand theory alone precludes systematic tests. Therefore, grand theories are non-scientific in nature. Hence, out of purview of present paper.

Steps involved in axiomatic theory building

Different authors have suggested different steps to be followed for constructing theory in an axiomatic format. Among these authors, Zetterberg is considered as the father of axiomatic theory building. According to him, the following steps need to be followed :

A. Basic concepts

These are definitions which we, strictly speaking, will introduce as undefined. However, by circumlocutions and examples, we may convey what we intend them to mean. There is nothing mystically fundamental or basic about these concepts. They only represent an assumption of an agreement that we may use certain words in certain ways.

B. Derived concepts

The derived concepts are obtained through combinations of the basic concepts. Granted the basic concepts, we accordingly can say that the derived concepts are precise. If we illustrate them with examples, we should remember that these examples do not add or substract anything form the definitions. The basic concepts from the nominal definitions of the theorem.

C. Hypotheses

These hypotheses must not contain any other concepts than the nominal definitions. We will formulate these hypotheses as we usually do in science. These hypotheses should be chosen in such a way that they do not contradict what we know is true and we will cite scientific research supporting these hypotheses or in the absence of the latter, we will cite informal observations in favour of them.

D. Postulates

We will select from among the hypotheses formulated a certain number to be the postulates of our theory. The postulates should be chosen so that all other hypotheses, the theorems, should be capable of derivation from these postulates. Specifically, the postulates should be chosen so that they become consistent and independent. In other words, no postulate should contradict any other, and no postulate should follow from any other postulate as a theorem.

Schwirian and Prehn (1962) recommend the following steps to be followed :

I. Substantive Domain

II. Basic Hypotheses

III. Frame of Reference

A. Assumptions

B. Basic concepts

C. Derived concepts

D. Operational definitions

IV. Derived Hypotheses

Substantive domain : It consists of the class of variables with which the theory is concerned. There are two types of domains as measured by their scope - the macro-domain and the micro-domain. The macro-domain consists of all variables of a given class. Hence, the macro-domain of communication behaviour includes all variables dealing with it. The micro-domain deals with a group of variables and their inter-relationship which are less in number and narrower in scope than these of the macro-domain of which they are a part.

Basic hypotheses : The basic hypotheses of the axiomatic model are those assertions of relationship between the variables of the system which are chosen on the basis of existing empirical knowledge, or some justifiable deductive process. The basic hypothesis are chosen and stated in such a manner that all the logical possibilities of relationships among all of the variables in the system may be deduced from them. Also, the basic hypotheses must be combination of any two other basic hypotheses.

Frame of reference : The frame of reference of a theory is the foundation upon which the substantive assertions are made, and is composed of the following sub-sections, assumptions, basic concepts derived concepts, and operational definitions.

Assumptians : Assumptions are logical assertions about the basic concepts, their nature of inter-relationships. Assumptions are considered to be self-evident and are not subject to empirical test within the theory. They are included in the frame of reference to aid an nurturing the explicitness, inclusiveness, and clarity of the theoretical system.

Basic concepts : Basic concepts are terms which are undefined in the frame of reference since this is assumed to be a general consensus of understanding about their meaning.

Derived concepts : Derived concepts are those terms which are defined through the combination of basic concepts and assumptions.

Operational definition : Two definitions of each concept are provided in the frame of reference. The formal definition is that which is specially deduced by the logical process. The seconp definition is the operational definition and it recasts the formal definition in terms of the data and procedures of the individual empirical situation.

Derived hypotheses : Hypotheses derived form the basic hypotheses are referred to as derived hypotheses.

Hage (1972) proposed the following steps to be followed : Theoretical concepts : Theoretical concepts are the foundation of any theory. The first task in constructing a theory is to find some concepts to use in our theoretical statements. The most helpful kind of theoretical concept is the general variable, a continuum that applies to any culture and at any point of times-to societies that have ceased to exist, that presently exist, and that have yet to come into existence.

These criteria culture free and timeless, are easy to apply and they make our task of recognizing general variables a simple one.

Theoretical statements

Our task of finding theoretical statements is facilitated considerably if we recognise that concepts can be connected in a either or way, such as with the verb to be and connected in a continuous way such as in the formulation, the greater the X, the greater the Y, for want of a better words we have called the either-or statements and the latter continuous statements. The latter formulation allows us to be more complex and informative in our analysis of reality.

Specifying the definitions

Instead of only one definition, it is better to have two for each concept. In this instance both theoretical and an operational definition provides much more than twice one definition. The two kinds of definitions provide two ways of perceiving our concept. The theoretical definition gives us meaning and the operational definition measurement. A dialectic between meaning and measurement makes each definition easier to specify and produces a creative evolution towards better measures and sharper meanings. In turn, this facilitates the development of theories and especially theoretical statements.

Specifying the linkages

If our theoretical statements are to be put on a sound intellectual basis, they must have two linkages, one theoretical and the other operational. The former helps us explaining social phenomena and gives our statements some testability.

Together they can be a winning combination because we have seriously taken our time to specify both linkages, we increase the probability of making our research more productive.

Ordering concepts and definitions

The techniques rest on the basis of logical reasoning.

Ordering statements and linkages

Theoretical order is more than just arranging our concepts into primitive and derived terms. We must also order our theoretical statements into premises and equations.

Singh (1975) advocated the following steps to be followed :

(i) Select an idea or social phenomena

(ii) Make prepositional statements in abstract terms

(iii) Propositions in axiomatic hierarchy i.e, most abstract or more concrete propositions.

(iv) Indicate existence statements

(v) Indicate relational statements

(vi) Indicate operational measures of concepts used in propositions

(vii) Hypotheses of relationship or difference based on variables i.e., variables corresponding to concepts used

(viii) Collect data

(ix) Test hypotheses with appropriate statistical or other techniques

(x) Examine the relational and existence statements as stated earlier

(xi) Relate these to proposition in their abstract forms

(xii) What can be said and what have you said about the idea or social phenomena. From the above analytical review it can be inferred that the following steps to be followed in constructing Axiomatic theory of communication behaviour.

1. Select an idea **or Social Phenomena : As for example** communication **Behaviour of** small **marginal farmer or farmers in general.**

2. Selection of variables/concept :

   (i) Basic concepts, and

   (ii) Derived concepts,

Derived concepts are deduced from Basic concepts with the help of assumptions. Selection of concepts/variables will be (i) based on the past researches on the area of the study, (ii) appropriate statistical method to be used, and (iii) according to Hage (1972), the No. of pf variables within the theoretical system has to be limited and must not go beyond ten variables The telationahty behind this suggestion is that even with the ten variables the No. of hypotheses would be n(n-1)/2 i.e. (10 x 9)/2 = 45 hypotheses are to be tasted empirically which by any standard is a very difficult task. With an example it is explained now the derived concepts are deduced from basic concepts with the help of assumptions.

S. No.	Basic concepts	Assumptions	Derived concepts
1.	Space	Space can arbitrarily bounded;	Spatial (Najafgarth block derived from concept (1) and assumption (1) and (2)
2.	Human individual (he may communication or receivers of the communication process)	The area of the arbitrarily bounded space can be identified;	Socio-economic status sources of information utilization, education, knowledge, innovative orientation, progressive values, entrepreneurship these characteristics are derived from basic concepts (2) and (3) and assumptions (3), (4), (5), (6), (7). The selection of these characteristics must be based on the above-mentioned criteria
3.	Characteristics of communicator or receiver of the communication process.	Communicator or receiver live in a space.	Scientist (source) farms (receiver) derived from basic concepts (2)
4.		The have definite no. of characteristics	
5.		They too have similar no. of characteristics	
6.		These characteristics very in degree	
7.		These characteristics can be observed either directly or indirectly	
8.		These characteristics can be identified.	
9.		These observed characteristics can be measured	
10.		Measured characteristics can be put for statistical analysis	

3. Definition of Concepts :

A clear concept offers the same meaning to all those who use it Hence, it is essential to define the concepts both theoretically and operationally.

4. Formulation of Postulates and hypotheses :

A fundamental principle of an axiomatic mode) is that given (n-1) postulates where 'n' equals the number of varibles and given that all variables appear in atleast one of the (n-1} hypotheses then all other hypotheses within the theoretical system may be derived. The total number of postulates and hypotheses is. Therefore, n(n-1)/2. Thus, in the present example with seven predictor variables and eighth response i.e., Communication behaviour there will be 28 hypotheses to be tested 8((8-1)/2 = 28.

Postulates : Postulates of the axiomatic model are those assertions of relationships between the variables of the system which are chosen on the basis of existing empirical knowledge, or some justifiable deductive process. The postulates are chosen and stated in such a manner that all the logical possibilities of relationships among all of the variables in the system may be deduced from them. Also the postulates must be such that no one postulate could be deduced from a combination of any two other postulates (Zetterberg, 1965).

Hypotheses : Relational statements derived **from** the postulates are referred **to as** hypotheses.

To exemplify the deductive technique as applied to derivation of hypotheses, let us suppose that we have two postulates :

(i) the higher the socio-economic **status, the higher the education;**

(ii) the higher the education, the higher the sources of information utilization

From those two postulates, we may derive a third one in the following manner (of course, it is necessary that two propositions contain one corresponding assertion, If a third one is to be derived from them).

if; I The higher the socio-economic status, the higher the education;

And : II The higher the education, the higher the sources of information utilization;

Then : III The higher the socio-economic status, the higher the source of information utilization.

In this manner, all logical possibilities of the relationships between the variables in the system will be deduced.

Postulates (n-1) = (8-1) = 7 : As the fundamental principle for formulating postulates statad in the formula (n-1 *t* where n = the number of variables. Formulated postulated are :

1. The higher the socio-economic status, the higher the education;
2. The higher the education, the higher the sources of information utilization
3. The higher **the sources of** information utilization, **the higher the knowledge**
4. **The higher the knowledge, the higher the** innovative **orientation;**
5. The higher the innovative orientation, the higher the progressive values;
6. The higher the progressive values, the higher the entrepreneurship;
7. The higher the entrepreneurship, the higher the Communication behaviour of an individual.

Hypotheses: By applying the deductive technique for formulating the hypotheses from postulates stated earlier all the logical possibilities of the relationship between the variables in the system are deduced and presented below :

1. · The higher the socio-economic status (S. E. S.), the higher the knowledge;
2. The higher the education, the higher the sources of Information utilization (S. I, U-);
3. **The higher knowledge, the** higher **the innovative orientation;**
4. The higher the SIU, the higher the progressive values;
5. The higher the innovative orientation, the higher the entrepeneurship;
6. The higher the progressive values, the higher the communication Behaviour;
7. The higher the SES, the higher the SIU;
8. The higher the education,, the higher the innovative orientation;

9. The higher the knowledge, ihe higher the progressive values;
10. The higher the SIU, the higher the entrepreneurship;
11. The higher the innovative orientation, the higher the Communication behaviour;
12. The higher the SES, the higher the innovative orientation;
13. The higher the education, the higher the progressive values;
14. The higher the knowledge, the higher the entrepreneurship;
15. The higher the SIU, the higher the communication behaviour;
16. The higher the SES, the higher the progressive values;
17. The higher the education the higher the entrepreneur-ship;
18. The higher the knowledge, the higher the Communication behaviour;
19. The higher the SES, the higher the entrepreneurship;
20. The higher the education, the higher the Communication behaviour;
21. The higher the SES, the higher the Communication behaviour;

These hypotheses are syllogistically deduced from postulates. Together, the postulates and the hypotheses are the propositions of the axiomatic theory which are subject to empirical tests.

5. Testing of hypotheses with appropriate statistical technique : To test the hypotheses an ex post facto research technique may be followed. A schedule questioned may be prepared with appropriate scale and scoring procedure for the selected variables for data collection. Both parametric and non-parametric statistical tests may be applied.

6. Validification of the axiomatic theory : The conditions for the validification of the axiomatic theory are : (i) empirical test must be favourable and (ii) no mistake has to be made in the derivation while formulating postulates and hypotheses.

Conclusion : In the discipline of Communication of the tradition of theory construction has been comparatively weak.. However, there is a need to develop empirically testable theories which could *be* cross culturally verified. An attempt has been made in this paper to various steps involved in exiomatic theory building in relation to Communication-Science. It is hoped that this paper will generate discussions at the profession level and enthuse the researchers for making a positive attempt in this direction.

References :

Biesanz, M, H. and Biesanz J. (1973). Introduction to Sociology. Prentice Hall, Inc. Engle Wood Cliffs, New Jersey.

Blalock. H. M. Jr. (1969). Theory construction from verbal to mathematical formulation. Engle-Wood Cliffs, N. J., Prentice Hall,

Blumer, H. (1969). Symbiolic Interactionism. Engtle Wood Cliffs. M.J. Prentice Hall, Inc.

Cohen, M.R. and Nagel, E. (1964). An introduction to logic and scientific method. Routledge and Kegen Paul Ltd, London.

Dubin, R. (1969). Theory Building. Free Press. New York.

Feigl, H. (1965). Some major issues and developments in the philosophy of science of logical empiricism. In Herbert Feigl and Michael Serivan (eds) Minnesota Studies in the Philosophy of Science. Vol. 1. Minneapolis Univ. of Minnesota Press.

Galtung, J. (1967). Theory and Methods of Social Research. London, Allen and Unwin.

Gibbs, J. (1972). Sociological theory construction. The Dryden Press Inc., 901, North Elm Street, Hiensdale, Illinois.

Greenberg, J H. (1959), An axiomatization of the phonologic aspect of language. In Symposium on sociological theory, Ed. by L. Gross, Row, Feterson and Company, Evanston, Illinois, White Plains, New York,

Hage, (1972). Techniques and problems of theory construction in sociology. A Wiley-Interscience Publication, John Wiley and Sons, New York.

John, D. et al (1954). An introduction to social research. Harrisburg, Pa. Stackpole Press.

Kckhardt, K.W. and Erman, D.M. (1977). Social Research Methods : Perspectives, theory and analysis, Random House, New York.

Kerlinger, F.N. (1978). Foundations of behavioural research, Surjeet Publications, Delhi.

Northrop, F.S.C (1947). The logic of the science and the humanities, Crowell-Collier-MacMillan, New York.

Popper, K.R, (1965). The logic scientific discovery. Harper and Row, New York.

Schrag, C. (1967). Elements of theoretical analysis in sociology. In Llewellyn Gross (ed.). Sociological theory. Inquiries and paradigms. Allied publications pvt. Ltd., Bombay, New Delhi, Calcutta, Madras, Banglore.

Schwirian, K. P. and Prehn. J. W. (1962). An axiomatic theory of urbanization, American Sociological Review, Vol. 27, December.

Singh, B. K. (1975). Theory and theory construction in Applied social sciences. Indian Journal of Extension Education, Vol. XI, July-December No, 3. and No. 4.

Smelser, N. J. (1969). The optimum scope of sociology. In Robert Bierstedt (ed.). A design for sociology : Scope, objectives and methods, Monograph 9, Philadelphia : American Academy of political and Social Science.

Stinchcombe, A. L (1968). Constructing social theories, Huscourt, Brace and Borld, Inc., New York

Wieks, D. R. (1872). Glossary of sociological concept.

Winton, C. A, (I974). The ory and measurement in sociology, A Halsted press Book, Schenkman Publishing Company, John Willey and Sons Inc., New York.

Zetterberg, H. (1965). On Theory and Verification in sociology, Totowa, N. J Bediminster Press.

Znaniecki, F. (1934), The method of Sociology, Holt, Rine-hart and Winston, New York.

Kckhardt, K. W. and Erman, D. M. (1977). Social Research Methods : Perspectives, theory and analysis, Random . House, New York.

Kerlinger, F. N. (1978). Foundations of behavioral research, Surjeet Publications, Delhi.

Northrop, F. S. C. (1947). The logic of the sciences and the humanities. Crowell-Collier-MacMillan, New York.

Popper, K. R. (1965) The logic scientific discovery, Harper and Row, New York

Schrag, C, (1967). Elements of theoretical analysis in sociology. In Llewellyn Gross (ed.). Sociological theory Inquiries and Paradigms, Allied Publications Pvt, Ltd., Bombay, New Delhi, Calcutta, Madras, Bangalore.

Schwirian, K. P. and Prehn, J. W. (1962). An axiomatic theory of urbanization, American Sociological Review, Vol. 27, December.

Singh, B. K. (1975). Theory and theory construction in applied social sciences. Indian Journal of Extension Education, Vol. XI, July-December, No. 3 and 4.

Smelser, N. J. (1969) The optimum scope of sociology. In Robert Bierstedt (ed.). A design for sociology Scope, objectives and methods. Monograph 9. Philadelphia : American Academy of Political and Social Science.

Stinchcombe, A. L. (1968) Constructing social theories, Huscourt, Brace and Borld Inc., New York.

Wieks, D. R. (1972). Glossary of sociological concept.

Winton, C. A. (1974). Theory and measurement in sociology, A Halsted Press Book, Schenkman Publishing Company,

John Wiley and Sons, Inc., New York.

Zetterberg, H. L. (1965). On Theory and Verification in sociology, Totowa, N. J. Bediminster Press.

Znaniecki, F. (1934). The method of Sociology. Holt, Rineh-art and Winston, New York.

Chapter 37

Principal Component Analysis : A Tool for Analysing Impact of a Communication Programme

S. Chakraborti and D. Das gupta

Isolating the good from the bad or the desirable from the undesirable has been a human activity. This is essential to society itself. Although institutions have often been accorded authority for making certain value judgments, only in recent times has evaluation of public programmes, not to speak of a technology transfer programmes alone, becomes so formalised as to constitute a recognizable separate activity and even a new discipline.

Evaluation leads to a settled opinion or to an objective judgment about the value/worth of something. This judgment is arrived at by grading or by set of standards or criteria. This is again essentially comparative and usually explicitly so. This obviously means that there must be a set of standards and a class to which the object is compared. Hence, in its essence, evaluation entails adopting a set of standards, defining the standards, specifying the class of comparison, and deducing the degree to which the object meets the standards.

With the valid and objective criteria of study and various designs of experiments in possession one may get into the investigation of proper dimensions in research in communication/technology transfer. It is not possible to make an exhaustive list of such dimensions. However, some of the major dimensions as suggested by Das Gupta (1991) are as follows :

- technology adoption, retention, dissemination;
- change/shift in overt/covert behaviour, value, interest, knowledge, attitude, etc;
- change/shift in production and productivity, land use pattern, employment and income :
- Constraints analysis.

In a large number of situations in social science research we are interested in a device with the help of which we can estimate (or predict) the unknown values of one variable (explained variable) from the known values of other variables (explanatory variables). The statistical device meant for this purpose is known as regression analysis. In functional notation we can write as $X_1 = F(X_2, X_3 \ldots\ldots)$, where X_1 is the explained variable and X_2, X_3 .... are explanatory variables which are either independent or at best moderately inter-correlated, when the functional form is linear we have multiple linear regression of the form (with 3 variables X_1, X_2, and X_3).

$$X_{1.23} = a + b_{12.3} X_2 + b_{13.2} X_3$$

where b's are unknown parameters that measure the respective weights assigned to each of the independent explanatory variables.

In social science research where we often interested in explaining some phenomena (effects) observed as the result of some previous phenomena (causes). These causes (explanatory variable) which produce effects are often highly interrelated variables. Thus straightforward multiple regression analysis is inappropriate under such circumstances. In this situation of high interrelationship among explanatory variables alternative methods of analysis that one can employ depending on the specific situation are :

(1) Principal component Analysis (PCA) : (2) Factor Analysis : (3) Image Analysis; and (4) Path Analysis.

Principal Component Analysis : In the Principal Component Analysis (PCA) if the explanatory variables are highly inter-correlated, we can transform the correlated variables linearly into a relatively small number of uncorrelated variables. Such derived variables, if considered as independent variables, will maximise the prediction of the original variables. These derived variables are called principal components.

Factor analysis : assumes that the observed variables are linear combinations of some underlying source variables called factors, some of which are common to two or more variables. That is, factor analysis assumes the existence of correspondence between underlying factors and observed variables and use this correspondence to arrive at conclusions about the factors. Thus factor analysis may be used as a means of ascertaining the minimum number of hypothetical factors that can account for the observed covariance.

The basic difference between *Factor Analysis & PCA,* is that the principal components are certain mathematical functions of the observed variables while common factors are *not* expressible by the combination of the observed variables, (both used for exploring interdependence among variables).

An alternative in initial factoring is image factoring. *Image Analysis* is different from common factor analysis in that the observed variables are considered a sample from the potentially infinite universe of variables in which the image factors are defined as linear combinations of the variables.

Path Analysis is a method of estimating the linkages between variables to provide information about the underlying causal processes. The simplest way to obtain the path coefficients is to employ ordinary regression techniques, provided the regression assumptions are met. Through path analysis one can measure the direct and indirect effects that one variable has on another. This technique also enables into simple and compound paths representing direct and indirect effects.

We take up Principal Component Analysis here for a detailed discussion.

PCA is a statistical technique that linearly transforms an original set of correlated variables into smaller set of uncorrelated variables that represents most of the information in the original set of variables. The goal of PCA is to try to explain part of the variation in a set of observed variables on the basis of a few underlying dimensions. However, it has no underlying statistical model of the observed variables and it involves linear combinations of correlated variables whose variable weights in the linear combination are derived on the basis of maximising the variance accounted for in the original variables. Hence as a statistical tool, PCA is, perhaps, best be regarded as an exploratory instrument to enable one to see what is the effective number of dimensions, or how dominant are certain linear combinations of the original variables.

Let us consider a set of p variables x_1, x_2, x_p which can be linearly transformed by

$$Y = \sum_{i=1}^{p} a_1 x_1$$

Here y is a one dimensional index. In PCA, the weights a_1, a_2, ... a_p are determined to maximise the variation of the linear composite (principal component), or equivalently, to maximise the sum of squared correlations of the principal component with the original variables. The principal components are ordered with respect to their variation so that the first few principal components together have, over all, the highest possible squared multiple correlations with each of the original variables.

Geometrically, the first principal component is the line of closest fit to the n observations in the p dimensional variable space. The second principal component is a line of closest fit to the residuals from the first principal component, so on. It there are discrepancy among variables, the number of principal components will be fewer than the number of variables.

Algebraically, the first principal component

$$Y1 = \sum_{i=1}^{p} a_1 x_i$$

is such that the variance of Y1 is mamximised

given the constraints

$$\text{that} \quad \sum_{i=1}^{p} a_1^2 x_i = 1$$

The random variables X_1, can be either deviation from mean or standardised variables. If the variance of Y_1 is maximised, then the sum of the squared correlations of Y_i, with x_1, x_2, ...x_3 (that is

$$\sum_{i=1}^{p} r^2_y x_i$$

) will also be maximized. PCA finds these optimal

weights a_{11}, a_{12}, ,,,,,,,,,,a_{1p} which satisfies above condition. PCA also finds the associated variance of Y_1 given by λ_1 the first latent root.

The second principal component is defined as

$$Y_2 = \sum_{i=1}^{p} a_{21}\, x_i$$

Here the weights a21, a22, ... A2p are determined in such a way that l_2 the variance of Y_2 is mamimised subject to the condition that is uncorrelated with the first principal component and also

$$\sum_{i=1}^{p} a^2_{2i} = 1$$

suits in Y_2 having the next largest sum of squared correlations with the original variables. However, the sum of squared correlations with the original variables (or equivalently the variances of the principal components) get small as successive principal components are extracted. The first two principal components together have the highest possible sum of squared multiple correlations

$$\text{(that is } \sum_{i=1}^{p} R^2 a_1\, y_i\, y_2$$

with the p variables. This process can be continued upto the desired level.

The relative sizes of variable weights for a particular principal component indicate the relative contribution of the variable to the variance of the principal component, or equivalently the relative amounts of variation explained in the variables by the principal components. It can be shown that the correlations of the variables with a particular principal component are proportional to these weights. They can be obtained by multiplying all these weights by the square-root of the variance $\sqrt{\lambda}$ of the associated principal component. This result is obtained due to the fact that j[th] principal component with weight vector (a_{j1}, a_{j2}, ... a_{jp}) has a variance l_j. Hence when we want to have unit variance we need a transformation of weight vector as $\left(\frac{a_{j1}}{\sqrt{\lambda_j}}, \frac{a_{j2}}{\sqrt{\lambda_j}} \cdots \frac{a_{jp}}{\sqrt{\lambda_j}}\right)$

and consequently multiply the coefficients of weight vector of jth principal component by $\sqrt{\lambda j}$. The correlations of the variables with the principal components are sometimes referred as factor loadings. Variables that correlate highly with a particular principal component give meaning to that component.

The basic statistics for PCA are p variances $\lambda_1, \lambda_2, \ldots\ldots\lambda_p \ldots\ldots\ldots$ (latent roots) ordered by size and the associated variable weight vectors $a_1, a_2, \ldots\ldots a_p$ (latent vectors). The variables considered are assumed to be such that for empirical data no variable will have a multiple correlation of 1 with the remaining (p-1) variables. If the latent roots are all distinet, then there are p distinct associated latent vectors. If all the points will exactly fall on a straight line and only one principal component will capture all information in the original variables.

The sum of variance of the principal components is equal to the sum of the variances of the original variables.

That is, $\sum_{i=1}^{p} \lambda i = \sum_{i}^{p} a^2$ when the variables are assumed to be standardized, the

variance for each variable is equal to 1. Hence the total variance in the system with p variables is equal to p. that is, $\sum_{i=1}^{p} \lambda i$ = P. Also, the sum of squares of the p elements in each of the principal components is equal to 1, that is, $\sum_{i=1}^{p} a^2 ki = 1$ also, $\sum_{i=1}^{p} a_{ji} a_{jk} = 0$, implying that j^{th} principal component and k^{th} principal component are independent of one another.

The proportion of total variance of p variables explained by k^{th} principal component is $(\lambda k/p)$v Thus the proportion of variance in the original p variables that k principal components accounts for is

$$\sum_{i=1}^{k}(\lambda i/p) \text{ where } k<pv$$

If the sum of the variances of the first few principal components is close to p, then we have captured most of the information in the original variables by a few principal components.

We use principal components primarily as a tool to describe a sample of observations. Here we ate interested in testing the only hypothesis that last few latent roots (variances of principal components) are equal, that is, H_0 ($\lambda k = \lambda k + 1 =$ .... $= \lambda p$). If this null hypothesis is accepted, we can conclude that these last few latent roots are equal. If these latent roots are equal, this in turn will imply that the latent vectors (associated variable weight vectors) associated with them are arbitrary, and hence should be discarded. Thus this test can help us to decide how many principal components are to be retained.

How many and which principal components are to be retained depend partly on the objective of analysis, If only we only want to describe a variable set without regard to subsequent uses, then we can retain k largest principal components. But if we want to use the principal as predictors of a dependent variable, then we should consider their correlations with the dependent variable as well.

Kaiser (1960) recommended dropping those principal components of a correlation matrix with latent roots less than one. He argues that principal components with variances less than one contain less information than a single standardised variable whose variance is one.

Jolhiffe (1972), on the basis of simulation studies, has suggested a cutoff of 0.7 for correlation matrices. Part of his argument is based on the fact that a population latent root greater than or equal to one can result in a corresponding sample latent root considerably less than one because of sampling error.

This rule can be extended to covariance matrices by using cutoff 0.7l, where λ is the average size of the latent roots of the covariance matrix.

Author criterion is to retain enough principal components to account for a given percentage of variation, e.g. 80%. All these rules are arbitrary and should be applied with caution.

It is to be noted that if there is no correlation between variables, there will be no principal component, because every component then is as good or as bad as the other. Each will account for only a unit variance. As the interrelations among the variables increase, the proportion of variation explained by first few principal components will increase.

Properties of principal components :

(1) A variable x_1 can be decomposed into additive contributions by the principal components. That is, in order to approximate x_i, we would multiply the score on the first principal component by the value of x_i in first principal component, similarly multiply the score on second principal component by the value of x_i in second principal component and so on upto k retained principal components and add these products.

(2) Since the principal component can be written in matrix notation as y = A′ x and A′ is a non singular matrix of weights whose inverse is A, we have x = A y. The principal components can be expressed as linear function of the variables or the variables can be expressed as linear function of the principal components.

(3) For well-defined correlation structures, where variables falling into clearly defined clusters with high correlations within clusters and low correlations between clusters, the pattern of latent roots readily indicates the number of principal components to be retained. The retained principal components can easily be interpreted from the latent vectors or equivalently from the factor loadings. Thus if we have p variables of which k variablesare highly inter-correlated, but have relatively low correlations with the remaining (p-k) variables, also if the said (p-k) variables are atleast moderately correlated among themselves, then the pattern of correlations indicates two clusters of variables and suggests that two principal components are enough for accounting the variation in the p variables.

(4) PCA can be utilised in selecting a subset of variables to represent the total set of variables. Need for such consideration arises due to the fact that (a) In order to compute the principal component scores, measures of all the variables in the variable set are needed ; or principal components are sometimes un-interseperetabl. The basic argument for selecting a subset of variables to represent the total set of variables is as follows : If the correlations among the variables are

high, or if there are clusters of variables with high inter-correlations, then, we can represent the variation in the total variable set by a subset of variables, and these variables are known as principal variables. These are various methods of selecting principal variables. One way may be to use Jolliffc criteria of $\lambda = 0.70$ to determine which principal components to retain. Then the variable which has the highest loading in the principal component can be selected to represent the said retained principal components, provided it has not already been selected to represent a larger variance principal component. In that case, the variable with the next largest weight would be selected. The procedure would start with the largest principal component and proceed to the smallest retained component.

Alternatively, one can use the discarded principal components to discard variables. We start with the smallest discarded component and delete the variable with the largest weight on that component. The process is repeated for the second smallest component and so on. If any variable had been discarded earlier, then the variable with the next highest weight would be discarded. The process continues upto the largest discarded component. The rationale for this process is same as in earlier situation. Any principal component (Principal variable). Thus if any principal component can be represented by a variable with highest weight inthat component can be discarded then its principal variable can also be discarded,

(5) In regression analysis, if the independent variables are highly correlated, they can be transformed to principal components and then the principal components can be used as the independent variables.

(6) PCA can be used to detect multicollinearities among the independent variables in a regression analysis.

Interpretation of Principal Components

Let us now consider how the principal components are interpreted with the help of a hypothetical example of communication. Let us consider that we have 5 standardised communication variables X_1, X_2, X_3, X_4 and X_5 where X_1, X_2 and X_3 belong to communication fidelity (say) and X_4, X_5 belong to infrastructural lapses for communication (say). On the basis of the data on these variables let us for example consider that the first principal component obtained is $Y_1 = .44X_1 + .42X_2 + .45X_3 + 41X_4 + .43X_5$ with $\lambda = 300$.

Here the weights are almost equal implying that each variable is almost equally represented in Y_1. Thus in this example the first principal component could be interpreted as a measure of overall communication.

The variance of Y_1 is 3.00. Hence Y_1 explains

$$\left(\frac{a_{j1}}{\sqrt{\lambda_j}} \times 100\right)\% = \frac{3.00}{5} \times 100\% = 60\%$$

of the total variance of 5 variables

Again, let second principal component obtained is

$Y_2 = .39\, X_1 + .31\, X_2 + .22\, X_3 - .52\, X_4 - .62\, X_5$ with $\lambda = 1.25$

Since all the 3 variables X_1, X_2 and X_3 which belong to the broad classification of communication fidelity have positive weights; and X_4 and X_5 which belong to the broad classification of infrastructure lapses for communication have negative weights, thus the second principal component can b_3 interpreted as a contrast between communication fidelity and infrastructural lapses for communication.

The Y_2 accounts for 1.25/5 X 100% = 25% of total variance of the five variables. First two principal components together account for (60 + 25)% = 85% of the variance in the five variables. Thus most part of the variance is explained. Hence we can say that we have reduced the dimensionality of our observations from 5 to 2.

The correlations of the variables with principal components (factor loadings) obtained through multiplication of all these weights by square root of variance $\sqrt{\lambda}$ of the associated principal components are as follows :

Principal components Loadings

Variables	Principal	Components
X_1	.762	.436
X_2	.727	.346
X_3	.779	.246
X_4	.710	-.581
X_5	.745	-.693

The proportion variance in X_1 explained by the first two principal components retained is $(.762)^2 + (.436)^2$ which is equal to 0.771. In this way we can get the proportion of variance that the first two principal components retained explain for each of the five variables-Scores on any principal component is obtained by putting standardised scores for the original variables in the said principal component. It may be noted that mean score for each principal component is zero. An individual having high score in first principal component in the above example will have a high level of communication. An individual having a high score in the second principal component will have high scores in communication fidelity variables and low scores on communication variables.

With the above example it is now evident that Principal Components Analysis can really be considered as a powerful tool for social science research in general and research in communication in particular. With the availability of computer facility this tool can really be used, effectively.

Suggested Readings

1. Das Gupta, D (1990). Pragmatic Approach to Evaluation of Impact of communication Programme. Interaction Vol 8, No. 1 pp 149-172.
2. Dunteman, G. H. (1989), "Principal Components Analysis", Sage University Paper Series on Quantitative Applications in the Social Sciences, Beverly Hills, Sage Publications.
3. Jolliffe (1972), "Discarding variables in principal component Analysis", I : Artificial data, Applied Statistics, 21 : 160-173.
4. Jolliffe (1986), "Principal Component Analysis", New York Springer-Varlag.
5. Kaiser, H. F. (1963), 'The application of electronic computers to factor analysis'. Educational and Psychological Measurements, 20 : 141-151.

Chapter **38**

The Language of the Mass Media with Special Reference in Radio Broadcasts in Regional Languages

J.V. Vilanilam

Language is the basis of most human communication and communication is the foundation of human society.

Language can make or mer human relations. It can soothe feelings and engender understanding, affection and love, mutual respect and national as well as regional cohesion. It can inflame people and even make them mad. It can make people laugh or cry. It can disseminate truth, but it can also disseminate untruths and halftruths. It can use and abuse people, appoint and disappoint them. It can manipulate condition or control their thoughts. It can inform, misinform, disinform, lead or mislead people. It can brief and debrief them. It raises expectations as much as it lowers them. It can activate and deactivate, motivate and demotivate people. Language can entertain as well as bore people. All these contradictions of language can be seen in the various mass media the newspapers, magazines and other periodicals, radio, television, films, theatre, music and every branch of the media used in communicating ideas to large or small groups of people.

Language and Society

Language is as much shaped by society as it shapes society. There is constant interaction between society and language and this interaction is reflected in the media of communication. If we accept the concept that society is constantly changing or growing, we have to concede that language in general and hence the language of the media also changes or grows or becomes different.

Linguists are not just language scientists they are social scientists too, because there is a symbiotic relationship between society and language. Since society is dynamic, language is dynamic too. Linguists have to keep themselves abreast with the ever-changing social phenomena because language is an important indicator of changes that come about in social organization and structure.

Social change thus gets reflected in the language of a given society and the mass media cannot escape the influence of this change. But social change or the lack of it depends on many factors, the most important of which is how society is organized on the bases of economic and production relations. Economics, politics, and technology influence society and hence its language. Communication does not occur in a vacuum. Communicators are people who have ideas to spread and the nature of their ideas will depend on their own cultural background and socio-economic goals. What kind of society they want to shape will influence their messages, and hence their language.

It is, therefore, of the utmost importance that the users of language take some time off their daily routine for a fresh look at their techniques.

Each medium has its own audience. When there is no consonance between the medium and its language, or between the language used in the medium and the language used by the audience, communication suffers because the act of communication is consummated only when the communicator's language fits in with that of the communicatee's language.

Communication involves the sharing of concepts and in order to determine how the sharing is effective, a close study of the language of the media of communication is essential. If mass communicators and the masses of people whom they try to reach through the various media do not speak the same language (not only literally but metaphorically), there will be a gap between the senders of messages and their recipients. However, communicators work under certain constraints. There is no single, unified, homogeneous mass of people in any part of the world. The task of the mass communicator is to reach the largest number of people and establish proper understanding with them. For achieving this, the communicator tries to use as universal a language as is possible. Research in language use and linguistics becomes essential for this purpose.

The purpose here is to examine the constraints on the use of language in the broadcast media, particularly the radio (since television use is still quite limited in our country), with specific references to All India Radio which is the main source of radio broadcasts here. Little research has been done so far in India, especially in Kerela, on the language of the mass media. And precious little has been done on the language of radio broadcasts. Although the ideas expressed here are not the result of empirical research, some references will be made to a recent study of broadcast news in Malayalam done by some researchers in the University of Kerala.

The language of the mass media

Before touching upon the language of the mass media, let us clarify what we mean by mass media and mass communication.

Ever since the invention of movable types by Johann Gutenberg in the mid-15th century, knowledge escaped the confines of a closed circle of knowledgeable people who happened to be priests and noblemen. Books began to reach a general audience and the process of the democratisation of communication started in a small way through printed books. But many advancement in the technology of printing had to occur before books reached large masses of people in any society. Some societies in the West had a technological edge over those in the East. Therefore, the imbalance in the flow of ideas started at a very early stage in modern civilisation. The technology of printing was closely followed by or it synchronised with, the process of colonisation through trade and military superiority. The industrial revolution in the West changed ordinary people's lifestyle and work-leisure activities. The newspaper became a daily phenomenon in the West in the latter part of the 18th century. Cheap newspapers or what is known as the penny press became a powerful force in society in the 19th century. Coupled with the spread of knowledge was the spread of sensationalism in newspapers and periodicals. The immense possibility of using the print medium for agitating people's political affiliation was also seen during this period. In almost all freedom movements in the West the newspaper played an important role. These things happened in the East slowly because of the economic and political situation prevailing there. However, whatever technological improvement came in the West had its reflection in the East as well.

With the advent of the telegraph, the telephone and the wireless radio, a new vista was opened for people with ideas to communicate to others. The radio became a powerful force in the third decade of the present century.

What has all this to do with the language of the print medium and the broadcast medium? When the process of communication became democratised, or when it became popular, the high-flown language of the books of olden days was replaced by simple language. Latin and Sanskrit lost their importance. Regional languages began to flourish. Prose became the main vehicle of thought and it replaced poetry. The masses of people who were reached by the print and broadcast media of this period were reached by a new language the popular language, which could be understood clearly and easily by large masses of people with average education.

In India, sweeping changes in mass media usage have not yet occurred. The total daily circulation of newspapers today may not be much above the total circulation of dailies in the England of the 18th century. This big gap has to be remembered when we discuss the phenomenon of mass communication in India. There is strictly speaking, no national mass medium in our country. Firstly, there are many languages and secondly, literacy is still very low. However, our broadcast media can reach larger sections of people who are illiterate, although the language of the broadcast media has to be adjusted to the needs of specific linguistic groups. To say this is easy, but to get this done is beset with huge practical problems mainly because of the large number of languages and their dialects through which broadcast communication has to be effected. There are technological difficulties too. Most of our people live in remote villages. Yet, the AIR is to be congratulated on the speed with which it has been able to increase stations in rural areas and through them reach a vast audience. Despite practical problems, our broadcast medium is doing its best and succeeding in reaching millions of people every day. Without forgetting this great achievement, one has to point out that there is a great deal of room for improvement in language use. This matter is of the utmost importance especially in view of government's decision to go ahead at full speed with the implementation of satellite communication. The idea is to bring remote sections of our large country within the network of communication particularly development communication.

Special problems of the broadcast media

Unlike the print media (including newspapers, magazines and other publications) and the film and the television media, where concepts are put in symbols—that is words and pictures which the reader can experience through sight and sound, the medium of radio depends entirely on the voice of the broadcaster. The listener cannot see the message or hear it a second time unless he goes to the radio station and asks for a copy of the script or does a preplanned taping of the message. Usually, this does not happen. Only researchers go to this extent of studying messages a second time. Thus the large majority of listeners hear a message only once and if they miss any word, mis-hear it, or miss its meaning, the message becomes incomprehensible and they are left in midair. It is, therefore, essential that broadcasters use language (that is words) in the clearest possible way.

Mere clarity in pronunciation is not enough. Words may be clear, but still the listener may miss their meaning if the words are not within the frame of reference of the audience. Words are concept-carriers and inscrutable words cloud the concept leading to a total incomprehensibility of the message or its partial comprehension, both of which defeat the purpose of communication.

Broadcasters, as mentioned earlier, face the task of reaching a message to the largest possible audience effectively. For this purpose, words that are unambiguous to the largest number of people have to be used. Certain words are not used in certain areas of the same linguistic region in the same sense. For example, "poochi" means insect in central Travancore, but it is obscene in some northern districts. Similarly, "marutha" means a variety of demon in some parts of Kerala, but it does not mean anything in the city of Trivandrum. Coastal people may not understand some words used commonly by the hills people. What the broadcasters try to do is to use the standard, textbook language, assuming that most listeners have received those words and their meanings from the same education system. Although, this is not always a solution to the problem, it is the only workable solution now available to broadcasters. Another drastic solution is what Professor Kierstead of the A & M University of Florida suggested at a seminar held by the Department of Journalism early this year, namely, narrow-casting. This means the multiplying of broadcast stations in order to reach more specific than general audiences. In our present situation, multiplication of broadcast stations is not easy since private broadcast stations are not part of our broadcasting network or strategy. Thus what is practicable in our situation is for broadcasters to use words that are commonly understood by people who are educated on the basis of standard language, universally applicable. Since literacy is still low in most parts of India, the question of a received language or standard language is yet unsolved. However, Kerala's high literacy rate is a helpful factor in solving the question of standardisation of the language of Malayalam broadcasting to an appreciable extent.

Is there a need for the controversy over the language of the mass media?

Looking at the question from another angle, there need not be any need for a controversy over the language of the mass media, particularly, broadcast media. There is no doubt that language of the media has to be adjusted in reference to the functions of communication, and also the needs of the audience receiving the messages. As mentioned earlier, there is no universal broadcast language as such. Depending on the broadcast material and the type of audience for which material is intended, the language has to change. The controversy usually comes only in the case of certain popular programmes and especially news. Otherwise, one does not quarrel about the language of a scientific article where complex processes have to be explained through appropriate language which usually is not easily understood by people who do not have any background in science. But even there, simplicity should be the aim of the broadcaster, for science can be explained in simpler terms for the layman. However, no language can be simplified beyond a certain level.

Avoidance of English and Latin words and foreign phrases is sometimes suggested as a solution to language difficulty. But Sanskrit words are sometimes difficult to understand for users of the local languages. I surveyed one of my classes with some words from the editorials of Malayalam newspapers. At least half a dozen words in the editorials were not properly understood by the students. (It is definitely no reflection on my students!)

The Need for audience surveys

Audience survey is essential. The mass circulated press in Kerala seems to be happy where it is now. Even without such research, the big newspapers of Kerala are doing quite well in circulation. Why should there be more research when even without it, circulation can be improved. There may be special factors that are helpful in increasing the circulation of newspapers, factors that are well-known to the press proprietors. They have learned this by trial and error or by inspired divination. The need for systematic research among the audience is not felt among them. But through proper research of audience preferences for language style, the same newspapers can raise their circulation to the million marks before the turn of the next century.

As for audience research in the language of radio broadcasts, there are some studies and reports from seminars.

Shaji studied one aspect of radio broadcast language, namely, the syntax of the Malayalam news bulletins of the AIR, although it was not just word order that he studied. He found that there was ample room for a great deal of improvement. For example, sentences could be made shorter, with minimum number of words, clauses, and word syllables. This, he found, would enhance comprehensibility of and audience interest in news broadcasts. The length of words and hence the number of syllables could be reduced by eliminating the use of auxiliaries to sentence endings. Nearly two thirds of the audience sample surveyed preferred an informal style which is closer to conversational style for newscasts.

Newsreaders are likely to sound official for the sake of authenticity, but my own experience in the U.S. for the past twelve years tells me that are informal style never reduced the authenticity of news items broadcast. That the large majority in Shaji's sample indicated preference for a simple, colloquial, dialectic style in news broadcasting, is a clear indication of what the audience really would like to have, instead of hard, high-sounding Sanskritized words. One peculiarity of Malayalam editorials and broadcast news is the large percentage of high-sounding Sanskrit words used by the communicator.

Another problem is the compulsory use of more difficult equivalent words for avoiding monotony. For the word "said", "opined" is used sometimes in English. "Opined" is a stilted word. Even at the cost of variety, "opined" can be obliterated (Sorry for the big word). In Malayalam, the broadcaster may use abhiprayappettu, prasthavichu, velippetuthi, choondicatti, koottitcherthu, soochippichu, etc., in place of paranju. One has to remember that each of these equivalent words has a specific meaning. For example, velippetuthi or revealed need not be used when something that was said was not previously kept a secret.

Redundancy or illogical repetition of ideas or words in another area where the broadcaster has to be careful. For example, prasthavichirun-nituntu-why irunnitunte?

Many of these observations are applicable to language used in other mass media. Look at this sentence in Malayalam from a newspaper: "pattika vargakkarkku veetu nirmmichu kotukkunnathinulla bhavana nirmana padhatiyute visadamsangalilekku chuzhinjirangan addeham tuninjilla".

Some linguists are purists. They want to avoid all "foreign" words. Words are like immigrants to a foreign country. They stay in the country for some time and get absorbed as citizens, as part of the local language. Of course, some immigrants do not get absorbed. Thus those who get absorbed should be given the full rights of citizens. They need not be treated as second class citizens. We have in Malayalam plenty of such citizen words. It is preposterous (sorry, foolish) to translate them, usually with comic effect.

The relevance of dialectological research in India

Sociolinguistics, sociology of language, social anthropology of language, social psychology of language, social dialectology all these form the domain of many distinguished linguists all over the world. Till recently, most studies were confined to standardised varieties of language but now-a-days, attention is being given to the total linguistic rapertoires of individuals and communities. We too have to turn our attention to our dialects in order to absorb more words into the standard language of the broadcast media. Only then can mass media messages be understood-fully by the large majority of our people who live in the villages cut off from the mainstream of our present day standard language.

Since we have 16 or 17 well-developed languages and hundreds of dialects in India, the task of the linguist and the communicator is not an easy one. For the Western world, research need concentrate only on the received language or the standard language which is more or less uniformly understood and used by the people of the same culturally homogeneous and geographically united region. Since literacy is high and the use of the mass media, particularly, the broadcast media is high in those regions, the dissemination of the standard language is easy. However, dialectological research is carried out in Western countries, although dialects of English as used in Scotland, Wales or Ireland have very many words in common with those of the English language in its standard form, and through the intermingling of dialect users with the main language users through frequent travels and through literature (*e.g.* the language of Robert Burns, the Scottish poet, is well known to the English or in general the British people; Irish writers such as Swift, Shaw and others are national writers of Great Britain). Such a situation is not obtainable in India. Although there is a common cultural link among speakers of Bhojpuri, a dialect of Hindi, and Bengali, a well developed language with a great deal of standardisation, a Bhojpuri poet may not be understood or appreciated without a great deal of effort by a Bengali reader. Besides, the differences in script and words that exist among *the* major languages of India create additional language-based cultural problems among Indians on the national scene. How can there be an integrated approach to the problem of the language of the mass media in India with all these practical problems confronting us? Though dialectological research is relevant in India, much more urgent is the need for evolving standard languages in each linguistic area of India if at least regional cohesion is to be achieved over the next few decades. This is especially important since the use of the broadcast media is increasing in different regions with the expansion of radio and television broadcasting in regional language. It is hoped that media directors, managers, producers, researchers, writers, announcers, and all those who are communicators will give more attention to the language of the mass media.

Chapter **39**

Interactive Local Radio as a Tool for Development : An Analysis of Selected Community Radio Projects

H. Fisher

Many early radio stations began as a local community activity. The first radio station 'broadcast mainly within a department store' (Arnaldo, 1988, p,v). Most early stations were low-powered and reached only the immediate vicinity. But the need for a constant flow of programming, increased listener sophistication, and commercial and political demands soon made radio primarily a one-way medium. As Stephens (1989) observes : "Radio was traditionally an a'oof medium, a one-way transmission of professional sounding programme from studios in an urban center. 'Access' meant taking radio to the people through taped interviews or live outside broadcasts". In recent years, as Head (1985) shows, a rebirth of local radio has taken place. Thousands of community stations now exist. However, most are still one-way systems, mostly commercial, with the station tunneling information to a largely passive audience.

Early development communications theorists (Lerner, Schramm, Pye and others) saw mass media as "an independent variable in the development process" (McAnany, 1980). They believed; for example, that mere exposure to radio messages was enough to cause changes that would lead to development. Ascription to this theory led to the launching of numerous radios for development projects. In the earlier farm forum programmes, development messages came primarily from the government agricultural department or the extension agent down to the rural people. Only limited feedback from the farmers was invited. Melkote (1989) found that poor Indian farmers "ware not involved at any stage during the process of message prepraration" and that the messages were too prescriptive, complex and technical to be understood. A large proportion of the earlier radio programming for development was also devoted to-top-down government campaigns, such as Tanzania's National Radio Study campaigns.

Later theorists and researchers, such as Beltran, Golding, Grunig and Rogers, saw communication 'as both a dependent and an independent variable in a complex set of relationships with social, economic and political structures and processes' (McAnany, 1980). In their view, radio messages should interact with the entire social system. This approach led to strong criticism of the top-down concept. Awa (Over coming Impediments, 1988) identified its main restrictions : lack of listener participation in formulation and implementation of development programmes, failure to consult indigenous knowledge and considering Western media as the instruments of communication. The interactive approach focused on the listener and the community. Words like 'interaction' (Can Radio ?, 1986), 'decentralization' (Laflin, 1987), and –'access,' 'participation' and 'self-management' came into vogue, and greater efforts were made 'to involve rural people in development' (Stephens, 1989. In the process, a number of interactive community radio projects for grassroots development have recently come into being,

This article briefly describes and analyzes the strengths and limitations of five community radio projects and then attempts to draw from their experiences criteria vital to making community radio projects useful tools for development.

Five Community Radio Projects

Five community radio projects for development were chosen for this analysis because they are the best-known efforts' with an established record of experience in employing interactive local radio for development and because considerable literature on their policies and activities is available. Although the project differ, they all use some form of local interactive radio for development.

The Jamaican Community Radio Project

The Jamaican community radio station was born in 1979 as a device to get farmers in central Jamaica involved in the agricultural development of the region. As Bella Mody (1986), the project's advisor, indicates a USAID Jamaican government agreement provided funds and in kind services to build and operate a station for three years. The Jamaica Broadcasting Corporation (JBC) appointed a Station Manager, but failed to delegate him with adequate decision-making and budget management authority. He set up a local steering committee comprised of 16 social service agencies. But governmental delays to authorize start up activity led the manager to quit after two years.

With one year of the project left, JBC officials requested USAID to extend the project and chose a new station manager with a background of dramatic actress, radio producer and social worker. Despite her enthusiasm and skills, she faced difficulties selling to her staff the station's intended policy of making listeners the source of their own messages and the concept of production teams consisting of a content expert, a community researcher and writer-producers concept. Producers gave little credibility to the community researcher's findings and advice. The producer assigned to agricultural programmes knew little about farming and his adviser, an agricultural officer, tended to dictate what was good for the farmers until he accompanied the researcher on village visits.

Valuable lessons were learned about audience involvement, some of which were applied. Farmers' ideas began to set the agenda for the agricultural timeslot. Audience preference for colloquial Jamaican English and community suggestions for health script ideas and participation in the programmes were adopted. Local news, the most popular programme, was changed to a dramatic format presented in Jamaican dialect after a survey of audience preferences. A workshop produced a network of sources from the audience for the news. On advice from the villagers, the station resorted to indigenous entertainment music. Thus, the audience became the source of programmes through its direct & indirect inputs to the audience researcher hired to 'get audience input in the formative stages of programme development' (Mody, 1986). Formative research or evaluation has since became a vital corrective tool for community radio as it analyzes the environment, audience needs, problems and characteristics, delivery system constraints, communication Content and strategies, feedback channels and utilization factors in the planning stages of a project (Adkins, 1983).

However, the Jamaican project has soon scrapped. The USAID contract ended, leaving the station manager to deal with accounting and authorization problems. Producers tired of the extra effort required to make the audience the decision-maker for programme content while their peers in Jamaica's commercial stations simply spun discs. Adviser Bella Mody's contract ended and she was no longer present to guide. The nation entered a period of economic hardship and staff and supporters lost heart to fight for the station. Within a year, the Station Manager resigned and the station became a 'white elephant.' Three years later, when the government adopted its new media policy of profitability, the JBC made the former community station a repeater station for its Kingston commercial programming outlets.

As Mody observes, an innovative methodology at the micro-level was circumscribed by, macro-level political actors and economic factors' (1986), not considered in the project design. Despite the station's demise valuable lessons about community involvement formative research, production strategies, programming content, that need for on going financial support and staff and community determination to make a project continue had been learned.

Radio Baha'i (Ecuador)

Radio Baha'i, a community station located in Otavalo, Ecuador, was established in 1977 to serve the interests, needs and tastes of the indigenous rural Indian Baha'i community. The station was built (1) to promote and maintain the values, dignity and of the rural people and the their traditional culture, which is being eroded by twentieth-century technology and values; (2) to promote basic education, the delivery of social services, and the distribution of development information, and (3) to serve as a nexus for communication and exchange of news and information about the region with the local listening community (Hein, 1982).

The station broadcasts are popular because most of its programming is in Quichua, the home language of the audience. The station places priority on programme content. Its news and indigenous music programmes are especially germaine to campesino interests. It seeks to keep its broadcasts appropriate and relevant to audience needs. A 1982 survey revealed that 94 percent of the potential audience was listening occasionally or regularly to Radio Baha's, despite the fact that over 30 stations can be heard in Otavalo district. Most of the staff, including several women was recruited from the local peasants. Many were illiterates who had no prior broadcast training or experience until the station trained them (Hein, 1983). The equipment mostly used and very modest came predominantly from the United States. The antenna, for example, is made of ordinary aluminum irrigation pipe used in local agriculture; yet, Ecuadorian engineers found it to be the most efficient 1000 watt transmitter in the country. The station broadcasts 19 hours a day, seven days per week, on an annual budget of about & 50,000, much of which is apparently spent on transmission costs (Hein, 1982).

Although the listeners apparently have little input to station policies and the staff makes programming and production decisions, the staff regularly conducts formative evaluation on the programming. Each staff member regularly spends several days time with the audience gathering information about the listeners' needs and preferences. Final programming authority lies with a three-member Radio Commission, one of whom is an illiterate farm woman.

The station apparently inspires listener loyalty by being responsive to their needs and interests. The local news programme "El Noticiero Local," aired twice daily, helps listeners communicate local events to one another. Typically, the programme includes stories about lost children, livestock, or property and about community activities. Since the station broadcasts all messages free, some local listeners walk several hours to deliver their messages and interact with the station staff. Examples of this service include a request from Ecuador's Ministry of Agriculture to produce 3 spot announcement advising farmers about an outbreak of hoof-and-mouth disease and where they could get vaccines, a Civil Registry campaign announcement that it had produced identity cards for illiterate campesinos and assistance to-rural health center to promote their offer for free eye care. The last brought 450 people to the clinic the first day after the station announcement.

The station also works as an intermediary between government ministries and formers. It cooperates with the Ministries to farmers' needs. It has set aside a van and a team of producers to produce a daily 30-minute programme on the opinions of the farmers in which it records farmers' concerns then visits local Ministry offices for their responses. In this way, the station has eliminated some of the 'vertical' process by which the Ministries were providing information and information has become more relevant to the listeners (Hein, 1983).

While Western pop music was predominant on competing stations, Radio Baha'i has endeared itself to the campesinos by playing a significant role in bringing back their traditional Andean music. It sponsors traditional music festivals, encourages local music groups and plays their materials on air. This model is now being replicated in other regions of the Andes (Hein, 1983).

Radio Baha'i's strengths appear to lie predominantly in its loyal staff and its interaction with the audience, in its formative programme research, and in its activity to maintain listener loyalty. But staff controls programming and it remains partially a top-down project in that the community has relatively little input at the station policy level. While the station encourages some government agency inputs, it apparently controls the station's programme content. Both practices limit interaction and audience access to the station. Available literature also fails to indicate where the station gets its basic budget, possibly indicating outside support is sustaining the project.

Radio Nagercoil (India)

All India Radio (AIR) comprises one of the world's largest national radio networks. It also has huge regional stations. But about five years ago AIR began experimenting with the local radio station concept as it established its first local community radio project at Nagercoil in the rural Kanyakumari District in Tamilnadu. Nearly a million people inhabit, this area is rich in traditional Tamil literature and culture. AIR determined that Radio Nagercoil, as a local station, should "reach right into the heart of the community," should "use its microphone to reflect and enrich the life of the society," and must "function as the mouthpiece of the local community" (Anjaneyulu, 1989).

Today, many of Radio Nagercoil's programmes are field-based. A small number of producers spend time in the district conducting field studies of the people and their needs and taking the role of participant-observer to study the culture and life of the people of the district. Free lance part-time producers help with the reporting, editing and presentation of the tapes recorded in the field.

These local producers have produced access programmes, community service broadcasts, and specific programmes designed to encourage the development and to preserve the culture of the area. The access programmes are produced with the help of community listener participation. Local artists, speakers and musicians come to the station to perform. To date, over 50 local clubs have taken part in this type of programming.

For its community service programming, the station serves as a two-way channel of communication between development agencies and the local people. For example, when listeners complained about a poor drainage facility, the station aired the public grievance along with a reply from the concerned municipal commissioner. In another case, the station helped resolve a feud between two rival fishing groups by interviewing the fishermen from both sides and conveying their views to the local Fisheries Department. A meeting of government authorities. Radio Nagercoil personnel and both groups of fishermen was then arranged, during which an amicable settlement was reached. Radio Nagercoil then broadcast its "A Meeting at the Sea Coast" programme. In this process both listeners and service agencies become involved.

Specific programmes deal with matters of relevance to the people. Producers interview people at work and learn of their needs and problems. They interview experts on the area's history and culture. As Radio Nagercoil's Assistant Station Director Shri Anjaneyuiu indicates, all of the station's programmes have a direct bearing on listener's sense of belonging to the community (Anjaneyuiu 1989).

While Radio Nagercoil seeks to focus on the community, there appears to be little formal formative evaluation of its process of involving the community. Undoubtedly the station is doing much to bring government agencies and local people together to solve problems and meet needs; however, the station does represent the Indian government and it remains in control of broadcast content. Other than the fact that the station is charged to emphasize programming of value to the locality, the system remains centralized and top-down. The listeners are asked to participate, but apparently on the terms the producers deem best. It seems the local community also has little or no voice in station policy matters or in broadcast scheduling.

The Mahaweli Community Radio Project (Sri Lanka)

To speed development and to correct serious problems of poverty, unemployment and food shortages, the Sri Lankan government introduced the Accelerated Mahaweli Development Programme in 1977. It focused on the development of fee farming which brought settlers to several areas of the Island. In 1981, the Mahaweli Community Radio Project (MCR) was introduced as a complement to the rice improvement scheme. The community radio project was designed to be under but independent of the Sri Lanka Broadcasting Corporation (SLBC). MCR has received financial and other assistance from UNESCO and DANIDA the Danish assistance agency. Knud Ebbesen (1986) the Danish consultant to MCR says the project encourages villagers "to take an active part in the development process or, even better, to take their own initiative to improve living conditions in their communities" From its inception, the project was based on the belief that radio could be an effective tool for motivating the villagers to self-improvement.

From the outset, participation of the listeners was based on production techniques which involved them in the planning, recording and editing of programmes. First the production teams study the resources, social structure, demographics, economic levels and agricultural activities of a village. Equipped with this data, they camp out in a selected village four days, meeting and holding discussions with the leaders and residents from every walk of life. The success of broadcast programmes depends on the materials recorded at these sessions. Later, the production team returns to the village for a day to organize and to record a musical evening in which young and old participate.

MCR broadcasts a daily hour of rural programming called "Garni Horawa" every evening except Sunday from 7.00 to 8.00 p. m. (the listeners' choice of time), A continuity announcer ties together several short segments including a feature on a village problem with suggestions for a solution, a drama produced in the field by villagers, a show featuring village culture or traditional music, an interview programme in which local "experts" offer information and advice to farmers, another feature highlighting the accomplishments or experiences of a villager, a segment in which a successful farmer explains his experiences with a new technology and a short "how to" talk pitched at the level of the listener and often given by a villager. Programmes cover information on the Mahaweli Development Project, agricultural news and information, irrigation, inland fishing, animal husbandry, health, nutrition, employment, loans, education, marketing, moral behaviour, and village history, organization and folklore. Development is the theme of all programmes (Valbuena, 1988).

MCR is evaluated regularly. Besides continual formative evaluation for its programming, a listener survey was conducted in 1982, and evaluation of its impact in 1984 and another evaluation of the "Gram Horawa" programme in 1987. In 1985, the MCR was extended to three more regional stations with community radio production teams assigned to each. One of the three, the Girandurukotte Community Radio (GCR) is linked to a European Economic Community project to establish a development center in the area. The production teams actually live in the community, making it a true community radio project and a possible model for other similar projects. Since its founding, two evaluative studies of the operations of the GCR have already been conducted. In 1988, Valbuena conducted an impact study of the entire project on behalf of UNESCO and the Asian Mass Communication Research Center. Using document analysis, in-depth interviews and field observation, he measured audience awareness, amount of practice of MCR-promoted activities, quantity and quality of MCR services and amount of interaction between the settlers and MCR.

These evaluations have yielded highly useful data. For example, the 1984 study showed a very high percentage (56%) of frequent listeners, 58% of whom heard the programme from start to finish. But the GCR study showed even higher listenership, indicating the importance of producer identification with the community by living among them and involving them in all programme decisions.

The evaluations have produced invaluable information on MCR's impact, on needs for programmes and programme content changes and on settler attitudes. Perhaps more important, MCR has created awareness among its listeners of MCR, of the Mahaweli Authority development programmes and of how the settlers can participate and became involved in the development activities. The studies also showed that MCR programmes have motivated people to achieve self-actualization through their participation in radio programmes, community dialogues, letters and sharing of their experiences. However, the studies also suggest that, while the MCR has stimulated awareness of the need for improved farming, actual practice may be motivated more by interpersonal contacts with field extension agents or local farmers regarded as sources of reliable farming information (Valbuena, 1989). Nonetheless, the project has, in fact, animated settlers into participation in activities that not only encouraged self-actualization, but community identity and development. In tandem with Mahaweli development workers, it has motivated the settlers to try innovative practices in agriculture and health. It has likewise motivated local development workers to take the settlers and their problems seriously, ensuring a more palpable degree of service to the people. (Valbuena, 1988)

However, Valbuena says the GCR community-based station, which is linked to a local development center and where resident production personnel are an integral part of the community, is "an unbeatable combination".

The MCR undoubtedly has incorporated many of the values which lead to successful community radio for development-identification with listeners, involving them in programme planning and production, and encouraging participation in development. However, it also demonstrates, as MCR and its evaluators recognize the need for multi-media approaches and interpersonal interaction if the awareness created by radio is to be converted to development improvement practices. While it appears the project is increasingly successful with the passing of time, the observer must bear in mind that the project has depended on outside sources (UNESCO, DANIDA and the SLBC) for financial and other, support.

Liberia Rural Communication Network

The Liberian Rural Communication Network (LRCK) is a branch of the Liberian Broadcasting Service (IBS). It began as a government-to-government project to which the USAID and the Liberian government committed financial, personnel and other Infrastructural support. This author was part of the team when during its planning, early construction and training stages; USAID provided funds for transmitter and studio construction, research and training and the Liberian government furnished some funding, personnel, station locations and support of LBS and government ministry personnel.

Liberia, a country of about three million people with-diverse cultures and at least 16 languages, has isolated and relatively underdeveloped rural areas. Limited transportation infrastructures have contributed to a fragmented nation, with small chiefdoms and clans, and to areas where new ideas have not reached or taken root. In this setting, many development projects have been frustrated.

LRCN was designed to support rural social and economic development by promoting use of existing government services, providing aid for development to the rural people, increasing communication between villages and regions of the country, promoting self-help projects, distributing news, information and entertainment and involving the rural people in development activities. To carry out these functions, the LRCN's main goal became the provision of a radio network to support rural development and to increase sharing of ideas between villages and between branches of the government and the upcountry citizens.

Today, there is a network of three medium wave stations and a central production studio, each with a staff of about a dozen persons. The stations are linked administratively and by single-sideband radio, but each station exists primarily to serve the local community surrounding it. Each producer is responsible for programming in his own language, but is aided by volunteers from the community. The stations broadcast two hours each morning, from 6:00 to 8:00 a.m. and four hours each evening, seven days a week. To date the stations have considerable autonomy. Participation is encouraged in several ways : local steering committees advise each station; village listening groups meet to listen and react to programs; local volunteers aid with translation and programming; program development groups made up of health workers and agricultural agents advise the station staffs; local music and drama groups regularly take part in broadcasts; and professionals from development agencies are trained in radio broadcasting (Foote et al, 1988). Laflin (1987) also notes that, because the network structure is decentralized with responsibility delegated to the field units, "each rural station must generate some local funds for the network by finding sponsors for its development information and local language broadcasting services". Because the Librarian government has often faced major financial problems, the raising of local support has become essential to maintain LRCN.

Development programming has focused on agriculture, fish farming, health, child survival, immunization, oral rehydration, family planning and village development councils. The network concentrates on specific development activities aimed at particular audiences. Spot announcements are used effectively. News is provided by the LBS. Local music and drama groups furnish entertainment to the daily schedule.

Through its daily 30 minute educational program, "Network Liberia," a four month campaign has been waged against the diarrhea and resultant dehydration in children which is prevalent in the country. LRCN staff and producers cooperated with medical specialists, Ministry of Health officials and mothers to mount the campaign in local languages. The campaign included prior spot announcements, drama, interviews and even a poster coloring contest to encourage mothers to visit clinical oral dehydration centers. According to Florida Kweekeh, LRCN Director, moderate dehydration cases dropped from 70% to 17%,, during the campaign, while mild cases increased by a similar proportion; however, she notes the trends may be attributed in part to increased use of the rehydration centers (1987).

In Bong Country, the station has carried numerous programs about the country's Agricultural Research Institute (CARI) and its work. The station in Lofa Country has centered its attention on fish ponds, which are a rich source of food and income. In both cases, post-tests have shown increased awareness and information on the part of listeners.

Since its inception, the LRCN has been subject to annual evaluations. One of the earliest was a study of the audience and its characteristics. Formative evaluation regularly plays a part in program conceptualization, planning and execution. For the most recent baseline before-after impact survey, made after the first full year of network broadcasts, a sample of 200 adults was drawn from each of the three station areas. Awareness, knowledge and change in practice were the key dependent variables. Results of each evaluation are feedback to LRCN personnel for adjustments to programming.

According to Dennis Foote (1989), who directed the evaluation, "LRCN listenership is large. Fifty five percent of those questioned ..... listen daily, and at least 80 percent listen several times a week" (p. 3). Foote also found that LRCN's popularity stems from producers whose program content comes from the community, from use of local languages and strong station signals. The study showed that LRCN has become a significant force in creating awareness about the importance of immunization, better family planning, the benefits of agricultural research, the value of fish ponds, and the role of village development councils and parent-teacher associations in the society. But far less evidence exists that there have been significant changes in behavior. This may be due to the fact that the stations have only been on air a year or that programs fail to identify small, manageable steps for action or that there is insufficient support infrastructure to bolster behavioral changes (Foote, 1989).

The keys to LRCN success include a corps of well-trained network indigenous leadership experts who plan, conduct research and give direction to the project. Also, the USAID has been patient and generous with funding and advisory personnel. Essentially, it has built a communications infrastructure of three 10 KW medium wave stations that did not exist prior to the project. As its support is withdrawn, LRCN will have greater funding problems; fortunately, the network has already begun to resort to local sources of support and increased volunteer help. But transmission expenses will continue to be a heavy burden. Thus far local and national government ministries and agencies have been cooperative and have not interfered with station programming; in fact, LRCN has made their work easier and more effective.

Although LRCN has made a slow, but good start, some significant challenges lie ahead : how to turn listener enthusiasm and heightened awareness of development ideas to behavioral changes; how to sustain sufficient budgetary support, especially since the nation faces a continuing financial crisis; and how to maintain the semi-independent decentralized status relatively free from political pressures which it has enjoyed to date.

Summary and Conclusions

This article has reviewed five community radio projects and analyzed the strengths and limitations of each. All of the projects exhibit similarities, yet each are a distinctive entity. The fact that each function in a nexus of differing political, economic, social and cultural variables invalidates direct comparison generalizations. However, the experiences of all five projects do point to certain criteria which appear to be essential if community radio is to serve as a useful and effective tool for development. Those commonalities can be enumerated.

1. First, as the experiences of all five projects indicate, there should be a program of baseline research and formative evaluation to determine community characteristics, interests, potentials and needs. Some of the projects studied, as LRCN and MCR, demonstrate the value of continuing research and evaluation throughout the project as a means of keeping it fully effective and relevant to the community it serves.
2. Second, program topics should be relevant community interests and development needs. If the programs are relevant, listeners will be interested and open to ideas expressed in the programming. Radio Jamaica, Radio Baha'i and MRC all experienced listener enthusiasm for relevant community news brought in by listeners. Radio Nagercoil has found that program of relevance to the community enhance a sense of belonging. LRCN and MCR concluded that topics should be especially relevant to the community's development needs.
3. Third, there should be community participation at all levels from the administration down, but particularly in program planning and production. In a continuum that ranges from simple

producer interviewing of local subjects to full involvement of community personnel in planning, producing, evaluating and editing programs for broadcast, all five projects have experienced the importance of involving listeners. One station of the MCR network. GCR found that audiences gave highest credibility to program producers who came tdive among them. Community interest and participation appears directly related to the degree these station staff identifies with its listeners. As Burke (1989) points out, '...radio producers should make it abundantly clear that the station does not belong exclusively to the radio staff. The radio station exists to give the community its own voice ..'

4. All projects found that use of the local language(s) is vital to the success of a community radio project.
5. Use of indigenous cultural materials, such as traditional music and folk drama, stimulates community interest, participation and loyalty to the station. Such materials provide the necessary entertainment side of community station programming. In all five projects, indigenous materials have helped the local community listeners feel the stations are truly theirs.
6. To differing degrees, the projects all found that radio in itself is insufficient for effective development communication. Interpersonal interaction and other media must also play a role. As Laflin (1989) indicates, 'People still want and need traditional face-to-face sources..' (p.15). The LRCN and MCR experiences seem to be demonstrating that radio is especially effective for stimulating awareness of a problem or a need, white behavioral changes appear to be more dependent on interpersonal and other factors.
7. Jemaican Radio illustrated the importance of gaining and maintaining the support of macro-level political actors. In various ways, ail the projects relate to local and national government structures and agencies. Radio Baha'i and Radio Nagercoil work closely with ministries in fact, they appear to be almost extensions of the agencies. MCR and LRCN, meantime, have a benign relationship with ministries, sometimes they have had to encourage the ministries to become involved. The projects should be working cooperatively with government agencies for development however, since they are within the system they could be in danger of top-down control or politicization.

8. Finally, establishment, operating and maintaining a community radio project costs money. Every one of the five projects studied is indebted to outside support for construction, transmission and maintenance costs, daily operations, research and training. In fact, none of them could have come into being without an infusion of outside support. The support agencies have been patient. A local station needs to become financially independent as soon as possible if it wishes to continue as a viable development tool if outside funding is withdrawn. As some of the projects studied have already done, a community station should encourage local funding and volunteer assistance. Both provide additional avenues for local community participation in the project.

In conclusion, it appears interactive local radio can be a significant tool for aiding the development process, especially if it encourages community participation.

Notes

1. The team was composed of five trained Liberians and five expatriate technical experts. The latter served as advisors. Thc author acted as chief coordinator for the advisory team.
2. Originally, a 10 KW MW station was planned for each of Liberia's nine countries, with a central production studio in the capital city of Monrovia. Togelher, they werc to comprise an interconnected network. Because financial constraints limited Liberian input to its share of funds for the project, the network was reduced three to stations, one each in Bong, Lofa and Grand Geddah countries, plus the central production studio in Monrovia.

References

Adkins, G. (1983). "Formative Evaluation in Educational Radio and Television : A Fundamental Need in Developing Countries." *Development Communication Report,* 44 (December), 3, 14.

Anjaneyulu, S. (1989). "Scaling Down: Local Radio in India." *Development Communication Report.* 64 (1), 1-2.

Arnaldo, C. (1988). "Preface." In Valbuena, V. *Mahaweli Community Radio Project: An Evaluation,* pp. v-vi. Singapore : Asian Mass Communication Research and Information Centre.

Burke, R. (1989). "Expectations for Local Radio : Looking Beyond Immediate Impact." *Development Communication Report,* 64(1), 7-8.

"Can Radio be 'Reinvented' for National Education?"' (1986). *Action*. 105:2, 6.

Ebbesen, K. (1986). "Mahaweli Community Radio." *Media in Education and Development*, June, p. 97.

Foote, D. (1989). "Local Radio for Development : Haba Na Haba." *Development Communication Report*, 64 (1). 3-6.

Foote, D., McDivitt, J. & Cassady, D. (December, 1988). *Final Report of the Evaluation of the Librarian Rural Communication Network*.. Menlo Park, CA : Applied Communication Technology.

Head, S. (1985). *World Broadcasting Systems : A Comparative Analysis*. Belmont, CA : Wardsworth Publishing Company.

Hein, K. (1983). "Community Radio in Ecuador : Playing Local Music, Strengthening Cultural Ties." *Development Communication Report*, 44 (December), 4, 6.

Hein, K. (1983) "Community Radio in Ecuador: Meeting People's Needs." *Development Communication Report*, 42 (June) 15-16.

Hein, K. (1988). "Community Radio Thriving In Ecuador : Otavalo Indians Running Their Own Show." *Development Communication Report*, 40 (December), 11, 13.

Kweekah, F. (1987). "Liberia's New Health Vehicle : Radio." *Development Communication Report*, 56 (1), 5-7.

Laflin, M. (1989). "Local Radio : Hard Questions." *Development Communication Report*, 64 (1), 15-16.

Laflin, M. (1987). "A Study in Decentralization : The Liberian Rural Communication Network." *Development Communication Report*. 58 (3), 1, 6.

McAnany, E. (ed.). (1980). *Communications in the Rural Third World : The Role of Information in Development*. New York : Praeger Publishers.

Melkote, S. (1989). "Effectiveness of Development Radio Programming Among Poor Farmers; A Case Study." Gazette 43: 17-30.

Mody, B. (1986). "The Receiver as Sender : Formative Evaluation in Jamaican Radio." *Gazette* 38 : 147-160.

"Overcoming Impediments to Communication for Rural Development in the Third World." *Development Communication Report* 62 (3), 5.

Stephens, D. (1989). "Power to the People; Not Just in Developing Countries." *Development Communication Report*, 64 (1), 9-11.

Valbuena, V. (1988). *Mahaweli Community Radio Project : An Evaluation*. Singapore : Asian Mass Communication Research and Information Centre.

Chapter **40**

Importance of Visual Literacy in Distance Education: An Analysis of Satellite Broadcasts

Valerie A.C. Collins and Peter J. Murphy

The professional literature on learning by television and the experimental projects undertaken at the University of Victorial all indicate the importance of 'good' quality productions. Educational programmes do not have to compete with sophisticated commercial television broadcasts. Students, instructors and administrators all emphasize this fact. The new distance education programmes must be designed to satisfy the educational needs. Learning styles, individual differences and career aspirations of the target audience. By instructors integrating verbal language a learning environment of this quality can be produced.

The launching of Sputnik in the late fifties heralded the beginning of the new world in which, we now live. Prior to this historical event, society continued to evolve at a slow but a regular rate. International conflict, unemployment and increased urbanization had characterized the first half of the twentieth century. Many new 'labour-saving' devices had been introduced into homes, offices and factories. Suddenly people found they had more leisure time than ever before.

Most of the changes which occurred in the period being reviewed tended to be incremental in nature. This happened for many reasons, but maintaining the status quo was undoubtedly a principal factor. People tend to dislike drastic changes in their lives, preferring to know with some degree of certainty what the future holds for them. Incrementalism, as Dye (1978) astutely notes : "is important in reducing conflict, maintaining stability, and preserving the political system itself" (p.34).

Often individuals who are responsible for introducing or maintaining a new innovation are frequently inadequately prepared for these responsibilities. For decades, 'on-the-job' training has tended to be the process used for acquainting individuals with the knowledge and skills needed to facilitate change. Consequently, 'muddling through' has characterized administrative behaviour in educational institutions.

The launching of the Sputnik was a 'mile stone' in man's technological and social development because it transformed the world into a 'global village'. It was clearly evident that events occurring in one part of the world would soon be transmitted to another part in a matter of minutes. A communication system had been created which would inform people of political, economic and cultural situations in other countries and provide them with an assessment of the consequences these conditions might have for them.

Frequently, important events do not occur in isolation but happen simultaneously and have a 'multiple effect' on society. As Toffler (1970) points out, innovations tend to feed on each other. While scholars and government officials were endeavoring to assess the consequences Sputnik might have for Western Europe and North America, advances in communication technology made it possible for television sets to be produced relatively inexpensively. By the end of the sixties, television sets could be found in most homes in the Western World. Simply turning a knob provided people with entertainment, education and information. Our society would never be the same again.

Once a television is switched on, it is difficult to prevent the eye from being drawn to look at the screen, and for the mind to become interested in the visual signals, even with the sound turned off. Television as we are all aware has an 'attention grabbing' quality (Fisher, 1978).

Television is a highly visual medium which has a substantial capacity for influencing people's values, standard of conduct and beliefs. To better understand the impact television has on people's lives and society in general more research must be undertaken to clarify important concepts such as visual literacy, visual language, visual thinking and visual learning. Possessing a more comprehensive knowledge base of this kind will enable those involved in distance education to use television more effectively.

Visual Concepts

What is visual literacy? According to Heck-scher (1975) visual-literacy is as essential to distance teaching as verbal literacy. He further submits that many distance educators are usually visually illiterate and as a result do not fully utilize visual literacy in the design, development and delivery of courses.

The term 'visual literacy' was first expressed by Debes (1969) who listed the skills of a visually literate person as :

To read visuals with skill.

To write with visuals expressing oneself effectively.

To know the grammar and syntax of
visual language and be able to apply them.

To be familiar with the tools of visual literacy and their use.

To appreciate the master works of visual literacy.

To be able to translate from visual
language to verbal language and
vice versa, (p.27).

A more precise definition has been put forth by Hortin (1978) who notes :

Visual literacy is the ability to understand (read) and use (write) images and to think and learn in terms of images (p.202).

Both these definitions indicate that visual literacy is composed of two components; competency as a consumer of other people's created images and competency as a producer of one's own images in order to communicate one's ideas.

Verbal language is 'generative' in that infinite numbers of meaningful utterance or statements can be generated from the application of a finite number of rules to a finite number of elements. Visual language is similarly generative, but, whereas in verbal language linguistic elements are morphemes, phonemes, syllables, words, phrases, sentences and paragraphs, the 'building blocks' of visual language are 1ine, dot perspective, tone, shape, colour, grain etc. Therefore, a visually literate person should be able to understand (read) images and use (write) visual language, and to be able to process information visually in order to think visually.

Is visual thinking possible? The answer to this question appears to be simple. If a visual language exists, then this language can be employed for reflective thinking. The ability to process information in a visual mode permits a learner to think by visual imagery. Visual thinking according to McKim (1972) is characterized by what we see, what we imagine and what we perceive.

Visual learning is known to be an important component of the complex human learning process. According to Dwyer (1979) there appears to be evidence which suggests visual umagery aids memory and facilitates learning. However, the relationship between the imagery process and the verbal process is still unclear. Several investigators (Dubin & Hedley 1969; Fleming & Hutton 1983; Haber 1970; Paivo 1969; Salomon 1979; Tidhar 1973; Lorayne & Lucas, 1974) have demonstrated that imagery improves memory and learning. These findings should be applied with caution.

The education system of British Columbia, at all levels, relies heavily upon printed matter for disseminating information and for nurturing learning. Yet, there is ample evidence to suggest that visual modes of communication may be more appropriate for transmitting certain concepts and for examining specific issues. If the potential capacity of television is to be fully utilized for delivering distance education courses, more consideration must be given to the visual aspects of the medium.

Methodology

The primary purpose of the investigation described in this report was to examine the impact of the new distance education system on the provincial universities. To facilitate analysis, the focus of the investigation was delineated to satellite-based, live interactive television broadcasts offered by the University of Victoria through the Knowledge Network of the West Communications Authority. Since a multitude of consequences are generated by this mode of communication, the scope of the study was further delineated by addressing only two principal issues, namely : (1) changes occurring in the educational experiences; and (2) changes occurring in the organizational structure of institutions.

The professional literature reviewed for undertaking this investigation indicated that visual literacy, competency in thinking, learning and communicating through the use of visual images are essential competencies for educators and learners involved with the new distance education systems. If modern communication technology is to be utilized effectively, distance educators must be familiar with as well as utilize more extensively, the elements, structure and style of visual language. Of the areas studied in the investigation, to satisfy the primary purpose noted, the issue concerning the use of visual information in satellite broadcasts will be commented upon in this document.

The data needed to complete the investigation were obtained from data banks established for two studies undertaken at the University of Victoria to explore the application of modern communication technology at the post-secondary level in British Columbia (Haughey & Potter, 1983; Collins & Forsythe, 1983). Data contained in the banks was obtained from a variety of sources to ensure a comprehensive knowledge base was available for analysis and discussion.

A review of the professional literature on learning theory, communication theory, learning from media, visual literacy, the adult distant learner, interaction in the mediated environment and experiments with modern communication technology was undertaken and information relevant to the purposes of the investigation recorded. Government documents, research studies and unpublished manuscripts concerning the application of modern technology in the British Columbia post-secondary system were carefully studied for important information. A selection of similar material from other regions of Canada, the United States, the United Kingdom and various countries of the world were reviewed so as to place the studies in a global context.

Face-to face interviews were conducted with senior government officials, educational administrators, university scholars, distance education coordinators and course instructors to obtain their perceptions and comments about the new modes of communication. The questions discussed during these interviews were derived from the professional literature and government documents reviewed as part of the data-gathering process. Telephone interviews with a number of individuals who had been involved in distance education for several years provided a users' perspective of the new distance education system. An examination of evaluation questionnaires completed in a group of students provided additional information of the users' assessment of the learning experiences offered by contemporary communication technology. A selection of video-taped broadcasts and supporting printed material analyzed in terms of a set of criteria established from the review of professional literature.

After the data had been gathered, it was analyzed by an illuminative evaluation approach developed by Parlett and Hamilton (1972) for studying Innovative programmes. This approach enables specific facets of an innovation to be discerned more clearly. Precision is achieved by a process of triangulation which 'illuminates' the issue by crosschecking data obtained from a variety of sources. In illuminative evaluation, there are three characteristic stages, these being : (1) investigators observe; (2) enquire further, and (3) then seek to ex-plain. The information derived from this analysis is used to establish profiles for the issues being defined. This form of evaluation, according to Parlett and Hamilton (1972), sharpens discussion, disentangles complexities, isolates the significant from the trivial and raises the level of sophistication of debate.

The creation of the Knowledge Network, as one might expect, had a significant, though somewhat delayed, impact on institutional relationships at the post-secondary level. One of the Networks mandates was to encourage greater co-operation, collaboration and communication among post-secondary institutions which it endeavoured to achieve through the delivery of the new distance education courses. Pask's (1980) Conversation Theory served as the conceptual framework for studyingwhether organizational structures and institutional relationships had been changed by the new communication technology.

Results

The professional literature suggested that students vary in their ability to learn through the different media. This observation was supported by the findings of this investigation. Students, instructors and administrators acknowledged that students differ in their response to various modes of communication. Some students indicated that they found television stimulated their thinking more than traditional correspondence courses, while others prefer to study printed material at their own pace and used television as a complementary reference source.

Many of the programmes offered by the Knowledge Network were found to be multi-media. Consequently, a series of broadcasts might be delivered by a variety of modes of communication including Satellite, printed matter, television, audio-tape, video tape, or computer. The challenge, according to instructors and distance education coordinators, is determining which mode is the most appropriate for delivering a specific item of information. Most students appeared to appreciate knowledge being delivered in different ways. There was a general consensus among the respondents that a multi-media approach enriched the learning environment.

Although there was ample opportunity for including visual information in the new distance education courses, there was less use of stimulating and informative graphics than was expected. Generally information was presented to students by verbal language, the mode of communication normally found in the traditional university lecture hall. Minimal considerations was given to visual literacy, visual language and visual thinking, which should have been integrated into presentation.

A distance educator coordinator reflecting the thoughts of many respondents maintained that there was an increased need for more visual information in the new courses. The visual techniques often used by instructors to stimulate thinking and to clarify concepts were perceived to be unimaginative and of limited impact. More recently, instructors have used video-inserts, role play and demonstrations with greater success but the visual component of most programmes still needs to be improved.

Several respondents questioned whether a sophisticated visual programme was necessary since highly motivated students would acquire the information even if a presentation was lacking in visual stimulation. The visual component of a programme obviously needs to be selected with due care and thought being given to the unique needs of the target audience. Students do not expect to receive a broadcast that is as 'attention grabbing' as many commercial television programmes but they do expect broadcasts to be of 'good' quality.

Some scholars investigating learning in the new multi-media environment have become concerned about the risk of students becoming passive recipients of knowledge, rather than active participants in learning. The data gathered in this investigation seemed to indicate that instructors tended to dominate the learning environment. Generally, communication was one-way, instructor to students, which supported the concerns being expressed regarding students becoming passive learners. Instructors were perceived as needing special in-service courses of instruction which would enable them to develop strategies for keeping students actively involved in the learning process.

The broadcasts reviewed revealed that a significant disparity existed among instructors and courses in making the most appropriate use of the medium of instruction. In some cases, imaginative use was made of techniques such as video-inserts or role play which enhanced the subject matter; while at other times a technique was totally inappropriate. A 'gangster' routine was used to teach a particular point of grammar in a programme teaching English as a Second Language. Such criminal activity is not an appropriate scenario for acquainting new Canadians to the nation's social and cultural milieu.

On occasion, television was used to convey material of a non-visual nature which would have been more effectively presented by a different medium. Such as audio-cassette. The practice of conveying the bulk of the information on a subject in printed packages and reserving television broadcasts for highlighting, demonstrating, exemplifying and discussing key points was in accordance with the professional literature on the effective use of media. At present, research is scarce on many aspects of contemporary communication systems, a deficiency which many respondents noted. This situation must change, if advantage is to be taken of the potential offered by modern technology.

Discussion

As the 21st Century approaches, modern communication technology will be introduced more extensively in post-secondary education. Universities, community colleges and technical institutes will discover new audiences for the services and programmes they offer. These new students will enrich post-secondary education in many ways by bringing with them a wider variety of lifestyles, a diversity of interests, numerous career expectations, extensive experiences and a maturity which is not usually found among undergraduate students.

Experimental applications of modern communication technology, such as those undertaken at the University of Victoria, have shown that we have much to learn about using the technology effectively. Instructors tend to employ, as one might expect, teaching methods, evaluation procedures and reference material which they found successful in the traditional university lecture hall. Many of these procedures and techniques, however, are inappropriate for the new media. The class being instructed may be dispersed over a large region which produces numerous unique problems for the instructor.

Analysis of Satellite broadcasts, undertaken as part of this investigation, revealed that instructors tended to use verbal language for presenting concepts, facts and issues. Yet, interactive television offers instructors an opportunity to present many aspects of a subject in a visual form. Often a graphic, role play or demonstration may describe or clarify an issue much more effectively than the spoken or written word. Visual information appeared to be used minimally in most broadcasts. This situation exists mainly because the instructors are usually visually illiterate.

The professional deficiency identified could be reduced by offering instructors special in service programmes to improve their visual literacy and understanding of the techniques of televison production. In the present period of fiscal restraint, in-service training is frequently perceived as being an exceptional luxury. If modern communication technology is going to be applied effectively in post-secondary education special in-service training for distance education instructors is a necessity. To expect scholars to transfer from one learning environment to another without any adequate preparation is unrealistic.

REFERENCES

Collins, V., & Forsythe, K., British Columbia - higher education and the integration of new technology. In organization for Economic Cooperation and Development. Education and new information technologies, area B. Changes in curricula and institutional structures. Paris: Centre for Educational Research and Development, Organization for Economic Cooperation and Development, 1983.

Debes, J. L., 'The Loom of visual Literacy'. An overview of a speech given at the First Annual Conference on Visual Literacy. Rochester : New York, March, 1969.

Dublin, R., & Hedley, R. A., 'The Medium may be related to the message', Eugene: Centre for the Advanced study of Educational Administration, University of Oregon, 1969.

Dwyer, F. Ma., 'Strategies for improving visual learning: A handbook for the effective selection, design and use of visual materials', Pennsylvania State University, 1976.

Dye, T. R., 'Understanding public policy'. Third edition. Englewood Cliffs; Prentice Hall, Inc., 1978.

Fisher, K. M., Idiosyncratic learning with T.V. Journal of Personalized Instruction, 3(3), 157-161, 1978.

Fleming, M. L. & Hutton, D.W., 'Mental Imagery and learning'. Englewood Cliffs: Educational Technology Publications, 1983.

Haber, R. M., How we remember what we see Scientific American, 222 (5), 104-112, 1970.

Haughey, M. & Potter, G., 'The principles and practices of good television teaching'. Unpublished manuscript available from M. Haughey, Univ. of Extn. Divn., Univ. of Victoria, P.O. Box 1700,Victoria, British Columbia, 1983.

Heckscher E., Visual literacy and distance education in E, Ljosa (Ed.). The System of distance education Brighton : International Council for Correspondence Education, 1975.

Hortin, J., 'Visual Literacy and visual thinking'. Manhattan College of Education, Kansas State Univ.,1978.

Lorayne, H., & Luces, 3., 'The memory book', New York : Ballantine Books, 1974.

MnKim, R. H., 'Experiences in Visual thinking'. Montereys Brooks - Cole., 1972.

Paivo, A., Mental imagery in associative learning and memory, Psychological Review, 76(-3) 241-263, 1960.

Parlett, N. & Hamilton, D. 'Evaluation as illumination'. Occasional paper. Edinburgh: Dept. of Comparative Education. University of Edinburgh.

Pask, G., The limits of togetherness. In S. H. Lavingtont (Ed.) Information Processing 80. Amsterdam : North Holland Publishing Co., 1980.

Salomon, G., Interaction of media, cognition and learning. San 'Francisco : Jossey - Bass. 1979.

Tidhar, C., Can Visual reminders increase learning from T.V. British Journal of Educational Technology, 4(2). 142-148, 1973.

Toffler, A., 'Future shock'. New York : Bantam Books, 1970 **Dr. A. K. Majumdar** is Professor, Dept. of Agril. Extn., Bidhan Chandra Krishi Vishwavidyalaya, Kalyani, Nadla, West Bengal, India.

Dr. Anantha S. Babbili is in Department of Journalism, Texas Christian University, Fort worth, Texas 76129, USA.

Dr. D. K. Gogoi is Associate Professor of Extension Education, Faculty of Agriculture, Assam Agril. Univ., Jorhat-13 (Assam) India.

Dr. (Mrs.) I. Graver is in Department of Home Science Education 4 Extn., College of Home Science, Haryana Agil. Univ., Hisaar (Haryana), India.

Dr. J. J. S. Yadava is Prof. & Head, Communication Research, Indian Instt. of Mass Communication, D-1 3, Ring Rd, South Extn. Part II, New Delhi-49, India.

Dr. Peter J. Murphy is in Faculty of Education, Univ. of Victoria, Victoria, British Columbia, Canada.

Dr. Raghbir Singh is Professor of Extn. Education, Punjab Agril. University, Ludhiana, (Punjab), India.

Dr. Srinivas R. Melkote is in Dept. of Radio-TV-Film, Bowling Green State Univ., Bowling Green, Ohio 43403-0235, USA.

Dr. S. Pathak, Scientist (S-2), Agril. Extn., Jute Agril.). Research Instt., Barrackpore, West Bengal, India.

Dr. (Mrs.) T. Verma is Associate Prof. cum Head, Deptt. of Home Science Education & Extn., College of Home Science Haryana Agril. University, Hissar (Haryana), India.

Dr. Valeria A. C. Collins is in Faculty of Education, Univ. of Victoria, Victoria, British Columbia, Canada.

Dr.Vighnesh N. Bhat is Project Director, Newspapers and the Reader, B 13(A), Kailas Quarters, Manipal-19 India.

Chapter **41**

Effective Dissemination of Information

B. P. Sinha

Dissemination of information does not mean mere spread of information nor does it mean unloading of information package on other persons a one way affair. In fact, it initiates a dynamic process which is ongoing, ever changing, at a large number of points and the largeness of points measures the extent of dissemination.

As we know there are multiplicities of information generation agencies existing in the modern world and so is the case of means and media of dissemination of this information. The net result of this is that the whole scenario of information dissemination has become quite complex. Several international organisations and agencies, agricultural universities and their various departments, research stations, marketing organisations, input manufacturers, commercial farms, cooperatives, voluntary organisations and a score of others are engaged in generation and dissemination of information in a big way. But increasing availability of information does not in itself mean that the clients are better informed. An examination of a simple pathway of information dissemination will make the issue clearer.

As depicted in Fig. 1, the distinction is first made between and data that are existing in records or in human minds and those which are yet to be identified, thus signifying the continuing nature of data generation. The data already identified are also anormous and the stock goes on piling up everyday, but the individual is aware of only a small part of it. Out of even this small part of the data that he receives, he comprehends only a part of it and the other part goes uncomprehended. Out of the data he has comprehended he finds that only a part of it is relevant to him under the situation he is. The relevant data is further divided into functional and non-functional. Functional data are those which are applied in some way to get or achieve something *i.e.* they are the means to an end. Non-functional data are those which are end in themselves they only provide the satisfaction of gaining knowledge *i.e.* knowledge for the sake of knowledge. Further, both the functional and non-functional data provide satisfaction either on their own, independent of other data or in combination with other data and hence become dependent on that.

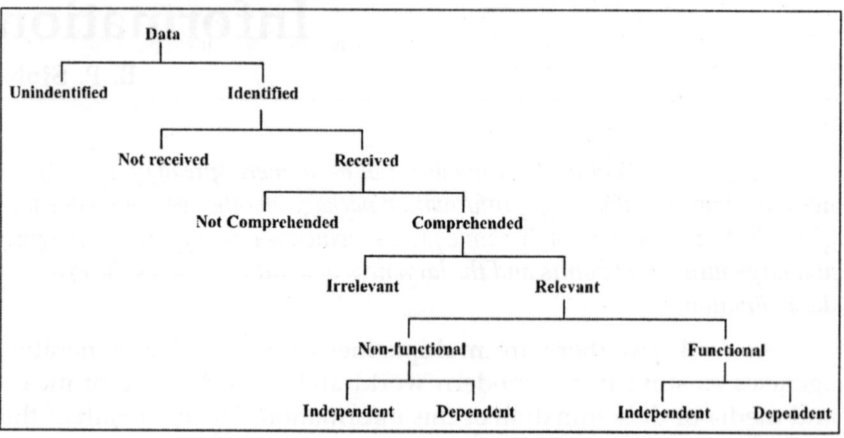

The pathway of information dissemination appears to be static, but in fact, is highly dynamic in nature. Data that is irrelevant at one point of time become relevant at another and vice-versa. Similarly, the facts that are non-functional may become functional over a period of time and vice-versa.

The Information

In agriculture, most of the data are not end in themselves rather they are means to an end playing significant role in problem-solving and decision-making. But it would be appropriate at this point to clarify that data as such is not information. To become information, data require analysis and interpretation to place them in a decision making context. Raw data or even semi-processed data are rarely used directly by decision-makers. In this sense an information system is a process which imposes form and gives meaning to data. Except in cases of some trivial and uncomplicated problems where raw data can be immediately translated into information, the simplest information system needing some form of processing will have at least three basic elements.

(If Q is the quantities of discharge for say concrete channel and Q^1 that for the earth channel, L is the length of concrete channel and L^1 that of the earthen channel)

Input ———→ **Processor** ———→ **Output**

(Raw data) **(Process)** **(Information)**

A simple example will make it more clear.

Objective : To have a water conveyance device with minimal water loss in transit.

Problem : Which of the two available devices - the concrete made irrigation channel or the earthen irrigation channel results inless water loss.

Analytical

Frame work : Estimate water loss per 100 metre of channel length.

Raw data : Quantity of water discharged into the channel (Q1), quantity of water discharged into the field (Q2) and channel length (L).

Process : Calculate and compare $\frac{Q_1 - Q_2}{Q_1 x L}$ **x 100 of the two devices.**

Information : $\left(\frac{Q_1 - Q_2}{Q_1 x L} \text{ x } 100\right) < \left(\frac{Q_1^{'} - Q_2^{'}}{Q_1^{'} x L^{'}} \text{ x } 100\right)$

Two other important elements of information system are *Control* and *Feedback*. The farmer involves comparison of output with some predetermined standard say for example in this case cost of lifting water, and the latter passing back into input of information about any deviation that occurs, so that corrections can be made, for example cost of the irrigation channel. Now on the basis of the cost of lifting water one can take a decision go for a concrete channel only if the channel is not more than 20% costlier than the earthen channel. Thus the information system will now be as under:

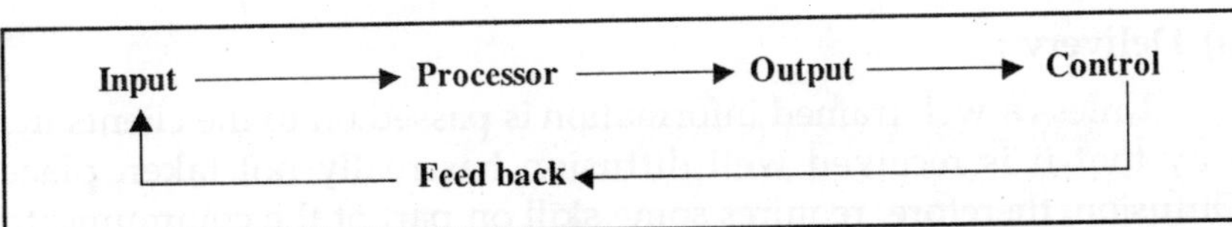

The soundness of information is extremely important in agriculture. The above information appears to be quite sound where cost in lifting water is the only or atleast very major variable constraining use of water. But in a situation like Punjab where the rent per pump is fixed irrespective of the quantity of electricity consumed, the soundness of this information will vanish. Contrarily, in a situation where water is not available in abundance *i.e.* a larger area of cropped land is to be irrigated with relatively less amount of water available for the purpose, this information would be inadequate and taking the scarcity factor into account a sounder information can be produced from these data. This however, does not mean that only physical and situational factors should be taken into account. In fact, psycho-personal factors of the clients are equally important. For example, in a study on effectiveness of farm school-on-the AIR programme, Ram (1981) reported that the motivating forces to adoption of the practices advocated by radio were (1) the promise made by the practice to increase yield (earning) and (2) easiness in adoption of the practices *i.e.* cumbursomeness of a practice should be so spelled out that it is not perceived as really combursome by the audience.

It can thus be said that information must be very sound and that in order to make the information sound, a critical analysis of the clients' situation must be made. This further implies that the information is client specific and not data specific. Thus like a good farmer who will sow his seeds only in good land, a farm communicator should give his information to a properly identified client rather than anybody and every body only to earn a frustrating experience.

Dissemination

The above mentioned aspect of information primarily relate to framing of information keeping the potential or targeted clients in view. The other equally important aspect of this is dissemination of the information to the target client. Dissemination has two aspects delivery and reach out. Let us first take the delivery aspect.

(i) Delivery :

Unless a well framed information is passed on to the clients in a way that it is received well diffusion has really not taken place. Diffusion, therefore, requires some skill on part of the communicator for effective communication of the information to his clients. A brief discussion of these follows as under :

(a) Frame of Reference : A frame of reference is the context of a person's thought while he is communicating. His context must correspond with that of the audience/clients, otherwise the communication will break down. For example, if a communicator while passing on information about a particular water conveyance device keeps his research farm in his context which is quite different from the fanners' farms, he is bound to fail in his communication, even if his information is sound.

(b) Empathy : Empathy is one's ability to understand the inner states of others. This enables him to encode messages which will overlap the receiver's field of experience resulting in high fidelity of the communication. As a communicator we carry in our minds an image of our receivers. The accuracy of that image (empathy) will determine how effective we shall be in our communication. Total empathy is probably impossible because of our peculiar personalities, but this can be developed through opening ourselves to others and practicing imaginary role taking.

(c) Authenticity : Here authenticity does not relate to the information, but to the informer, not to the message but to the communicator. To be authentic means to be genuine. Sometimes we try to protect our image rather than making an honest effort to communicate. We mask our genuine self and loose our authenticity in the process. We frequently become defensive when we feel inferior or feel that others are judging us. Sometimes our communication behaviours make others become defensive particularly when we start making judgments of them. In both the cases our genuine selves are hidden and communication is blocked.

(d) Trust : When we feel trust or can generate trust in others, we have developed a necessary ingredient for fidelity in communication. Trust is primarily communicated in the relationship between what we do and what we say. If our language and behaviour consist of conflicting messages, trust is reduced. Displaying authenticity generates trust, which in turn makes it easier for us to be authentic.

(e) The feeling content : While giving information we also communicate our feelings, many times quite pronouncing. This can help accept as well reject our message depending on what emotional state we are in at the time of communication. If we communicate the feeling of happiness, enthusiasm, affection etc. this will strengthen the communication, but if we communicate anger, ridicule, rejection etc. the communication will be ineffective. A communicator, therefore, needs to be cautious while giving information to his client.

(f) Source credibility : This is the extent to which a source of communication a communicator or a channel is considered trustworthy and competent or expert. Source credibility is an important factor for making the information or message acceptable to the receiver. Association of the communicator with a research/teaching institute will earn high credibility for him or else the message may be associated with a credible source to make it more acceptable.

(ii) The reach out :

An important aspect of information dissemination is reaching the information to millions of clients as quickly as possible. This involves several issues which deserve serious considerations. Ours is a society characterized by pluralities. We have value plurality language plurality, technology plurality and somany other pluralities. We have about 1252 officially recognised dialects to which communication system has to respond. Information must be communicated in the language people know. Compounded with this diversity of language is the extent of illiteracy which puts a big limit on the use of printed medium. A further diversity in this respect is that out of 36.17% of literates (1981 census), 52.4% literates are city born and only 23.7% are ruralites. Under these circumstances one has to think of literacy-free communication modes like interpersonal communication, traditional media, and the modern media like film, radio and television.

Further, the communication system needs to respond to the economic status of our people. "Indian poverty condition is such that 350 million people have the capacity to spend only Rs. 2.00 per day to meet all their needs. Hence, we have to provide communication to a large mass of our population on a Zero-private-cost basis.

We have, therefore, to think how can we reach our information to such a vast, diversified and widely dispersed population of ours. Ours is a society where no technology fades out, all co-exist from bullock cart to jet, and so is the case with communication technology, we thus need to exploit all the communication technology from traditional to modern to reach as many people as possible.

We have a strong band of extension workers numbering about a lakh. Looking into the size of our population this number of extension workers is quite meagre, yet they can communicate to a sizable number of people under inter-personal situation which is in fact, the most effective mode of communication. Communication as we all know follows a multi-step flow resulting in a wider dissemination of information if the information is communicated well at the first step. It is, therefore, necessary that the extension workers employed to disseminate information must be trained well in production of message and acquisition of skills mentioned earlier in this paper, to be effective in communicating information.

Traditional media of communication are still quite potent to communicate development information to rural masses. It is unfortunate that they are not being judiciously employed for this purpose. The extension personnel need to make special efforts in this direction which will greatly improve their capability to communicate to a larger mass with greater effectiveness. This requires identification of practitioners of traditional media, persuading and motivating them to give frequent performances with development messages to be supplied by the extension worker or the development department. This needs a high degree of motivation on the part of extension personnel in the first place, and the ability on his part to get the village, composer-singer or other artists to internalize the message and give it a creative expression.

A pillip can be added to this endeavour by bringing modern mass media into the play. Instead of attempting to have live performances frequently in each of the villages, the performance can be once recorded and then relayed through media like radio, television, video recorder and short films repeatedly. Traditional media being highly specific to each cultural region even within the linguistic zones, and development information itself being area specific, the mass media programme using traditional forms would have to be made and delivered on a decentralised basis. Unfortunately the mass media available in the country (barring print media) are government owned and they are not developed on this pattern. Even then mass media can be used for this purpose atleast in a limited way.

So far pervasiveness of communication network in India is concerned; the mass media have the greatest reach. The annual report of the Ministry of Information and Broadcasting for the year 1981-82 says that the technical reach of AIR's signals stood at "over 89% of the population and about 78% of the total area of the country". However, access to these signals is a question which needs a serious ponder over.

Urban dwellers and rural rich by and large are the bulk owners of radio receiving sets. The precise number of radio sets in use now is not known since the requirement of a broadcast receiver licensed, for low-cost (upto 2 band) receivers was abolised in August, 1980. The number of licensed radio receivers was 10.2 million at the end of 1979 which rose to 20.7 million at the end of 1981 (after abolition of the license required for 2 band sets). If we take the unlicensed sets into account this figure may go up by say three to four times. Even then we have one radio sets for several thousands of our population.

Even the community listening of radio did not make much of dent. There are only about 90,000 community listening sets as against 5,76,000 villages in which the bulk of India's population live. Further, more than half of these sets are liable to be out of commission at any given time.

Even the spectacle of the farmer carrying a radio set to his field what has been called the transistor revolution is confined only to some of agriculturally progressive areas like Punjab and Haryana.

Listeners' preference of particular broadcast(s) is another aspect which further limits dissemination of development information. All these makes it amply clear that the picture is really not that rosy as it appears from the annual report of the Ministry of Information and Broadcasting. Nevertheless, its coverage as compared to other media/ mode of communication is quite impressive and hence it can not be ignored for diffusing information.

The state of affairs with television is no better. Not only that the coverage of its signals is comparatively very small, but its accessibility is greatly limited by the high cost of its receiving sets.

A content analysis of one month's TV programme for seven years (for the month of May from 1972 to 1978) carried out by Jha (1978) reveals that 36.25 to 48.4% of the transmission time of Delhi Doordarshan was devoted to the programme providing entertainment. This was followed by the Information giving programmes which consumed 15.4% to 27.8% of the transmission time. Educational programmes occupied the third position consuming 7% to 15.1% of the transmission time. A recent analysis (reported by Shri G.N.S. Raghavan in his paper entitled "Centralised and Decentralised Communication; Software & Hardware Implications presented in the seminar on Communication in India : Prospects and Policy held by I.I.M.C., in April, 1983) of 14 days of Delhi Doordarshan's offerings-from 25th March to 7th April, 1983, shows that feature films and film related programmes occupied the largest single chunk of telecast time 21.1%. This was followed by news and current affairs (18.9%) and development oriented programmes including civic affairs, consumer prices etc. consumed less than even 14% of the total telecast time. It thus becomes clear that Delhi Doordarshan fulfills the three widely accepted functions of communication media in the order of Entertainment first, Information second and Education third. This has probably created an impression in the minds of the people that it is predominently an entertainment medium which obviously limits

its role in diffusion of information.

Despite all limitations, the potential of mass media has to be well recognised. To quickly reach a large mass of the population which is an essential requirement of development information, they have to be employed as far as possible.

The policy guidelines for official media states, "media units should give effective coverage to Governments' programmes, policies and achievements. Particular attention should be given to programmes and dissemination of information relating to developmental activities in all their facets including, for example, extension work in Agriculture, Education, Health and Family Welfare and Science and Technology". (From letter issued by Ministry of Information and Broadcasting on July 2, 1980 and placed before the Lok Sabha on August 12, 1980 and before the Rajya Sabha on March 16, 1981). For this, it appears that they will be employed atleast to a reasonable extent (hopefully) to disseminate development information. They will perhaps provide more listening hours and also diversification in programmes. A mention may be made here of the recent experiment conducted by AIR in instituting several farm-schools on the AIR. A recent study conducted by Ram (1981) revealed that this school had been markedly effective in disseminating development information 28% of the improved practices in paddy cultivation disseminated through farm-schools-on-the-AIR programme were adopted by the registered listeners of the programme in the first year itself and some of the practices were adopted by 90% of the listeners. At least some of the information received by them was passed on to 46 non-listeners on an average by each of the registered farmers, which speaks of the extent of diffusion.

So far print media are concerned they have their limitations. Rural population in general and farmers as well as farm labourers in particular have very little education to make any extensive use of print media. The national newspapers have practically no readers coming from this chunk of the population and hence are not willing to provide much coverage to the development information meant for direct consumption of the from population. Yet we can't afford to ignore it totally. At least a few of the farming population is educated to the level that they can receive message through print, of course in their own language. Then why not to feed them to make them better informed and set the ball of words of month rolling further. Thus bulletins, leaflets etc. should be extensively printed and language newspapers should be used as far as possible to disseminate development information through print media.

REFERENCES

APPLABAUM, RONALD, L., ANATOL, KARL, HAYS. E.R., JENSON, O.O., POSTER, R.E., and MANDAL, J.E., "Fundamental Concepts in Human Communication," Cambuld Press, 1973.

BLACKLE, M.J. and DENT, J.B., "Information Systems; for Agriculture," Applied Science Publishers Ltd., London, 1379.

BURCH, J.C. (Jr.) & STRATER, F.R. (Jr.), "Information System, Theory and Practice," Hamilton Publishing Company, California, 1974.

KUPPUSWAMY, B., "Communication and Social Development in India," Sterling Pub. Pvt. Ltd., New Delhi, 1976.

LILLEY, G.P., "Information Sources in Agriculture and Food Science," Butterworth *& Co.*, 1981.

RAGHVAN, G.N.S., "Centralised and Decentralised Communication : Software and Hard Ware Implications" - a paper presented in the Seminar on Communication in India - Prospects and Policy, organised by IIMC, New Delhi, April, 1983.

SMITH, ALFRED, "Communication and Culture," Holt Rinehart and Winston Inc., 1966.

SETH, SATISH C., "Communications Science in India (2001 A.D.) Prospect and Policy," a paper presented in the seminar on Communication in India - Prospects and Policy, organised by IIMC, New Delhi, April, 1983.

Part III

Chapter **42**

Development Communication in Search of a Perspective

S. Pathak, A. K. Mazumdar

The objective of this paper is to emphasize the need for evolving a suitable perspective for the analysis of development communication process. This task can be carried out in the backdrop of the controversies relating to the concept of development, linkage between communication and development, review of different communication models and finally an endeavour to search for an appropriate development communication perspective.

I. Concept of Development : The Two Perspectives

There appears to be a number of different interpretations as to how development occurs in a society. Two broad schools of thought have, however, emerged from the varying approaches which are briefly discussed in the following pages :

(i) Some scholars suggested that the existence of certain modern attitudes is a precondition to the development. David Mc. Clelland strongly argued that effective entrepreneurship presumes the existence of a way of thinking that leads men to behave in a particularly energetic way a "mental virus" which psychologists call "need achievement". Other scholars have further suggested that family particularism, fatalism, laziness etc. are obstacles to modern entrepreneurial behaviour. Max Weber suggested that certain traditional societies had within them the seeds of modernity. He described the essential elements of the Protestant Ethic as a belief in the value of work, an emphasis upon savings and thrift, and a desire for upward social mobility. He further added that the value system of Calvinism contained the 'mental virus' (to use McClelland phrase) that made modern entrepreneurship possible. This position was, however, opposed by others on two grounds.

(a) Alexander Gerschenkron contended that Calvinism though facilitated development, yet many different value systems seem to exist in modernizing societies. One might note that Catholocism has apparently not impeded the extra-ordinary high rate of economic growth of many Latin American Countries. Moreover values described as 'traditional' may hasten development depending on social context and the use to which these values are put. Even family particularism need not be an obstacle to development as Marwari, Jain and Parsi though committed to family particularisms, have impressive records of successful entrepreneurship. S. P. Bose also reported that farmers manifested change in economic behaviour although they nurtured familistic values.

(b) Milton Singer, Bernard Cohn and M. N. Srinivas jointly suggested that even Hinduism described by Max Weber as an impediment to modernization (development) in India has not proved to be an obstacle. They pointed out that there is a variety of value systems within Hinduism and that religious belief and practices may themselves undergo major changes.

(ii) Social scientists of the other school of thought suggested that the appropriate attitudes and the appropriate behaviour will be forth coming, once 'opportunities' and 'incentives' are provided. This view has been supported by many economists who pointed to the existence of institutional impediments to productive activities in many developing countries. Myrdal pointed out that institutional deficiencies act as impediments in achieving economic development. Stanislaw Wellisz stated that values do not impede profitable economic activity, since people often described as indolent in one environment may prove to be effective entrepreneurs in another, for example, Indians and Chinese have been more productive outside their home societies. Alexander Gerschenkron suggested that development has often occurred in the absence of these 'preconditions' of dominant values in a social milieu.

Many economists, however, concluded that the institutional framework provides both "opportunities" and "incentives" which make development possible. It has been widely argued that many traditional peasants engaged in subsistence agriculture have a low level of aspiration and do not respond to economic incentives. But Clifton Wharton, an agricultural economist, demonstrated that traditional peasants are also responsive to higher prices and profits. He concluded that inputs in seeds, fertilizers and water that make it possible for peasants to reap profits from additional effort, higher more stable prices for agricultural commodities and a reduction in risk may be the most effective means of modernizing agriculture. In short, one need not accept the values and attitudes of peasants as the only necessary preconditions to make agriculture more productive.

From the above discussions it appears that there exist conflicting views on development. This is a complex issue in which a large number of variables are involved and it is, therefore, difficult to assign priority to any single proposition. Moreover, societies may differ so greatly in their nature of needs and change etc. and hence the/ priorities of one may not be those of another. Both the approaches in respect of development, therefore, appear to be significant and in many situations, complementary.

II. Linkage between Communication and Development :

Development is an intensely human phenomenon it necessitates changes in the attitudes and behaviour of individuals, groups and institutions. The main goal of development is the development of human being, the improvement in the well being of the individual and the amelioration in the quality of life and environment. Basic to this concept is the need for maximum involvement and participation by all the people in the development process. It is therefore, imperative that people are made aware of their role in overcoming their problem and are fully involved in designing and implementation of the development programmes. This approach necessarily emphasises the role of communication in the process of development. Communication and development are considered to be closely related. It is through the communication the free flow of information, ideas, facts, feelings and knowledge are circulated and maintained among the members of a society. The UN General Assembly resolutions of 1953, too, emphasised the role of information media in education, economic and social progress. Daniel Lerner (1958) also pointed out that, "mass communication serves as the great multiplier in development, the device that can spread the requisite knowledge and attitudes immeasurably more quickly and widely than ever before." Similarly Schramm and Carter (1959) reported high correlations between the scale of economic development and other scales representing the development of mass media. The studies of Leanard Doob in Africa (1961) and Y. V. L, Rao in India (1966) explained how extensive exposure to mass media helps in bringing about change in the traditional society.

The application of fast growing technology has thus substantially changed the social milieux and it is in this context that communication assumes a new significance. We should, therefore, take into account human communication needs throughout the whole design, planning and execution of development programmes. In the words of Balcomb (1975) "development is not a commodity that can be sold like a particular brand of soap, but a complex of attitudes and responses that can only be promoted by taking the people into one's confidence and carrying on a genuine exchange of views and information with them". Communication helps individuals, groups and organisations to exist and function effectively by enabling them to interact with themselves and their environment.

The interwoven link between communication and development thus calls for the examination of several models of communication social scientists have earlier developed and described. Accordingly, a number of communication models have been reviewed and some of their potential shortcomings have been discussed in the following paragraphs The analysis would enable the readers to gain a basic grasp of this vital component in the development process A general model of communication involves five elements : sender, message, channel, receiver and effects (S-M-C-R-E). Some models appear to be simple while others are relatively complicated. The existing models of communication are briefly described as follows : Shannon and Weaver designed a mathematical communication model which was largely adopted by behavioural scientists in describing human communication. The model includes ; (i) a sender, (ii) a transmitter, (iii) a signal, (iv) a receiver and (v) a destination. Ruesch and Batesan conceived of a communication as functioning simultaneously at four levels of analysis : intrapersonal, interpersonal, group-interaction and cultural. Each level of activity consists of four communicative functions : evaluating, sending, receiving and channeling. Elaborating the model of Shannon and Weaver, George Gerbner extended the components to include the notions of perception, reactions to a situation and message context.

Wilbur Schramm altered the mathematical model of Shannon and Weaver and reported six elements in communication process such as source, encoder, signal (message), channel, decoder and receiver. Westley and Mac Lean conceived a model having six elements as sender, encoding, channel, decoding, receiver and feed back.

Berlo described linear model of communication with six elements namely, source, encoder, message, channel, decoder and receiver. The source and receiver are considered mutually dependent. Dance advanced a concept of a helix to describe communication as dynamic, continuous, unrepeatable, additive and accumulative.

Becker suggested 'communication mosaic' which shows that most communicative acts link message elements from more than one social situation. This idea, however, was not incorporated in other models. Barnlund viewed communication as "the transactions in which communicators attribute meaning to eventsin ways that are dynamic, continuous, circular, unrepeatable, irreversible and complex". He suggested two models of communication : one at an intrapersonal level and the other at inter-personal level.

A comparison of the various models discussed above indicates the great similarities among them. However, these models suffer from some inherent basic weaknesses. According to Berlo, none of these models may be considered to be 'right' or 'true'. Some may be more useful than others, while some may correspond more than others to the current state of knowledge about communication. He reported that the difference among models may be due to inherent differences in disciplines and areas from which they have emerged, Miller (1966) while examining salient features of models pointed out that a model of speech communication cannot be viewed as an identical replica of the process itself on two grounds :

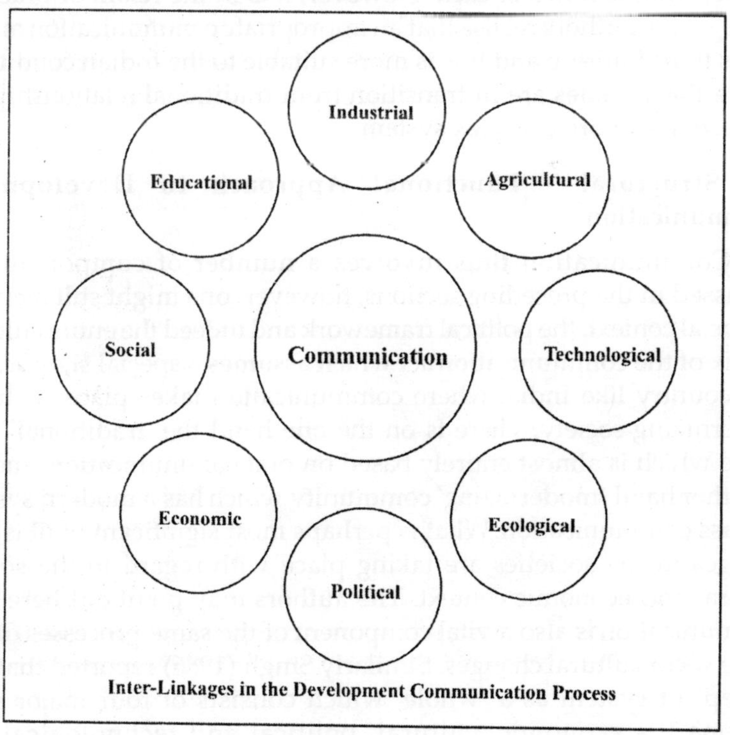

Inter-Linkages in the Development Communication Process

(I) Speech communication is a psychological phenomenon; it is, therefore, impossible to provide literal analogues for the many overt behaviours and internal responses present in any communicative situations. It employs words, numbers, or pictures rather than physical representations to depict potentially relevant aspects of the process.

(II) The very nature of process also prevents one from viewing a model of speech communication as an identical replica. Any model of speech communication must focus on only part of the total process, because it is impossible to construct a model containing all of the variables at work in the process.

A careful analysis of these models may reveal which has the greatest utility and this consideration should dictate the best selection once one's objectives have been determined. Miller (1966) pointed out that it would be fruitless to seek one true, correct model of communication but one should be aware that there are numerous models that can aid him in analysing and understanding the total process. Let him not limit his vistas by selecting anyone of these; but, rather, let him expand his horizons by making use of the appropriate and relevant features of each. However, it is in the realm of practical strategy, the authors realise that an appropriate communication model needs to be built-up and this is more suitable to the Indian conditions where the societies are in transition from traditional relationships to modern and more complex system.

III. Structural - Functional Approach to Development Communication :

Communication thus involves a number of components as discussed in the preceding sections, however, one might still mention the social context, the political framework and indeed the entire cultural matrix of the communication act which assumes a special significance in a country like India, where communication takes place within a modernizing society. There is on the one hand the 'traditional' way of life which is almost entirely based on oral communication, and on the other hand 'modernizing' community which has a modern system of mass communication. What is perhaps most significant ofall is that changes in the societies are taking place with regard to the social, cultural and economic context. The authors may point out here that communication is also a vital component of the same processes of the entire socio-cultural changes. Similarly Singh (1986) reported that the concept of system as a 'whole' which consists of four major sub-systems i. e economic, cultural, political and technological are integrated by communication as *an* interactive force of communication through linkage function. In the context of a changing socio-economic system, 'traditional' or 'informal' systems of communication are also undergoing transformations.

In the wake of changes occurring in the society due to modern knowledge and technological evolutions the role of all the components of the society are also changing. So is the change in the interactional relationships of communication with other components of the society.

The Asian Regional Seminar (1979) on the theme of Rural Development and Communication Policies jointly sponsored by the Indian Institute of Mass Communication and the International Institute of communications observed, "Communication cannot be Seen in isolation, but as a part of the total development system. Thus, communication cannot be dealt with by itself but only in relation to the political and socio-economic structures and cultural patterns. It is important to avoid a simplistic relation between communication and development effort". Singh (196) also reported that if development communication is indeed considered as a total process, the inter-relationships among the components must be investigated as well as relevant environment in which communication system is embedded. This type of intellectual focus represents systematic approach to development communication.

Development in the society has many dimensions : economic, social, ecological, political, technical, educational etc. In the course of developmental changes, the inter-relations of dimensions are greatly altered. Communication assists all the dimensions of society; it thus envisages multi-dimensional linkages within the societal frame-work. Communication helps to establish the harmony that is vital to the proper functioning of the society. The authors attempt to present the essence of such interlinkages visually. Figure 1 thus provides a rudimentary illustration of the importance of interactive linkages within projected dimensions. These dimensions are obviously not exhaustive, One might still increase the number of dimensions under exigencies of situational needs as would be apparent from the observation of Miller (1966) mentioned earlier in section II and the following comments of Sereno and Mortensen (1970).

Sereno and Mortensen (1970) reported, "the notion of interaction entails the far more complex idea of interdependence a mutual influencing process among countless factors, each functioning conjointly so that changes in one set of forces affect the operations of all other constituents activity in a total field of experience". Changes through these mutually influencing forces in the society also entails functional differentiation in the components of communication. This functional differentiation obviously calls for a critical review of communication in the development nexus. In this connection, the observations of the Seminar on Communications and Information for Development Purposes in the Caribean Area, Georgetown, Guyana (1974) is significant. The Seminar Committee established working definition of 'development support communication' as, "The systematic use of communication techniques for the planning, decision making and execution process of development programmes, in order to ensure, at all stages, participation by all relevant groups of people and the success of the development objectives". While discussing different approaches as to how to promote development through use of communication techniques, Boyd (1975) reported that the main interest among the people in the school of 'social process' is to explore and promote the role of communication in society, that is in stimulating self-awareness at the community level, increasing community access to media, using media to help coalesce and express community interest etc. What kind of communication do we really need to promote development in the country? It is obviously the kind of communication that encourages people to think for themselves and to act on their own initiative to function as participants not merely beneficiaries. Hence, the importance lies on the fuller growth of 'man' in the development communication process. Therefore, the emphasis should be on the need for an approach encompassing the total environment of 'man' and on the need for comprehension of and respect for his values. Singh (1986) reported that development communication should be viewed as a total process that includes understanding the audience and its needs, communication planning around selected strategies, message production, dissemination, reception and feedback rather than just a one way direct communicator to-passive-receiver activity. The development communication process, therefore, proceeds from an individual centre of consideration and then radiates outward to embrace the family, the community and finally the nation.

In this context, it is interesting to note again the remark of Boyd (1975) who suggested that proper communication analysis begins at the level of real development action the project level, where participation, involvement and benefit by people is actually supposed to occur and where well informed awareness, understanding acceptance, and implementation of change by people is a fundamental pre-requisite for successful development. It is therefore at the project level that communication support can best be designed and carried out, attend to the whole range of human expectations, development potentials and social-cultural, psychological characteristics that are present in the project 'community'. Development communication, thus must be based on the understanding of social, cultural, motivational and communication factors within the particular community, that is when it becomes more ideally the 'web which knites the strands of development in the society together'.

IV : Values and Development :

The idea of development communication discussed in the preceding sections is apparently close to action perspective in the behaviourist school of social science. Normally, the sociology of development is considered to be associated with development of action systems in the social-action perspective.

In the dynamics of action systems the component like actor, goal etc. have important roles and the concept of development of action systems is based principally on the structural-functional theory; the basic postulate being that the system is boundary maintaining and it tries to maintain its status compatible with the environment. According to structural-functionalism, relevance for the effective functioning of the system is sought in structures and in this effort the system should satisfy some imperatives and eventually to meet these demands the system should develop appropriate structures. In the process of differentiation, values have been considered as points of reference in legitimizing the introduction of new ideas and in this process values are conceived as central in the monitoring of processes of social development. Although this formulation is vitally concerned with the role of values in social change, less attention has been paid to the conditions which govern the emergence of new system of values as far as social action perspective is concerned.

In this connection, it may be mentioned that certain theoretical systems demand that structural arrangements are mainly responsible for social change as we indicated at the beginning of the paper. There are also perspectives which claim ideas govern changes. Hence we provide below a statement of Neil Smelser, "....................................... as social scientists it would be of great help to attempt to devise concepts that express relational state between symbols (of which values are one type). It is after all impossible to understand the importance of values in social change without at the same time inquiring into the relation between these values and the structural arrangements that presumably institutionalise these values; it is also impossible to inquire into the importance of structural factors in social change without simultaneously inquiring into the relations of these structural factors to the symbol systems that presumably legitimize them". Smelser has thus stressed on diminishing the force of old controversy over the relative importance of ideal and real factors and at the same time emphasized to generate more relational language.

SELECTED BIBLIOGRAPHY

Balcomb, John (1975) Communication for development : from propaganda to dialogue. Educational Broadcasting International, Vol. 8 (1) : 10-14.

Boyd, Paul D. (1975) Causes and cures of communication neglect in development planning. Educational Broadcasting International, Vol. 8 (1) : 5-9.

Doob, Leonard W. (1961) Communication in Africa : A Search for Boundaries. Yale University Press, New Haven.

Lerner, Daniel (1958) The Passing of Traditional Society. Free Press. Glenco.

Miller, Gerald B. (1966) Speech Communication : A Behavioural Approach. The Bobbs-Merrill Company Inc.

Rao, Y.V.L. (1966) Communication and Development. University of Minnesota Press, Mineapolis.

Schramm, W. and Carter, R. F (1959) Scales for describing National. Communication Systems. Institute for Communication Research Stanford.

Sereno, K. K. and Mortensen, C.D. (1970) Foundations of Communication Research. Harper and Row, New York.

Singh, K. (1986) Development Communication: Theory and Application in Third World in Sampreshan : 19-31, National Council of Development Communication, Varanasi, 221 005, U. P.

Weiner, Myron (1966) Modernization : The Dynamics of Growth (Ed). Voice of America Forum Lectures, 1966. P. 373.

Chapter **43**

Communication in Social Change

Malikhan S. Chauhan and Satish Verma

Taking off on the theme that "effective communication is essential for development i.e., social change", the authors identify major cultural, personal and communicative problems documented in the literature with respect to the developing world, particularly the Indian sub-continent. They argue that a communication strategy for accelerating development should encompass an interactive research extension teaching system and effective organization of people and programmes.

Communication is basic to all human interaction. It occurs in the most simple, oral exchange to extremely complex information networks. Societies depend on communication for transmitting and preserving the cultural heritage as well as for bringing about cultural and social change. The role of communication is all the more crucial in on going, world wide efforts to develop tradition bound, rural-oriented societies faced with problems of a limited (in many cases, primitive) infrastructure (including communications).

It is the thesis of this paper that effective communication is essential for development (or social change). The relationship between communication (used interchangeably with developmental communication) and social change is discussed first, followed by a description of problems faced by rural societies in the transfer of technological information for the purpose of development. The final section of the paper outlines a communication strategy to hasten development. Relevant work in developing countries, particularly the Indian sub-continent, is described to exemplify and, or support the thesis.

Social Change Communication Relationship

Social change is the process by which an alteration occurs in the structure and function of a social system. There are three sequential stages : (1) invention, the process by which new ideas and practices are created or developed, (2) diffusion, the process by which these new ideas and practices are communicated to the members of a given social system, and (3) consequences, the changes that occur within a social system as a result of the adoption or rejection of the innovation (Rogers, 1971). Change occurs when a new idea's use or rejection has an effect. Social change is, therefore, a consequence or effect of communication.

There are many definitions of communication in the literature. In the context of change, an appropriate description of communication is that it is a process by which a stimulus is sent from a sender to a receiver to bring about a desired change in the receiver's thinking, feeling or action. This interactive process, from a developmental standpoint, connotes motivation and confidence building, stimulating creativity and self-expression, and sharing ideas, information and experiences to be translated into actions leading to socio-economic and cultural development. A general model of the communication process involves live elements sender, message, channel, receiver, effects (S-M-C-R-E). A source (S) sends a message (M) via certain channels (C) to the receiver (R) with certain effects (E). These five elements cover the entire framework of the communication process. This is similar to Lass well's (1948) paradigm "... of who says what, in which channel, to whom, with what effect." The context in which communication occurs is also recognized as having an impact on the several elements of the communication process.

Communication elements are vitally involved in all aspects of the innovation decision process of social change. For example, when a farmer participates in a dairy cooperative extension programme, or a husband and wife decide to adopt family planning, a message (M) is conveyed to individuals (R) via communication channels (C) from a source (S), which causes the receivers to change (E) an existing pattern. Communication is, therefore, an important element throughout the social change process, and the societal and individual changes (consequences) that flow from the adoption or rejection of an innovation. A graphic representation of this relationship is presented in Figures I and 2.

Figure 1 shows the stages of the process of social change, namely invention, diffusion and consequences. At each stage, there are relevant social systems involved. Figure 2 is an example of this relationship in affecting change in agricultural technology. Agricultural research represents all social systems involved in the generation of new technology research scientists, technicians, public and private sector research interests. Agricultural Extension is the delivery system for transferring new technology.

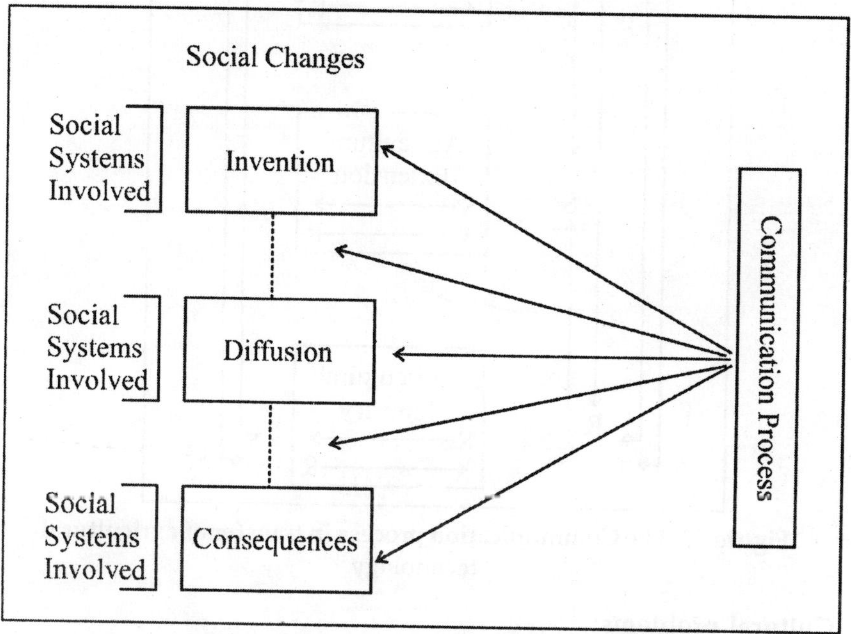

Figure 1. The Relationship between Social Change and Communication.

Involving the several social systems of Extension workers (specialists and generalize) and related agencies, both public and private. The third stage is represented by the several client systems in the agricultural industry farm families, credit agencies, marketing firms, etc. At each of these stages in the process of change in agriculture, communication elements are vitally involved, as messages are transmitted and feedback occurs between senders and receivers. Similarly, the social systems involved at all the three stages interact constantly.

Problems in development

The problems faced by peasant societies as they move towards development can be classified as cultural, personal and communicative.

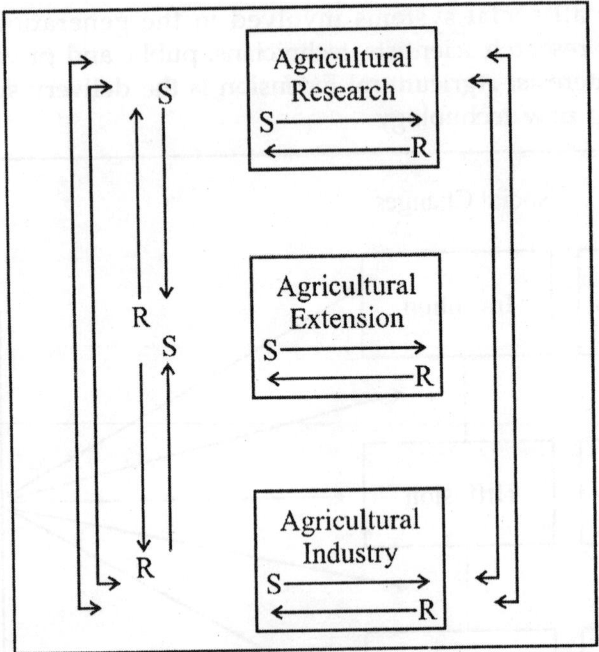

Figure - 2. The Communication process in transfer of agricultural technology

Cultural problems:

Basically, cultural problems stem from underlying value systems. According to a United Nations Report on the World Social Situation (1965), attitudes and values hindering progress in the developing countries were the causes of the tendency to think in absolute rather than relative terms, excessive reliance on personalities, heroes, etc., disrespect or disdain for manual labour, preference for civil service and white collar occupations, non-acquisitiveness, other worldliness, renunciation, fatalism, humility, non-aggressiveness and other non-economic values. These values, bolstered by the educational and religious systems, tended to paralyze initiative and, in general, were inimical to social and economic development.

Naturally, societies differ in their conformity to values. One would expect that less conforming societies would have a positive attitude towards change and, therefore, adopt innovations with lesser reluctance. For example, communities that place value on material gains, encourage and show prestige for innovative behaviour, eschew fatalism while promoting the scientific approach, reward individualism and personal success, are likely to change more rapidly than communities tied to custom and tradition.

Another significant cultural barrier to social change is the one imposed by class and caste. This is most evident in the diffusion stage of the social change process. Jha and Singh (1965), and Singh (1970) reported that higher castes in Indian society made greater use of institutionalised information sources than lower castes, causing further differentiation in access to information and innovative behaviour. Mathur (1967) demonstrated the strength of inter-caste communication of the communication behaviour of key communicators. He observed that 78 per cent of the recipients of information belonged to the same caste as the key communicators. He also found that the untouchables (Chamars) were completely isolated. This finding of poor inter-caste communication flow poses a barrier to the diffusion of information.

Poor credibility of the change agent in the eyes of the farming community limits the effectiveness of the change agent in promoting social change. Rogers (1969) views this situation as arising from peasants' generally negative attitude towards government. Such an attitude manifests itself in distrust for the agents of government. Dube (1958) reported that "... the relations between the common village people and government officials are characterised by considerable distance, reserve, and distrust...." while Foster (1962) alleged that "... the villager has been victimized by persons more knowledgeable" ever since.

Bailey (1966) commenting on the attitudes and beliefs of Indian peasants, maintained that, "... one justifies cheating government agencies by saying that the officials concerned are cheating you. This perception is often so firm that even behaviour which is patently not exploitative, but benevolent, is interpreted as hypocritical cover for some as yet undisclosed interest..."

Paradoxically, the dependence of peasants on local and national governments for the solution of rural problems is increasing as a result of governmental development programmes. Rao (1963) observed that because the perception of self-help was low, dependence upon the government was high. Similar perceptions were reported among Turkish peasants in regarding most village improvements as the job of government rather than their own (Rogers and Svenning 1969). The "help me" philosophy is more firmly embedded in peasants' minds than self-help.

Dependence on government may in some cases lead to over-reliance of peasants upon suggestions from official sources. Deutschmann and Fals Borda (1962) found that the predominant pattern for peasants in a Colombian village was to fully adopt farm innovations without first trying them on a limited scale. Similar results were reported by Rahim (1961) in Pakistan. The risk associated with this kind of adoption behaviour is evident. One possible reason that peasants do not adopt the installment plan may be that their fields are so small, often only a fraction of an acre that it would be impractical to try a new idea partially, because of the in-convenience of planting, cultivating and harvesting on such a diminutive scale.

Personal Characteristics as Problems

Diffusion studies have pointed out that peasant communities are characterized by a mentality of mutual distrust, suspiciousness and evasiveness in interpersonal relations. Carstairs (1958) reported that Indian peasants (Hindu) were characterized by paranoid reactions of mutual distrust, display of goodwill and flattery followed by dubious afterthoughts and an apparent lack of empathy regarding others' feelings. The motives of their fellows always seemed arbitrary, inscrutable and suspicious. Bailey (1966) also noted the significance of the distrust syndrome in Indian peasant behaviour, "Any peasant who adopts new ways and becomes rich must have cheated must have exploited his fellows ...". Such distrust serves as a powerful block to cooperatives and to most of the community development programmes, which are based on the notion that people can cooperate (perhaps with the help of some professional technical advice) to solve their social and economic problems.

The majority of traditional farmers generally do not react positively to innovations. This tendency of following prescribed ancestral ways could be attributed to lack of knowledge about available alternatives. However, even when innovations in agricultural production, health and marketing are presented to subsistence farmers; their record of adoption has seldom been enthusiastic.

What are some of the possible reasons for this lack of innovativeness? The farmer's life pattern inclines him to follow those ways he knows will produce positive, though small-scale results, rather than try a new idea that might end in failure and thereby endanger his existence. Strassman (1964) quotes a favourite peasant watchword,

"Sharks are dangerous only to those who go for swimming." Furthermore, peasant behaviour may not be oriented towards rational and economic considerations (Rogers, 1969). The degree to which peasants are efficiency-minded and economically rational depends largely on their level of modernization. Therefore, subsistence farmers may not be motivated to adopt agricultural innovations if the pecuniary advantages of such acceptance are only pointed out.

Scarce economic resources or inappropriate technology is another deterrent. Peasants are poor, and a lack of ready capital undoubtedly discourages the adoption of those new ideas which require cash outlay. Peasants are also poor in technological resources and know how. For example, since much agricultural research has been conducted in temperate climates, the innovations resulting from this experimentation are generally "rhore appropriate for the United States and Western-Europe rather than for Asia, Africa or Latin America. Many agricultural innovations from temperate climates have been introduced in tropical settings without adequate adaptation to the new conditions. The result has been failure and further negative conditioning of peasants' attitudes towards innovations.

Negative attitudes towards change are the result of generations of unfavourable conditioning towards new ideas. Inherited norms against innovativeness are firmly rooted by various, means of, social control on the part of the villagers. Deviation from these established, norms will gradually be encouraged as village isolation, breaks down and peasants gain new reference groups outside their village. One consequence will be a gradually more favourable reception of technological innovation by villagers but such change may be felt in the coming years.

Empathy is the ability of an individual to project himself into the role of another person. The majority of farmers find it difficult to empathise. A Turkish farmer, when asked, what he would do if he were made the President; murmured rather hesitatingly "My God! How can you ask such a "a thing? How can I cannot President of Turkey Master of the whole world?" Similar findings of low empathy of farmers were investigated in India (Rao, 1963) and in Pakistan (Eister, 1962), Rogers and Svenning (1969). Probable reasons for low empathy were cited as : Limited exposure to other ways of life and extreme psychological distance between peasants and elite urbanites make it difficult for villagers to imagine themselves in other roles. Yet there, is reason to believe that until peasants can empathize with more modern roles such as those depicted, in the mass media, they will have difficulty assuming a modern approach to life.

2. Geographical localities and limited mass-media exposure. A closely related facet of peasant mentality, which may partly account for interpersonal distrust, is the tendency for peasants to view the world as having only an absolute quantity of that which is really good. Foster (1965a, 1965b, 1967) terms this perceptual framework "the image of limited good". He contends that peasants commonly hold the notion that all the necessities of life including land, wealth, health, love, power, and safety exist in finite quantities, are always in short supply, and cannot be increased by any means within the peasant's power.

In the Indian context Mandelbaum (1963) states that Indian villagers hold "good things of the village are forever fixed in amount, and each person must manipulate constantly to garner a large .slice of his own."

Communication Problems Studies on communication fidelity have indicated the problems of language incompatibility between communicators and receivers. In India, it was shown that many Extension communicators in the high-yielding-varieties programmes were not posted in areas where their language was most compatible with that of their clientele, because of the administrative rule of-posting an employee out of his home district. This led to serious communication problems (Stngh, 1970).

Lack of knowledge of technical terms poses another serious problem. Singh and Modi (1974) reported that respondent knowledge about 13 technical terms commonly used in television programmes was put to test in an experiment for study. Programmes were selected through content analysis. These terms included words like "hectare." "centimetre." "metre," "litre," "acre," "kharif," "kilogram," "rabi." "October," "per cent," and so on.

It was reported that ignorance of the terms ranged from 23 to 92 per cent for the 48 farmers who were shown both programmes. An average farm viewer did not know the meaning of about 58 per cent of the technical terms that were used in selected programmes.

Singh and Modi (1970) also investigated message relevance and found that the level of interest in selected innovations was related to the perceived needs. They concluded that if messages were related to needs and interests, farmers could be expected to learn the content of messages.

The meaning placed by the intended receivers on the content of messages is important in the context of change. Lionberger (1965) remarked, "In India, a very nicely done poster intended to convey a 'drink milk' idea, created a great demand instead for the picture, because of the attractive 'mother cow'." The milk drinking message did not register. The meaning of messages must be judged from the vantage of those who are to be educated and not from the standpoint of the change agent.

Singh and Kumar (1976) also pointed out that "Multiple cropping messages has multiple meanings and this itself leads to a communication problem. Farmers do not understand whether multiple cropping is synonymous to double cropping or mixed cropping. We are multiplying verbosity and not crops."

The timing of messages for dissemination in a programme must be opportune. Dube (1963) described how inopportune timing ruined a malaria eradication programme in Bihar. DDT was sprayed in early October. Soon after the houses were white washed for the **Deepavali** Festival, and the efficacy of DDT was substantially lost. Naturally, the experiment did not succeed in eradicating mosquitoes. Apparently, the "message" of spraying was not completely told and or the timing of the message was inappropriate.

A Communication Strategy for Accelerating Development

Given an appropriate communications structure comprising information building and information-dissemination facilities, and working on the defensible premise that development will occur at a pace that matches the effective utilization of such a structure, it appears that a sound communication strategy revolves around two salient factors. These, in the opinion of the authors, are (a) the establishment and operation of an interactive research-extension-teaching system, and (b) the effective organization of people and programmes. Attention to these aspects will generate and transfer of technology in an effective manner while reducing the communication gap between information sources and users. Naturally, there has to be in place the necessary machinery, infrastructure and programmes of information exchange.

Interactive Research-Extension-Teaching System

A great deal has been said and written about the value of a strong research and teaching base to generate new technology to users. Technological sophistication has depended on the level of development and the organizational setup has been governed by the philosophy of administrative leadership in different countries. The integrated land-grant university system in the United States and the segregated government agency system in Asia, Africa and Latin America are the two major types of organizations for research, extension and teaching.

The U.S. system has been very effective, but its adaptability to other countries is limited. Limitations of the governmental setup have been cited as reasons for slower development. Regardless of the type of setup though, an important aspect in making it work is the extent to which communications structures and processes are effective in bringing about the desired end result, namely social change or development. In this task, some of the important concerns are coordination of departmental activities, establishing and maintaining communication links and feedback mechanics between various departments and agencies, and keeping open the lines of communication among various parts and the people. If these concerns are satisfied an interactive system could be installed.

Organization of people and programmes

Making programmes that are intended to benefit people work is dependent upon a host of factors cultural, social, technological, economic, political, etc. related to the objective of the programmes. Philosophically, Extension, for example has an educational mission. But in many countries Extension also has non-educational goals (regulation, inspection, service). Consequently, the communication strategies also tend to be different. In an educational setting, communications are persuasive, yet unbiased. Manipulation,bias and external pressure will be found in communications messages and techniques when the objective is to bring about change or to induce conformity to a prescribed behaviour education.

In summary, a communication strategy for development that will work is one that will reduce the communication gap between the system that is interested in bringing about change and the system that is the target of such change. To reduce this gap, one has to begin with the needs of change targets. Developmental programmes have to be based on these needs otherwise the change agency will lose credibility. Secondly, one has to use opinion leaders as a link to the general populace. Thirdly, the change agency has to use those persons to serve as change agents who will identify with change targets, in fact are or become one of them. Finally, people have to be organized so that they can grow on their own. In so doing, it is better to use an existing organization rather than install a new one.

References

BAILEY. F.C., "The Peasant View of the Bad Life", Journal of the British for the Advancement of Science. 23: 399-409, 1966.

CARSTAIRS, G.M. "The Twice Born : A Community of High Caste Hindu", Bloomington, University of Indiana Press, Indiana, 1958.

DEUTSCHMANN, PAUL J., "Communication and Adoption Patterns in an Andean Village", Programs Interamerican de Inforacion Poplar, Mimeo Report, San Jose, 1962.

DUBE, S.C., "India's Changing Villages; Human Factor in Community Development", Cornell University Press, Ithaca, New York, 1958. DUBE, S.C., "Communication Innovation and Planned Changes in India". Ford Foundation Mass Communication Study Team, Report, Sponsored by the Ford Foundation, Indian Ministry of Information and Broadcasting, New Delhi, 1963.

EISTER, A.W., "Critical Factors in the Modernization Process in West Pakistan", Year Book of the American Philosophical Society. 362-365, 1962. FOSTER, GEORGE M., "Traditional Cultures and the Impact of Technological Change," Harper & Row, New York, 1962.

FOSTER GEORGE M., "Cultural Responses to Expressions of Envy in Tzintzuntzan," South Journal of Anthropology, (22) : 24-35, 1965.

FOSTER, GEORGE M., "Peasant Society and the Image of Limited Good." American Anthropologist, (67) : 293-315, 1965.

JHA, P.N. and B.N, SINGH, "Utilization of Farm Information as Related to Characteristics of Farmer," Indian Journal of Extension Education, I(4): 294-302, 1965. LASSWELL, HAROLD D., "The Structure and Function of Communication in Society", in Lyman Bryson (ed.). The Communication of Ideas, Harper, New York, 1948.

LIONBERGER, H.F., "Some Implications from Diffusion Research for the Adoption of Farm Practices in India". Indian Journal of Extension Education (I); 18-25, 1965.

MANDELBAUM, DAVID G., "Forword." in William H. Wiser, "Behind Mud Walls", University of California Press, Berkley, California, 1963.

MATHUR, P.M., "Prestige Leaders and Family Friends as Communicators," Indian Journal of Extension Education 3(72) : 55-60, 1967."

RAHIM, S.A., "Diffusion and Adoption of Agricultural Practices : A Study of Pattern of Communications.." Diffusion and Adoption of Improved Agricultural Practices in a Village in East Pakistan, Academy for Village Development, Camilla, Pakistan, 1961.

RAO, Y.V. LAKSHMANA, "Communication and Development," A Study of Two Indian Villages, Ph.D. Dissertation: University of Minnesota Press, 1963.

ROGERS. E.VERETT M. and LYNNE SVENNING, "Modernization Among Peasantsr The Impact of Communication," Holt, Rinehart and Winston, New York, 1969.

ROGERS, EVERETT M. and F. FLOYD SHOEMAKER. "Communication of Innovations: A cross-Cultural Approach," Free Press. New York, 1971.

S1NGH, S.N., "Adoption and Diffusion of Agricultural Technology and Social Change" in Singh. K,N. and C.S.S. Rao and B.N. Sahay (eds). Research in Extension Education, Indian Society of Extension Education, New Delhi, 1970

SINGH, Y.P., "Studies in Communication in India" in Singh, K.N., C.S.S. Rao and B.N. Sahay (eds). Research in Extension Education, Indian Society of Extension Education. New Delhi 1970.

SINGH, Y.P, and KUMAR, "Multiple Cropping as a Communication; A Reflection", Ministry of Agriculture Report, New Delhi, 1976.

SINGH, PRAKASH AND BELLA MODI, "The Communication Effects Can: A Field Experiment on Television and Agricultural Ignorance in India," Communication and Development: Critical Perspectives, Sage Publications, London, 1976.

STRASSMA N. PAUL, "The Industrialist," in Johnson. John J. (ed). Continuity and Change in Latin America, Stanford University Press, Stanford, California, 1964.

UNITED NATIONS ECONOMIC AND SOCIAL COUNCIL, "Report on the World Social Situation." 1966.

Chapter **44**

The Mass Media and Economic Development

Samarendra Saraf

Mass media and economic development of a nation are interlinked and maintain a strong linkage. This paper brings into light the varying concept of economic development and perspective role of mass media in accelerated planned change in the society. The urgent task of mass media has been dealt in the form of deep probing original and individual thought. It has been advocated that the role of mass media can be effective and rewarding only when the entire strategy of mass media mobilization to the desired end is imaginatively planned, rationally integrated and progressively adapted, mainly in terms of their overt results and covert achievements. The paper finally creates sufficient scope to understand the strategies for the planning of the Mass Media.

Human, history has never witnessed such a global consensus for determined and concerted efforts towards betterment of mankind. This consensus is rooted in the ever-growing awareness that nearly three-fourths of humanity still continues to struggle amidst denial of amenities which the privileged one-fourth enjoys; and such awareness has resulted in mounting anxieties with growing concern for and confidence in the possible and effective measures towards alleviating human problems. With redoubled optimism, the pace of such efforts has been accelerated the entire world over.

The fast appearing independent nation/states rapidly changed the political map of the world during the post World War II period. Trade and commerce brought about great economic interdependence. Networks of modern means of transport and communication shortened distances and promoted closer contacts between different nations, thereby leading them not only towards a world community ".. where contending nations can grind ponderously for peace and war...." but where they could also " . . .*muster the moral and material forces of mankind. ..*" (Coyle, 1961 : 11). Such enlightened hope was charged with determination and optimism to help the people of the undeveloped world get more to eat and to free them from epidemics, ignorance and poverty; and the prospects, to start hoping for a better future through the instrumentality of national and international agencies became brighter.

Scourges of poverty, ill-health, illiteracy and primitive living standards of the three-fourths of humanity thus rendered it imperative for young democracies to conceive of programmes of planned change. Such programmes have proceeded from the many contributory factors and forces. "Growing interest in humanism, emergence of and subscription to the ideals of welfare state as against the eighteenth century political philosophy of *laissaiz faire,* new political ideologies promising to meet the challenge of changing time, desire for fuller participation in the world community through technological advancement, and the recognition of the dignity of the individual, have all had their respective shares in crystallizing the philosophy of planned change... (which has come to find) more coherent and articulate expression as a *philosophy* of life *and society in.* the present century, when many undeveloped and underdeveloped countries of the world (have chosen to commit) themselves to nation-building by initiating well laid-out programmes of change, under the guidance of qualified personnel. Development-minded ness has (thus), virtually become almost an obsession with every nation... the newly emerging free ones showing greater concern, reposing greater confidence, and relegating greater responsibility to such plans in their efforts at nation building" (Saraf, 1965:2). India is one of the many young but large democracies to have chosen planned change for national development.

Change is an on-going natural process, however slow and imperceptible, nevertheless never to be denied. During the entire course of human history, right from the unwritten vast space of time through the pre-historic era, through the historic periods and further down to the present, societies have witnessed many changes. But our contemporary society, better armed and equipped that it is, has given a new dimension to the process of change in the concept of planned change on a massive, nation-wide scale, diametrically opposed to the process of natural change. Planned change is a deliberate effort having a direction with targets fixed, resources stipulated and time-perspective defined.

Large-scale planned change has emerged as a philosophy and a science only because many factors have contributed to its growth.

Much of the economic thinking and researches have gone into the making of a rich body of literature on economic processes, economic organisation and the process of economic growth in different parts of the world. Scientific advances, particularly those of applied character, and the resulting mastery, relatively greater than ever before, over the physical world; have bred a sense of capability and have instilled confidence to take up the tasks of planned change. Growing familiarity with planning has opened up many more possibility casement for large-scale planning. Researches into the dynamics of human problems and the ever-increasing knowledge of social processes and institutions, through social and psychological studies, have offered insights to the students of planning. Moreover, national development as a process of arousal of new aspirations of producing an unblurred image of a better world, of motivation of people to action and finally of delivering tangible goods, has gathered some morale, good measure of public concern and great amount of reliance on such programmes, investing them with scientific respectability. And lastly, the growing skills in and the improved tools of communication have made it more feasible to effectively plan and carry out the message of such programmes to large numbers of people. For the barriers of physical inaccessibility now no longer pose serious a problem for communication.

Where should national development begin from?

It has been a rather perplexing question, begging a satisfactory and convincing answer, as to where should national development programme begin. The uncautious layman would enthusiastically argue : "Obviously, with the construction of dams and bridges and ...." The over-enthusiastic among the social Science students might well put-forth their counter proposition : "Obviously, at the human resource level, for it is this block which will provide personnel to take up effective roles in carrying forward the mission of planned change." Each of the contestants in the debate has his strong points in support of his hypothesis; each has his array of impressive logic to outwit the other and to convert his rival to his stand point. The issue of priority of *social overheads* over physical and material, and *vice versa,* may thus go on as an endless idle exercise, much the same way as the "chicken-or-the-egg" issue continues to baffle the philosophers.

To a cautious student of social sciences a development programme would mean a combination of *physical capital* and *human capital* in such proportions that neither suffers at the cost of the other.

Because, to him, the concept of *human investment* or *human resource development* is as important as the layman's logic in support of top priority to concrete and tangible attainments, due to the latter's having been over-powered and almost swept by their magnanimity.

It is now increasingly being recognized that apart from its natural and material resources, which constitute the *physical capital,* every nation has its *fixed capital* in the form of "the acquired and useful abilities of all the inhabitants or members of society". Elaborating his idea of *fixed capital* of a nation, Adam Smith says:

"The acquisition of such talents, by the maintenance of the acquirer during his education, study or apprenticeship, always costs a real expense, which is a capital fixed and realized, as it were, in his person. Those talents, as they make a part of his fortune, so do they likewise of that of the society to which he belongs." (Harbison & Myers, 1964:3).

So has Alfred Marshall's concept of *investment in man* gained very wide and popular currency, and has become almost a magical catch-word in the literature on development, in so far as Marshall has likewise emphasized the importance of education as "... .a national investment" and ".... the most valuable of all capital... invested on human beings" (Ibid). "For such investment," argues Marshall, "amply pays back its rich dividends". Marshall's phrase finds an echo in UNESCO Manual on Education Planning which defines education ".... not only (as) a fundamental human right but also a valuable investment and an indispensable factor in economic and social development" (UNESCO, 1965:1). Adam Curle has also voiced a similar opinion in his definition of the underdeveloped countries as those "... most of (whose) people are underdeveloped, having had no opportunity of expanding their potential capacities in the service of society" (Ibid, 13).

Schramm has put the entire philosophy of human meaning of development rather axiomatically:

"If national economic development is to occur, there must be a social transformation, and in order for this to happen, human resources must be mobilized and difficult human problems must be solved" (Schramm, 1954:9).

After defining the concept of social transformation as

"... a set of human transformations.... people to be educated and informed, attitudes and values to be changed, human relationships, customs, social behaviour to be reviewed and rethought" *(Ibid).*

Treating human resource development as "a more realistic and reliable indicator of modernization or development than any other single measure", Harbison and Myers have laid to rest the layman versus social science student controversy, by observing:

"It is one of the necessary conditions for all kinds of growth Social, political, cultural or economic. Admittedly, however, it does not explain everything. For example, a country may have well-developed human resources and its growth may be retarded by failure to develop the organizations and institutions which characterize a progressive society. Investment in man will not promote continuous economic growth unless it is accompanied by the investment of physical capital in productive enterprises" (Harbison and Myers, 1964 : 14).

The entire process of economic development has thus come to be equated with the process of physical capital investment and human resource development, the latter with its economic, political and socio-cultural emphases. Improving the fixed capital of its human resources by increasing the knowledge level of its people, a nation aims at "the accumulation of human capital and its effective investment" in the development of its economy; it seeks to prepare the people for adult participation in its political processes and democratic institutions; it focuses its interest on helping the people to lead fuller and richer lives, free from the shackles of ignorance, superstition and tradition. It is a social investment which will have its boomerang effect, in the long run, on the national economy by raising the GNP of its people. As a programme, planned change thus seeks to reorient new patterns of thought and action and to generate and release, by adding proper incentives, a self-motivated urge for further change. As a process, it aims at arousing newer aspirations, promoting perception of new goals, initiating motivation, spurring mass inertia into action, and enabling it to evaluate and assess its attainments. Such a programme, thus, necessarily implies patience and perseverance, imaginative planning and its effective translation into action.

To many, such a definition of the philosophy of planned change in terms of emphases on arousal of aspirations, goal perception, motivation and action promotion and their sustenance may appear to be both, an impressive paradigm built on some logically close-knit but *a priori* assumptions in an inevitable and invariable time sequence, and an over-simplification of the entire process in operation. True, it appears like that.

Quite a few others, might well argue that the entire process of change does not operate in a vacuum but has a solid human base an anonymity of mass that is not an unworked clay but which is a modeled clay a mass of clay modeled after its traditions, chiseled by its norms and painted in accordance with the native hues of its ideology. And out of this relatively inert mass has to be reshaped, as Nehru has put it, "... a dynamic and vital.... an adventurous and new" society (Schramm, 1964 : 19). Figuratively speaking, the proverbial ball that is relatively static because of its slow, imperceptible rate of change, has got to be set rolling so that it gathers an accelerated pace, partly through the force applied from without but largely drawing upon the sheer force of its momentum its new dynamics.

Moreover, the human base with which the planners have to deal is always embedded in a socio-cultural milieu which may be conducive to change, if the forces of modernity are effective and persuasive enough to evoke proneness or susceptibility to change; or it may inhibit change, should the forces of conservatism outweigh the appeal and urge for change. Degrees of proneness and resistance to change thus depend on the degree of integration to and provision of permissible range for variation through acceptable cultural alternatives within its socio-cultural setting on one hand; on the other, they are determined in terms of perception of goals and meaning of change as interpreted by such factors as education, tradition, values, norms and ideological make-up of the individuals, to whom such a programme is addressed.

Programme of planned change thus seeks to deal with and invade this 'vast hinterland of tradition' only to help the rural communities emerge as a modern society. Planned change is more or less an ideological warfare between the conservative forces of tradition and the progressive forces of modernity a passage from the dark recesses of infra or near-infra-backwardness on to the threshold of modernity an awakening from age-old traditional slumbers; in a nutshell, a choice between tradition and modernity. Tradition and modernity being polar categories, the latter inevitably claims the urgent attention of undeveloped and underdeveloped countries to prepare their people for some bold decisions, some new choices and adaptations.

Students of the process of modernization who have witnessed with social science interest, from rather intimate quarters, the old and traditional societies undergoing the entire process of transformation into transitional and further on into modern societies, have generally agreed that the process makes certain demands towards some basic adaptations, and that these adaptations become meaningful because they relate to the human context. And in doing this, it seeks to translate these new values into action based planned change should breed frustrations when new values might per chance fail to prove rewarding. An empathic orientation of mind helps in internalization of the new values. Modernization demands self directed, achievement oriented personality, open and responsive to innovations. Cultivation of scientific rationality, though complex, is the most desired. Such complexity of scene might well cast a shadow of the gloom, pessimism and defeatism, only forcing us to question as a sceptic : can anything be ever possibly done? How can the mass media of communication prepare the people for such desired values as empathy, mobility and high participation? What are the urgent tasks which the mass media can take up in order to accelerate the pace and tempo of modernisation?

The Mass Media

Their Conventional Roles and Some Urgent Tasks the conventional roles of the mass media of communication, as envisaged by one of the founding fathers of communications study and research, still continue unchanged : surveillance of the environment (both physical and social), interpretation and correlation of events, transmission of cultural heritage, and entertainment. But the emphasis on these roles differ from country to country. For mass communication may be largely occupied with surveillance and interpretation functions in one country, in the other, it may be engrossed with the entertainment function.

Rapidly multiplying authentic researches on both the effects of mass-mediated communication and the sociology of audience as well as growing acquaintance with the dynamics of change have offered us some insights and some "bases of hope", as Klapper would have it. As a result, communication specialists, as applied scientists, have found out three-fold roles for the mass media as an aid to national development. They are: the Watchman role, the Decision-making role and the Teacher role (Schramm, 1964 : 126).

Nearly every independent country that has hitherto been largely traditionally oriented is thus faced with the first and the foremost need to instill nation-awareness in the minds of its people as a preparatory step toward moderanization. The cabinet minister of a country (that expected its political freedom and sovereignty shortly) was right in his placing too high a premium on the mobilization of his country's mass media resources toward arousal of nation-awareness when he confided to Schramm; "You know what we are going to have to use radio for....? We are going to have to convince our people that they are a nation!" (Schramm, 1964:13). And Schramm too has recognized the urgency of such a task when he continues; "The idea of nationness and unity must be implanted; and, so far as possible, communication must help to control the divisive instincts that are always about to break the surface" (Ibid). Dube has voiced a similar opinion in not too dissimilar a vein : "Modernization presupposes organized national effort. A feeling of nationness is therefore essential. Anthems and flags cannot create a true nation; certain levels of social integration and interest articulation are necessary to provide the bases for genuine nationness (1967 : 42).

The mass media can be mobilized through well-laid-out, imaginatively planned and organized programmes, as an aid to arousing nation-awareness. As Schramm observes, while dwelling upon the flow of information through the mass media: "It is generally the increasing flow of information that plants the seed of change. It is also the widened background of information that furnishes the climate for nationness itself. By making one part of a country aware of other parts, their people, arts, customs, and politics; by permitting the national leaders to talk to the people, and the people to the leaders, and to each other; by making possible a nation-wide dialogue on national policy; by keeping the national goals and the national accomplishments always before the public thus modern communication, *wisely used,* can help to weld together isolated communities, disparate sub-cultures, self-centred individuals and groups, and separate developments into a truly national development" (Schramm, 1964:44).

People in traditional societies thus need be informed about the possibility of 'better ways' before they could be induced to change. Information relating to better living conditions, better health and better medical care, more food and better yield through modern agricultural practices in other parts of the globe will on one hand broaden the outlook (Schramm, 1964 : 127); on the other, an urge for such 'better ways' will germinate. Information regarding better possibilities lure of the promise for betterment gives rise to felt need, and a set of expectations will begin to take form if the mass media presented such possibilities as well within human reach (Ibid: 130).

By bringing into sharp focus the desired innovations, the mass media can be of help in their promotion through constant flow of relevant information relating thereto (ibid, 129). For a mere general awareness of possible alternatives is of little consequence unless they are adequately presented *(stimulus value)* in order to awaken interest *(felt need)*. By broadening mental horizons, the mass media can thus awaken interest by stimulating aspirations for change, and can sustain such interest by feeding it with more and more reliable information. Mass media should be cautiously used so that it does not make frontal attack on deep-set or strongly held social customs, norms and value systems. Since it only helps to change indirectly, otherwise, it will breed aversion to the mass media and reinforce efforts to build defense mechanism for traditionally bold value systems and customs.

Decision making is a complex phenomenon. Acceptance or rejection of a simple innovation inoculation or vaccination dramatizes the complex interplay and, sometimes, the decisive roleof the satellite issues that must be considered while arriving at any major decision. To have the prophylactic or not to have it no longer remains a simple matter for a moment's reflection and quick decision. It implies a Hamlet-like struggle "to be or not to be". The joint family ethics, the authority structure, the other-directed personality orientation, the religious dogmas and a good measure of cynical scepticism towards the therapy are some of the satellite issues that are at play as soon as the basic issue is faced with.

Thus, people directly faced with the adaptive demands of planned change are constantly exposed to vexing situations of decision-making. For under planned change, agricultural practices, food habits, hygiene, sanitation and the like, necessitate decisions for change.

Moreover, communication researches have established that exposure to communication through the mass media in general tends to be often selective (Lazarsfeld, Berelson and Gaudet, 1948). Perception of the message, interpretation of the contents and distortions thereof are also selective in nature (Allport and Postman, 1945). And, lastly, exposure to and perception of the mass-mediated communication are further conditioned by selective retention (Levine and Murphy, 1943).

Furthermore, the study of the sociology of audience has revealed the inadequacy of the "hypodermic needle model" of communication effect, forcing us to view mass communication effect from a "phenomenist" angle (Klapper, 1953:5). The resultant two-step or it may be a multi-step (Wright, 1964:65) flow of information well-nigh established the presence of an intermediary in between the mass media and the mass audience. Thus, the presence and functioning of opinion leaders the reference individuals or groups further complicate the situation.

In the midst of such established broad generalizations about the dynamics of mass-mediated communication, which limits its scope more than it liberates, can the mass media ever do anything positive in the furtherance of the decision-making process?

Viewing the Indian scene through communication spectacles, the mass media appear to promise little that can contribute to help accelerate the process of acceptance of innovations. For the norms and values of Indian society have given its people the bases for formation of predispositions. As Klapper has somewhat unequivocally put it: "The audience members do not present themselves to (the mass media) in a state of *psychological nudity,* they are, instead, clothed and protected by existing predispositions, by selective processes, and by other factors" (Klapper, 4). Equipped with the armour of predispositions as the people are, one might well question, "will not the mass media find such armour impenetrable and impregnable?"

Moreover, one's predispositions are ..."... ostensibly individual opinions and attitudes ... (which are) primarily social (in) character" (Katz and Lazarsfeld, 1955). For they represent the group norms which an individual has internalized. These group anchored values and norms are prized more often than not. Emotional identification with the group is a characteristic of social life. Group, thus, envelopes of individual members in "... a kind of psychological membrane..." (Goldschmidt, 1959 : 67). Joint family, larger kin group and the village community are, thus, the different group affiliations, each with its norms and values that cast their 'psychological membrane', as it were, to envelop an average Indian.

Schramm is thus right in maintaining that "Mass communication has never proved very effective in attacking attitudes and values or social customs that are deep-set or strongly held" (Schramm, 1964 : 132). Any frontal attack on these might only result in breeding aversion to the mass media and a reinforcing effort to build up defense mechanism. Defining the prospects for the mass media in the decision process, Schramm makes a cautious proposition : "The mass media can help only indirectly to change strongly held attitudes or valued practices" (Ibid).

However, the mass media can feed the interpersonal channels, thereby regimenting the progressively-oriented influential and opinion leaders, and) repairing them for opinion-influencing tasks through interpersonal communication channels. Media could hold the diverse interest groups in country across to share national efforts and establish dialogue with the policy planner. Thus beginning in its initial stages as a programme of planned change *for the people,* it is expected to shift to a stage of planning *with the people* and further on to that of *planning by the people* (Dube, 1965:52). The onerous task which the mass media can take up at the initial phase is to prepare the populace for the next phases to sow the seeds of empathy, enthusiastic response and determined involvement.

There is need for a shift in emphasis, the mass media have hitherto had, rather than diverting too much of the media resources to national leaders and their prosaic and monotonous routine, local influential, regional opinion leaders, change-prone individuals and open personalities and their entrepreneurship have to be focused and legitimized.

Alongside of the status conferral role, the mass media can perform the enforcement of social norms role. As an agency of social control in the urban anonymity, the mass media can highlight deviant behaviour thereby promoting public ethics. Lazarsfeld and Merton have driven this point home when they hold: "Publicity closes the gap between 'private attitudes' and 'public morality'" (Ibid, 137).

By familiarizing the unfamiliar, the mass media can help in the formation of new tastes and their stabilization. Consider how folk songs and dances that once appeared to be rustic and unsophisticated to the urban high-brow, have now become almost a craze. We have little realized how the invasion of cosmetics and dress fashions largely owes to the mass media 'ad' techniques and displays which are slowly, perhaps imperceptibly, making their headway in changing the rural tastes to the more "refined" ones. Urban impact through the mass media, is, thus, beginning to be felt (Rao, 1963:58 ff.).

Viewing education as a process of formation of new type of persons for developmental tasks, one notes that the mass media find a great virgin land of immense promise and prospects for them to play their teacher role. Schramm has, therefore, rightly made a bold proposition and loud claim: "Mass media can help substantially in all types of education and training" (Ibid, 140). Thus, from agriculture to literacy, or sanitation to modern therapy, the mass media cannot only widely disseminate new ideas and information to broaden mental horizons of the people they can effectively play the teacher role as well.

India is a largely agricultural country. National development programme seeks to promote the primary sector of agriculture to modernity whereby mechanized farming may replace age old plough and the bullocks. Chemical fertilizers may be used instead of the traditional manure. Improved seeds, disease-resistant and more productive, may replace the degenerated traditional seed variety. Eugenic principles may be implanted to improve the cattle breed. Insecticides and pesticides may be used to combat crop failure resulting from parasitic pests and insects. And modern devices of irrigation may take over from the traditional ones.

It is an ambitious, well-intentioned plan to introduce all these innovations. But all these involve new skills that have to be mastered and new information that has to be gathered. Through film strips and documentaries and the Rural Radio Forum programmes (Ibid, 6-8), the mass media can present such information where the audio-visual devices can combine information with demonstration to help the people learn new skills. For, each of these innovations involves unlearning of old practices and relearning of new ones in their stead. Demonstration through the movie will function as an aid to the extension services personnel in their tasks at mass-mediating the innovations. It will also ease their persuasion tasks through subsequent inter-personal channels. Looking to the wide-spread illiteracy, the audiovisual form of the mass media will be more welcome and rewarding, for, the mass media in printed form will be of little or no avail, where illiteracy looms large on the rural horizons.

Some Strategies for the Planning of the Mass Media

As has been outlined earlier, the mass media can be employed as a dependable device in the furtherance of modernization and nation-building campaigns. But their three-fold role-playing could be effective and rewarding only when the entire strategy of the mass media mobilization to this desired end has been imaginatively planned, rationally integrated and progressively adapted mainly in terms of their overt results and covert achievements. When we advocate the urgency and importance of mobilization of the mass media of communication in sharing modernization tasks, we mean to refer to the effective communication role of the media, for, the mass-mediated communications can have their tremendous persuasive impact on and appeal to the audience.

Explorations in effective inter-personal communication and its dynamics have long been made ever since the Greek philosopher denned "rhetoric" as the "winning of men's minds by words" or, as his disciple had it, as "the faculty of discerning the possible means of persuation in each particular case" (Maccoby : 1). Reformers, crusaders and war heroes, among others, have always employed such rhetoric skills the magic of words only to cast wide-spread hyponotic trance on their followers. Christ and the Buddha, Luther and Gandhi, Napoleon and Hitler are some of the best known experts of rhetorics who had worked wonders with the magical spells of their words. Communication, thus, has its own force; words have their own romance, next to that of none.

Old modes of communication have long been supplemented with the mass media. Park has characterized this transition very aptly : "... the aeroplane has well-nigh abolished the distance that once separated the nations and people, and the radio has converted the world into *one vast whispering gallery,* the great world inter-tribal, inter-racial, and international the world of business and politics has grown at the expanse of the little world, the world of intimate, personal loyalties in which men were bound together by tradition, custom, and natural piety" (Park 1962:945). And so has the Platonic image of rhetoric undergone metamorphosis only to evolve into "a new scientific rhetoric", as Maccoby would have it. For as many as twelve disciplines are pooling their academic resources together only ".... to cut across that shadowy boundary between natural and social science...." (Chase 1963 : 280), and to define the new rhetoric and its dynamics as a scientific tool, effective and efficacious through the mass media.

In planning out the strategy of the mass media as an effective scientific tool for rhetoric purposes, as an aid to nation-building, Harold Lasswell's classic maxim *who says, what, to whom, how* and *with what effect*? Has got to be borne in mind. For many factors and forces cut across each other during the passage of communication from its source to its destination only to add further complexity to the entire process.

Effective communication or, to borrow a phrase from telecommunication engineering, 'fidelity of communication' largely depends on the positive attributes of all the factors that go into the process of communication the source, the content, the mode or style, and the receiver; and a high-fidelity effect can, of course, be ensured only if the nuisance of 'noise', another telecommunication engineering concept, is overcome. For "noise is universal and insidious", observe Shannon and Weaver (Chase, 1955 : 16). And "noise and fidelity are two sides of the same coin. Eliminating noise increases fidelity; the production of noise reduces fidelity" (Berlo, 1966 : 40-1). Negation of one, thus, ensures the presence of the other.

In a programme of planned change, every new message, every novel idea, every new skill or mode of operation, when mass-mediated, has to cross a series of hurdles before it reaches its destination the mass audience; and has to subject itself to filtration through many a sieve before it attains the status of an effective communication. Fidelity-promoting factors have, thus, to operate in the midst of and wrest against the noise-producing factors. The fidelity-noise concept, with its extension and transplantation in sociological context, may, by analogy, mean caste affiliations, family and kin orientations and local community pressures and obligations as the major factors, among others, that may demote fidelity and increase the nuisance of noise element in the mass media communications.

Effective leadership, capable of opinion formation and opinion influencing, largely rests, at least in rural India, on the age-experience-status-wealth complex (Saraf, 1965:13). From amongst such potential influential, progressively oriented open personalities have to be identified, and status investitureon them as change-prone-multipliers has to be done in order to register their cooperation. To give a fillip to their opinion-formation and opinion-influencing roles with a redoubled tempo, these key individuals need be first baptised to the new faith of modernization and nation-building; and the information flow through the mass media needs be largely directed to them. And the valuable go-between of a VLW, who serves as a link between the rural folk and the opinion leaders on the one hand, and the mass media on the other, needs must thenceforth activate and, if possible, feed the inter-personal communication channels that often lead to decision-making. Moreover, demonstration and discussion roles have to further supplement the mass-mediated communication, thereby sustaining and reinforcing such communications. For it is well-nigh established that certain things can be better and more effectively communicated through the audio-means, others through the visual means; but a combination of audio-visual devices would leave a more lasting impress, for it makes the entire communication more comprehensible.

Decision-making in traditional societies, where "men are bound by intimate personal loyalties, tradition, custom, and natural piety", is largely a group experience a community involvement. Collectivity orientation of the other directed personality prevailing, decision making on major issues can hardly be an individual experience and action. Demonstrations and discussions will, thus, promote group experience community involvement; and, given an effective and persuasive lead and direction, they may well usher in a favourable climate for positive decision making.

Rural norms and values often present a delicate and hyper-sensitive area, which the mass media cannot afford to leave virgin but which, at the same time, has to be handled with great caution and psychological gloves, as it were. For any frontal attack on the "strongly held beliefs" will doomed to be a futile action; it might well invite controversy, antipathy and aversion. But side attacks on such areas through the instrumentality of the change-prone multipliers may promise to yield positive results.

The image of the mass media goes a long way in determining the persuasibility or otherwise, of their communications. For instance, if the movie to an average rural folk means, by and large, a source of wonder and thrill and a medium which deals more with fiction than reality, such a poor and negative image of the mass medium of the movie prevailing, the movie message will largely evoke more of scepticism and lack of reliability or dependability for the message of the source. A positive image of the mass media has to be created and implanted. Such efforts will depend, as Dube observes, "on a high degree of skill in impression (making and) management" (Dube, 1965:46).

The mode of presentation and its message content the how, what and why claim equally urgent attention in planning communication strategy for mass media. The content of communication needs to be so planned as to make it understandable in terms of the experience of the people to whom it is addressed. Chosen symbols and themes should be native to the audience, in the absence of local idiom, such communications will present more of a confused mass of ideas than clarity and more of puzzlement than information. Lest the mass-mediated communication should appear to the audience as a message in a different tongue from a representative from Mars or Moon, the employment of local themes, symbols and idioms is rendered imperative. For only such communications will mark intimacy and emotional involvement of the audiences with the communication channel.

Slogans and posters which function as an aid to great social campaigns, have to be meaningful and understandable. For instance, the posters that depict a small family as happy and prosperous one are to be found almost in every clinic. Such posters read, in English Family Planning Advice. Here, from which it appears as if they were meant for a class rather than the mass. Gay and colourful, the poster remains just an enameled plate depicting a husband, a wife and two children—all neatly dressed, healthy and cheerful. The picture part of the poster is more meaningful and understandable than its caption part.

"Grow More Food" had once become a magical catchword for a slogan. Exhortations were made from the public platforms; gravity of food scarcity situation was solemnly dramatized; drought, famine, population explosion and the like fears were aroused. Loud and laudable lip-service was rendered to the cause, but the total outcome of such mass-mediated communications was more high-sounding words rather than more food, for, only awareness and an urgency to grow more food were not enough; such slogans did not tell the people the ways and means whereby more food could be grown.

Arousal of needs being an important step in planned change, the problem of promotion of need awareness arrests our attention and raises some poignant issues. Need arousal calls for need identification; and the latter necessitates determination of priorities, for, not only problems vary from region to region, their relative urgency also varies. For instance, an area often threatened by drought and famine may have one set of needs; another area largely swept by epidemics may present another set of needs.

In other words, the mass media have to first locate the needs on a region-wise priority scale before need arousal to alternatives innovations as better choices could be taken up by them. The need identification and priority fixation can be done meaningfully when the mass media develop regular feedback channels. This can be effectively done in three ways: *(a)* the mass audience directly feeding back to the mass media; *(b)* the field workers feeding back to the mass media; and *(c)* the opinion leaders feeding back to the mass media. Such feedback channels will also offer occasions for the mass media to retrospect on their role they have hitherto been performing, and to prospect on their role they have henceforth to perform.

References

ALLPORT, GORDON and POSTMAN, LEO J. (1945). "The Basic Psychology of Rumour." (Summary in Joseph Klapper: *The Effects of Mass Communication).*

BERLO, DAVID K. (1966). *The Process of Communication.* New York, Holt, Rinehart and Winston.

CHASE, STUART (1955). *Power of Words.* London, Phoenix House Ltd.

CHASE, STUART (1959). *The Tyranny of Words.* London, Methuen and Co. Lid,

CHASE, STUART (1963). *The Proper Study of Mankind.* New York, Harper and Row, Publishers.

COYLE, DAVID CUSHMAN (1961). *The United Nations and How it works.* New York, Columbia University Press.

DUBE, S. C. (1965). *"Strategies of Community Development."* Hawaii, East-West Center.

DUBE, S.C. (1967). "Modernization and its Adaptive Demands on Indian Community", in *Papers in the Sociology of Education in India,* Ed., M.S. Gore, I.P. Desai and Suma Chitnis. Delhi, National Council of Educational Research and Training.

GOLDSCHMIDT, WALTER (1959). *Understanding Human Society,* London, Rout ledge and Kegan Paul.

HARBISON, FREDERICK and MYERS, CHARLES A, (1964). *Education, Manpower and Economic Growth,* New York, McGraw-Hill) Book Company.

KATZ, ELIHU and LAZARSFELD, PAUL F. (1955). "Personal Influence: The Part Played by People in the Flow of Mass Communication." (Summary in Joseph Klapper: *The Effects of Mass Communication).*

KLAPPER, JOSEPH "Mass Media and Personal Influence" in *Voice of America Forum Lecture.* Washington' U.S. Information Agency.

KLAPPER, JOSEPH (1963). *The Effects of Mass Communication.* Glencoe, The Free Press.

LAZARSFELD, PAUL F.; BERELSON, BERNARD: and GAUDET HAZEL (1948., "The People's Choice" (Summaries in Joseph Klapper; *The Effects of Mass Communication* and Charles R. Wright: *Mass Communication).*

LEVINE, JEROME M. and MURPHY, GARDNER (1943). "The Learning and Forgetting of Controversial Material" (Summary in Joseph Klapper: *The Effects of Mass Communication),*

MACCOBY, NATHAN. "The New "Scientific" Rhetoric", in *Voice of America Forum Lecture.* Washington, U.S. Information Agency.

PARK, ROBERT E. (1962), "Cultural Conflict and the Marginal Man," in *The Theories of Society,* Ed. Talcott Parsons, Edward Shils, Kasper D. Naegele and Jesse R. Pitts. Glencoe, The Free Press.

RAO, Y.V.L. (1963). "The Role of Information in Economic and Social Change: Report of a Field Study in Two Indian Villages" (Summary in Wilber Schramm: *Mass Media and National Development).*

SARAF, SAMARENDRA (1965). "Socio-Cultural Barriers in Adult Education Programme in Rural India" (Mimeographed). Hawaii, East-West Center.

Ibid (1967). "Some Observations on Our Education System" (Unpublished).

SCHRAMM, WILBUR (1964). *Mass Media and National Development.* Standford, Stanford University Press,

Ibid (1964). "The Uses of Communication to National Development" (Mimeographed). Hawaii, East-West Centre.

UNESCO Publication (1965). *UNESCO & Educational Planning.* Paris, UNESCO.

WRIGHT, CHARLES R. (1964). *Mass Communication : A Sociological Perspective.* New York, Random House.

Chapter **45**

Semiotics and Process: Communication Theory and New Technologies of Instruction in Western Industrialized Countries

Geoff Potter

Advances in communication technology have revolutionarised the globe. Classroom instruction manuals are being modified gradually as more and more new technologies of communication are being incorporated into them. Here author talks about new technologies of Communication and their impact of instruction methodology in Western Countries.

It is very interesting in this age of technological revolution to see how rapidly the embryonic field of communication theory is becoming an integral component of the systems that govern our daily lives. It is as though art, which the Greeks called "teckne" has found its ultimate expression in the organization of our increasingly electronic routines. As television and the computer carve deep inroads into information systems previously monopolized by print so the rules and conventions that govern communication and interaction alter.

Nowhere is this more critically felt than in the field of education. Not the traditional, print-based school system, but the new field of instruction that is called "Distance Education" and uses for its tools, television, telephones, satellites and computers.

The late Canadian philosopher Dr. Marshall Mcluhan wrote extensive warnings about the effect these new technologies would have upon the human condition. And of all the problems which he predicted might beset those of us projected into the technological age, none would affect us more than "culture lag" - our capacity for treating something new as though it were something old and familiar.

Education systems are, of course, well known as society's finest examples of culture lag. And in Western nations certainly, new technologies have presented educators with as many problems as solutions. The directions that Education may take over the next decade are fascinating to contemplate. Ho doubt for the young, the classroom will prevail. But for that enormous body of adults for whom lifelong access to information is becoming a necessity the traditional correspondence course or evening class may well be supplemented and eventually replaced by the electronic instructional systems which utilize home and micro computers, and of most significance to citizens of developing countries, interactive television, particularly that which is based upon transmission via satellite.

The challenge of these new technologies is partly one of mechanics and engineering but to the educator it is one of communication. Never before has the monopoly of the linear, curriculum-based instructional process been so challenged. And yet, it is now becoming clear that new information media can not be treated as though they are extensions of a classroom. The instructional methods employed by the teacher who stands at the front of his congregation and lectures with or without supporting resources, do not work with the new media. The difference is enormous between the experience of students sitting in a classroom with a teacher, listening, talking and absorbing as much non-verbally as they are verbally, and that of "external" students whose disembodied instructors come to them on a screen, and for whom "learning" has an entirely different connotation.

Since the arrival of the new communication technologies educators have quite naturally attempted to use them as extensions of traditional practice. Classroom sessions have been televised to external students, videotaped lectures have been packaged and mailed to isolated learners; cassette taped instructions attached to sets of photographs or xeroxed diagrams have been dispatched to individual students all over the world. At their best they are a reasonable alternative to the potentially intense interaction of the seminar; at their worst they sink to depths unimagined by the most incompetent instructor. I recall, for example, a friend in Malaysia receiving the material for an art course, consisting of one Polaroid photograph of Wu-Dynasty etching on the back of which was scribbled the instruction "Write an essay about this!"

To assume that new communication technologies extend old instructional processes is to beg the question. The new media are not of themselves instructional, they are neutral information in distribution systems the inevitable manifestation of collective mind and as natural a part of evaluation as is any flower or bird. They may extend certain aspects of the traditional work of the teacher, but if they do it will be the information his work contained, not him or his style. The more educators explore new media the clearer it becomes that a medium influences and changes the information it carries often dramatically and with far-reaching results.

Efficient utilization of new media by an educator is possible if he is able to step out of his traditional role of teacher aid into the new role of facilitator of the delivery of information maintaining the fundamental neutrality of information and distinguishing between three separate processes : (i) those occasions when he will preach; (ii) those when he will assist the learner to be taught by his own interaction with the information; and (iii) those during which it is his expectation that through interaction about the subject, he will learn as much as his student. The new media extend and amplify aspects of our humanity, but not all of our humanity. What they permit at one level, they prevent at another. Thus in using them the educator is obliged to be more controlled, more specific about his purpose and activities than may be desirable in a classroom. In short, he has to decide upon the type of instruction required before he embarks upon it. This requires a considerable knowledge of five things : (i) communication processes, (ii) types of learning, (iii) models of teaching, (iv) principles of adult learners behaviour, and (v) the characteristics of the system.

The extent to which these factors need to be understood was illustrated in a 1983 study conducted at the University of Victoria in Canada.

The University of Victoria had become as much a part of the adult education as any institution could :

1. It had offered and evaluated about fifteen courses via satellite based interactive video (one way video, two-way audio).
2. It had developed an adequate production and delivery facility.
3. It had approximately forty instructors on campus who had taught via the system.
4. Analysis of the course design processes and videotapes revealed that several different instructional designs and teaching styles had been used.

At the time each course was being produced individual decisions regarding instructional methodology had been made in the light of contemporary conditions. Retrospectively however, those decisions can be seen as explorations in a relatively unknown field. Subsequent reflections by instructors comparing distance education methods with more scientifically-based classroom methods revealed misgivings about the effectiveness of the satellite-based instructional process. Misgivings which are natural enough given the instructors familiarity with the classroom and unfamiliarity with these new media. There were, for example, differences between what an instructor considered to be his "message" and the apparent "meaning" derived from it by some of his students. Five aspects of interactive instruction were identified by instructors as "problem areas" i.e., aspects of the course in which conflicts arose :

(i) Inflexibility of the script :

The necessity of working from a rigidly defined script and the need to observe strict limitations on the allocation of time was a discipline that instructors found unnerving, and they regretted the loss of flexibility and spontaneity that this entailed.

(ii) Conflict between instructor and technical crew

On more than one occasion instructors commented on the conflict of interests between technicians and educators, the former being concerned primarily with professional production, the latter being concerned primarily with informing students. It was felt that the medium of television dictated that instructors should be fluent, professional and "slick". This slickness is not necessary or even desirable in classroom teaching. Instructors felt compelled to talk constantly while "on-air"; there was no place for silence. All this prevented instructors from adopting their natural style in teaching and produced feelings of discomfort and even resentment. They felt that they had become part of a "show", and at times powerless.

(iii) The Interactive component

Concern was expressed surrounding the value of interaction with students. The dissatisfaction focused on the incomplete nature of the communication process, and the one-sided nature of the flow of communication involved in a system where provision for feedback is severely restricted. Generally instructors entered the system of expecting fairly large amounts of meaningful dialogue, which was not forthcoming, partly because of the students' reluctance to speak and partly because of technical constraints which caused the student to have to wait for as much as ten minutes before he could ask a question.

(iv) The lack of non-verbal feedback

This was highly disconcerting to instructors. The lack of two way visual communication was seen as an enormous barrier and placed a strain on instructors who are traditionally quite dependent on non-verbal feedback from class and seminar members. Talking into a camera lens was considered unnatural. Many instructors felt constrained by having to sit still in a chair for the duration of a programme.

(v) Scheduling

Where students were receiving instruction in large audiences, instructors noted the impossibility of relating to individuals. The general attitude of most students towards satellite-based instruction is that given their isolation, it is acceptable.

However, many of the points made above reflect dissatisfaction with the process. The instructors knew their subject-matter and the students wanted to learn; but the system designed to bring them together has many components, each of which is capable of modifying both the intent of the instructor and the hopes of the student. Theoretical models of human communication processes have always noted the capacity of transmitting and receiving media to modify information transfer, and have warned people involved in interaction that other various and sometimes unexpected "noise" factors will frequently interfere also. Similarly, current learning theories and categories define learning and performance in terms which acknowledge the complex interactions of information, media and individual personality. A frequently made observation by some instructors surveyed at The University of Victoria reflected differences between the message they had assumed was being delivered and the meaning derived by their students.

Approaches to Communication and Learning Theory

The challenge of distance education to the instructor frequently begins when he has to indicate to those with whom he will be working, his own approach to communication and learning. Since as an academic he may always have considered himself to be a student, he may never have considered defining his approach, in which case it may, somewhat expensively, only be discovered by his co-workers and students as time passes. Thus should his own learning begin at this point, and he should familiarize himself with the basic tenets of communication and learning.

There are considered to be two approaches to each. In Communication they are sometimes referred to as The Process Approach and Semiotics (the science of signs and meaning). The philosopher J. Krishnamurti has written extensively about the unique relationship between the observer and the observed. In Learning Theory they are usually described as two Schools of Thought: Behaviourist and Cognitive. The behaviourist is process-oriented. The adherent to the Cognitive School is not. Few of us however, are purists. We will bring process and semiotics together, structuring our message, as Gerbner, Becker and Barnlund describe in order to facilitate our receiver's derivation of meaning.

The purpose of having the instructor study these approaches is so that he knows with which one or combination he feels most comfortable, thereby increasing his sensitivity towards the nature of his students' receipt of his message. In concrete terms it means this : if one is to adopt a process approach to the communication of information, one is concentrating one's energies on perfecting the message, i.e., instructing, making one's point. Under such circumstances one assumes that the student is receiving one's message. If that is not the case, one assumes that a communication breakdown has occurred and takes responsibility for correcting it. The behaviourist reinforces the process approach and measures the effectiveness of his transmission by the closeness of the students' response to the essence of his message. "Interaction" in such a situation may well comprise carefully-worded questions and monitored responses whose complexity will gradually increase. Evaluation of the students will be based upon the measurement of their retention of the message delivered. In a broadcast session supporting materials, graphics and other resources will have been utilized in terms of their ability to clarify the message.

An instructor taking a semiotic approach to communication may well be primarily interested in some quite different matters. Certainly he will be concerned to deliver each of his messages well, and may make use of a mathematical or functional communication model to do so. But he will regard the whole instructional process as interpersonal rather than intrapersonal, and will expect discussion on and around the content of his message. He may well expect to learn from the discussion himself, and will be prepared to modify his initial message as he learns of the meaning each of his students derives from it. He will also be aware that each of his students may derive different meanings from the message. Interaction may therefore be an integral part of his programme. In fact, he may well embark upon each session armed with a number of "key" points around which a discussion will take place. One can imagine that such a person may use supporting materials differently from his process-oriented colleague. Whereas the former may systematically integrate items into a presentation, the latter may wish to keep them on hand for use if and when the need arises. He may trust their veracity somewhat less than his colleagues, and be content if he and his students decide that they are not required. His outward casualness is dependent upon his mastery of his subject matter and its qualities. Since he is interested primarily in signs and meaning, he may well prefer to perfect the linguistic components of his message transmission and make only occasional use of graphics.

There are of course variations on the two themes. An instructor who divides his course into modules and units may decide that some units require one approach and others something different. Such a decision may even be made for segments of a single unit. In team-taught courses, individual instructors will each have their own unique approach to instruction.

One can well imagine the potential for misunderstanding between instructor and production staff or between instructor and student if there is disagreement about basic approaches to communication, or a lack of understanding of instructional intent. An example involves the instructor who is used to only small classroom and seminar groups, and the producer whose previous television experience has focused on broadcasting. The one may have grown comfortable with a semiotic, cognitive approach to instruction, while the letter's primary experience of communication is the broadcasting of messages. Unless each understands the other, at least one usually the instructor will become frustrated.

Instructors at universities and colleges are frequently subject specialists rather than trained teachers. Their knowledge of communication models, types of learning, and teaching models is limited. However, as many professional educators know, although such models are far from perfect, understanding of them provides a teacher with two critical things : insight into the structure and form of his preferred instructional method; and a selection of alternatives.

Comparative Communication Models :

Understanding how one approaches communication and learning is a strong first step and must be followed by consideration of the various models of communication. There are considered to be nine main models :

(i) The Mathematical Model (Shannon and Weaver)

A process-oriented model that sees communication as the transmission of messages.

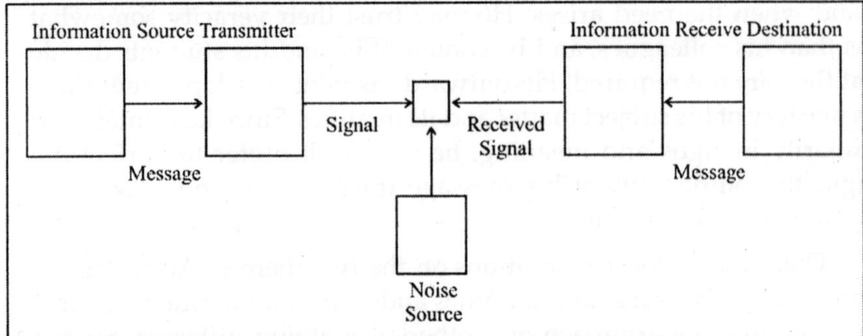

(ii) The Functional Model (Reusch and Bateson)

A conception of communication as functioning simultaneously at four levels of analysis: interpersonal, intrapersonal, group and cultural.

(iii) The Triangular Model (Newcomb)

An A.B.X system in which communication functions to maintain equilibrium within the societal system

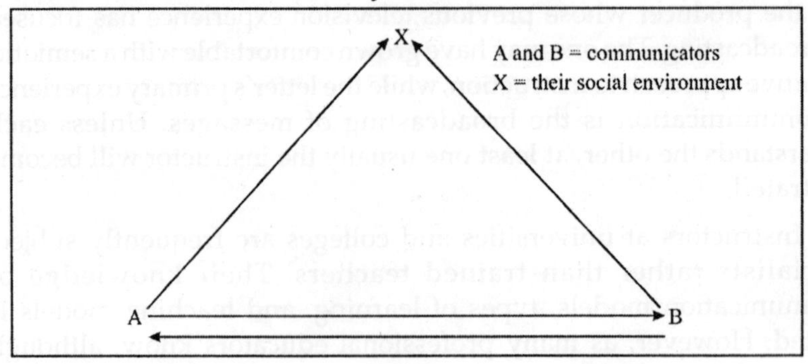

(iv) Schramm's Model

A mathematical model in which encoding and decoding occur simultaneously.

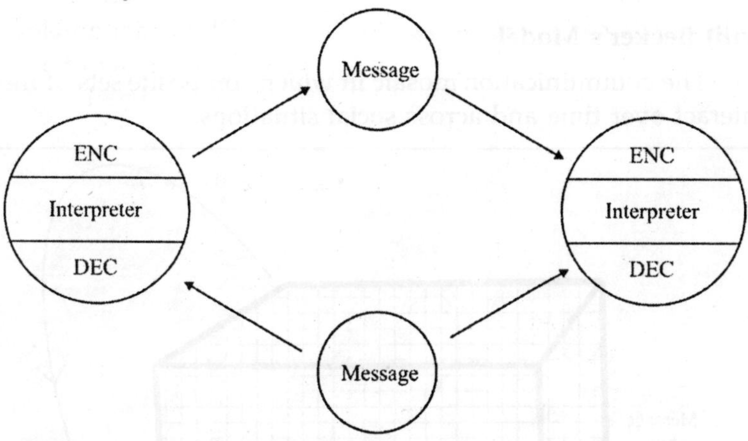

(v) Westley and McLean's Model

Newcomb's model with a 'gatekeeper' who serves an editorial function, allowing for feedback.

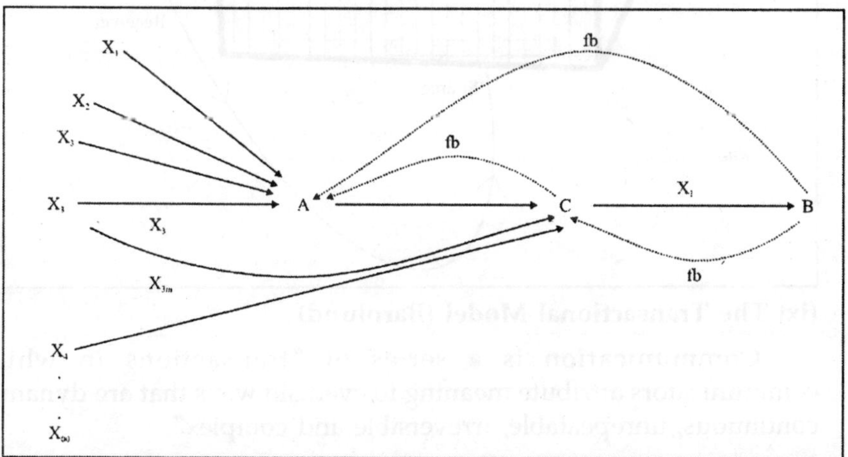

(vi) The Linear Model of Interpersonal Communication (Berlo)

In which source and receiver are mutually dependent. Encoding and decoding occurs according to the skills, knowledge and environment of both.

(vii) Dance's Helix

Implies that acts of communication are continuous unrepeatable and accumulative. Information has no beginning and no end.

(viii) Becker's Model

The communication mosaic in which composite sets of messages interact over time and across social situations.

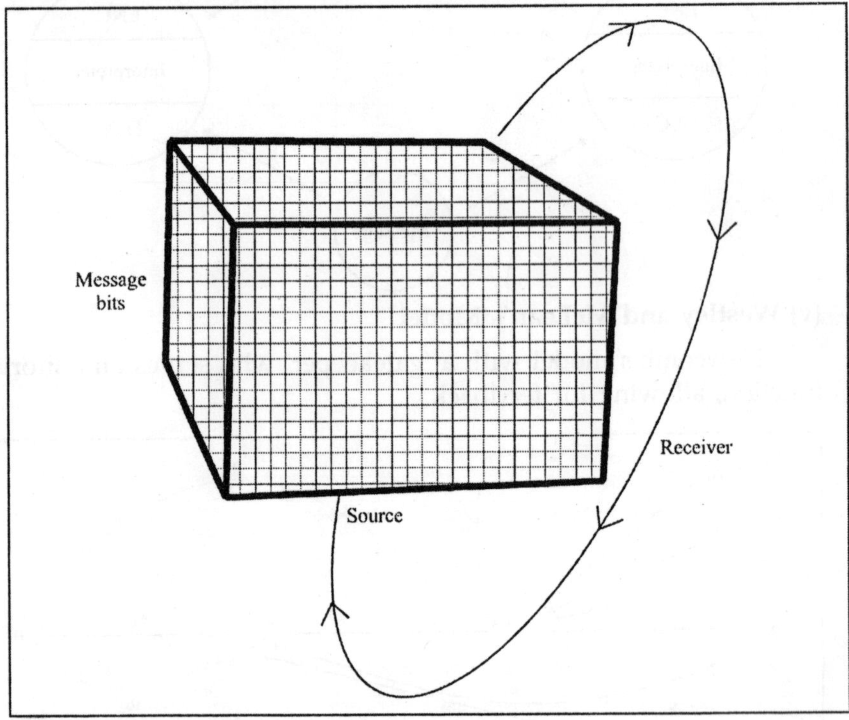

(ix) The Transactional Model (Barnlund)

Communication is a series of "transactions in which communicators attribute meaning to eventsin ways that are dynamic, continuous, unrepeatable, irreversible and complex".

Consider for example the difference in instructional method implicit in a comparison of Shannon and Weaver's mathematical model - one which is considered the epitome of the message-oriented flow-chart and Becker's "communication mosaic". An instructor adhering to the former would assume that it was his business to transmit messages. In seeking efficient channels via which to do so, he would concentrate on the capacity of a channel to carry the message, and on the clarity of the message. He would regard failure by the student to receive the message as a communication breakdown.

The mathematical concept of information has nothing directly to do with content. It is primarily concerned with the predictability of the signal, that is, the number of choice open to the instructor and the reduction of uncertainty in the student. Thus the emphasis is on efficiency the business of utilizing the channel to its maximum capacity with a minimum of input. Since natural language has what is called "in-built redundancy" or "high predictability" (i.e. it is not normally necessary for one to hear 100% of a message in order to understand the meaning), it is often the first choice as a delivery system in education (i.e. teachers talk a lot). And since the "talking head" or "tele-lecture" may be developed fairly easily into a relatively noise-free instructional method, an instructional team operating according to the mathematical model, may well select the satellite-based instructional mode because of its auditory efficiency. They may prefer the straightforward lecture over more elaborate graphic processes.

Becker (1968) has a different view of the communication process, and would regard it as very unlikely that a student receives only the message emanating from the instructor. Becker likens complex communicative events to the activity of a receiver who moves through a constantly changing cube or mosaic of information. Each section of that cube represents a potential source of information. Communication is semiotics an interaction between two mosaics, one comprising the student's private store of information and the other comprising a mixture of the instructor's message and the social information surrounding its appearance in his life.

An instructor designing a course, or a module or even a unit according to Becker's model may be just as anxious to deliver his message as anyone else may be, but may expect there to be a difference between the significance of that message to himself, and its significance to each student. He would understand that each student may priorize the content of the message uniquely. Consequently he may regard the "talking head" lecture as inadequate, and seek instead more variety in his own presentation and more interaction with individual students. Instructional techniques for teleconferencing designed by Monson (1980) and Potter (1982) recognized the significance of the cognitive approach and recommended the incorporation of reference points within discussions, tutorials and lectures that encouraged students to place received messages in personal, local contexts.

Thus the approach one takes to distance education will influence one's attitude towards the delivery of one's message and towards one's relationship with one's students in what are, in the final analysis very practical ways.

Comparative Learning Categories and Teaching Models

Knowledge of communication models and their differences arms an instructor particularly in his dealings with television production personnel who are usually very well acquainted with several models. However, what is ultimately being delivered is information. Someone, hopefully, is going to learn something. But how does learning occur, (or how do psychologists think it occurs)? Knowledge of the categories of learning and teaching models will give selected communication models and approaches to learning, some direction.

Gage and Berliner (1974) classify the different types of learning into five major categories :

i) respondent learning, elicited by a known stimulus;

ii) contiguity learning, accomplished through the pairing of events;

iii) Operant learning : a spontaneous reaction to the environment:

iv) Observational learning, or imitation; and

v) Cognitive learning : the learning of logical, rational relationships.

Respondent conditioning occurs as students learn to associate an "unconditioned stimulus", which might be a particular learning task, with an emotional response such as pleasure or dissatisfaction. The implication is that teaching tasks should be accompanied with pleasure arousing stimuli such as praise, or reward.

Contiguity learning takes place as the classroom discourse repeatedly presents two stimuli together so that the appearance of one calls forth the appearance of the other. This is the kind of process involved in rote learning or memorization.

Operant learning occurs through the sequence of discriminative stimuli, student response and reinforcement. One person's remark serves as the discriminative stimulus for another's responses which are then reinforced by the behaviour, verbal or non-verbal, of the first person.

Observational learning happens as the students see demonstrated some behaviour which will be imitated, such as how to conduct an experiment or write in the sonnet form. Thus the teacher becomes concerned with modeling appropriate behaviours and expertise.

Cognitive learning has been described indirectly earlier. It occurs through the students perceiving ideas presented to them which are then coded and stored in memory. The cognitive theorist demands that the teacher pay attention to the organization of material and the way it is logically connected to what the student already knows.

A subject specialist who has pursued his interests to a high degree is probably aware of how he learned. As a student he would have run the gamut of instructional experiences from the dynamic lectures and highly organized exemplars of personal discovery down to the middle head ed and disinterested few whose names he can scarcely remember but whose memory he resents. When such a specialist becomes an instructor himself, unless he has studied educational categories, he may well adopt the style from which he learned successfully as a student. This will probably work well enough in seminar, laboratory or tutorial situations.

Should this person find himself instructing a distance education course via an interactive system which imposes its own particular set of constraints upon his usual teaching style, he will be obliged to select from several options :

i) he can carry on with his usual approach, accepting the limitations the system imposes upon such things as spontaneity, feedback and non-verbal communication; or

ii) he can allow the producers of his programme to determine its style, and fit his message to their superior knowledge of the medium; or

iii) he can make himself aware of the various types of learning and models of teaching, in which case he may decide to alter his style.

Gage and Berliner's five categories of types of learning are available to him. So is the Joyce-Weil Classification of Models of Teaching which divides seventeen models into four groups :

i. Information Processing

ii. Social Interaction

iii. Person Orientation

iv. Behaviour Modification

Should the instructor and those working with him elect to make themselves aware of these classifications and groups, they may discover some specific model or combination of models which, modified perhaps, would be ideal for both themselves and their students.

According to Bruce Joyce and Marsha Weil (1972) teaching is "a process by which teacher and students create a shared environment including sets of values and beliefs which in turn colure their view of reality." A model for teaching is, therefore, "a pattern or plan which can be used to shape a curriculum or course, to select instructional materials and to guide a teacher's actions."

Information Processing embraces models of teaching designed to affect the ways in which students develop intellectually, use deductive and inductive processes, solve problems, form concepts and use the processes of a scholarly discipline. Nearly all models for this family are also concerned with social relationships and the development of an integrated, functioning self. It includes models based on the work of Taba (1967), Suchman (1966) and Schwab (1965).

The hierarchical formulations of Ausubel (1963) and Gagne (1962, 1970) fit with this group. Ausubel stressed the importance of meaningful verbal learning of firm anchoring ideas to which new concepts can be related, and also the need for teachers to introduce "advance organizers" to prepare students for the introduction of new material and concepts. Gagne proposed that intensive task analysis of educational objectives be conducted.

The developmental notions of Piaget (1963) and Bruner (1965) fall within this grouping. Bruner is concerned with the conceptual changes that occur during learning. He stresses the active role of the student in the acquisition of knowledge and advocates discovery learning.

The Social Interaction group includes models designed to teach ways of working cooperatively and democratically with other persons. Here, problem solving is aimed at improving interpersonal relations. This group includes the Oliver-Shaver (1966) Juris prudential teaching model, Thelen's (1960) group investigation model, the Massialas-Cox (1966) social enquiry model, and the sensitivity-group or T-group methods of the Esalen Institute.

The Person Orientation group emphasis meeting individual needs for the development of personal identity and emotional life. Rogers'

(1969) theories of non-directive teaching belong in this group. So do Glasser's (1965) classroom meeting model, Gordon's synetics procedures for the development of creativity, Schutz's (1967) awareness training and Brown's (1971) humanistic classroom. In this group education is seen as part of a self-actualizing process.

The Behaviour Modification approach has developed from attempts to create efficient systems for sequencing learning activities and shaping behaviour by manipulating reinforcement. It includes both programmed instruction which uses printed materials, and face-to-face interaction which uses the stimulus and reinforcing capabilities of the teacher interacting with the student. B. F. Skinner is identified with this approach. This model relies upon changing the student's external behaviour rather than his underlying and unobservable behaviour.

At this point of selecting learning and teaching models the instructor has to decide on the appropriateness of satellite-based interactive television for his particular subject. Several questions have to be considered at this point, for example :

(i) What is the purpose of the course?

- Preparation for professional certification or a credential?
- As a guide for personal study (laboratory experiments, testing, reading)?
- To provide a forum for debate and discussion of issues?
- To impart a skill or craft?
- To provide guidelines for appreciation (of art, music, film, photography)?

(ii) What is known about the students' learning styles?

- Are they "at a distance" from personal preference?
- Will they work alone?
- Will they need to interact with an instructor?
- What are the course's objectives?

These matters must be known before media selection can begin and before the considerable variety of delivery methods normally available to instructors can be examined.

Characteristics of the System

Assuming that the satellite-based interaction mode has been selected, its possibilities must be examined. In Western Canada six delivery formats are either in use of being designed :

(i) The Extended Classroom

In this format the instructor teaches a lesson in a classroom to a group of students, which is simultaneously televised to external student who may participate in the interactive segments of the lesson via audio links with the classroom.

(ii) The Tele-lecture

Here the instructor delivers a lecture to external students using traditional classroom lecture methods (i.e., blackboard, text book, transparencies). Students respond during a prescribed discussion period, usually towards the end of the programme.

(iii) The Illustrated Presentation

In this format the instructor determines the content of the programme and manages its design. While participating in the programme in an editorial or co-ordinating role, he makes extensive use of other resources (i.e., videotapes, films, guests, etc.). Students interact with the instructor during prescribed period.

(iv) The Integrated Telidon-Teleconferencing System

In this format programmes are designed by an instructional team and a selected instructor conducts each course session via teleconferencing. Structural and inter-concept relationships in the programme content are shown via Telidon,(videotext) broadcast.

(v) The Computer-Managed or Video text-based Self-Instructional Library

In this format course materials are designed by an instructional team and converted into videotext and computer formats which accessed as required by students.

(vi) The Tele-seminar

In this format instructor and students use most of the available time discussing issues raised either in texts or by the instructor. The format eventually used for a course is not necessarily selected by the instructor alone. Economics, time and political constraints often determine which will be initially selected and modified.

THE PRINCIPLES OF ADULT INSTRUCTION

Communication theories and models are not specifically directed at any particular age-group. Learning theories on the other hand, tend to be more about children than about adults. Similarly teaching models describe processes which are usually conducted with children. This is partly due, no doubt to assumptions about education being a childhood experience. Lately however, there has been a demand for lifelong learning and distance education for adults. So the instructor of adults armed only with knowledge of how children are taught, must be able to relate what he knows to the world of the adult learners.

Donald Brundage and Dorothy Mackeracher identify the following nine "characteristics of the adult learner" from which "learning principles" may be derived :

i) Physiological characteristics

ii) Self-concept

iii) Emotions, stress and anxiety

iv) Past experience

v) Time

iv) Motivation

v) Paradox

vi) Learning styles and abilities

vii) Developmental stages and transitions

Familiarity with these characteristics, plus others he may know of himself, will enable an instructor to mold his selected instructional styles to suit his students.

COURSE DESIGN

After considering one's approach and selecting one's learning and teaching model, and modifying it to fit the delivery system and to suit the particular characteristics of adult learners, bow then does one develop the appropriate course design?

Consider for example an instructor intending to offer a credit course in Literature. His path identifying and selecting an appropriate instructional design might be as follows :

Step 1: The identification of his own preferred approach to communication: His choice: The Process School or The Semiotic School. He selects the Semiotic School.

Step 2: Determination of which of the two approaches to the understanding of learning he prefers: His choice: The Behaviourist Approach or The Cognitive Approach. He selects the Cognitive Approach.

In making these selections he reflects natural, personal biases which will be reflected in his selection and organization of his materials and his working relationship with his students.

Step 3: So at this early stage a little fine tuning will help. He should now determine which communication model will bring him closest to the semiotic, cognitive style he desires. His choice: the nine major models of communication.

He selects Becker's model and in doing so indicates how he wishes to relate to his students. A student for this instructor is one who moves through a constantly changing cube of information; who derives meaning from information by comparing what is presented or discovered with what he already knows. The conclusions he draws are, like everything in his life, always changing. As he considers them", so they change. He assumes that this will be happening to his instructor and fellow students also.

The selection of this model as representative of his personal style indicates further the instructor's sense of responsibility for the clarity of the messages he transmits to his students. As he sees it, the course is a chance for mutual exploration from which all should learn, including himself. He is the instigator, the one who presents the ideas that must be discussed.

Step 4: How best to do this? Examination of currently-accepted learning categories will give direction. His choice : Respondent learning; Contiguity learning; Operant learning; Observational learning; Cognitive learning. He selects, primarily, Cognitive learning, knowing that he may also turn to Observational learning at times.

Step 5: Is there an appropriate teaching model? Joyce and Weil divided seventeen models into four groups. His choice: Information Processing; Social instruction Person Orientation; Behaviour Modification. Selection is difficult because each of the first three has value. So, with the understanding that his course is only a small part of each student's life and that he is teaching a specific part of a larger curriculum, he selects Information Processing.

By making this selection he is again acknowledging his students' need to develop intellectually, to use deductive and inductive reasoning.

Armed thus, with a clearer understanding of his own style and its practical implications, instructor may now find it relatively easy to select his media. His final course design may include the following:

- a wide selection of recommended readings mailed to the student before the course begins;
- a series of brief but carefully delineated introductory statements about central themes, presented at the beginning of each session;
- long, open-line discussion periods; or
- time allocated during each session for students located in groups to discuss a particular matter before airing their collective and individual opinions during an interactive session.

Thus does the instructor identify himself? He can now approach the production team to explain his requirements. He knows that he will depend primarily upon himself and his students for the majority of each session. , He knows he will require pre-taped segments and a few diagrams and other graphics for early use in each session. He knows the he will be making the most demands upon the production crew during the first ten or fifteen minutes of his hour-long session. He may indicate that he requires certain materials to be available for reference, if required, during a discussion. Predictably, if he requests that some resource materials be available for use on call, or if he asks for off-air time for student discussions in the middle of a session, he may be informed by a slightly irritated producer that such practices are "poor t.v.". But he must not be intimidated by a reaction that merely reflects the producer's style and values. In the final analysis he is the instructor and they are his support system.

The instructor who is not master of the theories and models behind his preferred designs may well find himself working with a producer who understands perfectly the capacity of his studio and the most effective ways of using it for broadcasting. Under such circumstances the medium frequently dominates the message.

What is the pay-off after all of this consideration of theory, to the instructor and his support team?

- The instructor will approach his task with a clear idea of his personal preferences and teaching style.
- He will know that the people with whom he is working understand who he is and how he wishes to teach.
- In addition to the basic method he adopts, he will know of alternatives from which to select.
- He will be able to influence to some extent, his students' responses.
- He will be able to priories the various methods of instruction and learning available to him.
- He will be able to indicate clearly to the production crew, the use he intends to make of himself, his students and other resources.
- He will know upon what principles he wishes to base his evaluation.

SUMMARY

The basic assertion in this paper is therefore that educators intending to use new media should know how to do so. Lack of knowledge of the principles of communication and instruction can be frustrating and expensive. The major benefits of such knowledge are as follows:

- Instructors will be confident in this new and intimidating system.
- Course and production team members will understand each other's style and preferences in communication and instruction.
- Course and programme design will be streamlined because planning decisions will be accurate.
- Selection and use of resource materials will be efficient and defensible.
- Instructional methods will be varied and flexible.
- The relationship between instructor and students will be based upon a clear delineation of the instructor's approach and instructional style.
- Students will understand the significance of their participation in the interactive sessions.
- A basis for evaluation will be established.
- The academic credibility of distance education courses will be increased.

Chapter 46

Information Use and Its Impact on the New Millennium

T. Manonmani

The world in recent times has moved towards a knowledge based society where information and media are the crucial ingredients as much as land in the agricultural society and capital and labour in the industrial society. Knowledge is the key creating wealth and improving the quality of life. The technologies especially computer and its allies for gaining, sharing and applying knowledge are changing rapidly. Knowledge tends to grow at exponential rate.

The global system in the new millennium is clearly bringing the various media of communication together to form a universal communication together to from a universal network. The network holding multinational economic and political interests tends to develop a core to establish a super-ordinate and sub-ordinate relationship mainly (US base and developing nations) and also an unequal international system of information rich and information poor. It enhances the gap between the haves and have-nots.

The capacity to communicate and have access to the media greatly influence information use and its impact. Whenever the availability, accessibility and the capacity to manipulate media are assured the people themselves will also generate new vital information for their own development in particular and national development in general. But the availability and accessibility of media and information in the developing nations is below the level of fixed norms. According to 1999 Human Development Report, India is actually is doing quite badly in this regard. For every 1000 people there were in 1996 only 15 main

Telephone lines, 64 Television, 1 -5 Personal computers and .01 Internet connections where as even china has 45 Telephone lines, 252 Television, 3 P.Cs and .02 Internet connections. The report shows that the vast majority of Indians remain excluded from media consumption and development.

In the above ground realities, there is a need to have vision for strategy to the next millennium . Otherwise there would be no meaning in pouring new media on masses of developing nations like India. Hence, the new paradigm will have to have an expand communication system with the "inclusive participation" and involvement of people at all levels across the length and breadth of this vast country. It is also an essential pre-requisite for better enjoyment of human rights.

Where Does India Stand ?

Half country ago in India, Development was to be a project meant for the overall integrated well-being of the whole society which would provide an opportunity for a fulfilling and enriching life for every individual. But over two thirds of billion population still live only on the threshold of survival. At the dawn of the new millennium India's circumstances are unenviable. Life expectancy is around 65 years compared with 78 years in the high income countries. Literacy of adult women is notoriously low, at some 30%. Under 5 years mortality rates of children remain above 100 per 1000 births. The AIDS epidemic is gathering force and could gravely undermine many of the social and economic gains of recent years.

Thus, the most lamentable afflictions for the people in India are illiteracy, large families, ill health and poverty each reinforcing the other and therefore becoming responsible for the very poor index for human and economic development.

The dawn of the new millennium with the technologies especially IT and Biotechnology gives new opportunities for social development and opening of the economy has produced new dynamism, most dramatically in the Information

Technology sector. The world is working up to India's crucial role as the largest democracy and as a dynamic economy. If still a low-income one an average. India is inexorably drawn into the globalizing economy. The most educated enterprising and talented sections of the people in our country are seizing the global opportunities and are becoming prosperous and bringing wealth to the country through IT and software business. But it is stunningly reported that only 10% literate and talented becoming part of the global economy while India has a large mass of totally 40% illiterate and 50% some what literate.

If the increased use of information technology in every productive activity will be freeing a large number of unlettered and unskilled people from traditional employment. This will heighten tensions and inequalities in society and that may lead to social strife and un-sustainability of free enterprise and democracy itself. The most frightening prospects are as follows :

As the contribution of agriculture to the Gross Domestic Products (GDP) of the country declines from 62% in 1951 to under 25% in 1999 and as S&T are used for larger production of food stuffs with less and less of labour, huge massess of illiterate and un-skilled for anything other than elementary agriculture operations, will become jobless and migrate to urban areas, occupying all open spaces, leading to the slummification and unblivability. They will become the feed stock for the populist destructive politics, or otherwise they will be lead to commit mass suicides like in Andhra Pradesh. Dr. Radhakrishnau rightly mentioned that the greatest danger to any society is not so much the existence of physical slums but the alarming growth of mental slum.

In the context the development agenda with the old approach either top-down or the bottom-up or horizontal approaches have assured the unfortunate side effects, the short comings and inequalities were mere minor perturbations which must be borne if India has to achieve developments.

To build a development praxis, it must be kneaded with the yeast of Justice, self -reliance, sustainability and participation. Building meaningful self-reliance at the national level will more beyond the realm of rhetoric, only when it is coupled with a careful and truthful search for community level resources and local knowledge. It should envision an idea of India contemporaneously at the global and micro level with the ingredients of sustainable economic development, socio-cultural pluralism, religious tolerance and vibrant democracy. Development is therefore the growth of the individual as well as that of the society in all aspects (John Kurien. 1998:1)

In order to make India a "Knowledge centered society there is a need for a detailed frameworks, which will consist of a lot of sub-sets and focus on areas of strategies resources generation to help individual to identify different knowledge areas and priorities for knowledge creation and exploitation" as advanced by Dr. A.P.J. Abdul Kalam, Principal Scientific Advisor and a member of the knowledge Task Force.

In this regard, it is suggested to envision on mixing up conventional technology with modern Internet and setting up an access, content and 'inclusive' participation model assuring the combination of top-down and horizontal and bottom-up approach for 'equal development' without swellings.

Access

Knowledge society must be inclusive and for that inclusion to be a reality, every one must have access to participation in the decision-making process. The information and communication infra-structure provides the means for inclusive because it allows for timely, inexpensive and broad dissemination of information from a multiplicity of sources to the majority of the people. It also allows for immediate assessment and internal ization of the information that is provided because of its built in capacity for interactivity (Wijayananda jayaveera, 1999:2).

As India enters the era of knowledge society and knowledge economy, access to the infra-structure to share knowledge is essential for social and economic progress. Any individualin the society would not be excluded to benefit from knowledge resources. Of course the importance of the Internet, telematics and multimedia for global knowledge endeavors is indisputable. But technology is not itself a determining factor to change any society into an inclusive knowledge society. (Wijayananda Jayaveera, 1999:3). Hence not only the physical access but also the mental accessibility of technology and any other knowledge resources might be the determining factors. The concept of access must be expanded to include 'interactive and inclusive participation'.

The first condition for the knowledge society is universal access. In the case of physical access both availability and affordability are paramount. The accelerated growth of the Internet for example has not lead to the equal penetration even among the citizens of industrialized countries, but they are incomparable with the divide seen in developing countries. The current penetration is insufficient to prevent a division between information haves and have-nots and particularly between urban and rural populations in developing countries.

Secondary, universal access would remain unachievable if physical access facilities are not available at an affordable price. Although telecom privatization and deregulation have made traditional operations more efficient, they are not a guarantee for local universal access to the Internet. Further more the transferred nature of telecom industry is more favourable to those who can impose revenue terms because of their advanced technology, high speed internet backbones and net concentration, developed over a period of time. This advantage has allowed few dominating operators to exert pressure on others to shoulder their access costs, making it even more difficult to provide the most basic services in developing countries to acquire and share knowledge resources.

For example, International call rates have been so far negotiated between the countries of call origin and reception, however the pattern of call generation has changed particularly between the carriers of United States and the rest of the world. The increasing amount of call back telephone services from the U.S. has substantially increased the volume of traffic originating from that country. Given this back drop there is an enormous pressure from the US operators to eliminate the cost sharing mechanism. To this effect it has been made a unilateral benchmark order for rate re-balancing which would effectively reduce the settlement rates benefiting US carriers at the expense of operators in developing countries. With this mounting pressure to abolish bilaterally negotiated cost sharing arrangements, developing countries will face with an unprecedented burden to maintain their tele-communication systems. The net resulting of this rate re-balancing is that the operators in developing countries will be forced to offset. Costs by increasing their local call charges. (Wijayananda Jayaveera, 1999:2)

Moreover Infra-Asian networks are linked to the US operators. This is even true with the most Infra-European networks, which often use Virginia as their international Internet hub. Thus positioning of the USA as the world internet hub; whereby operators in the US benefit far more than the connected peripheries. Ironically under these circumstances, both the Internet providers and the users in the US gain free Internet access to rest of the world at the expenses of other countries, including the poorest of them trying to join information highways despite many other pressing priorities like India. This has led to a sort of exploitation, a situation which is, indeed, in the long run counter productive to internet penetration and eventually will prevent many in developing countries from tapping knowledge resources.

Finally, the continuing state monopoly of Videsh Sanchar Nigam Limited in international telegraphy as well as in internet provision with in the Indian market raises the costs of telephones and IT services and is doing considerable damage to

India's international competitiveness in the IT sector. India's telephone density is abysmally low, international telephone calls organizing in India are among the costliest in the world, largely due to lack of competition. Physical infra-structure for data transmission with in India (eg. Optic fibre cables) remains under developed despite some recent progress. Restrictive policies on FDI have kept international chipmakers out of India and have in directly raised the prices of PCS in the Indian market.

To counter this, Wijaynanda Jayaveera puts forth the following suggestions :

1. The issue of strengthening infra-region networks has to be brought forward to be high in the agendas of regional forums. Serious thought should be given to the possibility 10 establish high capacity regional backbones to connect each country within a multi-hub global network in which nobody dominates consecutively. If this is achieved, operators in developing countries will be in a position to reduce cosis and extend telecommunication and Internet access well beyond the present elite enclaves of access. The civil society in developing countries could sensitize parliamentarian and regulatory authorities to address this important issues and to mobilize the relevant regional political bodies to act with determination.

2. ICT equipment, peripherals and access fees for communications and delivery systems have to be made affordable to ensure competitiveness in the E-World. For educational and community based institutions these items should be made affordable through the provision of tax exemption, subsidy schemes and special discounted fixed rates for net access charges. These fiscal and often incentives to be continuously reviewed and realigned to promote IT usage.

Physical Access

Regarding the physical access, the fact that the rural areas as such villages and small towns are getting left behind in terms of access to the world of computers and internet. It is true, that there are a few significant efforts, in a village in Punjab which is enthusiastically "computerizing" and even has a home page on the web ; in the village warranna co-operative sugar factory iii Maharashtra which has wired up 6 to 7 villages; and M.S. Swaminathan's village information project in Tamil Nadu. Constant promises from the ministers about telephones for every village turning STD booths into "information kiosks" and so on are examples. Yet these are meaningless if it excludes the majority from the scene. Each and every citizen should be connected with universal access. For that India and other Asian Nations must have sufficient infrastructures which mean the following :

Reliable, stable electricity and efficient telephone are essential to use computers anil to access the internet. Even when electricity and telephone services do exist, it is bad service in India. Without good telephone services or without constant disconnection from the server, use of Internet becomes frustrating and expensive. Without good electricity and telephone service entry into the information, age is impossible.

Mental ability

For the Universal access, mental ability to seeking the needed information instead of accepting the provided information is crucial in the knowledge society. With the advent or knowledge society, the opportunities for life-long education will become the most important requirement, l-irst to attain literacy, good education and excellence to global standards among a large proportion of the people. Every individual needs to make the commitment to life-long learning. Then we can make some one, somewhere information rich by sharing our tacit knowledge on a daily basis. So the Prime Minister of India Mr. A.13. Bajpayee also in his recent five point agenda stresses the education of developing a learning society and E. World. The agenda has three fold objectives.

(1) To build the knowledge and skills of individual to live and work in the knowledge society.

(2) To make people aware to understand the emerging new society. Programmes for diffusion and acculturation of ICT at the grass root level to be undertaken using public, private and community based "learning" institutions.

(3) To create vibrant Government, industry, academia and public interaction in Policy making and implementation.

Education systems with their traditional approach of fixed courses to make the society ready for adult careers will no longer sufficient to meet the demands of sufficient to meet the demands of knowledge society and economy. Due to accelerating rate of knowledge accumulation, the education with 3 to 4 years in a University will become obsolete shortly after recruited for a job. The objective is to make the education system an open learning system for to cater for the needs of the knowledge society and the changing E-world. A comprehensive review to re-engineer the national education system, specifically in terms of flexibility in curriculum and content development, teachers, training, examination system, university entrance requirements should be undertaken. A programme to educate the educators for the new millennium must be implemented to mass produce ICT fluent educators. Schools, colleges. Universities, ICT training institutions, mobile ICT training units and other virtual institutions of learning are to be identified as walk in log on ICT training centres for the educators. Educators here means all that consider themselves as such including parents.

Finally, in building knowledge and skills, 'value creators' should be encouraged to take the frontline activities to create new values. This inducement will achieve the diffusion and acculturation of ICT at the grass root level. Major commitment to Universities should serve as a core to knowledge based development strategy.

Content

In regard to content, the language, and availability of information is a matter of concern. The efforts would remain ineffective if users cannot obtain needed information because most information is not in an accessible language and in the public domain.

Sharing as knowledge content is equally important since knowledge is expected to be the driving force of post-industrial economy, to which many countries are expected to leapfrog. It is predicted that the country, company and industry with the most productive knowledge works will win. The knowledge base for the future knowledge economy is being developed largely through publicly funded ventures such as Universities and research grants. The publicly funded academic institutes continue to be the fountain of knowledge. Then who should own the knowledge ? There should be an arrangement to ensure at least that all research grants of public funds are issued on the condition that the research information is made available for fair use in the public domain and also in the accessible language. This is more important strategy for the volume of public domain information available on the internet like rural

Internet project of Tamilnadu. The national policies to promote public domain information and to ensure that they provide information and applications to improve education, health and environment and government functions should be considered as a priority. (Wijayananda Jayaveera, 1999:4)

One drastically approach to working towards the needed balance of priorities is to encourage pilot activities involving government, civil society, Non-Governmental organization. Universities and private sectors to collaborate in providing Universal access to knowledge. In this direction UNESCO has initiated a number of pilot learning activities. Using libraries in rural Africa as gateways to information highways through multi - purpose tele - centres in six African countries and the

Kothmale Internet Community Radio, Sri Lanka, in which the matching and ihe mixing up of conventional technologies with ICTs are being tested, arc some examples. The experiences gained advance solutions on a number of policy issues on access, tariffs, public domain information, content relevance, training, ownership of broadcast spectrum and participation of public and private institutions.

Participation

To ensure the combination of Top-down and Bottom-up or horizontal approach or in other words being the marginal into the mainstream development participation of the beneficiary in the programme and content (message production) should be given priority. Experiences gained through many pilot projects in Latin America, Philippines, Sri Lanka like Internet Community Radio have assured the possibilities of participation by making use of existing radio narrow cast networks.

In most part of the South Asia, it is the only common device that most people have access to. Radio in India is acknowledged by media professionals, NGOs and communication specialists as having a far broader reach (98.5% of listeners) than Television. Although many predicted it's decline following the influx of satellite television, radio can change its orientation to focus on areas in which it can do better than television by providing interactive, participatory and locally based broadcasting called Community Radio, an alternative to today's highly centralised and hierarchical media structures has the ability to adapt to different needs. The community itself decides what it's priorities and needs are in terms of information provision. Since more than 70% as ruralities have poor access to media source for decision making processes, it would seem appropriate to initiate community Radio in rural areas. Dr. M.S. Swaminalhan Foundation has been approved to own P.M. station for its rural information networks in Tamilnadu for example. Thus radio might become the gateway to Internet and the Universal access and participation.

Mixing up radio narrow casting and the Internet to mutually enhance their development impact. There are models like Kothmale Internet community Radio, Sri Lanka.

REFERENCES:

Herbert, I. and Schiller, 1985, *"Strengths and Weaknesses of the New International Information Empire"*, Lee, Philipp. (Ed.) Communication for all, Orbis, Newyork.

Jaynth, A, 1999, *"Information Highways : Paths lo prosperity or poverty"?* Paper presented at the AMIC's 8th Annual conference, Singapore.

Jayaweera, Wijayananda, 1999, *"Including the excluded - for common good of all"*, paper presented at TASKNET conference, New Delhi.

Kurien, John, July, 1998, *"Development - The micro global quest"*. The Hindu Magazine, Chennai.

Reddi, lisha V., 1991, *"Leap fragging the Industrial Revolution "* Kichael Traber (Ed.), The Myth of the Information Revolution, Sage Publ., New Delhi.

Singhal, Arvind and Rogers, Everett M. 1989, *India's Information Revolution,* Sage Publications, New Delhi.

Venkatasu bra mania n, K., September 5, 2000, *"India's Development acknowledge Society"* - An approach paper, The Hindu open page, Chennai.

Chapter **47**

Political Transformation and Communication Technology: Retrospects and Prospects

R. K. Madhvi

The dissemination of information has come a long way in the last few years but the concept of communication has been known from very early times. Indian mythology refers to two eminent characters as communicators. Narad Muni and Sanjay, depicted in their legendary roles. The former, a bachelor saint, an omnipresent oral communication activist who sought to Liase between the rulers and people. He enjoyed a sort of journalistic immunity in society. The latter was a mythological war reporter who is believed to have given a running account of the great epic battle, Mahabharata, to the blind kind Dhritarastra.

The ancient Monarchs in India had various means of political communication which helped them to administer their kingdom and protect it from internal and external aggression common in those days. Kautilya's Arthasastra' (300 B.C. 300 A.D.) a text of which was discovered in 1909 is regarded as the oldest known works on State craft and deals extensively with all aspects of administration of a state. A close examination of the Arthasastra reveals that political communication was given great importance by ancient policy makers for the administration and protection of the state (Nair, 1998). The 'Arthasastra' described how news was gathered through spies and secret agents broadly divided into chares and Sancharas. During the reign of Emperor Ashoka there was also a system of visual dissemination of the news of the day. These were published in small pictures drawn on temple walls; the laws of dharma or moral codes were inscribed in pillars, which can be seen even today. Although the first attempt to establish newspapers in India did not occur until 1766, rulers like Aurangazeb's methods of collecting information through newsletters on social and political activities in his empire are some of the important events to gather information. The British attitudes toward the press in India were influenced by some important personalities who occupied the positions of Governor-Generals and Viceroys. Despite long periods of repression, the Indian press took advantage of wavering British policies to rouse Indian sentiments against the British and to sow the seeds of freedom among the people. With the vastness of its territories, the complexities of its languages, and the varieties of its cultures. India posed difficult problems with regard to the implementation of press controls.

In the twentieth century, more and more Indian Newspapers began to speak out against the British rule and support freedom for India. Their impact began to be felt throughout India which enhanced support for the national struggle for independence. Prior to Independence, however, Nehru found out that the press was contributing to the already difficult situation created by the partition. Although Nehru supported the amendment to Article 19 (2) of the constitution and helped the passage of the 1951 press objectionable matters act, he did not implement them strenuously. With Nehru's unsurpassed popularity and the over whelming acceptance of the congress party's leadership, the Nehru erapress played a rather small role either in the conduct of national elections or as a unified force of opposition to Nehru's policies. However, the impact was greater with the arrival of vernacular press. The subsequent Prime Minister Indira Gandhi added a new dimension to this concept. In contrast to the Nehru period, the period of Mrs. Indira Gandhi's rule was characterized by confrontation between the Government and the press. The press was also uncomfortable with Mrs. Indira Gandhi's limitations regarding freedom of the press during emergency (Kharkhanis, 1981).

The first General elections in 1951-52 attracted worldwide attention. The press carried the news of the campaign in detail. Due to diversity in language, administrative units based on linguistic divisions were recognized. The fourth General elections did not help congress because of the reports and complaints of food shortages, rising prices and violence in certain pockets etc.

Election campaigns have how begun to act as occasions for civic education. Political awareness is increasing among the people with increased exposure to the mass media. The role of media in the 1998 Indian deletions is clearer in helping the people to choose their candidate from the wide array of parties in the fray. "Election Watch", a programme conducted by former I.A.S. personalities Mr. B.P.R. Vitahl & N. Jaya Prakash Narayan, is a "Voter education programme" monitoring party lists for candidates with criminal backgrounds, suggesting safeguards against bogus voting and organizing public debates between rival candidates. Such media programmes have activated public interest and led to greater pursuance of political news and information.

Media's Role in Political Communication

The Elections of 1988 have demonstrated the superficiality of our newspapers. It was treated as contest between political personalities, with little or no attempt to understand and describe the underlying socio-economic forces that determine the outcome. Much of the reporting gave the impression of being preconceived as if. the reporter had decided in advance what to write, adding local colour gained from last minute hurried tours to provide a touch of authenticity. The results showed that hurried chats of roadside corners were no substitute for serious study of a constituency.

The Indian press has been accused of being obsessed with politics and concerning themselves insufficiently with social and cultural issues. News from the provinces and small towns is ignored. Newspaper managements are to a great extent responsible for superficial coverage. Most of the newspapers restrict non-metropolitan reporting. With fewer whole-time correspondents stationed in state capitals, and those in the rural countryside are usually starting out in the profession and not completely able to interpret grass root social trends which takes deeper study and more experience (Bhattacharya, 1998). It is the quality of field reporting that distinguishes one paper from another. Similarly may be disguised by differences in make-up in the print media, but typography cannot match the rapidity and dramatic impact of the visual media. However television cannot equal thoughtful, well written depth coverage that goes with learning of political events. The press has been more vibrant in coverage of political news than the controlled electronic media. But this trend is undergoing a vast change with the telecommunication and satellite revolution. The television appears more authentic. The anxiety of candidates to be seen on the screen provided evidence of its reach and impact. Illiteracy is no bar to communication in the audio visual media. They respond smilingly to the toughest questioning for the privilege what they said on television made news; what they said on the press seldom did (Bhattacharya, 1998).

The media and Political Performance

The political opportunism, shifting loyalties and total abandonment of basic ethics that has convulsed Indian political scene since the elections in 1996, resulting in the college of two governments in quick succession, is as much a lesson to the congress and the various constituent parties of the united front as to the constituents of the 13 party BJP coalition. Gone are the days when regional parties were feared as contributing to the centrifugal forces of the country. Infact the regional parties today contribute to the strengthening of federalism while underscoring the need for fulfilling the regional aspirations.

India has been witness in recent times to the only dominant news in both the electronic media and the print media, the antics of the politicians which include the stories of their corruption, frequent change of governments at the centre and the states and the growing power of money and muscle to get elected into the Lok Sabha and the Vidhan Sabha. Electrical reforms of the kind people want for a clean policy are still a far cry. The media report more about the misdeeds of the politicians, bureaucrats and others holding various levels of power than about any positive developments and projects completed for the welfare of the people. At the International level, the national Government plays a vital role in restoring the country's pre-eminence in various regional and international affairs. It cannot be denied that any country's international standing arises from its herent internal strength, political, economic, social technological, and cultural. The cases of Singapore, USA, Germany and the Scandinavian countries all show that they have a reputation on the world stage based on what they have built-up within. Infighting politicians, dehumanizing poverty, a rickety infrastructure, malnutrition, disease, illiteracy, criminalized politics, corruption at every stratum, lack of dedication lack of a work ethic, etc, convey the picture of a sick society and such societies seldom evoke any kind of respect abroad. The media must conscientize the citizens on the living conditions and factors that contribute to it in various other nations and urge them to demand a better deal from their own national governments (Sachdeva, 1998).

Communication Technology and Political Transformation

At the time of independence in 1947, apart from the press, All India Radio was the only electronic media available and also among these mass media Radio, was the only channel that had the potential to be quickly developed into an effective and truly national service in disseminating political news. AIR proved its worth and utility both as an informal channel and a development and cultural activist.

But with the coming of television it has become the most popular, glamorous and acclaimed medium of communication. It combines the best elements of spoken word communication with the potency of visual communication. It has fruitful impact in terms of educational, informational and entertainment needs of rural and urban population in modern India. According to the findings of IRS (Indian Readership Survey, 1998) conducted between July 1997 & June 1998 by the Media Research user's council in collaboration with ORG-MARG, as many as 44.5 percent of the people have access to television when compared to 33.3 percent to the print media, 19.7 percent to radio and a mere 19 percent to cinema. The survey was spread over 2,965 towns and villages covering a population of 2,07,284 lakhs in India. Regarding television, the survey pointed out that DDI national network was watched by 39.6 percent in the rural as well as urban areas, DD (Metro) by 9 percent and Zee TV by 6.9 percent. All the other channels account for less than 5 percent. The progress in Satellite technology has made tremendous impact and its uses span a wide variety of applications. It has resulted in satellite channels which is a major boom for political parties to transmit their message through these channels. The expansion of communication services brought tremendous change in political transformation in the present day. The transition from electricity, telegraph, cable and telephone to electronic, microwave, digital, laser, optic fiber, microchip and geo-stationary satellite technologies based systems of communication on the one hand and tele-text, view data, micro processor based electronic exchanges and network control processors and multi-plexus on the other, marks an unconceivable leap frogging in the fields of communications and telecommunications. Thanks to the latest technologies, the world today has become a small global town and has entered into the age of videophones computer or teleconferences, facsimile transmission of data, voice and images, miniaturization and computerization (Malhan, 1985). Political campaigns using a multi-media strategy play a dominant role in disseminating political news and views of various parties to the people.

Political Advertising for Election Campaigns

Advertising used by political parties before election is called political advertising. The election campaigns mainly use banners, posters, pamphlets, hoarding, wall writings, badges etc. This kind of political advertising furthers theparty's popularity among the people. The election campaigns help in reinforcing and mobilizing the opinions of the electorate. An important feature of an election campaign is the use of media campaign material and face-to-face platform speeches. Political leaders generally start their campaign with heavily printed material carrying the 'motto' of their party. They exhibit those banners and posters with a specific symbol in every nook and corner of the state. With the same 'motto' in mind the party men use imagination and select slogans captions and cartoons and exhibit them in vantage public places. The ultimate, gain of such effort is psychological. These visual pictures get imprinted in the minds of the people especially among those people who are undecided and have kept their options open. While the conventional modes of electioneering are very expensive, audio and video cassettes which are specially prepared are able to reach more houses. Highly dramatic, poetic literature with music in the background has immediate impact on the minds of the people. But even this is a costly affair. A poor man cannot afford the luxury of a cassette player. Likewise public meetings, which necessitate a political leader to reach places in person, is through common, a difficult task. The personal charisma is indeed a plus point in this regard. But this is also expensive.

Today election campaigns are carried on through the mass media of communications, especially the press, radio and television. These election campaigns are carried on in a mass scale by the newspapers. Just before elections, radio and television allow election campaigning under a special programme called the "party political broadcasts". Except this, the broadcast media in our country does not allow any further election campaigning. In other countries especially in the United States it is most intensive on television considering television's mass appeal and mass reach, care is taken to make these campaigns interesting and effective.

During the recent Indian elections video cassettes and video conferencing was increasingly being used in state wide and national campaigns, to allow campaign leaders to meet 'face to face' without being taxed by already over burdened travel schedules. Electronic mail, though not widely, is available to some individuals. Personal computers and specialized campaign software allow even the smallest campaign to main list supporters and to target mail to them. Behind the usual posters, video films, songs and T-shirts of the election campaign lies a well -planned strategy designed for maximum impact. During the recent 1998 elections, the BJP took one step further, the campaign appeared on the Internet; the new medium of the Internet makes little difference to what the BJP really wants to talk about. The Website was launched with much fanfare, with party president L.K. Advani claiming that it showed how "forward looking" the BJP was. The real news from the Website depends on one's point of view. The Website is not for laggards in software; you need the recent net escape 3.0 or higher and no text only browsers. The opening page is a picture-heavy affair guaranteed to strain a terminal without sufficient memory, 'Vandemataram', scrolls in several Indian languages while a map of India appears super imposed with a Durga like figure. The jingoistic overtone in the banner proclaims the BJP has as its goal 'transforming' the world's oldest civilization into a strong, prosperous and confident nation. The internet users are greeted with pictures of prime Minister A.B. Vajpayee and L.K. Advani on the main page and rest of the site includes a piece on women's issues, the economic affairs page, the national security and Defence page, the foreign policy page and K.R. Malkani's page on the BJP's history.

From the 1998 elections 'Ditties' (parody songs) of hit times have been used. The campaign trail on video was also used during these elections. Audio cassettes, balloons, computer chips, etc made their debut in even Orissa's interiors during the recent political campaigns (Bhandare, 1998). In a campaign report by the "India Today", the correspondents and photographers who closely followed the candidates on their campaign trails discovered that India is not just a vibrant democracy but a heterogeneous one too. Democratic verdict is not born out impulses but after a boisterous carnival. A look at the recent elections belies the reality that India is a poor country. Parties hire planes for their leaders and advertising agencies for their campaigns. It comes as no surprise that we are facing India's first true television election.

The national broadcasting television channels had the monopoly the covering the elections a few years ago. Advertising agencies were used as far back as 1985, but there was little need to package leaders and parties. This time, a host of private channels have joined the fray bombarding the viewer with news programmes and special election bulletins. It is no wonder that jockeying for visibility among political parties is at its most intense. Television has given nation wide recognition to marginal leaders of marginal parties. In the age of the television election, politicians are being coached on how to speak, when to speak, and even on what clothes look good on television; politicians are becoming brands. It will be a new age election campaign as elaborately designed rathams (chariots) make way for helicopters and staid bullet proof ambassadors, traditional propaganda techniques give way to satellite and cable TV networks (Thomas, 1998).

Many factors influence the effects of political communication on diverse segments of society since India's population constitutes of a large rural populace. The differential effects of media campaigns on the rural and urban population deserve special attention.

Retrospect and Prospects : Some Issues

In political communication, the country has a long way from using radio as a dominant mass media in the earlier days to the present times of a television oriented political campaign. This has been possible mainly due to the expansion in the communication media especially television and the Satellite channels which give an opportunity to transmit messages of various political parties and their ideology with greater availability of time. Though the press scores over the television in detailed and factual reporting, people are usually impressed by the direct coverage of various political campaigns all over the country. Today the voter is not only aware of the mainstream parties but is also aware of most of regional and local parties which have come to stay in politics with the era of the coalition government. It would be interesting to examine the use of various media in political campaigning in future elections as they would provide information on the most effective multi-media strategy combining both interpersonal and mass media in political communication. Apart from actually covering the campaign trail, the mass media could also look at some issues that concern the people in political communication. The code for journalists covering political campaigns and elections envisages the following guidelines.

It will be the duty of the press to give objective reports about elections-and the candidates. The newspapers are not expected to indulge in unhealthy election campaigns, exaggerated reports about any candidate, party or incident during the elections. In practice, two or three closely contesting candidates attract all the media attention. While reporting on the point raised by a candidate and make an attack on his or her opponent. Election campaign along communal or caste lines is banned under the election rules. Hence the press should eschew reports which tend to promote feelings of enmity or hatred between people on the found of religion, race, caste, community or language. The press should refrain from publishing false or critical statements in regard to the personal character and conduct of any candidate or in relation to the candidature or withdrawal of any candidate or his candidature, to prejudice the prospects of that candidate in the elections. The press shall not publish unverified allegations against any candidate or party. The press council also advise guidelines on pre-poll survey and 'Exit-polls' survey that in the event of staggered poll dates, the media is seen to carry exit-poll surveys of the polls already held. This is likely to influence the voters where the polling is yet to commence. With a view to ensure that the electoral process is kept pure and the voter's minds are not influenced by any external factors, it is necessary that the media does not publish the exit-poll surveys till the last poll held.

The rapid expansion in computer assisted communication Technology has largely made its presence felt in the urban areas. It would be quite sometime before they begin to have any influence in the rural areas.

Today elections have become telephonic with private channels competing with Doordarshan to bring the results live to the people. The channels caught the politicians at their best and worst moments and tracked them down in the remotest pans of the country as well as in the Internet at international level. The internet is often described as a large city, with no boundaries, where one can just wander about, meet people and chat with them, know about the weather, read books, preview films and the latest music albums, know what one's favourite film stars are doing and of course, walk around the shopping malls buy things. One can go to the university, if one want, register for a course and take lessons of talk to customers and sell the latest wares. The internet has also come up with news stands where one take a look at the latest issues of all the major newspapers and magazines. People can go through most of them, without subscribing to them no matter where they are in Trivandrum, Tobago, or Tokyo. They don't have to worry about the news back home while they are traveling in any part of the globe, because they can read their favourite newspaper everyday on the internet. The era of what is being called on line Journalism' has truly arrived.

Today millions of people worldwide are turning to the Internet for news. One doesn't have to wait for the next television bulletin or for the newspaper next morning. All one has to do is log on to the net and get the latest news, views and much more including opportunity to comment on it and make his or her views known to other. Publishers, television channels and all those involved in the news business worldwide are realizing the importance of the internet as a new medium for disseminating news instantaneously to audience across the globe. In the past two years or so, the online media scene has virtually exploded and is reaching new heights with every major news event appearing on it (Sharma, 1998).

It is hoped that the expansion in communication technology would benefit the vast population of India and provide them with more political information to assess the ideology and performance of governments at the local level, state and centre.

References :

*Ajit **Bhattacharya** 1998,* Superficial coverage. *Vidura, April-June, New Delhi.*

Bhandare, Namita 1998, Poll Hard Sell, India Today Feb. 2nd

***Karljamos.** Sharad 1981,* Indian Politics and the Role of the Press, *Vikas Publishing House, New Delhi.*

Malhan, P.N.. 1985, Communication **Media. Yesterday, Today and Tomorrow.** Publications Division Ministry of Information and Broadcasting, Govt. of India. New Delhi.

***Nair, Kiran R.**, 1998,* Arthasastra and the Foundation of Political Communication. Prabuddha Bharata, *Vol.101 January.*

***Sachedeva,** Surendar **Kumar,** 1998,* A coalition that can survive on consensus. *Competition success review, May, New Delhi.*

Thomas, Sunil K., 1998, The Polished Propaganda. The week, Feb. 1.

Chapter **48**

Micro-Media : A Tool for an Activist

Umashanker

The most important criticism against the mass medial micro media is that it breaks people, fragments them into very individualized independent entities. The folk media brought the entire village into a common place and initiated a discussion. But macro-media like TV and computers push the people into their homes or shells and give a centralised programme for them curtailing the whole possibility of discussion : Walkman is still worse : It has literally cut-off the person from this world. He and his walkman is his only world. It is here that the gap between person to person is widened. It is in this act that people become less sensitive, less concerned and less human. From a man being a social animal he is now cut off from other human being in an island. These media have an inbuilt ideology which is inhuman and reactionary. The media does not help people to move towards a better humane society through collective action. In fact it forces them to be self-centered and very unconcerned individuals.

If the world is what it is today, it is because of communication. Can any body disagree with the fact that, what we eat, what we wear and the way we live are determined by the media around us? Probably, the most drastic and alarming effect of the media are on our psyche the way we think.

Communication is one of the basic needs of the human beings. If we trace the human evolution most of our inventions had been in the field of communication. From the basic qestures and simple sounds, today, we have moved to sophisticated communication involving satellites and computers. The journey had been very long from communicating with the known for functional use to communication with the unknown to satisfy our inner abstract feelings.

"Global Village" a concept so beautiful can be visualized only by the people in the field of communication. But in reality has the world come closer? The answer is surely No; and unfortunately the trend is towards more fragmentation and polarization.

Has the increase in the food production in the world in any way decreased the number of people dying of starvation? No; Has the giant leaps in the medical science stopped people dying of want of basic medication help? No; Has the breakthrough inventions in technology made our peasants and common man's life more humane and comfortable 7 No; Then, what has happened to the inventions, to the inputs? Where have the fruits gone? Who has benefited from ail these developments? We don't need a very big analysis to see that most of the fruits are being enjoyed by a very small minority, a few rich and powerful a small section of the vested interest. They not only enjoy it, ensure the continuous enjoyment they control it. We see this pattern more clearly in the field of communication.

All the popular media, the mass media are today controlled by the rich powerful rulers. They saw to it that this media is the only 'media' effective in this society. To achieve this they have systematically and subtly destroyed the people media folk media. Hence today we see only T.V., cinema, radio, press and drama becoming more and more popular. Here the audio and the visual is so strong that it bombards the senses of common man to such an extent that he becomes impaired to any other form or media which is less violent and less forceful. In total this process has created a vacuum, a sense of powerlessness among the common people.

This condition is not only hampering and hindering the development and growth of the common man, but it also creates a very problematic situation for the person who works with the people a change agent working for the community. The person needs to communicate and he sees that there is no media he can make use of effectively. To communicate through the macro-media he needs lots of funds and it is totally out of reach for him. Hence he is left struggling hard and he miserably fails to communicate through that media without the know how, the infrastructure and the funds to use it effectively.

Can the situation be left like this? No, if we are visualizing change. Any change agent to be effective will have to be a good communicator. Looking back to our great leaders, without exception we see that they were very effective communicators. Some of them had voluntarily imbibed the skill but others have acquired the skill by training and hard work. All of them have done one thing which is common and that is, taking the contemporary popular media into their control. They reframed the media according to their needs. At present, the change agent is far from reaching media leave alone the aspect of controlling it.

Is the condition of the change agent totally hopeless? If we analyse the situation we see that the skill have some hope. The T.V., the Cinema, the Radio, the Computers and the Press have not really reached all the remote villages of India. The statistics reveal that out of the 80% rural India only 50% have been exposed to one or the other forms of mass media. The remaining 50% of the rural population is totally out of this viscious fold. No wonder that in one of the remote Tribal hamlet of Northern India they had organised an official function on the third day of the assassination of Sri Rajiv Gandhi, totally unaware of this international Tragic news.

The only way left to the change agent who wants to work with the rural poor is to build his own media. A media in which not only he the people are also skilled the know - how should be very simple and easily available in villages. The finance involved should be very less and the infrastructure should be readily available in the village. It should be light and flexible without many and heavy equipments so that he or his troupe can travel to any of the inaccessible remote villages without any difficulty, taking these things into consideration a committed Communication activist or a cultural activist can use micro media to communicate very effectively within the rural poor. Talking, public speech, group discussion, public meetings, Role plays' Story telling song, exhibition, demonstration, puppetry, street. Theatre and Jathas are the powerful forms of micro-media.

These Micro-media; can be broadly divided into three categories,

1. Speech and Meetings
2. Demonstration and exhibition, and
3. Performing arts.

The culmination and integration of all these media is the powerful concept of Jatha. Let us look in to each of these categories.

Speech and Meetings

India is a nation with very rich and developed oral tradition, talk and speech is the life of rural India. It is one media, which is totally in control of the Common man.

Talk : Talking to people is the first and foremost step in communication. Though it is the simplest technique it is the most powerful of all the media. The key word here is "LISTEN" by listening we allow them to speak and this activates them. It is by listening that we pick up their words and their language. The art of talking should be such that most of the talking is done by them with the minimum of intervention we should be able to communicate whatever we want to communicate, probably, talking to learn from them will give the needed humbleness to us.

People like talking and they enjoy it. They can talk very artistically and with total command over their language. They talk with confidence since it is their language and that is why we have to learn from the technique and the art.

For this media you do not need any infrastructure, any founds and any equipments. All you have to do is use your Body, Cultivate you; Voice and develop gestures. It is one media which is totally in control of the individual rich or poor.

Speech

A slightly concentrated form of talking is speech. While talking is a dialogue speech is monologue. On a particular subject you prepare what ever you want to speak and very powerfully in a concentrated manner deliver the speech to the audience either you have collected or to an already gathered audience.

Speech can be delivered to a very small group or to a very huge gathering. To be a powerful speaker you have to cultivate the habit of speaking in public. You should plan your speech well in advance and practice it very meticulously. Observing your audience, being sensitive to their reactions are the fundamentals of public speech.

It you are using mike, you should be able to use it comfortably adjusting it to suit your voice.

Here too no sophisticated equipment is needed. It is you, your voice, your gesture and your preparation that is important. Any village is accessible to you by walk. For a small gathering you can speak without any outside aids, for a large gathering what you need is a simple sound system. Today for a very meager amount you can hire a sound system in any part of the rural India.

Meetings and Group discussions :

If we observe our villages we see people discussing amongst themselves in different parts in different sizes regarding a variety of things. These informal meetings are places where a lot of information is shared, feelings are ventilated, doubts are raised, decisions are taken and problems are solved. We have seen that once the meetings are formally organized they become impotent and ritualistic. Hence we should plan meetings which are very informal. It is they who should speak. It is they who should discuss. It is they who should decide. It is they who should act. Our job is to facilitate. The best would be to get ourselves in to a meeting of theirs and do the job effectively. This needs skill and talent. To conduct a meeting effectively there is no shortcut. You have to get the skill only by conducting a lot of meetings. You have to be alert, open and learn by trial and error.

For different issues and different gatherings different types of meetings should be planned. Role plays in the meeting helps you to initiate a discussion or to move the group from a blockade. Public meetings are effective to involve the government officials and demythyfy them.

This media does not need any equipment unless you use a simple sound system for the large, gathering. You can walk into any village and deliver a powerful speech without any special arrangements.

Demonstration and Exhibition :

To give a specific input to the rural people one of the powerful media used today is demonstration and exhibition. Especially in the medical field exhibitions and demonstrations have been used very extensively to create awareness in the rural people. Simple Charts, paintings and models can be easily carried in to the rural areas and with the help of these aids the symptoms, the causes, the curing process and the prevention can be very successfully demonstrated.

I have seen some groups working on dehydration and the preparation of the ORS solution very successfully using the skill of demonstration. With very little properties and materials we can put up very meaningful exhibitions. It is true that this media needs a lot of preparation. It is a group work with different talents used at different levels, with local resources and very less funds. This media can be used meaningfully and effectively.

Performing Arts :

The rural India still has the colour, the rhythm and the drama. We have to activate and work on this taste that they still possess.

We had the culture of Community theatre and it is this aspect we have to concentrate on. Theatre was never alienated from people. It happened in the midst of the people giving scope for the people to participate in it. Songs and story telling Puppetry and Street theatre are the forms which can be successfully used by an activist.

Songs and Story Telling :

Every work, every act, every festival and every incident has a song or a set of songs giving more meaning and life to it. These songs are sung communally with each and every one participating in it. The tunes are very pleasant and simple. The method of developing a song is very simple by continuously adding bits of matter in to it. Songs and story telling in prose is inter-woven into each other to make a interesting form.

Songs and stories have to be re-written. Keeping the tunes the content has to be worked out. Basing on some of the tunes new tunes can be created to send our messages. Except for one or two musical instruments the group does not need any other equipment. Even if the group has a member who can use the indigenous instruments of the village it would make the group closer to the villagers. The infrastructure and the funds needed are very minimum. The group can straight away walk in to any village and give a song session or tell a story or give a performance integrating both these forms.

Puppetry

This is a very exciting media. It can attract any body from a child to the aged. We have a long and rich history of Puppetry. When other countries had not even thought of puppetry in India was flourishing. But I am not speaking of these Traditional and sophisticated forms, I am speaking of educational puppetry which is very simple to prepare and perform. Glove puppets are very simple to prepare and perform.

A small group of six to ten can take up this media and master it. A script on the issue has to be prepared understanding the scope and the limitation of this media for the script different characters (Puppets) has to be prepared. The paper miche Puppets are the easiest and functional. The raw material needed is just old paper, some sort of gum (maida) paints (or any colours) and old clothes. A very simple stage can be built by a single bedsheet.

Themes on alchoholism, units sangha building, Environment, and on health education have been successfully performed by many groups. But this media is best when performed to children.

Small amount of funds is needed to build the unit with puppets, musical instruments services and lights. It can be minimized by organising the local resources. But most important thing for this group is the practice to manipulate the puppets. By investing time and energy in to practice and the skill to mobilize local resources the Puppetry troupe can be very popular and still be managed with minimum funds.

Street Theatre

Drama performed on the street, street corner, or in the open ground can be broadly termed as Street theatre. In contrast to the drama performed by the professional groups where there is costume, make-up, stage, lights. Orchestra and Publicity, here we have only human beings and their talents nothing more nothing less. We work as our body, our voice and our active ability. An issue which is very relevant (local burning issue) and a drama developed on the issue. Most of the time it improvised on a few facts that are available

In street theatre troupe practices for some time (from three days to six months) and stages it in a place supposed to create the most effective impact. Here the objective, the target area; the target group and the result are very clear and specific. Normally a discussion and an action follow a street theatre presentation.

This media had been very popular and powerful in organizing and mobilizing people towards an issue. It needs a sincere and hard working team with some basic ideology inclined towards the poor, No funds. No infrastructure. No equipments are needed. Normally the villagers have a culture of looking after the performing troupes, of course, when they like that performance.

Jathas

When a problem is connected to more than 10 to 15 villages a jatha is more effective to mobilize people than any one single media. There are different forms of jathas. The most easiest one which is very common to villagers and effective is the jatha by walk. A group of people with one of the media or as a combination of two or more media move from one village to other mobilizing people on the issue and building awareness. The size of the group could vary from six to sixty. The duration of the entire jatha could be from five days to six months. The stay in one particular village could range from one hour to three days.

It is a community effort where different groups join together and conduct jatha. This media touches each and every person in the villages. The jatha could be on one particular issue with all media taking up the same issue, or it can be a very broad based issue with different media taking up different issues or different aspects of the same issue. This media needs a lot of planning, public-relations mobilizationand publicity. But when the entire community is involved the mobilization and the organization does not really become a problem.

Jatha with the integration of all the Micro-media can be very powerful and hence it needs a cohersive core group to manage, activate, exite and execute the programme.

Micro-media is a very potential tool but unfortunately untapped.

To conclude Micro-media is a very powerful tool in the hands of a committed cultural activist.

Chapter **49**

Rituals as Theatrical Communication Amongst Jenukruba Tribals

K. Narayana

Jenukruba is a food gathering semi nomadic tribe living in the jungle of Mysore and Coorg districts of Karnataka. Their main occupation is honey collection. 99% of the tribal population is illiterate. Out of 40,000 only 5 students have reached S. S.L.C. for the first time. No modern media of communication has penetrated this community. The means of communication is oral media.

The only significant festival in this community is 'Kont paje', 'Kont' literally means 'pestle', 'Puje', 'worship', metaphorically 'kont' stands for 'phallus'. So, this is a ritual regarding the worship of Phallus', regardless to say is a theatrical activity through performance.

This paper tries to show how different sets of signs expressing the sexual exploit that have found a place in phallic worship or fertility cult. The signifiers to be studied include motion or movement sequencing which create actions throughout. Studying the contexts of these actions will clarify how physical movements, which can be called gestures or mines, can yield meaning as well as how they transmit meaning. The concepts of space they make use of another signifying factor and the third being the concept of time and the fourth the Actor spectator relationship.

Movements

Movement 1: The first movement performed by an elderly man who is a priest of this community by swirling in the beginning a bunch of neem leaves around the phallus **'Kont'** anti-clockwise and the same bunch is swirled around his phallus, who is nacked for a number of times and the female of all age group. One after the other by touches on their heads. Then he starts his body movements according to the rhythm of the song sung by the women.

Most notable acts are a set of movements or short motion sequences made by the priest. Sometimes he ties a piece of wood hanging to the front of his body and swings back and forth according to the rhythm of the song. These movements involve slight spreading of his legs and a forward movement of the pelvis to expose the genitals. A more complex movement involves rotation of the hips, and concomitant movement of the knees. This is done infront of women there is no actively participating female counterpart in this.

After this movement is done, there is a break for sometime and meanwhile the women go back to ring folkhymn inviting ones again all the people of the settlement. This break suggests the dramatic scene.

Movement 2 : The second movement is the movement in theme, or a change in scene. From first scene it moves to the second by the entry of a character with stripes on his front portion of the body and his physical movements resemble that of a tiger symbolizing manhood. He enters the circle yelling with great vigour followed by a woman character enacted by a male. Though he symbolizes manly-hood in his physical movement and appearance, his speech symbolizes that he is not able to be get a single child to continue his lineage. He sold his wife as a barren 'woman'. At the same time he is afraid of honeybees, wild animals of the forest, symbolizing his impotency and thus he enacts as a clown, a comic relief on the contrary, to mockat this tiger enters a pregnant lady with more than 10 issues. The language used by the tiger has metaphoric, matonymic expressions and revolved around sexual act and sex organs. He takes examples from live characters and situations.

Throughout this scene a dog is brought behind him sniffing his arise. Thus, symbolizing the human sexual instinct. This scene is acted by married people.

Movement 3 : Now it is the turn of young boys and girls, unmarried to occupy the stage. They come with all modern non-Jenukruba dresses. The girls and boys sing love songs, in duet and dance. They appear to be modern and emitate non tribal language. The boys go round the girls liking their faces and thus symbolizing 'sucking the youth' as though they lick honey with great contentment. In this scene, some boys dress like girls and some girls like boys and dance and thus there is a role reversal.

Movement 4 : An old man, an 'ascetic' comes to the stage in search of an young girl to marry her exhibiting his muscularity. He comes with a pot on his head and sways it, tempting the girl.

Movement 5: An actor comes covering his front portion of his body and leaving the back nacked. He goes on defecating and the people behind him go on collecting the excrement. This is enacted only through mime, symbolizing the 'perishability' of the body.

Movement 6: Finally towards the closure of the ritual, another song is sung requesting 'konti' to comeback next year without fail, while this song is going on the unmarried girls put 'sobalakki' the puffer rice. The idol is taken away from the backside of the canopy and immersed in the same lake from where the mud was collected.

Space : Physical and Psycological

Space in the context of the ritual is not distinguished from its surroundings by visible markers. But, has the ever shifting relations of proximity and distance between individuals. This has an intimate distance at the same time the people follow some conversion on their sitting arrange-merits. Usually, the women who chant the folk hymn sit near the male Konti' next sit the elderly women, and girls covering half of the circle. The menfolk also sit in the same hierarchy starting from female 'Konti' side, and leave an empty space and that the conventional distance is created signaling the boundary between the acting area potentially tillable visually and acoustically. Thus they maintain a sort of male-female, old-young, married unmarried oppositions before the commencement of the ritual. This is a sort of distancing the people based on psychological factors also.

Role of Binary Oppositions in the Ritual

In this ritualistic theatre one can find following types of oppositions :

I. Binary Opposition in Space : (i) Male Konti female Konti, (ii) Installed from the front taken out from the back, (iii) Male Konti female audience, (iv) Female Konti male audience, (v) Opposition in age is like Opposition in colour terms as well as opposition is also found in the circle.

II. Opposition in characters and sex. (i) elderly priest absence of the female counterpart, (ii) Tiger man.

(a) outward manly appearance toward impotancy,

(b) Tigerman barran woman,

(c) Barranness potentiality.

III. Boys-Girls
This is in opposition to (i) married-unmarried,
(ii) unmarried boys, unmarried girls.
IV. Traditional-modern
V. Ascetic-longing for sex.
VI. Strength of productivity-purposelessness of the produced.
VII. Actor-spectator.

Merger of Oppositions

At many points the oppositions merge together

1, *Actor Spectator :* As the action moves, the folk of an actor and spectators starts exchanging. A spectator can take the role of an actor in the subsequent scene or seenes, except that they cannot act as a priest once again.

There can be an actor-spectator involvement in due-course of dialogue exchange. A spectator can involve in providing clues to act according to the situation. He can join the actors at a particular juncture. Thus, the actor-spectator distinction is merged. They can sit anywhere in the circle after their performance. The physical space is mixed up irrespective of age and sex and gets merged even in their costumes sometimes. Thus the psychological distance is nullified.

Role Reversal: As soon as the clown (tiger) enters, the strict sanctity of theatre loosened. Boys dressed like girls and girls dressed like boys suggest the role reversal. The distinction between the sex comes to a naught.

Duality of Roles : Tiger, which is a symbol of strength and vigour for outward appearance finally attains the symbol of potency.

The ascetic longing for a girl also symbolizes the quality of the personality.

Thus, both the physical and mental distances and differences in terms of opposition collapsed. The multidimensional aspects of fertility is depicted throughout and both verbal and non verbal ritualistic theatrical themes revolves around fertility and its consequence, from birth to death.

Verbal and non-Verbal Code : Inter-textual Relations and Decodification

When a spectator wants to witness a theatre he needs to have more knowledge of that culture. Mastery of the rules of the game by the ritualistic theatre is in large part, a matter of experience. The spectator is bound to master the organizational principles of the performance inductively, that is, by experiencing different texts and inferring the common rules. Appropriate decodification of a given text derives above all from the spectators familiarity with other texts. An ideal spectator is in this sense endowed with a sufficiently detailed and judiciously employed textual background to identify all relevant relations and use them in a grid for a correspondingly rich decodification. The theatrical frame is never in this regard, pure. The performance is liable to draw upon any number of cultural, topical and popular references assuming various kinds of the extra theatrical competence on the part of the spectator.

Since there is no written text, in the case of Jenukuruba fertility cult performance, dialogues are spontaneous. Spontaneous conversation can be considered as a text to the extent that it forms a distinct unit more or less coherently structured semantically and pragmatically. The speakers charge the same object and topics of discourse; their references are linked that the textual constraints governing the progress are fair than those regulating dramatic dialogue. One can find internal coherence in the text. Since the performance is repeated, every year, there may not be a particular kind of texts used in its repeated performance every year. Since I have seen these festivals number of time I have tried to draw my inferences.

The change in the scene, context and topic look unconnected at first instance; but when they attain symbolic types, in the mind of the spectator, they provide contest to the disjointed sub-performances. Thus they enter a context of action, they posses the property of order in accordance with their own internally consistant form. In this way symbolic types can determine and provide context.

Thus human ritual has a unique characteristic both as a stage and is a disseminating media of messages. These is a common conceptual theme linking the range of meaning included in concepts about ritualization. This kind of ritual is a specialized communication sequence of complex discursive and non-discursive behavious. Thus it becomes very important to note the coding of the messages conveyed during the rituals make us think that however primitive the society is it is very complex one and it needs a special code to understand them.

Thus, the fertility cult plays a special role in securing regeneration to provide the path to going life. This kind of performance not only provides happiness for the community, but it acts as a means of attaining regeneration. This also introduces a novel and humerous solutions to the perplexed psychological problems.

Also, the mystery of creation is celebrated in a sexual form, because sex is the obvious symbol of new creation of birth. This is enacted and celebrated together. Thus man and nature are related. Coming of age sexually, socially and politically are dramatized in this single ritual.

Thus ritual does not dramatize only events in the lives of individual. It often commemorates historic events in the life of the group. These are commemorated, kept alive in memory, as traditions that foster the continuity and uniqueness of the group.

Ritual chiefly expresses celebration, congregation, dedication and commemoration. Mere celebration, might be an expression of joy in the event, but dedication, and communication place the event in a larger context, and affirms shared purposes of greater significance than any private ends in the primitive communities like Jenukruba.

To conclude, it is the rituals that create myth. In my opinion when the concept of ritual and myth are brought in to the area of media, as a dissemination of knowledge it has to have communication value. Languages as human artifact is a collective product of human skill and can easily be regarded as 'mass media'. But when we consider the arrest of spoken language in the form of print, the audible is transferred to visible. Similarly, myth or ritual is an arrest of the complex socio-cultural process. Then we can consider the fertility myth as a process of dissemination of message, in a highly codified, symbolically loaded communication. Similarly, after the advent of T. V. in this country we are returning to the oral narration via electronic media; which abridge both time and space. Cinema is another best media which has symbolized our concept of 'hero' and it has shown us what sort of hero' the common mass aspires for.

Our modern media should realize that whatever the programme they make should not be excluded from the rituals in the sense of community participation.

Our modern media should not give emphasis only to express celebration, but it should inculcate conservation, dedication and communication, and it should place the event in a larger context.

Chapter **50**

Technology and Social Behaviour in Hand Pump Programmes for Rural Drinking Water Supply-The Widening Communication Gap

Dilip Fouzdar

Natural calamities keep on claiming sizeable chunk of human resources of our society. Drought is one of the important components of this process. With his rich practical experience the author has made an attempt to highlight human aspect of misery minimisation. A case has been made to emphasise the need for educating the users and professionals. There is urgent need to plan for communication support for making this national effort a success.

The Indian drought of 1967, widely known as Bihar drought, happens to be a distinct land mark in the history of rural drinking water supply. This was the occasion when a new generation of man and machinery invaded the rural scene for displaying a set of ingredients unknown to its beneficiaries thus far, to fight against the drinking water shortage.

We have come a long way after 1967. Following Bihar drought there had been in a sequence several other drought events substantiated with similar relief type operations in various states including Bihar the home ground for the drought. Yet, handpumps in drinking water wells stand in the rural areas without any commitments to provide reliable safe drinking water for the community. Meanwhile, drought seems to have institutionalised itself in the name of 'International Drinking Water and Sanitation Decade' of the U.N. in which India happens to be a signatory.

The new device, the tubewell brought along a set of tools the hand pump and the drilling machines to name a few. These were like miracle boxes and were being employed during the following years by a group of multi-disciplinary professionals who had merely some partial knowledge of the subjects and tools they have been handling though in some other context.

Improper Device Hangover of the Past

During the drought of 1967, in Bihar, there existed tremendous confusion regarding the use of hand pumps. Diamond drilling machines were summoned which were just not the devices to help the occasion. Water (the ground water) has always caused problems in the progress of such drilling techniques. Drillers, most of whom came from the mineral exploration organizations like the Geological Survey of India (GSI) and the State Government Departments of Mines and Geology, out of their ignorance went on closing the fractures that were absorbing all the drilling water (or the drilling gel this is injected from the surface to clean the drill cutting) just out of their eagerness to attain drilling deeper and deeper. The ultimate result was the production of dry bores.

The dry bores fitted with hand pumps resulted in people 'mishandling' them. People were naturally angry at this mockery though things were not clear by that time either to the people or to the professionals who were rendering their bit in order to determine as to where lay the fault. Was it the bore that was dry or was it the pump that was not capable of drawing water? At some other locations where pneumatic blast hole drills (later to be known as down the hole DIH hammer drills) were employed,the water was visible. Often it soiled all the onlookers by its intermittent 'flush' of muddy water gradually grading into transparency. It created merry moments in the drought stricken villagers and the newspapers carried impressive photographs of villagers bathing near the drilling machine boring a hole which was yet to be commissioned as a water well.

Each pump had to be repaired frequently, at times 19-20 times within a span of 3-4 weeks. Soon it was discovered that the pumps were not study enough for the community handling. Children playing and eventually breaking them was an alarmingly common feature. Often domestic buffaloes and bulls did the job. By and by, the commercial hand pumps lost their credibility and by 1972-73 when Maharashtra was passing through another spell of severe drought, the local voluntary organisations had come out with a study hand pump by the name of JALVAD. Two closely related, closely situated voluntary organisations each located respectively at Jalna (WAR ON

WANT) and Vadala (M.M. Water Development and Mechanical Training) Ahmednagar District had been responsible for designing and producing it. This pump was later modified as Sholapur model and the yet later model of India Mark II, which became extremely popular throughout India are most identical due to their genetically relationships.

Having won the major battle of hand pumps with the help of India Mark II, which is strong and study enough to sustain community handling, it was apparent that the programme of rural water supply should have gone rather smoothly. The real situation however, does not support this view.

Whose Tubewein

Table 1 indicates the status of utilization of hand-pumps in 3 neighbouring districts. In the district of Purulia where the programme is running for 5 years, the number of break-downs (in a given time) are more than the other areas (for example Keonjhor where programme has just completed) because the India Mark 11 pumps generally give trouble free service for about a year. Again in Purulia out of 43 wells, frequent breakdowns and infrequent breakdowns were noticed in 16 borewells each, whereas 9 borewells reported no breakdowns. A point worth noting here is that out of 43 wells, 13 and 24 wells were drilled during the years 1978 and 1979 respectively and the following was the response to its repairs which is exclusively carried out by PHED crew (Table II).

Table 1. Condition of hand pumps and the tube wells in three neighbouring districts*

	Period of operation	Tubewells		Pump condition		
		Total commi-ssioned	Sampled	Good	Out of order	Unsatis-factory
Mayurbhanj (Orrisa)	1980-82	187	47	36	8	3
Keonjhor (Orissa)	1981	14	5	5	-	-
Purulia (West Bengal)	177-82	-	43	29	12	2

*Samples from work done by Lutheran World Service, Calcutta surveyed in April 1982.

Table 2

Time taken for repairs	No. of cases
Duration	
7-15 days	14
1-2 months	3
More than 2 months	12
Not applicable	11
No information	3
Total	43

The situation in Mayurbhanj is yet to show up since here the oldest bores were commissioned during 1980, but even here, out of 42 wells sampled, 8 sites had break downs with one well showing frequent breakdowns. The repair of these took 20 days to 2 months, while 2 of the pumps were not repaired for more than 6 months time.

The situation has developed clearly because of the centralised approach of the three tier infrastructure for the hand-pump maintenance which is being pushed through a strong UNICEF backing. The three tier system consists of a village level caretaker, a block level team for minor repairs and a district level superteam for major repairs and spare supply. This district level (3 tier) maintenance system rather threatens to jeopardise the entire self-help attitude instead of allowing the 'system' to work. By and large, the rural people are developing the tendency to believe that the tubewell installed in their village belongs to the Government of PHED and though they are the users they indeed have no responsibility of repairing it.

Unreached Villages

The objective of the Drinking Water and Sanitation Decade is to provide at least one safe drinking water source within a distance of 1 mile (1.6 kms). Although a water source could mean any thing from a dugwell spring or open pond the emphasis presently rests on the drilling rig and the hand pumps. It in a map of India of the size of a 70 mm screen (in cartographic terms this should mean a map of 1:1000,000 scale) each of the well points were marked by a small blue dot in the entire country would look as blue as the Bay of Bengal. But a number of villages that could perhaps be marked in the map with precise accuracy, could not be reached with a drilling rig mounted on the heavy truck. Table 3 gives an account of the reached and unreachcd villages under a planned programme in the district of Mayurbhanj, Orissa. This situation is despite of the fact that both PHED and the voluntary agency has been in operation in the district for over a year.

Table 3. The list of scarcity villages reached and unreached in Mayurbhanj district, Orissa.

Name of the block	Total no. of scarcity village	Vol. agency programme		PHED programme		No. of village not covered
		No. of village covered	No. of hand-pumps installed	No. of village covered	No. of handpumps installed	
1	2	3	4	5	6	7
Jamda	73	33	57	37	44	12 (16%)
Bahalda	97	23	31	30	93	57 (58%)
Tiring	80	27	30	58	125	13 (16%)
Bijatola	129	19	30	12	25	99 (77%)

The figure in column (2) is not equal to the total of column (3)+(5)+(7) because the voluntary agency and the PHED had worked in common villages. The example of 4 blocks of Mayurbhanj district should be at least indicative of the gravity of the problem we are advancing to tackle with only one type of drilling rig in the hard rock areas which forms almost the entire peninsular region of Ganga-Brahmaputra plains.

Now taking the country as a whole, if drinking water is to be brought within the reach of people the total number of tubewells to be installed is roughly 3,000,000 which should take about 30,000 machine years to complete. This implies that a regiment of 3000 fast mechanical drilling units must work continuously for 10 years to fulfil the task.

Now, if the requirement of machine is projected taking the irrigation needs into account the figure should be five to ten times higher. Obviously, the magnitude and the dimensions of the problem is such that the actions should be self-generating, economically speaking, and therefore, decentralized approach would perhaps be more appropriate rather than the imposed centralized programmes those are in vogue.

The Communication Gap

A number of questions emerge in the minds of the planners, administrators and operators of a drilling programme and these persist throughout the period of operation. For example, before the machine reaches a work site, the question may often be asked as to:

- What could be the depth of drilling?
- What (formations) are going to be encountered?
- How much water is expected?

and after the completion of the well it is often necessary to know that:

- Is the well going to support water supply on a long term basis?
- What should be the depth, capacity and make of the pump to be installed?
- What is the quality of water suitable for drinking and is it well protected from pollution threats?

These questions imply involvement of various functionaries at different levels who would obviously not be under one idealized umbrella to augment smooth operation of the programme. From village level workers to the national level planners it involves various agencies which in turn would include various sections of multidisciplinary scientists and technicians. There is the obvious difference in the knowledge level among the total cross section of involved groups and individuals. In the actual programme, various ranges of experts would take part starting from qualified geologists and engineers, to barefootians like water diviners and hand boring masters. Obviously their deployment would he in accordance with the priorities that develop and the available expertise at various stages and levels of the programme. The entire communication gap relating to drinking water programme can be broadly identified at two levels. Firstly, the communication gap within the system (establishment) handled by multidisciplinary professionals who carry out policy making/planning and implementation and secondly, the communication gap between the rural mass and the professionals.

There is a kind of mystry surrounding the occurrence and movement of groundwater and therefore in the total approach of selecting a well site to hand pump installation, professionals play a major role. Presently, dissemination of knowledge within the level of professionals is being considered but this may be with a determination. A part of the useful knowledge must percolate upto the level of the users. The people must not always treat the system from a distance due to the mystry that surrounds the total ambit of technology. The demystification is utmost necessary and it can be brought about it in the following ways:

(i) by promoting the slow labour intensive devices which may also be duly modified. This will provide adequate opportunities for the members of rural community to actually take part in the operation.

(ii) by creating local village level teams for repair and maintenance of pump. This can be derived out of the above exposure.

(iii) by disseminating the basic knowledge of groundwater occurrence (as may be appropriate to the area) and hand pump mechanism and also about the aspects of water resources and the sanitary protection of water.

(iv) by allowing the villagers to act with the tubewell like making a channel to save the village road from getting wet, using the water for vegetable gardening, organising the people to collect water in a disciplined manner etc.

These measures, to some extent, will help people understand the situation and adopt the new alternative drinking water system from within.

The authors of drinking water decade must realise the fact that the tube-well is yet to prove itself to be the workable alternative to the existing dug wells which are versatile enough to serve under various situations despite of their weaknesses (in terms of depleting water level rendering it to go dry and the lack of protection against faecal or bacterial contamination) they have the strength to serve. This basic strength of dug well must be translated to the system of tubewells as well in order that it works.

Communication plays a big role in bringing the technology and its interwoven components to the level of a remote village. Keeping in mind the fact that the majority of the beneficiaries would be unexposed to modern way of life. The basic strength of the dug well is that it has successfully and parenneally reached the level of a remote Indian village with its whole technological parafernalia and the total infrastructural fabric. It is so obvious and it is so much a part of life.

Behavioural Trends

Since the drinking water tubewell involves no cost to the community, the community hardly have any participation at the planning stage of the programme. Any alien proposal for the installation of the hand pump (tube well) would, therefore, at the initial stage receive an overwhelming support from the residents of a village full with enthusiasm, curiousity and amusement.

This is obviously because of the visual appeal of the total population and a kind of status value. Every powerful villager wishes that the tubewell should be located at his doorstep. No one at this stage would ever consider or realise that a tubewell on the roadside may eventually spoil the road by making passage extremely slashy causing increase in population of mosquitoes or flies threatening the hygiene and sanitary conditions. Due to the proposed tubewell the place may also turn noisy which is not uncommon specially during the rush hours of morning and afternoon,

But when the tubewell in the village becomes a reality it is not accepted by the community merely on its face value. If the taste of water differs even slightly from their existing traditional source, the people will have their reservation about accepting the new source of water. The fact that the tubewell water is protected from bacterial contamination would not convince the villagers to win their acceptance. Again, the new source must also pass the test of housewives who reject the water if it takes longer time to cook the lentils of grams.

This is depicted in Table V along with some other behavioural features. In Mayurbhanj, the percentage of acceptance is low partly because the programme is new whereas in Purulia it is higher as gradually the community sees the good points of accepting the new device. In Purulia the sample survey indicates the use of tubewell water for drinking on 2 occasions even though it has not been found suitable for cooking.

It is also observed that the people are protecting the sanctity of the tubewell site by not allowing the children toilet wash though taking bath or washing utensils are not considered a taboo except in a few locations in the district of Purulia where people have put restrictions against taking bath and washing clothes and utensils at the tubewell platform. In some of the blocks people have been able to stop these community habits totally.

Not so often, the excess water is being channelised to irrigate kitchen gardens (kitchen garden is rather a urban concept in India that was acclaimed by systematic official campaign. In rural areas, kitchen garden is a traditional vegetable farming in the backyards). This aspect is important because this helps in understanding peoples' spontaneous attitude towards water. The tendency is common throughout the country depicting a very hopeful trend. While the

Table 7

Area		Water quality		Water use		Behavioural trends					
		Good	Brakish	Drinking	Cooking	Kitchen Garden	Bathing	Children Toilet wash	Washing clothes	Washing utensils	Washing animals
Mayurbhanj (Orissa)	Sample 52	40	12	31	31	14	29	0	25	25	3
Total bores with hand pumps	Percenta ge 100	77	23	59.6	59.6	27	55.7	0	48	48	5.7
Purulia (West Bental)	Sample 43	28	13	36	34	15	21	0	19	23	8
(Total bores with hand pumps – 248)	Percenta ge 100	65	31	83.7	79	35	49	0	44	53.5	18.6

tubewell is marked exclusively for drinking water purposes, people have themselves derived further benefits from it which has eventually widened its scope. This should provide an important feedback to the planners of this huge national programme. The prospect for the entire drinking water programme may as well be thoroughly reviewed to see if there could be a programme of even a larger magnitude replacing the present drinking water programme. This broad based programme may serve both drinking water as well as irrigation needs. Awareness is gradually developing in the villages where tubewells have been installed for some time ranging between 5 and 15 years. This is after a prolonged mechanism of persuation truly demonstrative in nature because a number of hand pumps been have broken, left out and abondoned in the process a number of drilled wells have been written off. In the new villages the curious residents pass through all the experience that the earlier villages have undergone with the result in a number of interior villages hand pumps are still not functioning.

There is need to plan an educational programme keeping in view the experiences gained. Such a programme should be based on media mix for penetrating the heart and mind. Unfortunately even with heavy investment in technology, this dimension has been ignored. Consequently, many human problems are multiplying.

Chapter **51**

Communication Technologies in Indian Perspective

Bela Trivedi

Since communication in any form is the basic urge of human being, all mass media are extensions of his ideological existence. It is but natural that communication should be, and is, one of the areas which constantly undergo rapid changes bringing innovation to achieve the goals. I have tried to spell our significant aspects regards to :

1. Existing Communication Technologies.
2. Emerging Communication Technologies.
3. Existing and emerging areas of social applications of these communication technologies and
4. Communication strategies, approaches and skills.

With rapid technological advancement, the whole world is being converted into "Global Village" and India's achievement in communication technologies in short span of 50 years of Independence is a matter of pride.

Barring about 10% of geographical area and population, mass communication reaches to the people of this nation staying at the remotest areas. Newspapers and other print media have limitation to reach out the illiterates. India is such a vast country with diverse cultures, languages, dialects, religious and above all complex problems of illiteracy, poverty, unemployment etc., that without technological intervention, it is unimaginable to overcome these barriers.

Communication Scenario at a Glance

A cursory glance at communication technologies depicts that Print media has traveled from treadle to off set and now satellite based newspaper editions. All news agencies are varying with each other to have latest news collection in the least time consuming way using latest available communication media.

All India Radio is the official radio broadcasting organization in India. The number of radio stations has gone up from mere 6 in 1947 to 196 in 1998. In 1995, the estimated number of radio set was 111 million. The medium waves transmission covers 97 percent of the population and nearly about 90 percent of the area from analog to digital broadcasting.

Doordarshan is another official television broadcasting organization in India. It has 1034 transmitters with 18 channels that covers 72.7% geographical area and 84.4% population. It reaches to 63.2 million TV homes with approximately 331.2 million viewers. India has about 18.4 million cable and satellite homes. India has unique network of cables that work as distributors of television signals.

Satellite television channels like STAR Plus, STAR World, STAR Sports, ESPN, Sun TV, Sun News, ATN, AXN, Zee TV, Zee News, CNN, CNBC, BBC etc, and now the introduction of Direct To Home Services, about 200 channels will be available to the viewers.

Telephones, Chordless phones, m5obile phones, cellular phones and now global phones like IRIDIUM.

Broadcasting following with audio conferencing, one way video and two way audio teleconferencing using telephone lines, using DAMA, two way teleconferencing are the examples of induction of interactivity with a view to enhancing communication effectiveness.

VSAT and many more new technologies are in the offing. Computer has changed communication scenario radically with unimaginable pace. Its application in almost all possible areas of work has replaced traditional system. With increasing used of Internet. India is literally marching towards information and communication explosions. It is estimated that there are 80 million internet users. E-mail has become very popular because of speedy and accurate and cheap transmission of messages. Audio text, video text, hyper text and many more dimensions are being added everyday.

Constant Up-gradation in Technologies

Another interesting aspect to be noted is that even in these technologies, content up-gradation takes place like Microsoft introduces updated versions of Windows. More and more satellite channels provide news from different corners of the nation and the world using interactivity. DOORDARSHAN NEWS, STAR NEWS. ZEE NEWS, CNN, CNBC are examples of this.

Related to this is the change in the size of equipment. Earlier, a computer was of the size of a room. Now, it is of quite a small size. We know that a radio set used to occupy more space and now today transistors and pocket radios are available.

Compliance to Users' Demands

Third noteworthy aspect is compliance to the demands of users of communication technologies. With acceptance by increasingly large number of users, technology is forced to go innovative to accommodate more and more users. According to Prasar Bharathi Chief, "Now digital audio broadcasting is in offing". Digital broadcasting by satellite television channels is the future as it has potential to provide space for number of users at a same time (which is not possible in analog system) that too, with quality output.

Space Programes for Rural Development

In short, next millennium would find us amidst great communication technological revolution and information explosion. For this, credit goes to Dr. Vikram Sarabhai, a visionary founder of Indian Space Programmes who declared at UNISPACE-70 that India is not varying with other countries for supremacy in space programmes. India is developing space programmes for its social applications aiming at rural development. Subsequently, ISRO leadership Prof. Satish Dhawan, Prof. Yash Pal, Prof. U. R. Rao, Dr. Kasturirangan have continued this philosophy and have contributed meaningfully.

Pertinent Question

The pertinent question repeatedly asked is how a deprived section of society is going to have access to it? Central Government and State Governments and Non Government Organizations are making efforts. Government has made a provision for rural community radio sets and community Television sets, rural broadcasting mobile van to show films and video programmes where irrespective of class, caste, age, sex, religion, education all the people can have entertainment as well as information. NGOs reach out to interior places and unexplored areas and show them videos, films etc and conduct group discussion to impart information and make them aware. Need is to strengthen and intensify these efforts.

Indian Government's mandate is very clear and that is overall national development focusing on rural development through education, employment and healthy life span. All resources at the disposal of Government and society are geared up to achieve technological excellence and its most relevant social applications.

Media Complimentary to Each Others

Communication strategy is to reach out one and all and to provide information that can empower people for decision-making at all stages in all walks of life. No communication medium works in isolation. All are supplementary and complimentary to each other. With television, films have not disappeared. Along with films, television and radio, newspapers still thrive. In fact, Television incorporates films and film based programmes and newspaper journalists, radio critics. Newspapers do include Television and film based columns.

It will be appropriate if I mention here that Doordarshan started broadcasting in 1956 with Krishidarshan (agricultural programme) has now covered practically all aspects of rural development. The Satellite Instructional Television Experiment (SITE) in 1975-76 established is a milestone because of the utility of satellite communication which was later on confirmed by a field laboratory situation experiment Kheda Communication Project. Now onward effort is Jhabua Development Communications project where broadcast and interactive training for human resource development are being tested. In addition to this, DECU/ISRO runs a Training and Development Communication Channel for Distance Education. Rural Development, Women and Child Development, Panchayati Raj and Industrial Training. As we all know that ISRO is all set to launch one satellite GRAMSAT completely dedicated to rural development. A million-dollar question is how to make it the most effective in terms of rural development by devising appropriate communication technology, strategies and skills.

Though most of the areas have been given due weightage but some areas, according to me, are less explored and need more thrust like imparting literacy through broadcasting, gender sensitivity, legal literacy, skill imparting and upgrading, income generation avenues, communal harmony, patriotism, nation building etc.

Technology for What?

Coming to communication skills using these technologies, much depends on how a communicator uses that particular technology. For example, a journalist can use a computer for report writing and also for data storage. I have seen one journalist in change of current affairs she always keeps monthly data, important events in a data diary that helps her to trace historical events and how the events take turn. Camera is in our hand but for what purpose we use it is the crux of the matter. Is it for communicator's creativity or for fulfilling his/her social responsibility to contribute in nation building? Television channels are at our disposal, what a communicator is supposed to do? Is he/she to reinforce whatever is there or to use television channels as "information carrier", "motivator", "mobilizer of public opinion", weapon for social change, "mouth piece of people's voice" etc.?

Need For Well Defined and Well Orchestrated Efforts

At this juncture when India is facing such problems that are complex and along with educational institutions, mass communication requires well orchestrated efforts to pursue well defined philosophy for national development, particularly, rural development. A mass communication has to perform a role of link between people and the policy makers and should continue to contribute as a catalyst.

I am glad to tell you that we are using technology with proper understanding i.e. instead of using one way, vertical communication approach we are using the same with two ways horizontal and participatory approached. With Prasar Bharati, broadcasting will play more significant and more credible role.

Mass Communication with Vision and Mission

I would end up saying that, tillnow, it is commented that "There is hardware led software policy in India". I appeal to all of you to work in the direction with vision and mission to convert this comment into "Software led hardware policy is extremely successful in India". Indian scientists have immense capability to face whatever challenge communication experts will pose in terms of needs and I am sure that communication technology will go indigenous.

Chapter **52**

Development Communication through the Ages

G. N. S. Raghavan

The term 'Development Communication', with a narrow economic signification, has gained currency in the second half of the 20th century, in the wake of the decolonization following World War II. But the communication of values conducive to human development understood in the widest sense is as old as human civilization.

There have been, through the ages and in many parts of the world, outstanding men and women who have sought to propagate the equality of all persons irrespective of sex or race or other circumstance of birth. In ancient India, for example, the sages of the Upanishadic period (c. 600 B.C.) repudiated the hierarchy of caste which reinforced the power structure, with the Sudras (or working class) assigned to the bottom of the social scale. One of these sages said : "It is not our religion, still less the colour of our skin, that produces virtue; virtue must be practised. Therefore, let no one do to others what he would not have done to himself." The Buddha (563-483 B.C.) spoke in similar strain :

"No Brahmin is such by birth;

No outcaste is such by birth.

An outcaste is such by his deeds;

A Brahmin is such by his deeds."

In the West during the pre-Christian era, Socrates advocated as early as in the 5th century B.C. that women and men of the same aptitude should be given the same kind of education and the same responsibilities in public life. The Greek dramatist Aristophanes, a contemporary of Socrates, had the courage to write a play ridiculing the war party in his own native city state of Athens. In the comedy Lysistrata ('She who disbands armies') he depicts an international sex strike by women of the Greek states which were wasting themselves in the Peleponnesian War. They refuse their men sexual intercourse till peace is negotiated.

Jesus Christ's teaching, 'love thy neighbour as they self, was truly revolutionary. Soon, however, the deeply humanist teaching of Jesus was buried under papal institutions. Whereas Jesus had driven the money-changers out of the temple, the Church itself became the biggest of exploitative feudal landlords. In India, during this medieval period, the Padma Purana laid down that a wife should look upon her husband as a god and serve him, "be he debauched, immoral, a drunkard, a gambler."

In 1727 the Bishop of London wrote to the slaveowners of the southern American colonies to assure them with regard to the Negroes, that "as to their outward condition, whatever that was before, whether bond or free, their being baptised and becoming Christians makes no manner of change in them." And in 1934 the Archbishop of Canterbury, head of the English Church, criticised as hasty the Montagu-Chelmsford reforms which Indian nationalist opinion regarded as wholly inadequate. On the other hand, there was Charles F. Andrews, a Britisher and a practicing Christian, of whom Jawaharlal Nehru said : "India does not posses a mere-devoted friend than Charlie Andrews."

Thus there have been two streams of religious communication: one humanistic and promotive of a better society; and the other narrow, fanatical and anti-humanist. In the context of the tragic events of recent years in Punjab, and incidents of Hindu-Muslim tensions from time to time in different parts of the country, it is well to recall the humanist message of such great egalitarian communicators as Kabir and Nanak. Kabir, who on his death in 1518 was claimed both by Hindus and Muslims as one of their own, represented the confluence of the egalitarian Bhakti movement among Hindus and the mystical Sufi movement among Muslims. He proclaimed :

"Hariisin the East; Allah is in the West; Look within your heart, for there you will find both Karim and Ram. All the men and women of the world are his living forms."

The syncretist movement culminated in the emergence of a new religion, Sikhism, based on the teachings of Guru Nanak. The compilation of his teachings, known as the Adi Granth, includes a large selection from Kabir and other Bhakti and Sufi saints whose message was in consonance with Nanak's. Mahatma Gandhi invoked the Bhakti-Sufi strain in the India heritage when he popularised the verse: "Iswar and Allah are both Your names. Grant us all good mind."

It is a pity that the grand concept of humanist development expounded in our country by a long line of egalitarian thinkers from the Upanishadic sages to Mahatma Gandhi has been irivialised by the Ethno-Centric definition of 'development' by the industrially advanced countries of the West. They have encouraged a replication, in the former colonies, of their own pattern of industrialisation, urbanisation and mass communication. A strong motive behind this encouragement is export promotion of their machinery and manufactures, including the hardware and software of mass communication.

The Western patronage of development communication, through their scholarships and textbooks, has made the term synonymous in the main with agricultural and health extension including family planning. The governments of the former colonies have been led to think of the expansion of the mass media as something good in itself, irrespective of whether the media are intended for and reach the masses with content relevant to their needs and interests. The ruling elites of the developing countries have been only too willing to follow the advice of Western Communication gurus because mass media established and controlled by governments can be used to build up the image of those in power. Since people will not buy radio or television sets merely to hear and see their leaders all the time, entertainment is provided on a liberal scale-much of it meretricious, and much of it imported. It is a variant of the formula of the Roman emperors of old to keep their subjects supine: giving them bread and circuses. Bread may be in short supply for the masses in many developing countries. But circuses (of which the modern variants include Asiads and international tennis and cricket matches) are available in plenty.

If the Western model has influenced many former colonies, in the Communist World which has expanded since World War-11 alongside the attenuation of the old colonialism, the Soviet Union has sought to 'develop' the countries of Eastern Europe-and now Afghanistan in its own image. And communist China has likewise forced its helping hand on Tibet and, to the extent it could, in East Asia. The earliest example of 'development communication', narrowly conceived as an aid to increasing industrial and agricultural productivity, was Stakhanovism in the early years of the Soviet Union. The cult was promoted because the State as capitalist needs, as much as a private enterprenuer, to extract surplus value from the labour force to generate savings for investment in further economic growth.

Thirdly, it was in the Philippines under the authoritarian and corrupt Marcos regime—now discredited even in the U S.A. which propped him up for long years—that the expression 'development reporting' first gained currency. It became popular with the authorities in India during the internal emergency of 1975-76 when the democratic liberties of citizens were suppressed, free expression throttled, and censorship was imposed even on the proceedings of Parliament. But the trivialiaation or misuse of the concept of development communication does not invalidate the possibility of employing communication to promote human development in the broadest sense, Those who believe in a holistic approach to development should rescue the concept of development communication from the notion that it is an appendage of government departments of agriculture, health and family planning. It is this narrow concept that, in the context of grim events like those in Punjab, makes 'development communication' look somewhat like Nero fiddling while Rome burnt Development communication will gain in stature and enhance its role to the extent that the approach to development becomes comprehensive, It will entail critical examination of the value orientation of development and of the constitutional, political, educational and other frama-works in which development is attempted. It cannot avoid the issues of peace within the country and with neighbours, promotion of which is very much a task of the development communicator whether he is engaged in it as a communication professional or a political activist or as a social worker.

III

To illustrate what I mean by the value orientation of development, let me quote a passage from the first Maulana Azad Memorial Lecture delivered in February, 1959 by Jawaharlal Nehru, the initiator of planned development in India. Speaking of the manifest differences and of the rivalry between the Western Powers and the Communist Powers, as well as the similarity at the same time of the industrial culture of both, Nehru spoke of the challenge of preserving humanist values in an age of rapid technological change. He said : "I have no doubt that India will progress industrially and otherwise; that she will advance in science and technology; that our people's standards will rise; that education will spread and that health conditions will be better. But what I am concerned with is not merely our material progress but the quality and depth of our people. Gaining power through industrial processes, will they lose themselves in the quest of individual wealth and soft living ? We see the effects of rapid technological changes more especially in young men and women today. In extreme cases there is a tendency to criminality, alcoholism, des-tructiveness, eroticism, in addition to a cynical and negative attitude towards life and work. These tendencies are present more in the developed and advanced societies than in India or other under-developed countries. But it is important to note them because similar forces are likely to affect our life too."

The warning sounds prophetic a quarter century, later, when the amoral gospel of hedonistic consumerism is being spread by the mass media in countries with diverse political economics. In India the foremost disseminators of consumerism are the publicly owned and govern mentally run All Irtoia Radio and Doordarshan.

There is a close linkage between consumerism and the erosion of moral integrity. The late Prime Minister, Indira Gandhi, said in the course of an address to a conference of officers of the Central Bureau of Investigation and of State anti-corruption agencies in New Delhi on 23 March 1981. "The constant emphasis of the media on material goods, the vision of the comforts enjoyed in advanced societies, proliferation of regulations, the pressure of inflation, the breakdown of religious and ethical constraints, a general rise in permissiveness, all these contribute to corruption." It is another matter that it was during her tenure as Prime Minister that commercial advertising was allowed over All India Radio in 1967 and, subsequently, Doordarshan in 1976. Indian Television, introduced in the name of social education and being expanded with public funds to supplement the earning from commercials, has come to be dedicated to the provision of meretricious entertainment to the leisured class and to political image-building through the news and current affairs programmes.

The Working Group on Doordarshan Software, which gave its report in 1984, pointed out. "Unlike newspapers which carry a variety of advertisements besides consumer goods-ranging from industrial machinery to job opportunities, housing accommodation and matrimonials-the advertising on Doordarshan is overwhelmingly of consumer goods and services which should, in the Indian context, be classified as luxuries." Over the period since the above report was submitted to the government, commercial advertising over Doordarshan (as well as All India Radio) has gone up. The reason is obvious The government is bent on expanding the electronic media because they are directly useful to it for political purposes. Indirectly, too, they help to preserve the status quo by keeping the ears and eyes of the educated classes glued to Vividh Bharati's film music and to Doordarshan's film-based programmes and serials. The intelligentsia might, if they are not fed on these opiates, think seriously, discuss and even do some-thing about contemporary problems like corruption in public life and the administration; distribution of powers between the Centre, States and Panchayat institutions: electoral reform to correct the over-representation of the ruling parties at the Centre and in the states; and the persistence of widespread poverty.

It is ironic but true that the privately owned media of film and the Press, run for profit, have been making a greater contribution to development. From Garm Hawa to Phaniamma private film makers have from time to time produced films that qualify for description as consciousnessraising communication. But such worthwhile commercial films are few and far between.

More consistent and solid is the contribution made by the Press. The Second Press Commission acknowledges this handsomely in its report of 1982. "It is to the credit of the Indian Press that, though judged by readership or by ownership it is not necessary for most of our newspapers to highlight the issues of poverty, the Press has made a major contribution by reminding readers of those who live below the poverty line and giving ihe ruling middle and upper classes a feeling of guilt."

Press exposures of politicians' misdoings have brought down some State Chief Ministers, e. g. Mr. A.R. Antulay of Maharashtra and a successor of his. Newspapers have drawn atiention to the inhuman working conditions in quarries and in small factories to which for all practical purposes, laws on minimum wages and safety standards do not apply; to the persistence of bonded labour under various guises; to the problems of rural women for whom the process of industrial development, entailing the cutting down of forests, has increased the drudgery of fetching drinking water and fuel.

Even the sacred cows of defense expenditure and foreign policy, which are supposed to be exempt from critical scrutiny on the ground of 'national interest', have been probed by some of our newspapers. Every government knows that it has only to talk of a threat from across the borders to produce a Pavlovian reaction of 'national solidarity'. It is a tribute to the perspicacity of the Indian Press and its courageous exercise of freedom of expression that a cartoonist, some years ago, depicted the Indian Prime Minister and the Pakistan President, each seeking some relief from domestic pressures, discussing on the hotline, 'How about a little co-operative war ?'

National integration' should not be sought as a coming together to fight foreign devils. What should be promoted is harmony among our linguistically and ethnically diverse population on the basis of equality of citizenship rights and obligations. Recognition of the brotherhood of all people within a country will enable the perception of threpeople of other countries too as human beings with hopes and aspirations, prejudices and fragilities, like ourselves. This will lessen the heat of passion with which territorial claims are pressed. It will help governments, if they are wise enough, to negotiate in a non-hysterical atmosphere of public opinion and to arrive at peaceful resolution of border disputes on the basis of give and take.

While the Indian State should be secular in the sense of not identifying itself with any one religion, it is idle to advocate 'secularism' as an ideology which entails our people becoming non-religious, as atheists or agnostics. As Mrs. Gandhi once put it, "No body can expel religion through laws. Many countries have tried but they have not succeeded. And India is far more religion-oriented than other countries. Our aim should be to encourage the basic teachings which are common to all religious, like love and sacrifice and renunciation, service and brotherhood.

The development communicator should recognise and use the positive and humanist elements in our religious heritage, surveyed in the first section of this paper, as a development resource to unite men and women of goodwill and to isolate the narrow-minded fanatics.

Chapter **53**

Practicing Development Communication : Some Thoughts

B. N. Challopadhyay D. U. M. Rao

Since Independence of India, a large number of communication efforts have been put in, in many development programmes and specially so in agricultural development and other allied fields. Such communication efforts were normally tagged as 'Extension' rather than 'communication' by most of the academicians and practitioners of development. Probably the extension efforts heavily relied on research institutes and agricultural universities who generated improved crop varieties and new agricultural technologies that are transferred to villages to achieve higher crop production and higher level of economic growth. In the process, human resource development in developing societies suffered since the development input and their acceptance were not well understood by the people and the fruits of economic growth did not reflect upon the socio-psychological well-being of people and general quality of life. Thus, a concerted action by the people has been lacking in all our development efforts.

Development communication has come to mean more then just communicating information to the people. It is a circular process where the users are seen as participants in the communication process and are just as active as the sources and affect the outcome of communication. It is periodically renewed by continuous evaluation. The accent of development communication is not on the dissemination of information but on its reception and utilization by the users (Quebral, 1983).

It is only in the late eighties and nineties that communication for development purposes got a headway. Since it has been realised by development experts all over the world that development communication can be seen as a social process designed to seek a common understanding or consensus among all the participants of a development initiative leading to concerted action. Thus, development efforts should be geared up in such a way that would enable the rural people to take control over their own agriculture, health, nutrition, environment and other areas of development that critically affect their quality of life. As Theodore Schultz, the nobel laureate and economist puts it : "Population quality index of a developing society should be as high as possible in all our development communication projects so that every citizen contributes substantially to the development of the nation."

ENERGIZING GROUP INITIATIVES IN DEVELOPMENT COMMUNICATION PROJECT

Extensive studies have shown that individual approach adopted so far in India and elsewhere in Third World Countries could not take care of the quality of life for all in the villages as solving a number of social, agricultural and allied problems require concerted mobilized actions by groups of similar interests, values and culture. Probably human qualities like concern for others, liking for group actions for productive and constructive purposes, tolerance for accommodating others point of view etc. have to be imbibed in people of developing society by development communicators. Institutionbuilding activities around different groups of people have to be energized for maintaining continuity with change through not only a clear communication strategy but also through a vigorous management in all our development communication efforts.

ESSEMTIALS OF DEVELOPMENT COMMUNICATION

Much has been said about the development communication its nature, scope and mechanisms of operation in this Seminar "but one should not forget that all our development communication efforts must take into account four basic essentials in conceiving, organizing, managing and practicing communication efforts. The are :(a) Participatory communication, (b) Decentralized decision making, (c) Sustainable development, and (d) Development perspective in terms of conserving natural; resources, raising productivity and sustaining labour intensive development in a country like India and other Third World countries.

(a) Participatory Communication

History of communication research in India and elsewhere have made it amply clear that communication researchers have wasted nearly the last 30 years of research efforts by conducting mainly gross comparative studies of media and methods used for development. Such researches could not answer the simple and yet blunt question : "So what?" Probably it is better to investigate how best a particular method/media is unique in communicating a specific message keeping in view the aptitude and interest of the users.

Centralised media materials productions in the government and the private sector have failed miserably to influence the people at large because of this very reason. Thus it calls for developing media in a decentralized fashion and involving the people in the production of media materials. Experiments at Nagercoil, Tamil Nadu by All India Radio on local radio station have shown that people are competent enough to create, sustain, control and manage local programmes on their own with little assistance from All India Radio Staff. Local radio stations role in resolving a feud between two rival fishing groups reveals the confidence that people began to place on the local radio station at Nagercoil (Anjaneyulu, 1989). Experiences of Training and Demonstration (T and D) project in Kenya have shown participatory communication can make a real break through in developing and using all sources of communication materials in development projects.

(b) Decentralised decision making

Unlike conventional extension approaches, development communication efforts require a vigorous management of decision making at all levels of the project. If it is not done, communication will not be participatory and self sustaining but result in extension/ clientele dependency syndrome. More over the very conviction of communicating messages across people will suffer since conventional communication approaches do not take care of decentralised decision making process. Rather they stick to thrusting decisionsthat are taken at the top to middle or lower level of development communication projects. Besides decision by consensus gets a real meaning when all development efforts are decided in a decentralised fashion at grass root level. Decentralised decision by consensus encourages collective responsibility and accountability among the group members. Since such decisions are 'their decisions' they are likely to be implemented thoroughly.

(c) Sustainable Development

Experience of extension projects in India and elsewhere have shown that the changes brought about in agriculture, village industries, women's development, and family planning projects could not be sustained once the project is withdrawn from the area. This is probably because of the missionary approach of extension and knowledge generating user dependency syndrome. The people, for whom the project is said to bring about socio-economic change, will not continue the project unless the people themselves feel that it is for their own good such development projects have been taken up in the first place. Again if they don't feel that the project is their own and that their participation is quite important, then too, the project may fall flat with the withdrawal of the project staff. As a development practitioner one must not over emphasize on targets of crop productivity alone, but should find ways of mobilizing the people to sustain the same vigour and energy that they had when the project was in full operation. Institution building activities will help sustain such development efforts for a long time.

(d) Development Perspective

Development communication must not look at a narrow angle of development like giving a greater emphasis and effort on only one of issues like productivity of crops, dairy development communicators should make the people conscious of the development perspective in terms of conservation of natural resources, sustaining productivity levels and creating more jobs in other related areas of development in the area. Contingency communication planning has to be developed by development communicators in order to provide jobs to people in times of need and when no other conventional employment opportunities are available only to sustain the development process in the villages.

PRACTICING DEVELOPMENT COMMUNICATION

Now the basic questions remain do we need a special brand of development communicators? Do they require special skills to handle all development efforts? Should they concentrate more on "Extension" or "Communication"? All these questions need to be answered at this forum if we are really interested in practicing development communication.

If we feel development communication can be perceived differently from 'extension', we need professional development communicators and just not the communicators. In similar fashion we need, not just the broadcasters, but probable development broadcastors.

Since development communicators are committed to meaningful communication in all development efforts, it is essential that certain special process skills have to be nurtured pertaining to :

(i) special understanding of development process for a given area both at micro and macro levels;

(ii) generating suitable communication media technology at the project level through a participatory approach; and

(iii) developing need based, situation-specific linkage mechanisms among various development agencies through both formal and informal networks, not only at grassroots level but also at higher operational levels of the project.

Integrating media development with development project planning and execution.

The history of media development, availability and use in the last four decades in the country indicates very clearly that media development was never planned thoroughly by most of the development agencies. They were conceived and produced through ad hoc mechanism at a centralised place leading to ineffective impact. For example, agricultural films produced by Directorate of Extension in and around Delhi has little relevance and impact in Southern or Eastern states. Similarly poster designs on family planning messages in Kerala were misunderstood in other parts of the country. Local identity and credibility of communication messages were questioned quite often in all these communication situations. Therefore, conception and development of media generation lias to be integrated with development projects through a process of media planning which is otherwise non-existent in most of the development projects in the country today. A poor country like India cannot afford to produce costly and expensive media which do not serve the purpose for which they were produced. Development communicators should be conscious of the phenomenon that the new communication technology and media are just the tools and means of communication process and not an end by themselves. The real challenge before them is to design the communication message through the media and new communication technology in such a way that sustains interest in the communication content, among the people and the process stimulate conviction and action.

Rigorous media planning based on the essential and optional taxonomical characteristics of the media to be used, aptitude and interest of the people and diagnostic techniques of designing communication message have to be made at each phase of the development programme and also at each operational level of the programme. Media materials testing and further modifications in the media production has to go hand in hand at each phase of the programme and evaluation of the media is imperative for better results.

In all these efforts, a very clear perception of the communication message, to be designed for each of the media in use, ii required. Quality characteristics of communication message should get a clear emphasis based on the capability of the media for which the message has to be generated as well as the user's abilities of perception of illustrations and understanding and comprehension of communication content by the clientele.

DEVELOPING INFORMATION SUPPORT SYSTEM FOR DEVELOPMENT COMMUNICATION

FAO and other organizations have experienced intensively the need for sharing knowledge across the people in all development communication projects since the inception of Development Support Communication (DSC) Unit in Rome in 1969 Similar observations were also highlighted by experts in different for a from time to time.

If our development communication efforts have to be streamlined across the people within a specified time and budget, it is imperative that a strong information support system be developed at the project level so that information flow to people becomes a continuous one. The present system of providing information to the farmers with the help of large number of technicians and field staff is quite expensive and time consuming and should be discouraged in all our future development communication projects.

The new in built information support system within the development communication project not only should be in a position to provide answers for all sorts of personal queries but also document the information through a computer base of database acquisition and retrieval system. In a similar manner, permanent agricultural museum/ exhibition should project latest developments in all the areas of interest of the people.

ESTABLISHMENT OF DATABASE INFORMATION ON DEVELOPMENT COMMUNICATION PROJECTS

It is an interesting, at the same time, intriguing phenomenon that most of the development communication practitioners as well as extension researchers are not aware of the development projects that are in operation, or have been completed, in different areas throughout the country and also in the Third World. We heavily rely on clearing houses established in the west to know the experiences of development communication projects in our own country. In view of this, a centralized database information acquisition and retrieval system on all the development communication projects in the country and elsewhere should be established at an early date at the premier Institute/University level so that development communication practitioners and researchers get ready access to such information at one point.

NEW ELECTRONIC MEDIA AND DEVELOPMENT COMMUNICATION

Large number of new communication technologies has been pouring in our country in the last one decade and more are likely to come in near future. Number of institutions have tried out such communication technologies in development communication projects. To name a few, video for development, desk top publishing for printing handouts, leaflets etc. interactive video disc to train engineers in the Ministry of Water Resources, Government of India (Brodman, 1993) are some of the sporadic efforts made so far. Newer technologies like remote access to database through a computer terminal with modems linked with telecommunication networks, facsimile (FAX) which enables printing at the receiving end, of graphs, text and even hand-written messages to be transmitted over a telephone line is being used in the west by development professionals to send reports. Similarly posters and newsletters are being faxed to rural communities in France and Germany. Electronic mail (E-Mail) which enables person to person communication via computer is cheaper than sending messages via Telex or FAX. Computer Conferencing trough satellite is yet to make entry in development communication because of high cost. Probably, teletext and videotext are the two communication technologies which may be tried out in development communication projects in our country in near future. In France and other western countries, videotext is being largely used by development professionals and the people to get ready access to all sorts of database information in all spheres of life.

Though large number of communicators prefer to opt for new communication technologies (and probably upgrading them if budget permits) and fashion their communication strategy based on such expensive media, most of the development projects have yet to conceive such thinking, what to talk of using them in the project. Probably, communication specialists trained in the West as? Well as leading communication institutions in the country are prone to establishing new setup of sophisticated communication equipment judicious planning. There are three observations we would like to make in such endeavors. Firstly, media materials production and testing through a participatory communication approach is imperative in order to affect acceptance of communication messages across the people. Secondly, they should be reminded again and again that all these new communication technologies and the media are simply the tools or the means of communication process and not an end by itself. Educational component of communication has to be taken care properly and this requires overall comprehension of the subject of communication media, details and the people for whom the communication messages are to be generated. Thirdly, there are certain constraints and limitations in terms of software and course-ware, cost, use, and effective implementation that are commonly considered as inhibiting factors in the use of such sophisticated communication technologies.

These constraints become complex and visible in a country like India, where the real challenge before the development communicators is in communicating effectively with illiterate people. It is surprising that not enough information is available based on research as to how illiterate people of India perceive developmental visual information. Similarly, the dynamics of perception of multi-media presentations on development information has not been researched so far. More research and development work in the field of visual literacy and visual education is needed to in order to communicate effectively with poor and illiterate rural masses. By any stretch of imagination, development communicators cannot afford to overlook this phenomenon in the present and the future development projects.

REFERENCES

Anjaneyulu, K (1989). Scaling Down : Local Radio in India Development Communication Report 1989/1 No. 64 : 1-2,

Brodrnan, Janice (1993). Cutting Edge of Multimedia Technologies : Promises and Pitfalls. Development Communication Report 1993/2 No. 81 : 1-8.

Quebral, Nora C. (1983). Concepts and Practices of Development Communication, *Interaction* 1 (i) : 107-114.

Chapter **54**

Knowledge Gain through Selected Media Combinations according to Bloom's Taxonomy of Educational Objectives

I. S. Mangat, B. S. Hansra

Demonstration and discussion is the best media combination for gain in knowledge of specific facts, "Methodology" and "Principles and generalizations". Film and discussionwas the best in gain in knowledge of 'trends and sequences', ('classification and categories' and 'theories and structures'),

Bloom (1979) started "that as achievement testers and educational research workers, the major phenomena with which we are concerned are the changes produced in individuals as a result of educational experiences. Such changes may be represented by the global statements of the educational objectives". If we could know our educational objectives and design our educational/instructional material accordingly and also know about the effectiveness/suitability of media of instructions for achieving a particular class/sub-class of educational objectives, then most of the educational problems will be solved. Bloom et. al (1979) designed a taxonomy to classify these educational objectives or students behaviour which represent the intended outcomes of the educational processes. They emphasized on obtaining evidence on the extent to which desired and intended behaviours have been learned or the educational objectives achieved. Educational objectives have been classified in a hierarchical order from simple to complex. It contains six major class viz. knowledge, comprehension, application, analysis, synthesis and evaluation. Each of these classes is further subdivided into sub-classes. This study deals, with educational objectives from knowledge class only.

Research has established that different media perform different functions with varying degrees of success in transmitting knowledge.

A number of studies by different authors like Davis (1975), Singh and Singh (1976), Bhardwaj (1981), Manchanda (1981) and Sidhu (1983) have been conducted to ascertain the effectiveness of media viz. demonstration, discussion, film, slide-tape, lecture, printed material, tape-recorded message, radio and telephone etc, But the present study is a unique approach which no other investigation has tried earlier. Its specific objective is to study relative effectiveness of selected media combinations viz. Film, Demonstration, Slide-tape and Lecture all combined with discussion in gain in knowledge of farmers according to Bloom's Taxonomy of Educational objectives.

Bloom et. al (1979) have classified the educational objective related to knowledge into various sub-classes as following :

1. Knowledge of specifics :

It refers to symbols with concrete referents. It is further sub-divided into knowledge of terminology and knowledge of specific facts.

2. Knowledge of ways and means of dealing with specifics :

It refers to knowledge of ways of organizing, studying, judging and criticizing. It includes the methods of enquiry the chronological sequences and the standards of judgment within a field. It includes knowledge of conventions, trends and sequences, criteria and knowledge of methodology at sub-class level.

3. Knowledge of the universals and abstractions in field :

It refers lo knowledge of the major schemes and patterns by which phenomena and ideas are organized. These are the large structures, theories and generalization which dominate a subject field or which are quite generally used in studying phenomena or solving problems. These are at the highest level of abstraction and complexity. It is divided into knowledge of principles and generalizations, knowledge of theories and structures.

Methodology

The study was carried out on 100 farmers from four villages of Ludhiana District with 25 respondents from each village. These respondents were matched with the characteristics of age and education. This being an experimental study a pre and post-test control group design was used. Knowledge about techniques of grain storage was given through different media combinations viz. Film, Slide-tape,

Demonstration and Lecture, all combined with discussion. Contents of message were kept same for all media combinations. A knowledge test comprised of 30 items was developed on the basis of Bloom's Taxonomy of educational objectives. Four groups of respondents were randomly assigned to the treatments. Knowledge test was given to the respondents once before the treatment and then immediately after the treatment. Gain in knowledge was determined by subtracting the pre-treatment knowledge test scores from post-treatment knowledge test scores.

Findings and Discussion

The figures in table revealed that for gain in knowledge of specific facts like dates, events, persons, places, magnitude of a phenomenon and other approximate or relative information, demonstration and discussion was the most effective media combination with mean gain scores of 6.71 followed by film and discussion 5.57, slide-tape and discussion 4.71 and lecture and discussion 4.14.

In gaining knowledge of 'conventions' which refers to characteristic ways of training and presenting ideas and phenomena, usage of styles, practices and forms for consistency, demonstration and discussion and film and discussion were equally effective with mean scores of 12.00 each, this was followed by slide-tape and discussion with 11.00 mean scores and lecture and discussion with mean gain scores of 6.00.

In gaining knowledge of 'trends and sequences' which refers to knowledge of processes, directions and moments of phenomena with respect to time, film and discussion was the most effective with 10.67 mean scores followed by slide-tape and discussion (9.67) while demonstration and discussion had 8.00 mean score and lecture and discussion with 7.33 mean scores.

'Classification and categories' which refers to classes, sets, divisions and arrangements regarded fundamental to a field, film and discussion was the most effective with 11.50 mean scores followed by demonstration and discussion with 11.00 mean score and slide tape and discussion 10.50 and lecture and discussion with mean gain scores of 6.00 in gaining knowledge of classification and categories.

Gain in knowledge of criteria i.e. by which facts, principles, opinions and conduct are tested or judged, was the most effective through film and discussion with 11.00 mean scores followed by demonstration and discussion, 10.25; slide-tape 9.25 and lecture and discussion 7.75 mean scores respectively.

Demonstration and discussion was again the efficient medium in gaining knowledge by the respondents in methodology i.e. method of inquiry, techniques and procedures, with the 11.10 mean scores which were the highest followed by film and discussion with 7.80, slide-tape and discussion with 6.20 and lecture and discussion with 5.30 mean scores respectively.

Under knowledge of 'Universals and abstractions' that refers to knowledge of the major schemes and patterns by which phenomena and ideas are organized in a field, knowledge of principles and generalizations was dealt with most effectively through demonstration and discussion with 6.50 mean gain scores followed by films and discussion, lecture and discussion, and slide-tape and discussion with 6.00, 6.00 and 5.50 mean scores, respectively.

In dealing with the knowledge of 'theories and structures', film and discussion was the most effective of all the selected media combinations with mean gain scores of 8.00 followed by lecture and discussion and slide-tape and discussion with 7.00 mean scores each. Demonstration and discussion was the least effective in imparting knowledge of theories and structures with gain scores of 6.00 only.

Conclusions

It could be concluded that demonstration and discussion is the best media combination for gain in knowledge of specific facts, 'methodology, and principles and generalizations'. Film and discussion was the best in gain in knowledge of trends and sequences, (classification and categories and 'theories and structures').

S. No.	Knowledge levels and sub-levels	Mean gain (scores)				
		Frequencies of items	Demonstration and discussion	Film and discussion	Lecture and discussion	Slide tape and discussion
1.	Knowledge of specifics					
1.1	Knowledge of specific facts	7	6.71	5.57	4.14	4.71
2.	Knowledge of ways and means of dealing with specifics					
2.1	Knowledge of conventions	1	12.00	12.00	6.00	11.00

S. No.	Knowledge levels and sub-levels	Mean gain (scores)				
		Frequencies of items	Demonstration and discussion	Film and discussion	Lecture and discussion	Slide tape and discussion
2.2	Knowledge of trends and sequences	3	8.00	10.67	7.33	9.67
2.3	Knowledge of classification and categories	2	11.00	11.50	6.00	10.50
2.4	Knowledge of criteria	4	10.25	11.00	7.75	9.25
2.5	Knowledge of methodology	10	11.10	7.80	5.30	6.20
3.	Knowledge of universals and abstractions					
3.1	Knowledge of principles and generalizations	2	6.50	6.00	6.00	5.50
3.3	Knowledge of theories and structures	1	6.00	8.00	7.00	7.00

However, demonstration and discussionand film and discussion were at par for gain in knowledge of conventions. 'Slide-tape and discussion is second best for gain in knowledge of conventions and trends and sequences'. While film and discussion and lecture and discussion were second best and at par for gain in knowledge of 'principles and generalizations'. Similarly lecture and discussion and slide-tape and discussion were second best and at par for gain in knowledge of theories and structures. Lecture and discussion was found to be least effective for gain in knowledge of all other levels and sub-levels.

References :

Bhardwaj, N. 1981. Imparting Knowledge through selected modes of communication to the members of Ladies Charcha mandals of Ludhiana District. M.Sc. Thesis, Punjab Agricultural University, Ludhiana.

Bloom, S. D. 1979. Taxonomy of Educational Objectives, Longman Group Ltd., London, U.K.

Davis, M. L. 1975. A comparison of the effectiveness of Audio-Visual instruction and the Traditional Teaching Method of unit on the Radio-Telephone Third Class License with Broadcast Endorsement. *Diss. Abstr. Int.* 36 (4) : 1985 A.

Manchanda, R. K. 1981. The retention of nutritional knowledge gained by the members of Ladies *Charcha mandals* by the use of selected extension methods. M.Sc. Thesis, Punjab Agricultural University, Ludhiana.

Sidhu, J. S. 1983. Relative effectiveness of selected modes of communication in imparting knowledge to the members of *charcha mandals* of Ludhiana District. M.Sc. Thesis, Punjab Agricultural University, Ludhiana.

Singh, K. N. & S. N. Singh, 1976. Effective Communication media for rural audiences (An experimental study) project sponsored by the Dharma Upkrar Ji Chemical Company Ltd., Bombay, pp. 98.

Chapter 55

Communication Media in Non-Formal Education

S. P. Kanaga Anbuselvan

Communication is fundamentally an organic process, evolving in capacity, content, style and purpose, leading to changes and lagging in others. Although much has been achieved at various levels there are still gaps, imbalances and distortions in the development and use of communication everywhere. Illiteracy is still a basic impediment to the growth of communication. Communication technology despite of its rapid development is still costly, complex, difficult to operate, and even more difficult to maintain. The disparities in the world communication sources, the anomalies in needs and capacities, the confronting sophistication of technology and the philosophical contrasts in the use of communication will not be resolved by legislation or the evolution of an applicable consensus.

Media and Technology in Non-Formal Education :

Non-Formal education is "an attempt to bury the notion that persons lacking formal schooling must for ever be sentenced to an existence of poverty, misery and squalor". (Ved Prakash Gandhi, 1995:2)

The formal, non-formal, informal are the idea of "Learning systems". In each type of learning, environment is to be found mutually reinforcing and complementary elements which integrate to form a country's total educational system. Non-formal education is more oriented towards helping an individual solve problems has to absorb particular curriculum content. Instruction is carried out in co-operation with many kinds of teacher e.g. workers, health educators, community development assistants or agricultural supervisors who may use a wide range of communication media, which have included posters, games puppets, flash cards, charts, displays, exhibits, comic books, newspapers, video tapes, slides, films, filmstrips, radio and television. Non-formal education incorporating varied types of communication media for instruction has been on servicing since 1930's in many nations.

Television

With the exception of experimental attempts at using television in the Krishi Darshan Programme in India, television has not been used extensively in India as seen in other countries due to the following reasons.

- High costs
- Relatively small number of individuals owning televisions
- Available receivers primarily in the possession of those who have less need for non-formal education.
- Difficulty in providing relevant and interesting programmes for a rural audience. Some of the developing nations that were using television and some using radio for various educational activities are given below :

1. America Samoa, using ETV for the upgrading or instruction.
2. Niger, with its radio clubs to support literacy teaching and the use of television at the elementary level to diffuse educational opportunities and promote the use of national language.
3. Colombia, using television, to help teachers overcome restrictions of under qualification and limited teaching materials at the primary school level. Using radio for adult basic and literacy education.
4. Nigeria where television was intended to help solve problems of upgrading instruction at the primary & secondary level as well as for teacher's training.
5. Thailand is using radio to provide technical training as a home service to school children, teachers & general public.

Algeria using films, for discussion groups, correspondence courses and radio programmes to provide in-service teacher education.

Australia and Newzealand serving its "out back" sparsely settled central parts and families living in remote across with correspondence lessons supplemented by regular radio broadcasts and the flying Doctor two-way radio services.

Japan using radio broadcast, television to provide secondary education to scores of young workers.

India is using Radio Rural Forums for communicating with rural people, promoting innovations in community action development and disseminating information dealing with agriculture, health, literacy, education, local self-governmental and other aspects of economic and social development.

Kreimer has classified educational media utilization particularly television in over 50 nations. Most of these media have received extensive consideration as to their appropriateness for non- formal education in scores of articles, publications, research reports and documents.

1. The general experiences of countries, which have used Television for non-formal purposes, suggest that at feast in rural areas, television's usefulness have to be restricted to non-formal education purposes, as in the case of Salvador.
2. The cost-effective use of television in non-formal education programmes may be extending of the services of an existing educational system within a country to rural areas where electricity is available.
3. Recently Ivory Coast is following this approach by using its existing television broadcasting facilities to reach adult illiterates in the rural areas with programmes on infant nutrition, soil conservation, folk taboos, hygiene and the problem of rural exodus.
4. Radio is one of the best medium for non-formal education. The use of radio for training purposes, for practical instruction in agriculture and health, for talks on national policies and priorities for correspondence schooling, for community development action, for study groups, for literacy, and in training and development campaigns, such as family planning.
5. In terms of cost-effectiveness, radio can be given a favourable rating, but the evidences suggest that in and of itself radio or television, is insufficient to accomplish learning tasks. Rather it must be built into a teaching-learning system combining study/ discussion groups, printed materials, other media and some type of organizer/leader.
6. Radio is cited as being particularly helpful in accelerating a programme and in reaching "hard-to reach" rural audiences quickly and quiet inexpensively.
7. Among those countries using radio successfully in non-formal

education are Australia, Canada, Colombia, Pakistan, Taiwan and China where farmers plough their fields with transistors slung over their shoulders.

Video Taping : The use of Video taping and filming has been given extensive consideration in Alaska, Canada, India, Peru and Republic of South Vietnam and in rural areas of Appalachia in the United States. But the use of video tapping has not been systematically reached and there are many questions in need of answers. The problem areas in video tapping are :

(a) **Costs** - While the operational costs of video taping equipment are not too high, the initial investment can be expensive. In many cases the cost prohibits easy access by villages to this technology.

(b) **Compatibility** - The compatibility of the video playback equipment with the tape recorder must be insured, so that instant replays are possible. The lack of standardization of equipment is a real drawback.

In Tanzania, video tapping "year 16" project was an experiment in the use of communication for social change and for Tanzania National Development video taping and filming gave to illiterate communities the means to express themselves and to establish a communication between the villagers and their outside leaders. Video tapping stresses a process rather than a product and its real strength for development purposes stems from the fact that its use is oriented towards community problem-solving activities. The potential video taping "is that it is not a broadcaster but a narrow-caster - zeroing is on particular problems, communities and constituencies".

The skyriver project directed by Kim Kennedy is a good example of how the video taping works within community development control.

Audio Tape Technology - With the advent of audio cassettes - introduced in mid 1960's, the message has been given more control over the receiving and listening context than traditionally media and technology. Unlike the broadcast media, playing time can be arranged to local conditions, and in the instances where it is used with a discussion group, the convene of group as well as participants can listen its ahead of time and be better prepared for a group discussion session.

Time of exposure to the message, as well as frequency of exposure, is controlled by the person receiving the message so that questions of when, where and how are solved easily.

Audio cassettes have been used in the training of Para professionals, particularly midwives in Pakistan's family planning where manpower to train personnel was scarce. In Taiwan, New York's Yales Country, one of the poorest rural areas in United States with a population about 2000 where people still live in homes with dirt floors, no electrical power, no plumbing. But they were informed about medical help through audio cassette technology. In South Vietnam inexpensive battery operated audio cassettes were used to teach adult basic education in rural hamlets outside of Saigon.

Multi-Media Projects indicate that "the choice of medium is less important than the content and the way it is used" and that the more glamorous media at great expenses may be less effective than the careful integration of existing resources.

Some of the influencing factors have been given below as a point of reference and practical guide for those individuals considering the use of communication media and technology for non-formal education.

1. Non-formal education programmes are using media and technology but they are not always essential. The smaller projects, the less need is for media; the bigger, the more costly and complex project, the more difficult is to localize its efforts and therefore media and technology can play crucial role.
2. For successful utilization of media and technology in non-formal education the media must come to the learner, rather than the learner to the media. Put the media where the people are market places, on the job, in the field, buses and trains etc.
3. The use of communication media and technology such as radio, simulation and gaming, audio cassettes and video taping can overcome learning obstacles associated with illiteracy. The efforts in Vietnam, Brazil and Guatemala Sri Lanka are encouraging in this respect.
4. If there is one medium better than suited for non-formal education than others, it is radio because of its wide coverage, relatively low unit cost, and its ability to reach beyond literate as well as beyond the electrical power. It needs to be combined in a teaching-learning situation with some personal contact, some opportunity for response interaction and usually with visual stimuli such as printed texts, posters, newspapers and film scripts. This combination is particularly important in non-

formal education because the media are often used by other than a highly skilled trained teacher and far from an organized institution. Thus a combination of media and interpersonal communication is more effective than any one medium alone.

5. The use of communication media and technology in non-formal education should not be limited to dissemination of learning-teaching information, should also be used for tasks such as motivation, attitude change, reinforcement, community participation and sheer entertainment.
6. There is no one perfect medium for non-formal education. Each one on the variety of media has its own limitations. Their effective use requires that they meet the existing fell needs in a society, with particular attention being given to the local social and political contexts into which the media and technology must fit.
7. If a decision is reached to use media for non-formal education, the first focus should be on the possible use of indigenous folk media and low cost technologies that already exist within a country and not on the introduction of new media that are beyond the economic or technical grasp of poor nations. It is more realistic to concentrate on communication media already available within the country, introducing new media only when they are necessary for the attainment of particular educational goals.
8. Different non-formal programmes call for quite different types and combinations of media chosen because of differences in the audiences to be served, learning up objectives and most important, the conditions under which they operate.
9. Non-formal education programmes whether or not they use media and technology must be closely tied to "grass roots" level participation, and not be conceived and implemented only by the highest levels of the educational system or outside foreign agencies.
10. Group listening to broadcast media followed by group discussion is more influential in changing attitudes, beliefs and behavioral intentions towards adopting an innovation than is group listening without discussion. Thus the most potent element in the radio programme is group discussion and public commitment to action after the discussion. The same holds true for the low cost technology such as audio cassettes or gaming

techniques.

11. Finally, through country leaders and donor agencies must frequently be involved in the development of learning opportunities, non-formal educational programmes intended to be local in character and purpose, must involve the people they are intended to serve from the very beginning of the project. This local involvement must begin at the discussion planning stage and continue through the stages of implementation, evaluation and eventual modifications.

References

Ved Prakash Gandhi, 1995, Media and Communications Today Policy, Training and Development, *Vol.2, Kanishka Publishers, New Delhi.*

N. M. Ansari, 1992. Economics of Distance Higher Education, *Concept Publishing Company, New Delhi.*

Klapper, J. 1960 The Effects of Mass Communications, *Free Press New York.*

Mc. Quail, Denis 1994, Mass Communication Theory - An Introduction, *Sage Publications, New Delhi.*

Chapter 56

Role of Communication Technologies in Rural Development

Meenambigai, J, G. Tamilselvi

Communication is now one of the central issues in developing societies. Revolutionary changes in communication policies, techniques and strategies are needed in order to make it a fundamental right practiced by the poor rather than making communication a privilege enjoyed by the rich and powerful. Communication science in general and agricultural communication in particular has emerged as a profession. Here, one wants to transfer knowledge and skill about improved agricultural technology with a view to inducing change in the existing farm practices of farmers, for increasing production in rural areas. In this process both Information and Communication Technologies (ICT), play a major role. Communication Technology (CT), is galloping it is therefore necessary to be sensitised about it, ability to bend or twist it for social purposes especially to the advantage of the majority of people inhibiting the less developed countries ought to be more thrilling some important developments include microform files, high resolution large screen television, instant colour movies, computer controlled photo composition in printing and mobile radio communications.

Multimedia/Interactive Systems :

Multimedia have influenced the quality of extension work. It encompasses all the traditional forms of communication like video, stills, graphics, audio, animation and text is an excellent tool for extension communication. The computer integrates all these media into a single platform and provides interactivity to the system. It encourages feedback from extension workers to research and help to develop technologies of relevance to farmers.

Interactive computer video technology (ICVT) :

One of the most popular multimedia technology that has been used in the ICVT. It links the computer to audio-video replay in such a way as to provide the trainee with individualised truly interactive instruction. ICVT can contribute to problems of relevant farm information and improve the quality of extension services. The only constraints are inaccessibility to computers and illiteracy of farmers. Their most beneficial and practical use appears to be in training of agricultural extension staff to upgrade the quality of their services and to improve the relevance of technical information to farmers.

Interactive video disc :

IVD is one of the popular multimedia technology. There is a video disc player which access the video images stored on a two-channel audio disc.

This is specially valuable in the Indian Context, because it becomes a bilingual package, other media of information are text, graphics, animation and digitised audio which supplements the audio on the video disc. These are programs which are stored on the magnetic medium and the user interacts with the system through input devices like the mouse and the key board. The cost of the multimedia is probably the biggest issue that will be raised. However, the cost drops proportionately as the number of user get increased.

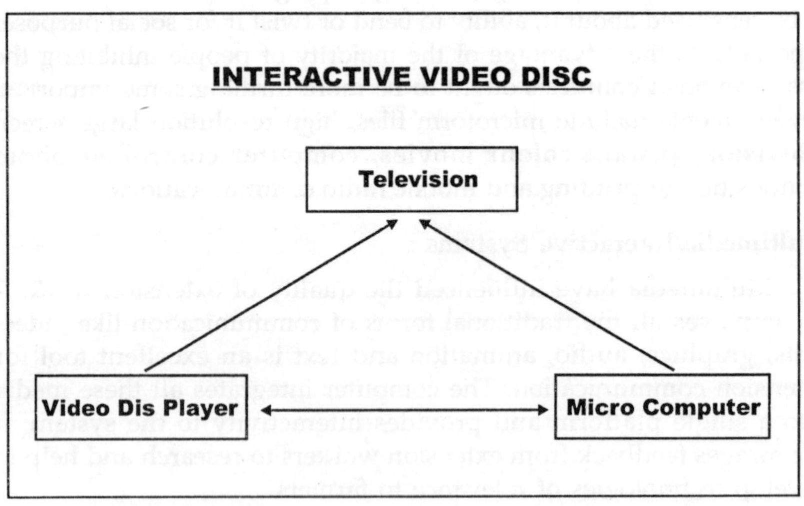

Video text :

Video text refers to an information system with which text and graphic information is transmitted and then received either by a videotext television or on an ordinary television set with an adopter unit. This concept has been developed to provide a method of low cost information delivery in offices and homes. A video coupled with a micro computer permits editing as well as retrieval of information on a videotext page. Videotext system is nearly first a computerised information storage and retrieval system. Videotext links TV set through a telephone line to mainframe computer which may have lot of information. It is a fully interactive system in that you can communicate with the computer.

Interactive videotext :

The videotext which is transmitted via telephone, cables, data line or private network is called interactive videotext. It is fully an interactive system and can even print a hardcopy of any page that appears. Its two way nature helps users access to thousands of pages stored in a number of computers that may be geographically dispersed, but are in communication with each other. Using the normal telephone network, the user dials the required number or the videotext device automatically dials the local videotext computer. The television set displays the videotext page and asks the user to use the available index or to key in the required page number. Its attributes makes it a potential teaching medium. It can be effectively used in training and in distance education.

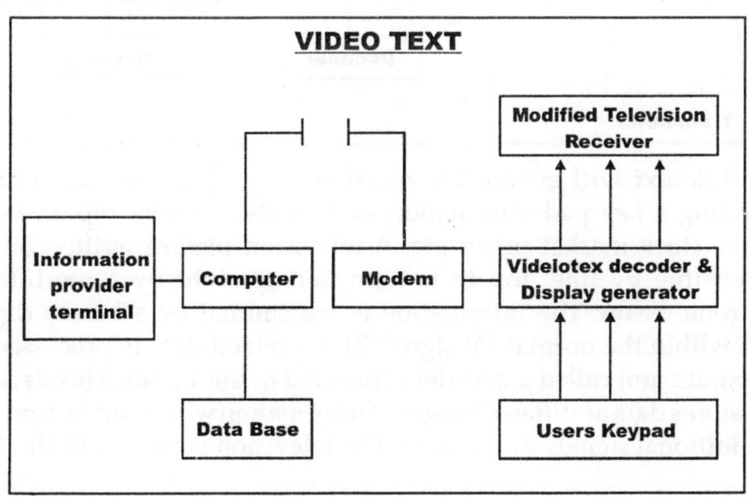

Teletext :

Teletext is a generic word for text-based information systems available on a public or network basis. Any one who installs a video presentation system may require one of the teletext variants. The best known is broadcast teletext. This is a relatively limited information library containing up-to-the minute public information including news, weather reports, current events, travel information and stock market prices. The information reaches the viewer via a hidden part of the normal broadcast television program and is even used to produce subtitles for the program. Teletext can send "digital pages" of information using scanned lines of the television picture not visible without an encoder. The information potential for teletext is almost limitless. From grocery lists to newspaper pages, the system promises still new frontiers in communicsation as television becomes more and more a part of our information society.

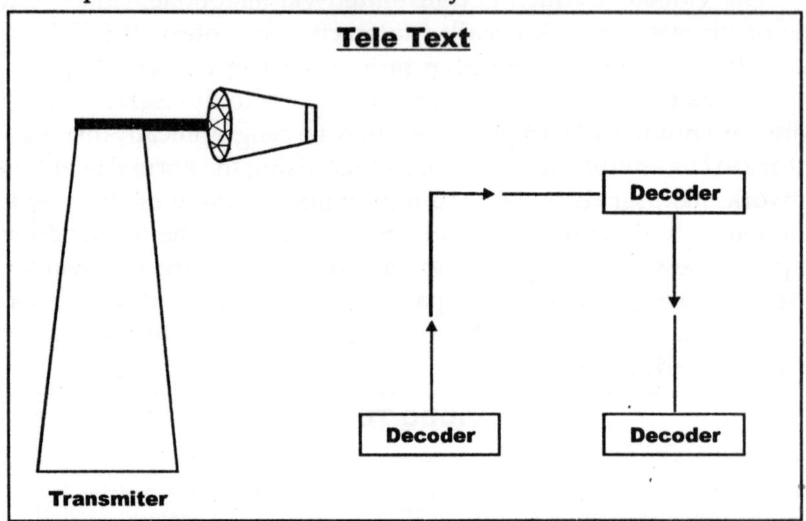

Teletext will enable TV receivers to display on demand by operating a key pad information such as the weather report, news, sports, stock market prices etc. A television picture written on the screen line by line, top to bottom, left to right by a modulated electronic beam. The information is transmitted by a binary digital code within the normal TV signal. The received data is processed by a separate unit called a decoder connected to the TV set. This decoder unit stores data as different pages of information which can be updated as additional signals as received. The television viewer with the help of the add on adopter keypad can select any one of these pages for display. The viewer can if he wishes hold the page indefinitely or for as much time as required to read the page at his own pace and may then switch over the next un-required page by pressing button. Existing TV sets can be converted to receive teletext either by internal modification or by the addition of an add on unit. Future TV sets can be made with built in record or with proper arrangements to plug in a decoder at a later date.

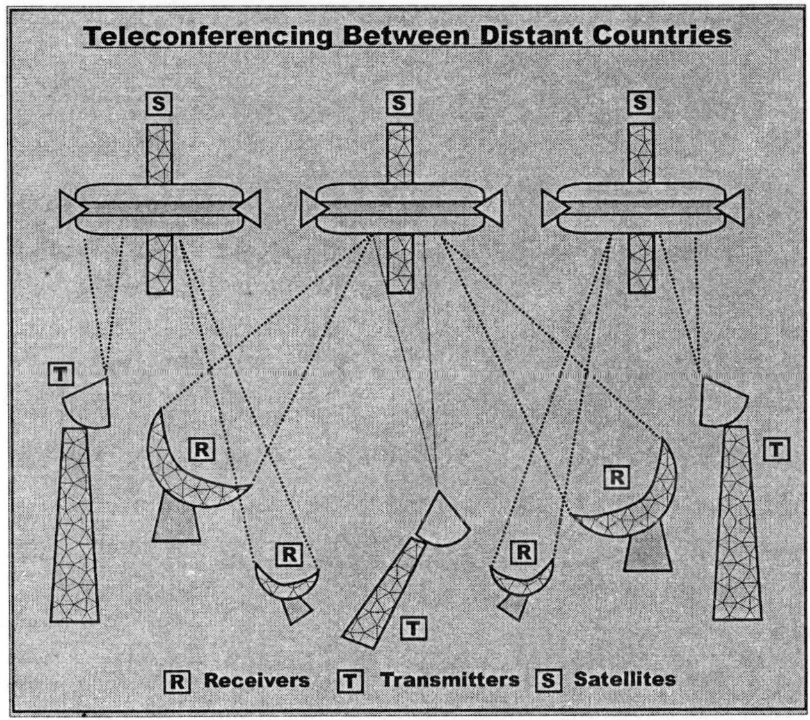

Applicability of teletext :

·Since is provides added display of information it can be used in transfer of technology. The farmers can get additional information about agricultural programme.

- Help the farmers to store the information memory and he can refer when it is needed. The teletext medium that catch the interest of news viewers provide a simpler less expensive enterprise than videotext and with high potentialities of success.
- Teletext makes use of broadcast facilities availing big

investments on computers and complex telecommunication facilities.

- Teletext brings information and news into the home through TV screen.

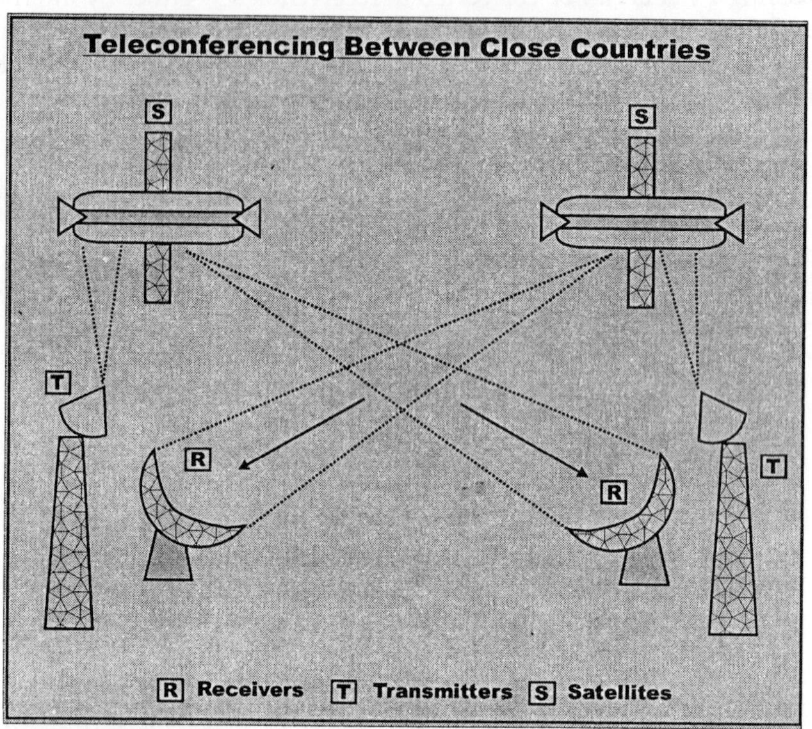

Teleconferencing :

"Teleconferencing is so rational, it will never succeed" says futurist Naisbitt in his best-selling book megatrends. In a state with 114 countries covering over 69,000 square miles, its often too costly in terms of time and money for extension personnel to travel for face to face meetings. On the other hand audio-teleconferencing permits rapid communication with less travel. The new integrated digital communication system is being used. This technique will add a new dimensions of control, timing and flexibility to communication. Instead of traveling miles for a meeting, conference rooms will be wired and cameras will be used to carry images and visuals. The time and cost for traditional day to day communication will thus be reduced.

The teleconferencing can effectively be carried out with the help of geo-centric satellites to cover the entire globe for communication purpose.

Teleconferencing between close countriesis comparatively easier than that of distant countries. For teleconferencing planning, timing and proper arrangements play important role. For close countries we need two geo-centric satellites or channels for teleconferencing. No need to have an intermediate agency of it is between Pakistan and India means, the satellite of India receive the signal from Indian TV transmitter and sent to the receiver of Pakistan TV Kendra.

For distant countries we need three geo-centric satellites. Two from teleconferencing countries and one from intermediate agency required. Intermediate agency is the link. It receives signal from both the countries and exchange this signal with the help of its transmitter and satellite. Then die teleconferencing paves way for international integration.

Applicability of teleconferencing :

· Even though it is a costly and international oriented effort it will be useful to strengthen the international communication network.

· This may also provide exchange of latest agricultural technologies available in different countries for the benefit of the farmers.

Some of the potential advantages of teleconferencing include savings in travel expenses and travel time. Also conferences can be held whenever necessary, since there is no need to make travel plans necessary, since there is no need to make travel plans long in advance. Because meetings can be held more frequently communication is improved between for example, headquarters and geographically scattered divisions.

There are also certain drawbacks to teleconferencing, because of the ease of arranging meetings in this manner, they may be held more often than necessary. Moreover, since this approach uses rather new technology, the equipment is subject to breakdowns. Most important perhaps, teleconferencing is still a poor substitute for meeting with other persons face to face. Despite these limitations in increased use of teleconferencing is likely in the future.

Video conferencing :

A telephone or radio network may be used to provide audio communication among groups at two or more locations. This is called audio teleconferencing. A video channel added to an audio link between groups by means of satellite transmission, microwave transmission or a two way cable television system is video teleconferencing. Video teleconferencing is also known as video conferencing. Experts sitting in the studios listen to the questions and answer live on television. The system reduces the need for travel and is of much help to the groups in remote areas.

Sometimes, two or more groups of people in separate locations use audio and video links to conduct a conference. The process requires specially equipped rooms. The participants must be arranged so that they can be seen by the cameras and a special sound system must be installed to ensure clear transmission of sound without acoustic feedback howl round. Arrangements must be made so that visual and material can be directly fed I into the system so it can be viewed by all the participants. Not only is the installation expensive, but the cost of renting the broad-band links between the locations is very high. Slow scan video systems can send pictures over ordinary telephone lines. Although superficially attractive for low cost video conferencing in practice they completely miss the point because conferences can only succeed if they work in real time. The enormous increase in international digital links in bringing the price per hour of link time down to more affordable level and within a few years the cost should no longer be a limitation.

Users of video conferencing need more money to make a success of it. They must have a commitment to making the technique work for all the participants. This can really only happen when the participants are regular users of the system. Therefore its main application is and will remain the support of regular management meetings.

Another use of video conferencing is distance teaching where one lecturer is able to speak to two or more audiences at once. The extra-dimension that the two-way link brings is that the remote audience can have a dialogue with the speaker and he can see his questioner. This method of teaching is likely to be economic and effective in high level training, where there are only a few expert teachers available, and in adult learning.

Satellite communication :

Long before positioning India's own communication satellite in the geostationary orbit, an intensive experiment was conducted to assess the technological as well as the sociological aspects of satellite TV broadcast. This was the satellite Instructional Television Experiment (SITE) which ran from August 1975 for a year. The object of this complex mass communication experiment was to demonstrate the use of direct broadcast of TV. programmes for rural development in the fields of agriculture, health, family, welfare, education and national integration.

For the eradication of illiteracy in the rural areas ISRO is working on a dramatically new concept of utilizing the most powerful audio-visual medium through the "GRAMSAT" satellites which will disseminate culture and region specific knowledge on health, hygiene, environment, family planning and better agricultural practices.

Application of information and communication technology for rural sector :

The communication revolution can be effective for reforming a numbers of sectors such as agriculture, animal husbandry, health, education, water conservation etc., to name few of them. It can certainly go a long way in obtaining a number of developmental aspects in women empowerment, proper utilization of funds by the government and implementation of various schemes, maintaining swift communication between the research institutes and village residents, for generation and conservation of energy resources, disaster management, revenue collection, land record maintenance and encouraging entrepreneurship development among the village artisans.

The communication revolutioncan be very effective in the health sector as well. Information about the various healthcare services can be effectively disseminated. Specialized campaigns like malaria, tuberculosis, polio, guinea worm, AIDS eradication etc, can be implemented very effectively using mass media based on communication technology. Effective monitoring of the major work carried out by the primary health centres can be done. The, family planning and welfare programmes can also be effectively implemented. Inaccessible and remote areas can be connected to the hospitals at urban areas and information can be provided to the people. Complications in health of individual patients can be discussed by doctors with specialist doctors from urban areas by use of video conferencing techniques. This can leads to effective diagnosis and medication of the patients in rural areas. The concept of telemedicine can also be possible by use of ICT in the health sector. The ICT will help to connect the rural part of India with the outside world and provide information to the students and villagers access to current information. The students from rural areas will get exposure regarding the happenings of the world through this networking which will help them to increase their knowledge and get information of various renowned education institutes for higher education. It can also play a crucial role in improving and changing the education sector in rural areas. Information technology based computer as well as other education can be provided to the students from rural areas.

Unemployment proves as major bane among the Indian youth, more so with that of the rural youth. Such youth trained properly in computer and other CT related aspects can be inculcated in the mainstream and employed for performing specialized jobs. This will be considerably helpful in using the talented human resources in national development. A Delhi based non-profit organisation, Vaancha ICT Association, a project that provides training and technical assistance in information and communication technology (ICT) to the disadvantaged youth. Similar projects can be implemented in the rural and semi-rural areas of the country which can be very effective in improving the employment potential in rural India.

The ICT sector can be a boon to the rural people by providing them. Informationa bout the availability of finance through various agencies the purpose of availability, the changed rates of interest and repayment periods. Proper networking of PACBs, DCCB and other co-operatives and nationalized Banks with the rural areas can go a long way in this direction. Proper monitoring of the funds disbursed by the finance sources and prevention of misutilization or diversion of funds for unproductive uses can be possible by using advanced communication technology means.

Information regarding the various schemes implemented by the government through the Panchayat Raj Institution like the Zilla parishad, Panchayat Samiti and Gram Panchayat can be disseminated to the villagers. This can be possible by connecting every village with the district headquarters through the Internet. Prompt and proper information about the schemes implemented by the government can be very useful particularly to the deprived sections of the society. Many projects and programmes initiated by the administrations and the local people have been put up in various parts of the country utilizing internet, local area networks etc., latest being the Lokmitra project launched in May 2001 in the Hamirpur district of Himachal Pradesh. The main object of the programme is to provide easy access to government information at the remotest corners of the state, redressal of complaints to the respective departments and getting their grievances addressed within ten days. This may help in transparency in working of government schemes, saving in time and responsible and responsive administration. The district administration in villages of Himachal Pradesh encourages people with computer system, modem and a telephone line for obtaining user ID from the district offices to start a citizen information centre in villages. The connectivity of the talathi offices (revenue office of the village) with the collectorate as also with the commissionrate can be very useful to the public in getting the land documents at any place within the state. This will result in increased transparency, considerable saving of time for the people as also lend to reduction in corruption, nepotism and other malpractice's indulged in by the revenue officials. It can also be helpful for women empowerment. Women being a vital part of Indian economy constitutes about one third of national labour force. The entire family is dependent on women. More than 90 per cent rural women in India are unskilled, restricting them to meagerly paid occupations. Proper training to the women and imparting skills through the use of CT will be considerably helpful. It can lead to increasing leadership abilities among the women thereby increasing their participation in local government and decision making process. Self-employment opportunities provided to the women through the use of CT can go a long way in reducing their dependence on the male members, making them self-reliant and thereby leading in their improvement.

In the present era of technology development can provide information to the village artisans and encourage entrepreneurship among them will lead to considerable self-employment. The information provided to artisans will be based on the needs of the modern society and changes to be made in the old pattern of work. It will also help in making provision of the market for their products. This will certainly lead to improvement in their standard of living and thereby in the upliftment of this lower stratum of the society.

ICT will also provide information about ways and means to save and create energy resources by use of biogas and gobar gas. This can lead to reduction in the energy crisis particularly in the rural areas and reduce wastage of valuable cow dung manure which is used as fuel. Information to the villagers about the use of non-conventional and renewable energy sources like solar and wind power for harnessing energy which certainly go a long way to tide over the impending and persistent energy crisis particularly in rural India.

ICT can also be useful to the farmers in the field of animal husbandry and dairy science. It can also be helpful for proper marketing of dairy products as also for providing information to the farmers regarding the same. The projects like the warana wired village in the Kolhapur District of south western Maharashtra has been trend setter in this direction. The 50 project connects the farmers in the area comprising of 70 villages with internet facilities available at the Primary Milk Co-operative societies. Detailed information regarding the animal husbandry and dairy practices, as also marketing of milk and milk products is provided to the farmers.

The information village project implemented by M.S. Swaminathan Research Foundation in collaboration with International Development Research Centre (IDRC) is aimed at bringing the benefits of modern Information and Communication Technologies (ICTs) to the rural population. The project objective is to access the impact of ICTs in fostering transition to sustainable agriculture and rural development and document their role in promoting the process of knowledge empowerment of rural families.

Information KIOSK is one such most advanced technology development which provides information about the dairy aspects to the farm women through touch screen technique. It found to be easy for the dairy farm women access to lot of information about dairy by simply operate the machine through touch the screen by their finger tips.

Information and communication technologies (ICT) have brought issues of communication and rural access to information to the forefront of the development agenda. Widespread access to ICTs in rural areas is still several years in the future, but wireless technologies may eventually end the reliance of ICTs on costly telephone infrastructure and helping to overcome rural isolation and enabling dispersed families stay in touch. At grass root level where information is scarce and communication difficulties and expensive.

ICTs have a lot to offer, but the high initial costs not only in hardware, but T in the cost of training and applying the technologies in day-to-day work can be an insurmountable obstacle.

The ICT revolution can thus be of considerable use for the underdeveloped countries like India and more so for the rural areas considerable lacking in development. The ushering in of this new technology can totally change the face of rural areas and possibly bring all round prosperity as envisioned by the forefathers of this great nation.

References :

Bjorklund, J. and I. Fredmayer. 1985. Keeping current via Teleconferencing. Journal of Extension *p: 21-24 Dhane, A.V., S.B. Gurav and J.B. Dhane. 2004. Applications of Information Technology in Rural India.*

AGROBIOS Newsletter- *Vol.2 No. 9 p. 47-49.*

Dugger, C. W. 2000. Connecting Rural India to the world. New York Times. *May 28 Harriman C. Lynda. 1989. Distance Education. Making video conferencing work.* Journal of Extension. *XXVII : 28-29.*

Harris, R. et. al. 1999. Success Stories of Rural ICTs in a Developing Country. PANTLEG-IDRC *Meera, N. Shaik, Anita Jhamtani and D.U.M. Rao. 2004. Information and communication Technology in Agricultural development : A comparative analysis of three projects from India, AgREN.* Network paper *1 35, London.*

Melkote, R.Srinivas. 1991. Commiunication for development in the third world - Theory & Practice, Sage Publication. *New Delhi.*

Morrow Katherine. 2002. The ICT agenda: global action plans and local solutions. LEISA INDIA *Vol. 4 No. 2 p. 9-10 MSSRF. 2002.*

Information Village Research project in Union Territory of Pondicherry. Second phase-first year report submitted to IDRC, *Canada.*

Sadanandah Nair, K and white A. Shirley. 1993. Perspectives in Development Communication. Sage Publications. *New Delhi p. 207-219.*

Venkataramani, G. 2004. Holistic extension service. The Hindu Survey of Indian Agriculture, *pi 67-169.*

Vilanilam, J.V. 1993. Science Communication and Development. Sage Publications. *New Delhi p. 102-104.*

Chapter **57**

Use of Video in Rural Development

V. K. Dubey and S. K. Bhanj

Communication research indicates that development takes place in a local context with a decentralized set-up in which people have scope for meaningful participation in controlling their own lives and environment. Unless the content of media originates at the community level and relates to the viewers' world of values and environment, it is bound to be irrelevant and ineffective. In this context, there is increasing consciousness in the present day world of the need to demystify and democratize media use at the small community level so that people can present their problems and explore possible solutions in the context of the realities of their situation. Besides increasing message credibility and acceptance, this effort can work equally well to motivate action. Communities thus can derive the strength and confidence to solve their own problems.

Community communication is based on the credo, "of, by, and for the community." The natural choice falls on video as the most suitable medium for generating interaction. Video has the ability to present the community in both moving images and sound for immediate reproduction. Unlike movies, it does not require processing. Video equipment does not require sophisticated training to operate, can be easily handled by illiterate rural folk, and is portable enough to be carried by bullock cart, by camel or horse, or by hand. The colour camera, portapack, and monitor all operate on batteries and are not dependent on electric power networks. India is making colour monitors and cassettes. Hence, only the camera and recorder need to be imported.

Video in Developing Countries

Experiments with small-gauge videotape have recently interested non-formal education instructors in the developing countries. Videotape is seen as an ideal medium for promoting audiovisual literacy for motivation, attitudinal change, behaviour reinforcement, community participation, and entertainment.

The approach used in the Tanzanian Year 16 project (1971-72) was based on the premise that communication is a necessary prerequisite for social change, that rural populations involved in the process have valuable information and knowledge to transmit, and that video was the right medium to choose for communicating this.

Video has been used in rural development both as a process and as a product. Its utility as a process was illustrated by the Fogo Island experiment in which the Memorial University of Newfoundland and the National Film Board of Canada participated; film and video screening brought problems of a political and economic nature into the open to be discussed, analyzed, and acted upon.

In Chile, video was used for training thousands of members of farmers' cooperatives. In Thailand, video was used in a National Family Planning Program, where it proved its utility in training rural field health personnel of all types. The uses varied with the context and place of use. Some of the fields and countries associated with video are indicated below:

- Agricultural development : Bangladesh, Gambia, Sri Lanka, Tunisia, Upper Volta, Peru and Venezuela.
- Child Care : Egypt, Gambia, Iran, Jamaica, Philippines, and Thailand.
- Literacy program : Egypt.
- Urban development : Pakistan.
- Generating community consensus and problem-solving with community effort and self-help growth : Alaskan region of U.S. and Newfoundland in Canada.

Video in India

The Indian experience with video as an interactive medium for community development is very limited. There are only a few examples of its use in a community context. The effort made by CENDIT (Centre for the Development of Instructional Technology) in its Saharanpur media experiment brought out vividly the possibility of using video for humanizing the process of communication.

A second experience was that of ISRO (Indian Space Research Organization, Ahmedabad), where video was used to produce programs for the local television network on such problems as bonded labourers, untouchability and other such evils, and local development. But the utility of the video medium has been demonstrated in other settings too. Recently, the National Dairy Research Institute, in collaboration with the Memorial University of Newfoundland, Canada, has been engaged in a rural development action research project in the village of Taparana, Karnal, using video extensively in various contexts.

The Taparana Experiment

Taparana is an insignificant entity in the midst of prosperous Haryana Villages. Essentially a village of low socio-economic status groups, Taparana consists of 209 families from seven different castes belonging to low and backward communities. Sixty four percent of the households are landless agricultural labourers. Only 13.4% of the families those who own more than two acres of land are able to live comfortably on agriculture. The average household income is about 250 rupees per month (for a family of three adults and four children). There are only three married women who can read and write and, in all, there are about 45 families with one or more members who can decipher printed letters.

The core component of this action research project (which has already lasted two years) is the faith that the ultimate users of an idea must have control over the what, why, where, and how components of the message, so that it reflects their own reactions and opinion, and are responsible for what they say.

This project is a departure from the conventional aid-oriented government programs made for the people. It aims at activating an individual's dormant potential to better exploit the available possibilities existing around him at a given point in time. The philosophical assumption is that human potential is unlimited and that ignorance of this potential is the basic cause of human suffering. Thus, the developmental efforts must aim at cultivating in the community the creative power to realize and estimate its own latent strength and resources from the perspective of available institutional provisions and at initiating self-action for growth. This means leading people gently to a gradual realization and assertion of their own hidden abilities and the progressive utilization of these abilities to achieve their desired goals through carefully regulated exercises.

Development from this perspective means fusion of individual and community needs. An order created whereby emerging individual growth is reinforced through newly created social networks. People join together to produce those resources that are missing but necessary for the well-being of the community. The inaccessible necessary resources are made available through mutual patterns of interaction. Achievement of this gear requires three things : (1) recognizing the importance of the individual; (2) building the community network; (3) creating a decentralized media use pattern. This amounts to the creation of a new culture. The project emphasizes creating a new order of self-governance. In general, it means cultivating personal and social responsibility, sensitivity to the environment, confidence, mutual trust, knowledge of one's own capacities and possibilities, and acquiring the skills to manage community affairs. This is achieved by democratization and media control by the people of the community, who are the ultimate beneficiaries of any development program. Thus, interactive media originated by the community should ultimately result in increased initiative, freedom of action, shouldering of responsibility, and acquiring of power to overcome the sense of helplessness. This socio-economic transformation of the individual and his community is the ultimate goal of the effort. Realizing these values in society is a time-consuming and hard-to-measure process. Hence, the usual yardsticks of roads built, hours of voluntary labor contributed, etc. need to be dropped.

The achievement of the above goals demands a new approach and a new outlook. At the same time, it demands patience and faith in the supremacy of human self-affirmation. In discussing the methods being followed in the project, we do not claim that these methods must be used in every situation. Variations are always possible and welcome.

Initiating Community Actualization

This is the process whereby the community identifies its own problems and their causes and explores possible alternatives for their solution. In this process, the individual or community may need help. The role of the media at this stage is to link the isolated individuals, groups, and communities and get them started on the process of positive interaction. To initiate this process, the community worker needs to be acquainted with surrounding realities and to have intimate knowledge of community culture. He should contact people individually and in functional informal group settings and engage in a series of open-ended discussions relating to general problems or topics of interest. Gradually, 'people can be led to speak out about their concerns. At this stage, the community worker needs to isolate those issues that are easy to solve and are also on the priority list of the individual, group, or community. Such matters then could be videotaped and presented in the community for discussion. Through this process, a few individuals in the community could be identified as "community actualized. Those are the persons who have a conscious concern for their community and the issues it is facing. There is seldom a community where no such persons are available. They posses a treasury of knowledge for rationally reasoning toward a solution. Such individuals generally have social acceptance, though they may not be from the elite group or ruling segment in the village. Before the taping is done, thorough discussion with the actualizes is a must. Afterwards a final recording can be made. This recording should be presented to the concerned group or community for generating or arriving at a consensus. Interactive community media help to spot the effective leaders.

Democratized media participation plays a major role in the community actualization process. This amounts to completely reversing the role of media from being top-down to being bottom-up. In other words, many of the messages originate at the community level and people have full control of what is presented, how, why, and to whom. This breaks the vicious cycle of dependency, and real power of decision-making and self-governance is achieved by the community.

The Community Worker as social Animator

The role of the community worker in traditional development programs has been that of an advocate and executor of programs for people. To the community, he is the state personified and a custodian of the people's interest. Such persons often find it difficult to win the confidence and trust of the community, and their motives are always suspected. Even the most concerned development workers find it difficult to identify with the problems faced by the community. Most of the time, their feeling of concern is one of pity and sympathy for people they consider to be inferior, and is marked by a condescending or paternalistic attitude. People sense this condescension and respond by not being responsive, since this hurts their self-respect.

A community worker is expected to be a catalyst to facilitate the development process. His conscious involvement with the villagers own desires, problems, and suffering identifies him with the community. He no longer remains above, but moves ahead to be of and for the community. Due to his intimate involvement, he becomes more knowledgeable about the community he serves. He learns how to engage people in constructive interaction and transfers the real power of decision-making to them 50 that his continued presence does not become a necessary precondition for progress.

In the Taparana community development project, the project workers cultivate a sense of empathy. Since the basic concern of this development project is improving the quality of human life, there is no effort to achieve a predetermined target. The project workers take their time to become part in and part out of the community. These workers try to identify issues and help in finding solutions by initiating interaction within the community, The community is thus assisted to become involved in problem-solving and to assume responsibility for, and control of, their own action for development. Once the action is begun, the proect workers withdraw from it and involve themselves in looking for some other issue.

Introducing Media

In the Taparana project, videois used as the community medium due to the advantages discussed earlier. The following steps for the gradual introduction of media into a community are suggested.

Step 1: Media demystification. The video units (half inch port pack, camera, and monitor are carried into the community. Scenes of common community interest, such as their fields, their school playground with children playing, village streets, etc., are recorded; in addition, the work of village artists may be recorded, with their consent. This tape is then played to community members. Care must be taken at this crucial stage not to record any sensitive subjects within the community such as village girls); community workers must be aware of local cultural values. It is essential that the social animator build a good rapport with the community prior to introducing the new media. At this stage, the community members come close to observe the operation of the video equipment and gradually come to understand media capabilities like instant reproduction, playback, erasing, recording, etc.

Step 2 : Recording and playback of an informal community discussion. As a second stage in the process, informal group discussions are recorded and played back instantly. At this stage, the social animator may also participate, engaging the group in constructive and positive discussion. Gradually, the community members core to identify some of their felt and unfelt needs. This could be described as problem identification and sharing by the community. Potential community actualizers are identified during this process.

Step 3 : Selection of community actualizers and small-group discussion to identify priority issues. Democratically-selected community actualizers are subsequently engaged in discussion of the broadly identified issues in a general informal group discussion. After viewing a recording of that informal discussion, the social animator must assist the small group to pinpoint some immediate but easy-to-solve issues through general consensus. An action plan is then prepared by the community actualizers. One or more individuals are then selected to help the social animator to video-record a particular issue.

Step 4 : Video recording of a community issue by one community actualizer, and discussion. The issue is discussed with a representative community actualizer an outline of the issue to be recorded is prepared with his help. Then the issue is video-recorded under his guidance. Time and again it is replayed to this community actua-lizer for modification. After his approval, the tape is played to the other previously selected community actualizers for their approval. Needed modifications are carried out as desired by the group. Thus the tape becomes ready for general community viewing to discuss the issue.

Step 5 : General community viewing and issue discussion; reediting if desired in cases where the issue must be presented outside the community. At this stage, the videotape is played to different groups in the community in their free time and at a convenient place. Since the community at this stage gets engaged in serious discussion of problem-solving, care must be taken to avoid any outside distractions or disturbances. Children should be kept away at this stage. In this process, the people in the community have begun mutual interaction, and consensus on an action plan commonly emerges.

If the issue involves an official agency or other outsider, the tape, after general community approval, is sent out for presentation to the concerned official. The response of the authority is likewise video-recorded and presented for viewing by the community in small groups. In this way, the community and the authority concerned are joined in mutual problem solving.

There are many other ways to use Video, such as for science education, transfer of technology, community entertainment, etc. Such tapes can be directly recorded by the social animator and may be viewed by the community in small groups.

A Case History : Owning a Rikshaw

Although the project was launched on January 1, 1980, we did not make any headway until April, because we did not know where to start. The community did not trust us; none of us was native to the region, and we were unaware of the cultural orientation of the community. But our repeated visits to the village gradually led tom rapport.

The workers were open minded about village issues. Mostly in informal chats, people used to ask for jobs. Attempts to discuss personal or community problems led to expressions of community helplessness, the blaming of others, and general indifference. Criticism of government village and state was a popular topic.

A ray of hope emerged when a rickshaw puller from the village met with one of the team members in the town. During an informal chat while Mr. Rajpal, the rickshaw puller, was pulling his rickshaw with the community worker on it, he asked for help in solving his problem of buying a rickshaw. He said he had been pulling a rickshaw for the past three years and was paying a rental charge of four rupees per day. He had already paid more than the cost of two rickshaws. (A new rickshaw cost about 1,000 rupees). He wanted to own a rickshaw. The worker came to know that there were a dozen rickshaw pullers in the village. He advised Mr. Rajpal to gather the others for a meeting and discussion. The group gathered one evening at the home of another rickshaw puller. The discussion focused on the modalities of the work and the possible consequences.

A second meeting was held two days later. The details of economics were discussed, and five members of the group were identified as articulators. These were the persons who could present their ideas to others clearly and systematically. The next morning the group was taped, and they presented their problems and possible solutions. The taped message was presented to the group for approval and modification, if necessary. Later the tape was shown to the manager of the Bank of India branch office in a neighbouring town.

The bank manager pleasantly surprised by the rationality of the group's problem analysis and by the alternatives put forth for the solution. The communication of the group convinced him of the rickshaw pullers' ability and willingness to act. The bank manager's response was taped. He expressed his agreement with the suggested terms of repayment for a loan and gave full details of how he could help. The tape was played back to the group on the next evening and discussions were held. Five rickshaw pullers decided to participate in a loan; they selected a group leader, who was literate, to represent them. His travel expenses to the neighbouring town were shared. In this way, the rickshaw pullers were able to acquire ownership of their rickshaws and take a vital step up the economic leader.

A Word of Caution

Video is a powerful communications tool in the hands of the development worker. But no tool, however powerful it may be, can ever replace the human element. The effectiveness of the tool depends on the people using it. Unless the community worker is committed, motivated, and prepared to play the ever-changing role of empathizer rather than that of preacher, the media alone can never deliver the goods. The support of national and local political authorities and a conscious desire to solve people's problems are other prerequisites for the successful use of video.

On the basis of the above discussion, we believe that the community communication approach attempted through the use of videotape-recording and playback offers definite possibilities for building people in relation to their community.

Chapter **58**

Community Video : Claims and Criticism

S. Talawar

Wide variety of communication aids have been used in consonance with different extension methods by extension workers throughout the world in agricultural and rural development. Recently, video is being - recommended for use in the wider areas of non-formal education for organising the individuals and communities, and to get them started on positive interaction to identify their own problems and find the solutions leading towards development. The proponents of community video have combined the chief advantage of user friendly technological video equipment and the concept of bottom-up, participative, more democratic, interactive, concientizing approach of communication that are not present in the broadcast media.

While agreeing that video has proved to be a powerful tool in community animation work in North America and Europe, its appropriateness for the Third World, in its own cost-effective terms and in comparison with alternative communication technologies, needs to be re-examined (Kidd, 1978 : 178). However, before embarking on to examine the concepts involved and claims made for the community video it would be interesting to take a stock of the nature and characteristics of community video.

(I) THE CONCEPT OF COMMUNITY VIDEO

Community video has also been referred as group media. Group media is, according to Klaus Muller, 'Any medium which can foster the process of group interaction through communication based on the life situation of the group members and sharing of personal experience that will lead to common endeavours and actions' (Muller, 1987).

According to Criticos (1989) 'Community media is community in media. Community takes priority over media. Community in media means that the community, i.e. a community of common struggle, has collective and democratic engagement in the contents, production and distribution of community media'.

Communication with a popular audience does not mean adopting simplistic forms. Besides being accustomed to watching sophisticated productions shown by big networks, this public demands imaginative efforts. The format has to be organised in such, a way that it corresponds with the language, the experiences and the real problems faced in their everyday lives. Such 'alternative video' attempts to provoke a critical attitude towards what is presented, in a process of information/reception in which feedback plays a fundamental part Group discussion produces new knowledge' resulting from collective participation. The videos presented are a codification of reality which is de-codified, in Paulo Freire's terms, by the people who participate in the process (Ceccon, 1989 : 32).

Peter (1977) referred to the video for local production and replay of AV material by non-broadcasting organization, groups or individuals. Further he pointed out that this kind of television is 'non-institutional' (in relation to broadcasting) and its operation is often by non-professionals (again in relation to broadcasting).

Features of community media include 'programming based on needs and interests of the community ... the involvement of non-professionals ... and an active relationship between the communicators and receivers' (Berrigan, 1974).

(a) Access and Control in Communication

One of the differences between video and TV or Film is the ease of use, and the degree of access and control which users can exercise (Norrish, 1992 : 1; Garforth, 1985 : 176). Many developing countries with the available limited one or two national channels, and also with limited air time, fail to address the diversified needs of communication of every section of the society. Mass communication through the national television consists heavily of top-down communication with very minimum feedback where the needs and problems of the individuals, groups and communities fail to get expressed. On the contrary, community video emphasizes the bottom-up, participatory, decentralized and more democratic way of communication, however difficult it is to operationalise these concepts in the actual situations. But there are a number of evidences indicating a fair degree of success in that direction.

(b) Participatory Communication

Community video is participatory in nature. Anderson (1988 : 163) defines video message as 'participatory' when those who are featured are not merely documented as objects, but are actively involved in structuring and then using the communication for their benefit.

Community video provides for horizontal, vertical and exchange communication. It gives the context for interaction among people and their adjacent communities and villages, providing the scope for effective organisation for collective effort. Even if there is a lack of print and electronic media, people do quench their thirst for communication by relying on oral tradition. Thus, community video could be more relevant as a means for community action than as one of the means of communication.

Anderson (1988 : 165) attributes more weight to the participatory nature of community video, while stating that the video itself may do nothing to influence a particular farmer to improve his or her practices, it is the interaction between viewers and those who produce the video that makes the message part of an active dialogue.

(c) The strength of Local Entity

National television represents national entities often irrelevant to the local circumstances of a large section of audience. Greatest strength of the community video is its local entity. Dubey and Bhanj, 1984; Besett and Tighe, 1988; Norrish, 1992; Garforth, 1985; McLellan, 1987; Anderson, 1985; Lewis, 1977; Fraser, 1980; Ogan. 1989; Gay, 1982; Kidd, 1978). McLellan (1987) pointed out that the video could be easily geared to local needs, languages and customs, as well as immediately involving people and providing them with a record of their activities and actions.

Local entity becomes very important for making the communication more purposive. People understand and comprehend when the communicator talks something that is very relevant to their day-to-day life and their vicinity, in their own language. It is more satisfying for the people to participate and know something that goes on around them. Further it could motivate them to become an effective agent in the system rather then a passive receiver, an object pushed around in the process of change.

(2) COMMUNITY VIDEO AS A PROCESS

It would be more interesting to go into the details of some of the typical experiments of community video with an intention of understanding how the communication tool, video is used as an integral component of the process of development of people, rather than as an aid in the top-down communication.

(a) Fogo Island Experiment

The first experiment in using the medium of video to help people help themselves was in an isolated fishing village on Fogo Island in the Canadian province of Newfoundland. The government had plans to resettle the community on the mainland because it was no longer economically viable. Many of the citizens had already left and morale was low.

The National Film Board of Canada started to film an ordinary documentary and recorded the views of the people who lived on Fogo Island. The film was shown to the community and a lively discussion ensued. The community turned out to be far from reticent and was reluctant to be uprooted. The discussions led to concrete plans of action. The film was shown to administrators and the community eventually received financial help to start a boat building company and reactivate the fishing industry (McLellan, 1987 : 145).

(b) Taparana village experiment of India

Taparana village experiment was started in India with the philosophical assumption that the human potential is unlimited and that ignorance of this potential is the basic cause of human suffering. Thus, the developmental efforts must aim at cultivating in the community the creative power to realize and estimate its own latent strength and resources from the perspective of available institutional provisions and at initiating self-action for growth (Dubey and Bhanj, 1984 : 84).

Dubey and Bhanj emphasize the initiation of community actualization with the help of video and community worker where the isolated individuals are linked together for the process of positive interaction. At this stage the problems and the 'community actualizers' are identified from the members of the community. Democratized media participation plays a major role in the community actualization where the role of the media is completely reversed to gear up from the grass roots.

Based on the work of Taparana experiment Dubey and Bhanj suggest some steps that help in gradual introduction of the media into the community.

Step I : Media demystification. Here the video is used to record common community scenes like fields, schools, children playing, villages artisans at work etc., taking care not to record any sensitive subjects that might agitate the members of the community. At this stage they get acquainted with media capabilities like recording, replaying and erasing etc.

Step 2 : Recording and playback of an informal community discussion. Informal group discussions are recorded and played back to the participants where the felt and unfelt needs are identified, apart from identifying the potential community actualizers.

Step 3 : Selection of community actualizers and small-group discussion to identify priority issues. Democratically selected community actualizers view the informal community discussion and involve themselves in a small-group discussion to pinpoint some immediate and easy-to-solve issues through general consensus.

Step 4 : Video recording of a community issue by one community actualizer, and discussion. The issue is discussed with a representative community actualizer and the outline of the issue to be recorded is prepared with his help. Then the issue is video recorded under his guidance. Time and again it is replayed to this community actualizer for modification. After his approval, the tape is played to the other previously selected community actualizers for their approval. Necessary modifications are made as desired by the group. Thus, the tape becomes ready for general community viewing to discuss the issue.

Step 5 : General community viewing and issue discussion; re-editing if desired in cases where the issue must be presented outside the community. At this stage the video tape is played before different groups at their convenient time and place, following which, the community gets engaged in serious problem-solving discussions leading to the emergence of common action plans. If the issue involves an outsider official agency, the tape is sent for presentation after the approval of the community. And likewise the response of the authorities of the outsider agency is recorded and presented to the community in different groups.

(c) Video for Resolving Conflict

In the summer of 1979, an exploratory study was conducted in Maine that used small-gauge video as an intervention tool in a herbicide spraying conflict. Video was used to promote dialogue between citizens protesting against the use of aerial spray and the blueberry growers and industry using the spray. Not only did video serve to document the environmental problem, it also enabled both parties to see themselves as well as the other side without direct threat or antagonism (Gay, 1982 : 21). The growers claimed it necessary for them to spray to comply with federal regulations & consumer demands, while the protesting group countered that long term effects of persistent pesticide use were questionable, and claimed that water supplies and the general health of communities adjacent to spraying operations were endangered. To resolve this conflict a video project was conducted in different phases.

Phase I : **Entry and pre tests -** A pre test was conducted to determine the attitude of the leaders involved in the conflict towards the environment and video technology, and with the assurance that any video segments they did not want shown to the opposing side would be destroyed.

Phase II : **Training, production -** Selection After the participants learned to script and operate the video equipment, each group produced its own tapes describing the conflict as they envisioned it and outlining possible solutions to the dispute. The tapes were then viewed and analysed. Facilitators helped participants in each group select segments to be shown to the other side.

Phase III : **Showing and feedback** - The first productions were shown to the opposing parties by the facilitator. Reactionsto the tapes were then videotaped, and selected portions of the reactions were shown to the opposing parties.

Phase IV : **Analysis -** Post tests were administered to determine what changes, if any, occurred in the thinking of the participants. Results were systematically analysed by the facilitator and participants.

When the opposing parties learned to handle the new communication technology, they could express their concerns in a way that helped them understand the effects of their behaviour on other people and the environment (Gay, 1982 : 22).

(d) Video for Agricultural Management and Soil Erosion Problem

Keith Anderson (1988) used the video to help farmers to confront their own problems of agricultural management and soil erosion. He makes a distinction between the outside individual organiser producer (who in other cases could include several people or a team) of the specific video communication project undertaken in the community and the local participants producers. And the different stages involved in the practical production were as follows :

I. Integration stage - The first stage is the beginning of a process of assimilation and understanding of the local situation for the producer of the video; it is a stage when a working script is developed from the ideas and personalities of the people. The producer arrives at a general understanding of the problems faced by the community, but with the full awareness that is his or her perspective. The role of an outsider was to organise the local knowledge on the issues of soil erosion and declining productivity in order to present it in a way useful to the community. And video was produced by recording the interviews that were exclusively semi-structured in nature.

II. Videotaping stage - The first videotaping activities involved recording introductory footage of familiar landmarks, the beauty of the region, and the people at work and play.

Further interviews were conducted at their convenience using semi-structured format with enough flexibility comprising general questions that stimulate farmers to show and reflect on their practices and lead them to discuss agriculture at the level of complexity they deem appropriate. The principal goal of participatory video was to solicit and respect what is community knowledge and minimize the role of outsiders and professionals.

III. Editing stage - After the interview, farmers edited their interviews by going through the recordings and indicating the important parts.

IV. Dissemination stage - The dissemination stage entails visiting both those who took part in the production of the video, and other groups that might be interested in seeing and discussing it. In the meetings people who showed emotional reactions at first, were asked to add their own experiences and situation to the issues presented in the video.

V. Evaluation - The evaluation has got to be integrated with the dynamics of semi-structured group meetings while people discuss how to solve the problems. The producer of the video can use the post-viewing discussion to understand better how video might have been used more effectively to address the issues.

(3) THE CLAIMS AND CRITICISMS OF COMMUNITY VIDEO

Fogo Experiment is considered as a mile stone in the use of community media methodology. The community media method gears the process of communication from the side of the people who are the subject of communication rather than making them the object of the whole process. Many of our development programmes originated without their knowledge and functioned by keeping them at the end of the process as clients. It has been realised that without the active participation of the people, development can not take place. People are better informed about their lives, surroundings and their problems than outside agents. Any intervention should spark off human abilities to function for the betterment of individuals and communities. Development services of many development programmes have led to the trap of continuous demand by the communities for non-stop flow of services. Development services mean the development of human abilities for generating and developing resources, and institutions for sustaining the process of development on a continuous basis.

Community video showed that the media could help stimulate, and even mobilise communities. But it also demonstrated that a development project could be approached in an open ended way. That is. rather than the educator or development worker trying to stimulate people (to participate in previously decided plans, the community itself could formulate developmental goals, provided they had enough information on which to base decisions (Berrigan, 1979 : 24).

While writing about the success of the Taparana experiment Dubey and Bhanj (1984) mentioned that the project was a departure from the conventional aid-oriented government programme made for the people.

(a) Novice Effect

Comparative research in communication over six decades has tried to find out the significant difference in learning due to the use of different media with an intention of finding the superiority of one media over another. However, a number of overview researches have consistently pointed out the lack of consistency of significant difference of a particular media over others. While taking stock of all the previous overview research and media analysis, Clark (1983 : 49-50) pointed out that the subjects tend to put in more efforts and attend to those media that are novel to them. The increased attention paid by the subject results in persistence which yields achievement gains. If they are due to a novelty effect, these gains tend to diminish as subjects become more familiar with the new medium.

Essentially, novice effect is an extraneous variable in the systematic method of research on media which needs to be properly quantified, measured and isolated to understand and use it advantageously or to eliminate it.

In Botsawana, video had to be shown twice to get over the initial novelty of the medium. Video was used in a conference situation to present the dramatised case studies (fictitious, but based on the factual situation) to spark off discussion among the villagers. In some other cases the initial enthusiasm could wear out with the passage of time as the participants get used to the new medium. Initially very successful programmes of rural radio forums in India became dull and unattractive to the participants with the passage of time. People fell out from participation, and the radios were used for listening to songs and music by the keepers.

Novelty effect could be tackled by message redundancy wherein the message is repeated in different forms of communication. Video conferences should be supplemented by discussions to make the process dynamic and purposeful.

Repeatedly drawing the attention of participants to the message of the media could help greatly to get over the initial effect of novelty while using new media over a period of time.

Are our few successful community video projects free from novelty effect? In problem-ridden societies, how many video projects can we afford to solve how many problems? To what extent are we confident that video will ensure the enthusiasm of people to participate continuously on long term basis?

(b) Power structure and community video

Extension repeats the principle of education for all like a parrot while accepting the intensive training and visit system which assumes that the benefits of technology and education shall trickle down from the contact farmers to the rest of the members of the social system. Training and Visit (T&V) system of extension involves intensive training of village level workers who, in turn, train the selected few contact farmers. Often the benefits of development programmes meant for those to be developed have been garnered by the well off who dominate the social system. It is also realised how difficult it is to reach out to the poor directly by passing those few well-offs acting as gate keepers to the social system. Development programmes of India and the sub-continent have been criticised for having caused sectorial and regional imbalances in the process of development.

India has to launch a number of corrective programmes as a component of the Integrated Rural Development Programme (IRDP), a mammoth rural development programme, to take care of the needs of specific target clients belonging to backward communities' balanced growth across the communities should be ensured for effective harmony. Equity is one of the important principles of sustainable agriculture and Rural Development.

Community denotes a group of members who compete, co-operate, and conflict among themselves and with other neighbouring communities. Community video, where community takes priority over media, should address the wide range of issues of power structure of the community. Community consist people of different classes, clans, castes and religions. How does video make sure that the views expressed come from the cross section of the community? How does it ensure the participation of the oppressed group on an equal footing against the dominant group of the community? What about the long term effect of intervention on the inter group relations? These are only a few of those questions which shall arise in the context of community video and need appropriate answers.

As the community is an embedment of power tensions, Tomaselli (1989) has raised a number of questions in relation to the power structure, factionalism etc., in communities while using community media.

(c) Video technology : The costs and reliability

The concept of community video is the product of hardware video technology and a method of using it in a particular fashion. Thus, the cost factor of the hardware becomes one of the important criteria to look into before investing on it. However, the literature on a community video does not provide clear idea of the cost effectiveness of using community video. Some experts back community video as low cost and reliable by comparing with the costs involved in broadcast media (Norrish, 1992; Garforth, 1985; Anderson, 1985), while others who criticise the community video for its costs and technological qualities, do so by comparing with the alternative small and culture based media that are equally effective (Kidd, 1978; McLellan, 1987). Interestingly some ignore the cost factor while glorifying some of the intrinsic qualities of community video.

Though the cost of video Equipment is coming down with the improvementin quality of the technology, it is still relatively expensive to purchase and maintain. Other media, such as community radio, might be cheaper, easier to use, more durable and can be just as effective (McLellan, 1987 : 148).

Ross Kidd (1978 : 131) making a comparative analysis of video and other communication media as tools, came out with conclusions based on work on video for non-formal education in Botsawana. He emphatically stated " .....however I feel it has no role in mass rural education in Botsawana. Its technical requirements and cost prevent it from being used on a massive scale. These factors force it to be used as a tool brought in by 'outsiders'; it is just too complex and expensive to become a 'people's tool' in the way it has in North America".

Mbughuni (1977) pointed out that video in the field has proved to be chronically unreliable. As a new and exciting medium it creates a certain mystique and builds up expectations which are quickly shattered when it doesn't work. Further, video equipment is vulnerable to bumps, dust, and other conditions which are difficult to avoid in rural use. Without regular servicing it can become a liability rather then a useful tool.

Gery Gay (1982 : 24) who demonstrated that video could be very effectively used in resolving conflicts warns that no one should use video without first analysing why and when to use it, and its cost effectiveness in a particular situation.

Heller (1978 : 135) mentioned that the video equipment is certainly going to become lighter, faster, more sensitive and coloured, but never cheaper, tougher or more reliable. Heller, further went on to state that the role of video has been very precisely to head the attack on public service control of broadcasting which stands in the way of this objective in most European countries, and to Soften up developing nations for infiltration and capture. Further, to quote the author, ".. It may be lamentable that so much good radical energy has gone into these games and that people who should know better have provided willing collaboration. What is worse, and ultimately dishonourable, is that we are now being asked to play the role of 'experts' in providing these fragile and expensive toys with a dressing of educational or professional respectability for export to the third world".

Slides have played a major role in social development programmes in Latin America; slide projectors are less expensive and have far fewer technical complications than video. Similarly, tape recorders provide the immediacy as video, are less expensive and more robust (Colle, 1977).

Video can be a waste of money, time and energy. The argument can be made that indigenous cultural media such as plays, puppet shows, songs and dances, or group discussions can be just as effective and much less expensive. 'Why have second hand discussions and animation when you can organise the real thing?' One social animator with experience in Botswana (McLellon, 1987 : 149).

When countries are facing global recession, their governments are freezing the salaries of public servants, and when the forced redundancies have become more frequent, it would be wise to seek quantified answers of social and material benefits of investing on community video.

(d) Rival media for video : Alternative small media and village drama

Video is a tool (however integrated it is), a means to generate interaction and seek action by the members of a community. The objective of the whole process is to facilitate the process of communication by giving an opportunity to participate, express and to respond. To achieve this objective of communication one can use different media. Media selection is guided by either the special ability of a particular medium to perform certain functions (media analysis) or by the need-cost-effective ness analysis. However, we can communicate the same message by using slides or video, although both differ in the manner in which they function to do the same job. The selection by need-cost-effectiveness is mainly guided by the availability of a wide range of media, costs involved in using a particular medium and feasibility for a particular task.

Any alternative medium however differently it may function, if it costs less, is found more reliable to use and yet works as efficiently as community video for social animation, it could well be used in preference to community video.

Often it is the cost factor that discourages people from relying on the video for 'community communication. Video could be cheaper to use in certain situations than others. We should identify those situations to harvest the dividends of investment on video. When it is required inevitably to exploit the special ability of video (moving image and sound etc.,) for the desirable effect, it makes sense to use video irrespective of costs.

Among all it was Ross Kidd who made conclusive remarks on the use of alternative media to community video. According to Williamsons (1987) it is a 'live theatre' which seemed a much more appropriate technology for the Fogo process.

The Canadian experience with video and film in community education had provided inspiration for the Botswana facsimile which combined drama and community discussion. Yet live theatre, specially in the rough and participatory style used, seemed much more manageable in a grassroots programme than film or video (Kidd, 1978 : 130).

Low Colin (1974) pointed to the great oral tradition of story telling, versifying singing and so on. People can be witty and colourful with the language sometimes richer than the homogenised textbook variety. While emphasising the indigenous forms of expression, J. C. Mathur (1968) mentioned that an abrupt denigration of traditional forms of culture means denial of access to a kind of literacy to which people have been used. As McLellan (1987 : 148) suggested, radio could be more reliable, less costing and equally democratic means of communication.

(e) The concept of non-professionalism

Community video is referred to as the local video production and replay of AV material by the non-broadcasting organisations, group or individuals. And it is pointed out that it is a kind of television that is non-institutional and non-professional in relation to the broadcast television (Lewis, 1977). Berrigan (1974) also mentioned about the involvement of non-professionals, and an active relationship between the communicators and receivers in community video.

Community communication using video starts with the intervention of outsider agent where he/she contribute to the process of communication till the end. Although many of the proponents of community video attribute a major part of the success to the human element present in the process, the role of outside agent is not very clear.

It is often very difficult to be in the process without influencing it. Colin Fraser (1980 : 27) mentioned that, in practice, however, it has not always proved possible to maintain the non-specialisation approach and it is premature to decide whether it is valid as a concept.

(f) Content vs. Media of communication

Very recently the content of communication has received the attention of experts over media. All these years researchers kept on producing and reproducing literature on the alternative use of media assuming that the content does not matter much as long as there are effective media to communicate. The symbol system used to communicate differ significantly with different media resulting in differential effect in information processing by the subject. It is not known whether it is more beneficial to use video for controversial or non-controversial issues in the community. In a complex power relation to the individuals and communities it is rarely possible to deal only with those that are non-controversial for all those affected. And it is inescapable for the development intervention to address the issue of inequity in the community, where there is always a bargain for resources and power.

When face-to-face communication becomes impossible due to threat and violence video could be the best way to mediate between conflicting groups. Video could be of great help when more facts and information are to be presented to the conflicting groups which can not be achieved by using alternative small media (low cost) to promote the understanding of opposite views for easy solution of the conflict (Gay, 1982).

(g) Learning from the video

Research on alternative media has always tried to grapple with the problems of using media discriminately for more and better learning. Using a combination of media to present the message through the most possible symbols for the benefit of the learner with wider choice of mode of learning has been the content of multi-media approach. Media symbol system theory emphasise on the symbol systems and the related attributes of a media (Saloman, 1976). Media characterisation in terms of its ability to deal with a particular communication situation, message, learner and task would help us in understanding better about a particular medium. The manners in which we wish to use a particular media (in terms of its symbols, and related attributes) influence the method of communication, whether it is in the class-room or on the field.

Many community video experts hold on to different opinion on using video effectively. Some say it is good for changing attitudes and behaviour, and for motivation (for example, Dubey and Bhanj, 1984; Fraser, I980). While others believe that it provides horizontal, vertical and exchange communication. Some others put it as a medium that builds the confidence of people and concientizes them (for example, Mac-Lellan, 1987). Some of the experts support the use of community video to improve merely the audio-visual literacy of illiterate masses of the many developing countries.

Ross Kidd who criticises community video for the costs, technological fragility and due to the availability of equally efficient alternative small and culture based media, emphasises that the video could be the best bet for training, and to offset the distance barrier in communication

Many of these observations are based on the few field experiences of the authors. Thus, there is a need for clearly stated learning benefits from video, and consensus on the effective and efficient use of video based on its communicative ability, costs and conventions.

(h) Organisation context of community video

The community worker is an interface; between the people and the government; through whom the government programmes are channelised. In a typical bureaucratic extension and development organisation the community worker is often regarded as a postman who delivers the packages of development to the clients. In an intensive system of T&V extension he has little time to devote to any activity other than his own training and training of farmers by regular visits to the villages. In many of the developing countries the village level worker (VLW) is characterised as less educated, poorly trained, low paid person having to deal with a large number of farm families.

It is such personality complex of the VLW that faces the intervention of community video as a part of the organisation traditionally looking after the developmental needs of the people. If such a VLW is to be involved in community communication as a social animator, it has to be seen over and over again whether he would be able to deliver the goods consistently. And if community video is launched without involving the VLW, he may view it suspiciously as a threat to his position and turn hostile to the whole idea. If we think of marrying community video to the existing mainstream extension, how many VLWs would be articulate enough to become an effective agent in community video as a social animators?

Sometimes it is said that the video could be used to by-pass the non-communicative channels in the bureaucracy, while often it is suggested that community video should work by taking into confidence the people in power.

Before investing in communication technology, it is important to define clearly how it will be used to support extension, including the adjustments that will be needed in budget and staff allocations. At the same time, it is clear that the existing communication facilities are often underutilised (Anderson, 1985 : 178). Thus, for the effective functioning of video for either community organisation or training, it is important to have the appropriate organisational structure so that it sustains on a long term basis.

(i) Per se media contribution and media-method dichotomy

Serious doubts have been expressed by a number of experts on the contributions of media to the overall change brought about by community video. The most important realisation people using participatory video for development work can make is that only a small portion of the success of video projects is related to the intrinsic quality of project concepts. Rather, most of the variation in success is related to the implementation process (Hornik, 1980).

Anderson (1988 : 165) pointed out that the video itself may do nothing to influence a particular farmer to improve his or her practices It is the interaction between viewers and those who produced the video that makes the message a part of the active dialogue.

Dubey and Bhanj (1984) are of the opinion that no communication tool, however powerful it may be, can ever replace the human element. The effectiveness of the tool depends on the people using it.

Colin Fraser (1980 : 28) who reported on the use of video for training pointed out that though it influenced the attitude it was still too early in experience to enable one to gather conclusive evidence on that.

Berrigan (1979 : 37) emphatically stated, '....video can not do development work. It can show the needs, open channels of dialogue, establish communication links and stimulate action but it is not a magic medium. It needs to be used with full field support'.

Where results have been claimed, (for example in the Fogo and Drumbheller projects of the famous Canadian Challenge for Change programme), it has proved impossible to isolate any particular video contribution from the general benefits flowing from programmes of investment in deprived areas. In other cases there are simply no results to record (Heller, 1978 : 34).

It has been reported that editing is time consuming and irrelevant to the process of communication in community video.

The literature on community video implies that video has been used as a tool or component of the whole process of social animation. Thus, video has been a component of the communication method. Video is used as a ground where community animation is a figure in the process. And it is also used to create the context for the interaction and participation of the people.

Since the purpose of communication is to spark the abilities of individuals and community for action, social animation process can dispense with video to use equally effective, cost-effective and feasible media like village drama, radio forums, slides, tape recorders etc. However, video on its own has communicative capabilities that are unique in it. Inspite of its similarity with low gauge film in terms of linguistic and pictorial symbols, it has the advantage of instant replay, re-use of tapes, and of not requiring processing etc.

Future research on video use, whether it is for community animation or class-room teaching, should be aimed at probing more on how best to use all the communicative capabilities of video. It needs thorough understanding on the content of a message, its relevance to the language, customs and the needs of the audience, costs and conventions of using it for a particular problem.

There is some consensus on using video for the purpose of training. UNICEF uses video for training regionally located staff members throughout the world (Ogan, 1989:3). Ross Kidd (1978:129-31) recommends video as a best tool for training, where the fragile technology can be handled safely under shelter, and also for reasons like cost. Further, he pointed out that video could be used to make information available to policy makers. Thus, it would be more appropriate to identify the specific uses of video that are cost-effective, feasible, and where unique abilities of the video symbols are required to be used.

References

Anderson, K. (1985) Participatory video in community video. *Media in Education and Development,* 2 (4): 163-168.

Berrigan, F. (1974). *Community media and development.* Middlesex Polytechnic, U.K.

Besett, G. and Tighe, D. (1988). The integration of video in development projects. *Media in Education and Development,* 21 (2) : 44-47.

Ceccon, C.S.P. (1989). Brazilian centre shows that video is an agent of change. *Media Development,* 4 : 30-32.

Clark, R. E. (1983). Reconsidering research on learning from media Review of *Educational Research,* 53 (4) : 445-459.

Colle, R. D. (1977). The traditional laundering place as a non-formal health education setting. *Convergence,* 10 (2).

Criticos, C. (1989).) *Community in Media* Mimeo.

Dubey, V. K and Bhanj, S. K. (19.84). Use of video in rural development. *Interaction,* 2 (3) : 82-90.

Fraser, C. (1980). Video in the field. *Ceres,* 13(1): 24-28.

Garforth, C. (1985). Mass media and Communication technology. *Media in Education and Development,* 18 (4) : 175-179.

Gay, G. (1982). Using video to resolve Community conflict. *Journal of Extension,* 20 : 21-24.

Heller, C. (1978). The resistible rise of video : some thoughts on technology and social change. *Educational Broadcasting International,* 11 (3) : 133-135.

Kidd R. (1978). The technology trap; A case study video use in non-formal education in Botsawana. *Educational Broad-casing International,* 11 (3) : I 24-132.

Lewis, P. M. (1977). Video in non-formal education. *Educational Broadcasting International,* 10 (2) : 64-67.

Low, C (1975). *Media as mirror.* In a resource for the active community. Canadian Radio-Television Commission, Ottawa,

Mathur, J. C. (1918). Cultural role of mass media in transitional societies. *Convergence,* Vol. 1, No 2.

Mbuguni, L. (1977). Appropriate audio-visual communications technology in Tanzania. Report (Unpubl.), UNESCO : University of Dar-Es-Salaam,

McLellan, I. (1987). *Video and narrowcasting* : TV for and by ordinary people. *Media in Education and Development,* 20 (4); 144-149.

Muller, K. (1987). The group media in the next five years. *Group Media Journal,* 6 : 8-10.

Norrish, P. (1992). *Uses of video.* Class-room notes. Department of Agricultural Extension and Rural Development, University of Reading, UK.

Ogan, C. (1989). Video's great advantage-decentralised control of technology. *Media Development,* 4 : 2-5.

Salomon, G, (1979). *Interaction of Media,* Cognition and Learning. San Francisco : Jossey-Bass.

Tomaselli, K. (1989). Transferring video skills to the community; the problem of power. *Media Development,* 4 : 11-15.

Williamsons, H. A (1987). The Fogo process, user-oriented communication systems and social development : A Canadian experience. Paper presented at the UNESCO Meeting on Planning and Management of New Communications systems, UNESCO, Paris.

Chapter **59**

Emergence of E - Commerce

Mahesh Patnaik

This paper intends to inspect the role played by e-commerce in today's communication technology. To give an illustration in Punjab an exporter of hosiery and sports goods business noticed an unexpectable downtrend in sales for no apparent reason. On deeper analysis it was discovered that they were losing their customers to their counter parts across border in Pakistan. The reason is not the quality or the service or the price, but Pakistani businesses were allowing international customers to transact over the web. Thus we can think about the plight of communicating with or without the web or e-commerce. With the expected transactions of e-commerce to rise over to $300 Billion no business can afford to ignore this kind of communication.

E-commerce encompasses the monetary transactions over the web in the form of business or any other financial transactions it may be used as Electronic Data Interchange (EDI) to transact between the industries. This may be used in three forms namely-Business-to-Business e-Commeree, Business-to-Consumer E-Commerce, Intra-Company (Enterprise) E-Commerce. All this transactional data has to be protected from being leaked to public this involves many security features of the interest. Like encryption alongwith the FTC rules about the different trade practices on the web, which is similar to the normal laws in practice? To help in understanding of the paradigms of e-business this paper tries to set up a small glass any on E-commerce at the end.

Electronic Commerce is the term used to describe another avenue available to small and medium sized merchants to successfully sell their products or services on the internet. The message has finally sunk in that the Internet is safe and now in addition to taking orders by phone, fax and mail you can take orders on line 24 hours a day, 365 days a year without opening a year without opening a door.

US MERCHANTS provide a "ground opening" for your new On Line Store with a Web Page that tells potential customers all about your Company. Next, visitors to your site will browse through listings of your products. They can view pictures and descriptive text for each item. When a visitor decides to purchase an item, all they need to do is clink one button to add the item to their "Shopping cart". The shopping cart passes through the check out line where the prices are totaled. Lastly, you receive payment information from your new order through our automatic response system. It's just that simple to become another satisfied customer of US MERCHANTS. E-commerce with US Merchants is very wise decision. The article featured in Network Computing confirms how important establishing your e-commerce site is.

Something to think about

Did you know that online retail sales revenue in the United States in expected to grow 404 percent from $2.4 billion in 1997 to $12.1 billion by the year 2000, the demand for electronic commerce software is increasing (Forester Research, April, 1998). The overall U.S. market for such tools is expected to reach $3.8 billion by 2002, up from $ 121 million in 1997 (Forester Research, May 1998).

Success Stories

"We opened our virtual doors in July, 1995 with a mission to use the Internet to offer products that educate, inform and inspire. Today we offer 3 million books, CDs, audio, books, DVDs, computer game and more. Though we've grown considerably since our early days, we maintain the same mission and commitment to our customers and we've proud to say that the 2.5 million people in more than 160 countries who have bought from us have made us the leading online shopping site". ....JFEFEREY P. BEZOS, FOUNDER & CEO, AMAZON.COM

Key Applications of e-Commerce KINDS :

AE Business-to-Business e-Commerce

AE Business-to-Consumer e-Commerce

AE Intra-Company (Enterprise) e-Commerce

These three strategic focal points for e-Commerce will be discussed and we need to determine which of the three can gain the greatest competitive advantage as a starting point for your programs. We will learn more about each one of these :

Business to-Business e-Commerce

The Internet can connect all businesses to each other, regardless of their location or position in the supply chain. This ability presents a huge threat to traditional intermediaries like wholesalers and brokers. Internet connections facilitate business ability to bargain directly with a range of suppliers-thereby eliminating the need for such intermediaries.

There are, however, tremendous opportunities as well as threats for companies regardless of their position in the supply chain. They include :

Providing information, McKesson is shifting from its position as a drug distribution business to a new position of value added provider in the health care industry. Its product is information, not drugs. By exploiting its massive database of information on sales and units for pharmacies, that information package is provided to improve the financial performance of pharmacies.

Purchasing and Selling General Electric's Trading Process Network allows suppliers to download GE's request for proposals, view part specification diagrams, and communicate with General Electric to complete the transaction. GE is expanding this network to help other companies buy and sell from each other and in investigating a transaction free business model.

Moving to an Internet platform, Banks, like Bank of America and Wells Fargo, are conducting trials to investigate whether EDI functions for business-to-business transactions can be moved to the lower cost Internet platform.

Supplying extract service companies with established Internet systems are beginning to provide subsets of their intranets to strategic customers; partners; or trading partners such as General Electric's Trading Process Net Work. Federal Express is leveraging its existing infrastructure for rapid package delivery to provide web-based order fulfillment services to businesses.

Business-to-Consumer e-Commerce

Kinds of business-to-consumer:

One-way-marketing : Corporate web sites are still prominent distribution mechanisms for corporate brochures, the push, one-way marketing strategy.

Purchasing over the Web : Availability of secure web transactions is enabling companies to allow consumers to purchase products directly over the web. Electronic catalogues and virtual malls are becoming common place.

Relationship Marketing : The most prominent of these new paradigms is that of relationship marketing. Because consumer actions can be tracked on the web, companies are experimenting with this commerce methodology as a tool for market research and relationship marketing.

- Consumer survey forms on the web
- Using web tracking and other technology to make inference about consumer buying profiles.
- Customizing products and services
- Achieving customer satisfaction and building long-term relationships.

An early example of a company that has leveraged the power of Internet interactivity to revolutionize customer relationships is Firefly. Firefly has created its products through web tracking technology. What would have been a conventional audio and video distribution company is now literally creating a custom product for each of its customers, based on the customer's profile and buying behaviour.

Intra-Company e-Commerce

The new tools of e-Commerce, such as TCP/IP connectivity and web browsers, have finally enabled the building, sharing and use of information within companies. Intra-Company applications of web-based technology are called Intranets.

Companies are embracing Intranets at a phenomenal growth rate because they achieve the following benefits.

Reducing Cost - Lowers print-intensive production processes, such as employee handbooks, phone books and policies and procedures.

Enhancing Communications - Effective communication and training of employees using web browsers builds a sense of belonging and community.

Kinds of business-to-Consumer - Upgrades and new software can be directly distributed over the web to employees.

Sharing Intellectual Properly - Provides a platform for sharing expertise and ideas as well as creating and updating content - "knowledge webs". This is common in organizations that value their intellectual capital as their competitive advantage.

Testing Products - Allows experimentation for applications that will be provided to customers on the external web.

Overview of Issues and Challenges

There are similar issues and challenges in planning and implementing an Intranet to our earlier lessons on business-to-business e-Commerce and on business-to-consumer programmes. It will be useful to reiterate these issues and relate them specifically to the challenges of building and maintain internal webs.

Security

Companies need strategies to secure sensitive data that should be share only internally or only with select employees on a need-to-know basis. Those strategies include :

- Firewall security systems
- Passwords
- Unpublished URLs outside the firewall for extranets
- Encryption

Corporate Identify

There's a need to strike a balance between freedom of expression and maintaining the corporate look and feel. Just as you've experienced receiving copies of emails that were intended for internal use, internal web pages are sometimes shared with the outside world. Make sure to involve corporate communication, at-least in an advisory capacity.

Business Intranet Applications

Why is there such a widespread adoption of web-based technologies inside corporations? The statistics are staggering :

Intranet Growth

- Forester predicts that the intranet server business will reach $1 billion by the year 2000. (Filed : 22-Jul-96 Source : Business Research Group, Newton MA).
- Zonal Research Inc. predicts that worldwide Intranet software sales will increase 16-fold in three years.

The Intranet is not just a corporate Internet behind a firewall, but a way of doing business. The intranet allows you to connect people inside your corporation as well as people you deal with outside your organization. Increasingly, companies are integrating their business-to-business and intra-company e-Commerce strategies. Applications that are valuable to employees may also be valuable to strategic partners, suppliers, distributors and customers. A new phrase has been coined to describe the provision of internal information to a group of users outside the organization - the extranet.

Electronic commerce, which has been defined as the buying and selling of goods and services on line, now is being redefined to mean new, emerging and resigned business processes that can take place online, done in a secure way.

Intranet Opportunities

Internet opportunities can be grouped into five general categories:

Internet Communication : creating an involved and communicative environment by making information accessible to organizational members.

Collaborative/Co-operative Work : Exploiting Web technology to create team oriented work environments that are more productive, higher quality, and enjoyable.

Knowledge management : Facilitating an environment where knowledge is shared and exploited by any group or division in an organization (e.g. best practices database). This involves the process of capturing and leveraging a company's intellectual capital, the most valuable asset of contemporary companies.

Process redesign : Transforming and dis-intermediating inefficient practices and processes inside the organization, from supply chain management to internet procurement to expense reports.

As was illustrated for business-to-consumer e-Commerce, Intranets are also in a rapidly evolving stage. Current implementations are primarily for basic communication and sharing-publishing existing documents in a web format and providing bulleting boards for sharing. Other stages of development are already in progress at many best practice companies. They include :

Creation or acquisition of new content : This content can be better, more systematic collection and sharing of customer information or it might be a competitive intelligence alerting system from an external publisher such as Dow-Jones. Collaboration, best practice and expertise sharing. Structured database for sharing and transferring individual knowledge to all employees (Booz, Alien & Hamilton and Hewlett-Packard are two examples of firms that have begun this process).

Advertising and Marketing On The Internet

The Rules of the Road

April 1998 : who's reaching a global market? Advertisers on the Internet a new frontier for advertisers and marketers, the Internet is connecting you to customers from Boston to Bati with text, interactive graphics and video and audio.

If you're thinking about advertising on the Internet, remember that many of the same rules that apply to other forms of advertising apply to electronic marketing. These rules and guidelines protect businesses and consumers-and help maintain the credibility of the Internet as an advertising medium, The Federal Trade Commission (FTC) has prepared this guide to give you an overview of some of the laws it enforces:

General Offers and Claims

The Federal Trade Commission Act gives the FTC the tools it needs to act in the interest of all consumers to prevent deceptive and unfair acts or practices. In interpreting Section 5 of the Act, the Commission has determined that a representation, omission, or practice is deceptive if it is likely to mislead consumers; and affect consumer's behaviour or decisions about the product or service, in addition, an act or practice is unfair if the injury it causes, or is likely to cause, is: substantial; not out weighted by other benefits; and not reasonably avoidable.

The FTC Act prohibits unfair or deceptive advertising in any medium. That is, advertising must tell the truth and not mislead consumers.

A claim can be misleading if relevant information is left out or if the claim implies something that's not true. For example, a lease advertisement for an automobile that promotes "$0 Down" may be misleading if significant and undisclosed charges are due at lease signing. In addition, claims must be substantiated, especially when they concern health, safety, or performance. The type of evidence may depend on the product, the claims, and what experts believe necessary. If your ad specifies a certain level of support for a claim- "tests show X" - you must have at least that level of support.

Sellers are responsible for claims they make about their products and services. Their parties-such as advertising agencies of web site designers and catalog marketers also may be liable for making or disseminating deceptive representations if they participate in the preparation or distribution of the advertising, or know about the deceptive claims.

Advertising agencies or web site designers are responsible for reviewing the information used to substantiate ad claims. They may not simply rely on an advertiser's assurance that the claims are substantiated. In determining whether an ad agency should be held liable, the FTC looks at the extent of the agency's participation in the preparation of the challenged ad, and whether the agency knew or should have known that the ad included false or deceptive claims.. To protect themselves, catalogue marketers should ask material to back up claims rather than repeat what the manufacturer says about the product. If the manufacturer doesn't come forward with proof or turns over proof that looks questionable, the catalogue marketer should see a yellow "caution light" and proceed appropriately, especially when it comes to extravagant performance claims, health or weight loss promises, or earnings guarantees. In writing ad copy, cataloguers should stick to claims that can be supported. Most important, catalog marketers should trust their instincts when a product sounds too good to be true.

Protecting Consumers Privacy Online

The Internet provides unprecedented opportunities for the collection and sharing of information from and about consumers. But studies show that consumers have very strong concerns about the security and confidentialityof their personal informationin the online marketplace. Many consumers also report being wary of engaging in online commerce, in part because they fear that their personal information can be misused.

These consumer concerns present an opportunity for you to build consumer trust by implementing effective voluntary industry-wide practices to protect consumers' information privacy. The FTC held a number of workshops for industry, consumer groups and advocates to explore industry guidelines to protect consumer's privacy online. In addition, the FTC is monitoring the World Wide Web to see whether commercial Web sites are posting privacy policies and honoring consumer's privacy preferences; the agency will report its findings to congress in June 1998. You can find our, more about information collection from children at the FTC's site. CARU also has information to help you develop a privacy policy for children's information.

Mail and Telephone Orders

According to the Mail or Telephone Order Merchandise Rule, you must have a reasonable basis for stating or implying that a product can be shipped within a certain time. If your ad doesn't include a shipping statement, you must have a reasonable basis to believe you can ship within 30 days.

If you can't ship when promised, you must send the customer a notice convincing him for the delay and his right to cancel. For definite delays of up to 30 day, you may treat the customer's silence agreement to the delay. For longer or indefinite delays, and second and subsequent delays, you must get the customer's consent. If you don't you must promptly refund all the money the customer paid you without being asked.

You can give updated shipping information over the phone if your Internet ad prompts customers to call to place an order. This information may differ from what you said or implied about the shipping time in your ad. The updated phone information supersedes any shipping representation made in your ad you still must have a reasonable basis for the update. See Complying with FTC's Mail or Telephone Order Merchandise Rule.

Telemarketing

Credit Repair, Advance Fee Loans & Investment Opportunities Advertisements promoting credit repair, promising loans for a fee in advance, or touting investment opportunities may trigger application of the FTC's Telemarketing sales Rule if the ad. Allows consumers to order goods or services by telephone, in general, this Rule does not apply to general media advertisements. If you're advertising credit repair, advance fee loans, or investment opportunities or offering to recover money paid in previous telemarketing transactions, however, the Rule likely applies to you. Among other things, the Rule requires that certain disclosures be made before a customer pays for the goods or services. The Rule also prohibits material misrepresentations.

See Complying with the Telemarketing Sales Rule

Testimonials and Endorsements must reflect the typical experiences of consumers, unless the ad clearly and conspicuously states otherwise. A statement that not all consumers will get the same results is not enough to qualify a claim. Testimonials and endorsements can't be used to make a claim that the advertiser itself cannot substantiate.

Connections between an endorser and the company that are unclear or unexpected to a customer also must be disclosed, whether they have to do with a financial arrangement for a favourable endorsement, a position with the company, or stock ownership. Expert endorsements must be based on appropriate tests or evaluations performed by people that have mastered the subject matter.

See FTC Guides Concerning Use of Endorsements and Testimonials in Advertising.

Warranties and Guarantees : The rule on Pre-Sale availability of written Warranty terms requires that warranties be available before purchase for consumer products that cost more than $15, If your ad. Mentions a warranty on a product that can be purchased by mail, phone or computer, it must tell consumers how to get a copy of the warranty.

Guarantees

If your ad uses phrases like "satisfaction guaranteed" or "money-back guarantee", you must be willing to give full refunds for any reason. You also must tell the consumer the terms of the offer.

See Guides for the Advertising of Warranties and Guarantees, A Business person's Guide, to Federal Warranty Law and Consumer Product Warranties.

Glossary of Terms

Commerce Server

Is web software that runs some of the main functions of an online storefront such as product display, online ordering and inventory management? Works in conjunction with online payment systems to process payments.

Cookies

Cookies are snippets of information delivered from a web site to the client's browser, and then stored on the hard drive. Examples are the time of the last visit, or the pages downloaded 'Cookies' can be ready by that web site on the next visit.

Credit Card Processors

Provide shared network systems and on-line credit processing systems for individual credit companies. They integrate transactions between merchants, financial institutions, and card companies.

Cyber Cash - is *a company that developed one of the early payment systems for the Internet. They give away a consumer "wallets". Online merchants using Cybercast software receive orders from shops with the "wallet". Orders are sent through the Cybercast server to bank networks for credit card verification.*

Debit Card - is a cash substitute for consumers. They look like credit cards, but don't provide credit. Amounts for purchases are debited immediately from the user's bank balance.

Digital Certificates *- are* digital Ads. used to present credentials online. Digital certificates are issued by companies which act as "trusted third parties". In a SET transaction, the buyer, the merchant and banks for these parties all have digital certificates.

Digital Coins - can be downloaded to a user's hard drive from an account at a bank. When the shopper wants to pay, a "wallet" pops open on their screen. "Coins" are transferred from the shopper's computer to the online merchant's server. The merchant deposits the "coins" in their bank.

Digital Wallet - is software that stays resident on the hard drive of an online shopper. When they are ready to make a purchase, the "wallet" pops open to reveal payment options. Some "wallets" hold credit cards with encrypted information. Other "wallets" hold digital coins.

ED - is Electronic Data Interchange. EDI provides electronic formats which allow for an exchange of business data between companies over networks.

Issuing Bank - Issues the credit to a credit card holder. When sale authorization is requested, the merchant's bank, requests the funds to be transferred from the credit card company, which in turn receives the funds from the issuing Bank.

Java - is a programming language originally developed at Sun Microsystems to create software for consumer electronic products. Java creates platform independent applications it can run on any operation system.

Java Script - is an object-based programming language that allows for the embedding of Java objects directly in an HTML, Page. Netscape 2.0 and all their subsequent browsers can interpret Java script.

Merchant Bank - When credit card authorization is processed, the first stop is the bank where the online store has a merchant account. This bank is the merchant's bank.

Micro Payments - E-commerce began with purchase paid by credit card-roughly $10 and higher. New Micro payment systems allow for purchases ranging anywhere from a fraction of a cent to $5.

Net Cheque - Can be set via e-mail. When the cheque in deposited, just as a paper cheque, funds are moved from the account of the cheque writer to the account of the receiver. The Net cheque has a digital signature mechanism.

Set - is an acronym for the Secure Electronic Transaction protocol. It's means for authenticating credit card purchases on the Net. Digital signatures are used by all parties. Transaction information is encrypted using 24 bit RSA encryption.

S-Mime encryption - Protects the privacy of e-mail. If the sender and receiver both have e-mail clients that support the S-Mime protocol, they can communicate with e-mail that is secure.

SSL encryption - was developed by Netscape to provide data encryption and authentication of servers or clients. SSL stands for Secure Sockets Layer. It can be used for any functions on the Internet. FTP-Usenet or the Web.

Scale ability - is the ability to grow incrementally. If an online commerce system is scaleable, it can grow in capacity as the demand requires.

Shopping Cart - is a piece of software that operates on an online storefront. The "shopping cart" keeps tract of all the items that a buyer wants to purchase, allowing the shopper to pay for the whole order at once.

***Smart Card** - is* a credit sized plastic card with an embedded microchip. The chip can be "recharged" with funds. The store of value on the card is debited as a transaction is made. The card can also store ID information, health care details and security information.

Need more? Check out PC Web media for definitions of more than 3,000 computer terms, plus links to thousand of other Web pages where you can find additional information.

Chapter **60**

Sacrality and Symbols : Some Notes on Changes in Urban Symbolism

Vighnesh N. Bhat

Our basic premise that television has the power to create issues and symbols carrying mythical significance are not new, for, it has long been recognised that there is a significant potential for myth analysis within the scope of communication studies as is evident from the existing literature. One of the outstanding studies in this direction which demands a special mention here is of Real's examination of distinct types of television programming in American network TV. However, it is apt to point out in this context that much of the recent literature on myth in the communication process has been piece-meal in its approach, and when it comes to television proper, It is all the more obvious. Television is looked in isolation from the total social context and the multilaceted presence of myth in both the form and the content of this medium are not linked to the day-to-day lives of the individual.

I remember reading Arnod Toynbee who said :

"Every city - or it might be more accurate to say, every city before the present age of mechanization-has been, among other things, a holy city in some degree". And that is what we as a civilization have learned in the course of human history.

Cities first took form around 500 B.C. in Egypt, India, Mesopotomia and above all ancient Rome. Despite their comparatively small populations, all these cities exerted a great influence upon their societies. It may not be too much to say that ancient human history until recently has been the history of cities and the rural area played a historical role only by virtue of their relations to these cities.

Cities are hardly a recent human development. The first urban conglomeration seems to have taken root about 6,000 years ago and since then city's influence is out of all proportion to the number of its actual inhabitants. Historically, city had always been both a product of as well as a vehicle for change. It had been, like modern ones, a place of diverse social interactions because of the fact that in the city, people lived close together and developed a diversity of skills and occupations. This, in turn, necessitated new patterns of organization, with different chains of command, and distinct forms of social stratification. All these multitude of factors required improved and diversified systems of communication skills and symbols.

Even a cursory look into the existing literature would provide various theoretical models in understanding the nature and development of the city as a basic form of human groups. As is well known, other such forms are collectivities like crowds, mobs and social movements, associations like clubs, pressure groups and service organisations, and institutions like universities, nations and large corporations.

For the purpose of this paper a working definition of the term city, whether ancient or recent, is needed. We propose that a city is a relatively densely and permanently settled area where a great variety of specialized services like manufacturing, administration and commerce are available and where new symbols of changing social life are found. There are several theories of urbanism but it may be wholly unnecessary and inappropriate in this context to review and critique these theories. Our scope here is limited in the sense that we would be arguing that cities had been the cradles of new symbols - distinctly, symbols relating different aspects of society such as commerce, leisure activities and communication. These symbols have been, and still are, sacred and sublime to the total way of socio cultural life. To do so, we will engage in analyzing and critically looking into historical information that is readily available to scholarship.

Let me begin by briefly looking into some of the great civilizations of the East as well as the west. Rigveda (before 1500 B.C.), the earliest 'of the Hindu holy texts, describes Indra (meaning the lord of the Heaven in a loose translation) as the possesser and the breaker of cities. Infact the word 'puram' used in this context literally means an organized city. It is of commonplace knowledge that many holy cities in India have their mythological origin in the hymns and prayers to gods for better life on the earth. Rishis (holy monks) asked for towns and cities to live-in while praying for a better life. The priests of Shang dynasty of China (about 1400 B.C.) wrote people's queries pertaining to all affairs of life on a piece of animal bone, burned the bone in fire, and read the oracle through the patterns of cracks which appeared on the surface of the bone. And among many questions, archeologists found that the people of Shang often asked Heaven about building a new city - big city. Only when the divination from Heaven was positive would people start to build a city. Quite similar developments are found in cities in other parts of the world. A 'tel' (city) of ancient Sumer was the place where a guardian god resided and such descriptions are found in the text of 'Ramayana' popular not only among Hindus in India but also among the Hindus of Bali, Sumatra and Srilanka. Both in Sumerian and Indian traditions, different gods had their private cities and protected different aspects of certain holy cities. Japanese urban civilization emerged in the early sixth century as the result of conscious efforts of Princess Shotoku, who built Shitennoji and Horyuji - clusters of Buddhist temples, urban culture flourished in the Hara basin. It is clear from available scripts that before the new site of the capital city at Nara were selected, Priests made divinations in regard to the direction and location of the city, and the Emperor followed old rituals of the Shang and Chou dynasties of China.

Greek cities, needless to say, were the headquarters of gods who were worshipped as the symbolic guardian spirits of each of these cities. For instance, Zeus, the most powerful of the Greek gods resided in Mount Olympus, and other deities had their own cities to live in. These same deities presided over Roman cities. As pointed out succinctly, "Romans had gods for everything and in this sense, had left nothing for chance[11]. Every aspect of Rome was related to, and in most cases protected by, some divine spirits.

In all these holy cities, the lives of the residents were full of rituals and ceremonies. Indian holy cities like Banaras and Rameshwaram are still known for their ritualistic lifestyle partly because rituals and religious ceremonies themselves make these cities what they are now. In other words, the holy cities themselves come to symbolize certain aspects of the religion which they are expected to represent. Historically, the rituals in these cities varied both in the format and in content and that most if not all of these urban forms encouraged their inhabitants to get engaged in rituals and ceremonies. 'Dionysis', for instance, was a festival initiated in Athens in the sixth century B.C. and as a part of this festival, competitive performances of tragedy and comedy were staged. The citizens of Greece and Rome were able to maintain spiritual solidarity with the gods through their participation in such festivals and ceremonies which seem to have given initial impetus to theatrical arts in the West. Both in Greece and in Rome, priests and religious leaders played vital roles not only in political and other socio-cultural activities of the state, but they had a considerable voice in deciding upon the nature of sports their society should have. Both in the East and in the West, most of the sports events which took place on holy days had as their primary purpose the entertainment of deities. In some cases, specially in the case of Far-Eastern cities, people dedicated their festivals to their ancestor spirits. A joint feast of gods followed by men was held in Indian as well as Chinese cities on all ceremonial occasions, and the people shared the sacrificed animals, products, and wine (Brahminical concept of "bali" among Hindus may be referred here). The third of the Vedas - Sama Veda - as also the holy texts of Shang describe the music and dance to be offered along with foods and drinks in city feasts. Ancient Japan was no exception and in Japanese townships, people gathered around the fire and exchanged wine cups with the gods.

Whether these similarities in the city life of East and West were the result of cultural diffusion of independent parallel development is an interesting historical and sociological question which can hardly be answered satisfactorily at this juncture. However, the vital point here is that in many areas of the globe cities have been perceived by human beings in separable from super-human deities. In this sense, a city belonged to its god (king often being the part of the spirit or incarnation), and the citizens of the city were the servants of that god in his temple. In brief, to be a citizen in an ancient city was to be continuously in holy communion with the guardian deity of that city.

Of course, we all know that ancient cities served many other functions - commercial, industrial, educational and so on - but their spiritual or religious functions were fundamental, and the architecture of many ancient cities demonstrate their religious nature through the very structure of temples, shrines, and other holy locales in the central parts of the city. A good example is the city of Cologne. In the middle of the thirteenth century, this city with a population of 25,000 decided to build the greatest church in the world but the people of Cologne could get started with this idea only in the early part of the fifteenth century and for a period of 400 years they worked continuously. Even today, a traveller along the Rhine can see the gigantic building of the Cathedral from a distance of several miles and this religious structure overwhelmingly dominates the whole area. Meenakshi temple in Tamil Nadu, South India, Jagan-natha temple of Puri, East India, could be cited as other such examples. The people of the - city of (Jim in Southern Germany worked 500 years for their Church and finally succeeded in building the highest church tower in the world. In all these cases, the history of the city has been integrally tied to the successive construction of sacred structures of ceremonial centers; the city grew with the increase in the number and size of temple 'shikharas' or church towers.

We are aware of the fact that the world's greatest religious structures - from pyramids to the Toji temple of Kyoto or Vijaya Vittala temple of Hampe - were designed and financed by the ruling elite class of the day, and the people of the cities were often exploited through taxes and their labour for such construction projects. We are here trying to carry home altogether a different point: the holy shrines had an absolute symbolic and spiritual value in the minds of these city dwellers, and thus, they were often happy to participate in such great ventures.

Cities before the modern time of industrialization grew up with their sacred rites and ceremonies. In Rome, for instance, there were as many as 135 "holy days" a year, and each such day was filled with particular ceremonial events. Traditional Hindu calender lists atleast 32 major festivals besides the minor ones which will run over dozens. The situation in Japan was hardly any different and archaeological evidences point out at similar facts in ancient China and Southeast Asian societies.

The holy tradition, however, has become transformed gradually since the medieval period and decisive changes in the meaning of such festivals took place with the emergence of "spectators" -the outsiders of the festival (and, therefore, of the community) - who visited these festive occasions , not to be involved in communication, but to have fun at the festivities of the city life. To put this phenomenon in ametatheoretical frame, as Weber notes, this shift from participation to "spectatorship" may be defined as the "secularization" of "holy days". Religious festivals began to lose their significance and became secular entertainments; "holy day" became a "holiday" in its contemporary sense. The more the cities developed, the less their religious functions were directly felt by the people and as the years and decades passed by, the basic integrating process in the city life communion- became secondary and in some cases even forgotten. As pointed out in earlier pages. in ancient times, city planners located the temple in the center of a city; often, commercial and industrial areas were on the periphery (as can be exemplified by the map of ancient Banaras city in India). But quite evidently, the priority has been reversed, both in the West and in the East after the dawn of industrial era, specially so after the seventeenth century. Thus "holy days" were transformed into "holidays" or "leisure", and "holy events" became "entertainments".

II

We would like to put forward one of the basic hypotheses of the paper at this juncture. We plan to develop in this section of the essay the argument on the very nature of modern time "leisure" activities. Our proposition is that contemporary "leisure" activities and "entertainments" in cities, almost without exception, have religious origins. In essence, "holy communication is not totally forgotten; rather it has changed its form. Old symbols gave way to the new ones. Old myths were replaced by new myths. New gods have arrived on the screen. Football in England is said to have developed out of religious competitions between communities. Japanese 'sumo' also came from sacred sports, and 'Kabuki', another Japanese traditional art, was founded in its holy relationship to Izumo Shrine. Indian traditional wrestling developed from the practices able bodied devotees had in their ritualistic performances to please the deities and so did develop Indian classical dance system, 'Bharatanatyam'. The .importance of religious practices in the historical development of music in general all over the world cannot be overemphasized and horse racing, a universal spectator game, has pious exordium.

The field of commerce is closely connected with sacred events and we do not have to refer to Weber's thesis of Protestant Ethic here. Both in East and in West, market place started as a part of religious festivals and even today the concept of "holy day market" - retail shops in the yards of temples - are quite popular in India and Japan.

Traveling is another area of great significance for my present purpose. The great expansion in world travel has its origin in theological sentiments. For instance visiting Mecca is the lifetime ambition of Muslims and visiting "Maha Kashi" is 'the be all and the end all' for Hindus. The faithful members of these and other religious would try to visit the main temple atleast once in their lifetime, thus fostering contacts and communications between far off societies. Gradually the element of "fun" and comfort were added to the sacrality of such pilgrimage and once the devotees finished their rituals in the temples, they would spend more time in sight seeing and other sources of entertainment. Soon there came a reversal of goals-and-means relationship and the pleasure became the major purpose of travel. Visiting the holy shrines or temples came as an honourable justification to sight seeing.

Through the past several decades, the secularization of holy experiences has been going on with amazing speed. Places of worship have been transformed into objects of sightseeing. The quality of time-space relationship in cities has completely changed. A visit to any holy place in the world - Notre Dame of Paris, Canterbury Cathedral of London, the Temple of Dawn in Bangkok, Gokarna Temple in South India, Chionnin of Kyoto - makes obvious the changes of sacred time and space in the last 300 years. Of course there are still monks, priests, and other people in holy professions with little changes in their services to the deities, but changes have taken place in the nature of people who visit them. It is evident that today, instead of participating in the holy communion, 'pilgrims' often engage in taking pictures, and are approached by vendors for selling souvenirs and other decorative items.

An observation of great curiosity, however, is that this process of "secularization" by no means complete and unambiguous. While many of the tourists are happy merely to see the holy places and take photographs of their pleasant tour, there remains an ambiguity in their minds. On, the one hand, the sacred places are perceived as "something to see" (as the Indian expression goes), but on the other hand, they are still regarded as somehow sacred, regardless of their affiliation or non-affiliation with particular religious sects. This ambiguit gives people an uneasy psychological experience and as a matter of fact, a careful examination might reveal certain continuities in the sacred time-place complex in contemporary cities.

One illustration of this fact can be seen in the architectural style of central parts of cities both in the East and in the West. If we take a ten-minute walk in the centre of any major city, we will notice that the major governmental buildings, courts, the capital structures, and so on are built in the architectural style non uncommon in that city's traditional forms. In the West the influence of Greco-Roman style is evident in all but few major public buildings. In the East, especially in

India, cities like Delhi and Hyderabad have similar resemblance to their old architectural styles of temples - wide, dignified steps in front of the buildings, giant pillars with decorative sculptures surrounding these buildings, and so on. In the United States, the Capital building in Washington D.C. is more similar to the Greek Pantheon than to any architecture in American style; and one can pick up many more such instances from different cultural settings from various parts of the globe in the same way.

III

Our second proposition is directly related to, or rather an offshoot of, the first proposition that we tried to argue out till now. This second hypothesis is that the modern cities of the world seem to have been very consciously replacing old gods with the new.

Let me try to muster some evidence before we go on any further with this argument. We shall attempt to demonstrate in the following instances that new sacred symbols in the forms of "memorials" are present in big cities in a way that is not qualitatively different from the historical process of deitification or sacralization of symbols. The Arc de Triomphe of Paris, in the very heart of the city, is a memorial in the true Roman tradition to Napolean Revolutionary Memorial, National Plaza, statues of Henri IV, or Joan of Arc. London is not less in any count. Admiral Nelson, Marble Arch at a corner of Hyde Park, is the ones which immediately come to our mind in this regard. Tokyo, which emerged as the new capital of Japan relatively recently (in the middle of the nineteenth century) has been trying to establish new sacred places following the models of Western Europe. Because of the sudden shift of the capital from Kyoto, new gods, - including the emperor himself - had to be invented instaneneously. Yasukuni Shrine was built to commemorate the souls of those who died in the successive war of 1860s to 1945. Emperor Meiji has been worshipped at the Meiji Shrine. Gateway of India Bombay, India Gate Delhi, Rashtrapati Bhavan and Nariman Point are but a few instances of this kind in India. Various memorials in Washington D.C. and the Empire State Building may also be cited in the some vein. Exactly as the ancient gods have been worshipped, these new gods and "ruptures in the space" have been perceived as the sacred core of the city.

This also has to be noted that cities traditionally have been places of symbolic interactions. They started as communion with the divine and continued in interpersonal human communication mediated by the sacred as well as the profane. At any stage of city development, to live in a city was to be exposed to an extremely complicated symbolic environment, and city people have been characterized by their readiness and willingness to participate in symbolic interaction rather than to deal with substantive environments. Indeed, historically speaking, there had been an increasing demand for symbolic sophistication all the time. Even today, cities are admired inspite of the difficulties they pose, for their "cultural" values - schools, museums, concert halls, theatre groups, and so on. This obvious point needs little evidence.

A city is an agent of for special types of socialization. "City people" are the people who are specialized in the production and utilisation of symbols more than substantive material objects. The nature of the symbols they work with has changed from sacred to quasi-sacred and to secular, but city life has remained filled with symbols for thousands of years.

For example, through the better part of this century, the nature of children's play in cities has undergone tremendous changes. When the modern city was in its developing stage, there were many small lots in the city when children could play various games. "Ride tiny carts, or simply run and jump around. In other words, children were physically mobile. But as the city gradually developed into the current "Westernized" form, children hardly could enjoy physical movement in play any more. Thus today it is our not so uncommon experience that children in "developed cities" of Western Europe or North America, engage in more symbolic indoor games rather than outdoor activities. Computer games, television cross-word puzzles, riddles - not to speak of simple games like scrabble - have pervaded' the early, years of life in these cities. As the physical density of the space gets tighter, require the skills of symbol manipulation.

With the tremendous power of sight, sound, and motion, television as medium of communicationhas set a new era of symbolic interaction and urban communication. As a matter of fact, there seems to be a mutual cause-and-effect relationship between high density, over-populated urban life and the success of television as a medium of mass communication. The TV screen is relatively small but there are constant movements going on all the time, with different kinds of sound arrangements. The TV watcher can experience on that tiny screen the gigantic buildings on this planet, American "steeler" football, spaceship to the Moon or Mars, C.N. -Tower of Toronto or almost anything that he would care to imagine just by manipulating a simple switching device. Living in a confined space made of steel and concrete, people can experience all kinds of thrills and experiences through this little/"magic box". TV has now become an instrument in the continuum of symbolic compensation to the loss of nature which the eighteenth century paintings did, although not as powerfully. Television goes to prove that there is little physical space left in the metropolitan areas but symbolic is plentiful.

Television is certainly a powerful authority, specially in the 'developed' societies. If we just try to standby and observe the unusual relationship that the ordinary viewer maintains with this 'almighty' box, - for instance a typical American middle class family gathered in, or rather drifting in and out of, the living room, watching the screen - we are bound to recognize its proximity with the day-to-day live experiences. In this sense, television is a 'real' member of the family and not just a medium of information in the typical sense of the term. Think how hard it is even to read the sports page in the morning newspaper while carrying on a conversation with one's wife about a vacation plan or a new purchase; how altogether difficult it is to work on office business or correcting a graduate assignment while watching a film, seated in a theatre, or attending a Bharata Natyam dance program; how virtually impossible it is to read a decent novel, or listen to a concert while at the same time talking sensibly with a student or a good, friendly colleague on the telephone. But on the most of the evenings of the year, a good number of families in industrialized societies engage in viewing the TV happily while such parallel activities go on. Television in the urban locales of the Western world has become a regular part of the individual life, an object which has the capacity and reliability of 'myth making'. In a real sense, people see what the television cameras see and the values, faiths and mores are redefined in the context of this new eye. Individuals, to an increasingly greater extent, while stay home, staring live lives in terms of what they think the television cameras tell them. No wonder if the words of Dan Rather of CBS carry more weight than these of a serious researcher on the topic.

It is significant to understand the complex fact that the new "monumental[1] gods of contemporary cities acquire a mythical significance due to societally shared experiences. The ultra-modern cathedral towers and 'shikharas' - the Eiffel Tower, the New London Tower, the Empire State Building, Vidhana Soudha of Bangalore, and

Tokyo Tower - are being watched today as our shared cultural experience instead of the past communication through the identification with churches and temples, and thus, acquire a mythical significance in the lives of citizens.

IV

In this modest attempt, we tried to reason out that the contemporary 'leisure' activities and 'entertainments' have their origin in sacred and mythical issues or events and that the modern cities consciously attempt to reestablish new symbols of mythological significance. We termed this process, in a very generic sense, the progress of replacing old gods with new ones.

Basic functions of myth in the societalization process are the following:

1. Myth functions as part of the perceptual, and hence communication, system of a culture through which unfamiliar situations, originating either within the culture or outside it are interpreted and fitted into old symbolic molds.
2. Myth helps create exemplary models for the whole society in a process that translates a single life history into what is popularly termed as an 'arche-type' thereby establishing regularities and patterns for imitation.
3. Myth provides the members of any society the power to resolve conflicts within their own culture as well as between cultures.

Our argument has been that the new leisure activities and entertainments in modern cities serve these basic functions so powerfully that they act to create mythical symbols as exemplified in the case of television programs in developed countries. Taking the example of the Western TV. We tried to reason that the very nature of television as a medium, not just the format and the content of programs, hides the fact that the mode of narration and the total TV context is arbitrarily structured. In other words, television realism presents itself not as one of the ways of perceiving and communicating symbols, but as the only way. Television 'realism' therefore, acts as a salient process in the extension of certain mythical symbols over all other sections of society. We hasten to add, however, that any such medium is only an integral part of a total process of mythologizing symbols in modern cities, other parts being "entertainment", and 'monumentation' of ruptures in the space.

References

1. **Toynbee, A. J.,** Cities on the Move, Oxford University Press, New York, 1970.
2. For an elaborate discussion on the Cities in the East and their historical development, refer **Singh, Y. and Sharma, A.P.,:** Urbanization of Eastern Tradition, Popular Prakashana, Bombay, 1966.
3. I am quoting these lines from Class lectures of **Professor Charles Long of Chicago** and thus do not feel the necessity of explicit reference.
4. The papers by **Jeffery Meyer & Michael Ong Twin** presented in the course of a Seminar at Pittsburgh, USA, during the Spring of 1983, drew my attention to this special phenomenon (which I was not previously aware of) very vividly.
5. Refer **Kato, H:** "Changes in Urban Symbolism", Journal of Communication, Vol.23, No.4, 1973 for an extended discussion of 'Gion' festival of Kyoto and other yearly ceremonial events in ancient arid medieval Japan.
6. See Weber, M. : The Protestant Ethic and the Spirit of Capitalism, (translated by Parsons, T) Scribner, New York, 1930, 1954.
7. See, for instance, **Huizinga, J,:** Homo Ludens: A Study of the Play-Element in Culture, Beacon Press, Boston. Huizinga argues in support of this point of view: "In Greek thought everything 'musical' was closely related to ritual, above all to the feasts where, of course, ritual had its proper function ... The gads, he (Platoi says, out of pity for the race of men born to sorrow, ordained the feasts of thanks giving as a respite from their troubles and gave them Apollo, lord of the Muses, and Dionysus to be their companions in the feast...".
8. Refer **Wheatey, P.** City as Symbol : An Integral Lecture delivered at University College, London. Wheatley refers at great length the cosmo-magical symbolism in the lives of city dwellers and the central importance of such cultural symbols in all aspects of city life including day-to-day communication.
9. **See Voltaire, F.M.A.de:** The Works of M.de Voltaire, Vol.23, (tr) Smollett, T.et.al., Newbery, London, 1802. In an interesting discussion on this point (though in a little different way) Voltaire observes that the citizens of Paris were happy with the presence of aristocracy because through the literature noble people, common men could learn better language and better manners.
10. **Huizinga, J.** Op-cit. As Huizinga points out, play in ancient cities, "dorens life, amplifies it" and is to that extent a necessary "life function" and by its quality of spatial freedom and physical mobility makes itself off from the course of the natural process in the human

life. Play like the ceremonial centers, is not "real" and "ordinary".

11. For a good discussion of this issue, see **Michael, J.A.** The Camera Age: Essays on Television, Penguin Books, New York, 1981; **Ewen, S., and Ewen, E. :** Channels of Desire, McGraw-Hill Book Company, New York, 1982; **At wan, R. et.al:** American Mass Media, Random House, New York, 1982.
12. See **Breen, M. and Corcoran, F:** Myth in the Television Discourse, Communication Monographs, Vol.49, No.2, June 1982.

Chapter **61**

Web-Based Education at Higher Level : Prospects, Challenges and Problems

Uma Joshi, Anjali Pahad, Avani Maniar

Internet is becoming very popular as means of accessing information and disseminating information quickly and economically. Therefore, the teaching strategies are being thought of in which world wide web can be used a mode of instruction, which can provide subject matter.

Information Technology (IT) is becoming the buzz world in almost all the areas concerning human life in 21st century. IT has brought about radical changes in the ways and speed with which peoples acquire knowledge, store, share, and disseminate information.

IT has enabled us to work quickly and has helped decision makers. The definition of information has broadened due to the role of non-print material that is stored electronically. The world wide web provides vast amount of information or data which can be accessed from any part of the world and at any point of time. Thus, there is telematic revolution. Realizing the growth of information technology today, it is an urgent call of the day modernize teaching, research and extension in higher education, otherwise, our higher education system will be unable to keep pace with the development of the world.

Today, web-based teaching is possible due to online technologies, which enable global connectivity that can support innovative and constructive classroom teaching practices.

Main features of web-based technology (WBT)

According to Shah (2002, 11), main features of web-based teaching are :

- Students can access course materials through internet/CD (web instruction emulated on CD).

- CD based video support.
- Contact programme may be organized at few identified study centers to give support for complicated content subject.
- Web access **through dial up** connections internet, **cable modems can be used.**
- Learners can send their queries to the tutor through internet.
- Mode of evaluation can be both summative and formative and can be carried out through study centers and web respectively.

How to design Good Web based Teaching Programme?

- Identify user needs and decide the expected outcomes out of teaching.
- Interaction through web-based teaching should lead to more information acquisition and advanced learning.
- Keep individual differences in intelligence and other activities in mind such as audio learners, visual learners and verbal learners.
- It should provide feedback and reinforcement to the student.
- It should not include unnecessary details and explanations.
- The web page should be clear and readable.

Web based Teaching should have Knowledge regarding

- Present trend and developments in hardware software and communication technologies.
- Exploring information and judging its relevance for a purpose.
- Means of preservation for information.
- Internet and its efficient use of quick access of desired kind of information.
- Technology for intellectual property protection.
- Information Management.

Paradigm for web-based instruction

Web-based instruction can be implemented in two ways; according to Mohansundaram and Kumaran (2001, 10).

a. Web assisted classroom instruction

In this method, the teacher will contact the suitable website and collect more and recent information related to a topic and present in the classroom teaching. Historical places, scientifically important places and pictures can be shown in the class by using this method. Abstract ideas can be explained easily with 3-D picture, animation and multimedia. In this method the teacher will handle the internet and use the website as a teaching aid and the classroom teaching will be more effective and will meet the individual differences of the learner.

In this method the teacher will first frame the instructional objectives and plan. Then he will teach the lesson with the help of the website. There will be interaction with the learners in the form of discussion, questioning demonstration and project/experiment. The teacher/ learners can also interact with the experts suggested in the website. Learners can be evaluated through formative and summative evaluation. This approach will be effective at school instruction.

b. Individual learning

In this approach the learner will contact the website, collect information and learn the lesson. The learners individually learn the concepts as per their self-pace. This method will be effective in higher education and in research. The teacher will act as a facilitator and create suitable for the learning process.

After learning the text available in the website, the learner can interact with the teacher and carry-out the teacher prepared follow-up activities. The learners can be evaluated through formative and summative evaluation. This approach will be effective at higher education and research level.

There are certain merits and demerits of website teaching according to Mohansundaram & Kumaran (2001, 11).

Merits

- More and recent information related to a topic can be accessed and delivered.
- Updated information is available in the website.
- Abstract concepts can be easily explained with the help of information and graphics.
- Self-pace learning is possible in this method.
- More information will be gathered in short duration of time. So energy and time will be saved.
- Web-based instruction will be more effective at higher education and research level.

Demerits

- Information related to all types may not be available.
- Information may not be available in all languages or regional language.
- Its' an expensive method.

Potentials of web-based technology in teaching

Teaching at higher education, mostly concentrates on giving information which is not the sole objective of higher education. The objectives are multidimensional in nature, so to achieve these, the appropriate methods should be used in an integrated fashion. Web technology can be of help. It is an accepted fact that, today a single teacher is not capable of giving up-to-date and complete information in one's own subject. We can fulfill this gap by providing the access to different sources of information. The information provided will be correct, comprehensive and in different formats. Web technology provides this facility so that learners can exchange their ideas and views and get clarification from experts. It also provides variety in the presentation of the content, which helps learners in better understanding of concepts and long retention of information, thus helping in qualitative improvement of teaching learning process.

In research

Web technology should be made available to research students and teachers. The networking of research basses and the creation of information highways have already changed the methodologies of research. Today, a researcher has access to unlimited information and is exposed to a wider spectrum of ideas and concepts. The research data sharing analysis and interpretation has improved the status of research finding. Consequently the acceptability also has increased with research contribution getting much greater exposure an improvement in quality can be foreseen.

Challenges of web-based technology

With the introduction of online teaching tools, some challenges have also come up for higher education. First challenge would be in the area of teacher's education or teacher training. Teaching will have to be ready and trained to use online facilities of teaching. Secondly, the sources of knowledge of student have become unlimited today. Teacher will have to be prepared to guide students to use these sources constructively.

The institutions of the learning will have to plan and work for implication of I.T., as there is a shift from the culture of print and paper to culture of digital technology. This can affect and require changes in all areas of higher education.

Teachers will face the challenges of redefining their role in web-based teaching. They will have to develop competence in teaching in the digital world order.

Universities will have to work towards network of university aids, redesigning of curriculum and learning experiences, which can homogenize higher education system in the country. This requires finances, modern attitude of the authorities and teachers and enthusiasm to implement web-based teaching.

Institution will have to make sure the access of IT to the learners otherwise it may become unaffordable to students from underdeveloped and remote areas or those from low socio-economic status.

Teachers will have to choose between various alternatives while planning web-based teaching. Such as, whether he/she wants to use online teaching when meeting together in class or when the learners are at a distance or a combination of both. Different strategies will be required depending on these choices. Teachers will have to be careful aware that in order to use online teaching they do not sideline the pedagogy.

The purpose of web-based teaching is defected if it is unable to provide learning opportunities to the large majority of learners. Therefore, the policies for online course at government level and institutional level have to create to provide for such facilities and service, such as free supply of equipment etc. to the institutions.

All those involved in planning for web-based teaching in development of software for teaching administrators and teachers will have to take all the necessary steps to sustain the relevant and necessary content on the web, because the content and course on the web change after sometime.

The research designs will have to be developed and implemented which accommodate nuances of teacher centered, non-technology teaching as well as to try out alternative instructional paradigms that optimize the potential benefits of web-based teaching. Finally the challenge thereon by web-based teaching to the education system is to maintain the learning sprite, maintain academic freedom, dignity of teachers, quality of education and access of web-based teaching to all the students.

Madhukar (2002) discussed the following problems related to web-based teaching, (i) At present, limited bandwidth and slow modems, hampers the delivery of sound video and graphics, although the technology is improving all the time, (ii) Learner success also depends on technical skills in computer operation and internet navigation as well as the ability to cope with technical difficulties. Technical problems were found to be an obstacle for student taking web-based courses, (iii) Information overload is also an issue, the volume of e-mail message to read, reflect on and respond to can be overwhelming, and the proliferation of databases and website demands information management skills.

(iv) Access to the internet skill is a problem for some rural areas and people with disabilities.

(v) Social isolation can be a drawback and the lack of nonverbal cues can hinder communication. Although the internet can promote active learning, some contend that, it can breed passivity.

(vi) This new format requires more time for preparation and maintenance than traditional on-campus teaching. This causes additional stress for instructors especially if they are put into action without proper training and support. Integrating unfamiliar technology into one's instructional strategies can be stressful at first.

(vii) Developing a web-based course is a complicated and time-consuming process. Without the help of support service, developing and implementing successful web-course can be an uphill battle. Poor course design conversely can foster shallow assessment procedures, restrict student freedom over activities and topics and cause students to be extrinsically motivated.

(viii) Some educators claim that the time involved in delivering a web-based course can be up to three times as much as face-to-face one.

(ix) In concentrating on the individual course interactions have often neglected the fuller academic resources support necessary to mount full-service web-based degree programmes.

(x) Computers are good at automating multiple choice matching and short fill-in-the-blank test items, but are usually not to grade essays and performance assessments. To use this latter type of assessment, a certain amount of personal interaction and communication are required.

Conclusion

In the new digital environment, adaptive capability is the key to survival and growth because traditional methods of functioning will guarantee neither success nor sustenance. Reddy (2001) feels that universities in the developing countries face similar obstacles in terms of paucity of funds and material resources, capital equipment and infrastructure. This requires careful planning that will certainly reveal the possibility of pooling resources and expertise in areas of shared interests.

Universities and colleges will have to be adaptive and restructure their education system and adopt IT based prospective in teaching, otherwise, they will not be able to compete with other higher education segment. This can increase the demand for IT based services from colleges and universities to be point the existing faculty resources remain fully employed and continue knowledge development and application.

References

Ahlawat P.; Khurana A.; Rajan A. and DhunnaM. (2001) Role of Information Technology in Enhancing Academic Productivity. University News, 39(42) October 15-21.

Ganchi D.A. (1999) The Cyberage, Education and the Paching Profession. University News, 37 (43) October 25.

Gill N., Dabas K. (1999) Challenges before Information Society. University News, 37 (12), March 22 p.g. 17-22.

Garg S.; Panda S. (2001) World Wide Web and Distance Education. University News, 39 (44), October 29-November 4.

Hawthrone E.M. (2001) Technology and Teaching - Why Change? In Media Asia,Vol20No.l.

Joshi. S. Khirwadkar. A., Tomar, A (2002) Information Technology - A Revolution in Higher Education. University News, 38 (30), July 24. Madhukar I. (2002) Internet Based Distance Learning. Himalaya Publishing House, New Delhi.

Mohansundaram K.; Kumaran D. (2001) Web-based instruction : An Innovative Teaching Strategy. University News, 39 (47) November 1925.

Shah B. (2000) Web-based Instruction - online learning. University News, 38 (51) December p.g. no. 10-13.

Venketeshwarlu D. and Shah B. (2002) "Information Technology and Teachers" in University News, February 2002, Vol. 40 No. 5.

Chapter 62

Communicating Culture in Times of Distress : Indian Media during the Kargil War

M. Neelamalar, D. Viswanathan, Ila Joshi

Indian culture is well known in the world for its uniqueness, greatness and sacredness. It is one thing to be civilized and cultured in ordinary times and quite another to possess the same quality in a war situation. In all the wars independent India has faced, the Indians - Army as well as public - have shown extraordinary glimpses of their culture. While in the first four wars, i.e., Indo-China War 1962, Indo-Pakistan wars 1947, 1965 and 1971, the media in India was still developing, but during the Kargil War (1999) the country had a developed electronic media and a vibrant print media vying with each other to bring out the greatness of Indian culture along with war information.

Kargil War

Kargil War (May-July 1999) was fought against the Indian Pakistani forces. The Pakistani forces were found to be composed of Pakistan army units, mercenaries and Afghan war veterans (Bahl, 2000). The metropolitan papers and electronic channels sought to rush their mainline correspondents to the scene of action. From then on, it was 'Kargil by the night and Kargil by the day' as there was a continuous flow of informations about Kargil from the warfront to the rest of the nation. Let us see in details the features of Indian culture evident during the Kargil War and how has the media communicated those aspects to the world.

Unity in Diversity

The basis for Indian culture is deeply rooted in the doctrine, 'Unity in Diversity'. Though internal disturbances occur here and there, in the face of an external threat, Indians have always stood united. And this was brought out by Indian media during the Kargil

War. During war, the Indian Government decided to handover the bodies of dead soldiers to their families, instead of burning them at the war front. The media captured the scenes of body bags of soldiers being brought from the war front to their hometown and their last rites being performed amidst military honour (Kargil Review committee, 1999). In front of the television, the whole nation expressed grief for the soldiers-irrespective of the region they belong to as well as saluted them. Blood Bank camps were conducted throughout the country to help save the injured soldiers. From Kashmir to Kanyakumari, Indians indulged in fund collection for the Kargil Relief Fund. From Children in Bangalore to prisoners in Tihar jail, people contributed their saving to the fund. Youngsters all over the country were eager to join the Army and longed for a chance to throw the infiltrators out of the country (Sidhu, 2001). The famous personalities like cricketers and film stars of the country hosted shows for the Kargil Relief fund. Thus, people in different regions, of various age groups and background were united in their will and wish to do something for the Indian Army (Verma, 1999). Coming together of the whole nation for serving the soldiers, who were ready to give up their lives for the nation was effectively brought out by the Indian media.

Valiance

From time immemorial, Indians were known for their valiant deeds and valiance is an established part of Indian culture. Indian literature eloquently described valiant deeds of ancient Indians. Indian history speaks in length about valour of various emperors who have ruled different parts of India from time to time. During Kargil War, the Indian Army exhibited an incomparable act of valiance and to know the full extent of their valiance, a vivid look at the background of the war is essential.

Kargil is a small town in Ladakh in Indian Kashmir, which lies on the old caravan route from Tibet to the Kashmir valley. The caravans carrying pass from Tibet, preceded to Leh and thereafter to Srinagar via Kargil and Zojila (Zoji pass). This pass, the only negotiable opening in the Great Himalayan Range, which separates the Kashmir Valley from Ladakh, remains snowbound for six months during winter every year. Kargil and the rest of Ladakh are cut off from the Kasmir valley during this period. The Indus Valley route from Gilgit-Skardu to Leh, remains open for most of the year and can be used by the Pakistan Army even in winter (Y. Bahl, 2000). Brass, a town near the entrance of Zojila, and Kargil a few miles to the east, remain desolate and deserted throughout the long winter. These habitations come to life again only when Zojila reopens for traffic in summer. The road from Srinagar to Leh crosses Zojila at an elevation of nearly 3,520m in one way, with steep gradients (Bahl, 2000).

The Line of Control in the Drass-Kargil sector passes uncomfortably close to Zojila, Drass and Kargil. Several Pakistani posts along the LOC overlooked National Highway IA in the Drass-Kargil sector. In the late 1990s, artillery barrages from Pakistan to disrupt the supply convoys en route to Leh have been common feature. Extensive armed intrusion in Drass, Kargil, Batalik (east of Kargil on the Indus) across the LOC created a war-like situation in Kashmir. The Pakistani forces were found to be composed of Pakistan army units, Muslim mercenaries from aboard and Afghan war veterans. They were well equipped with sophisticated weapons, including stringer missiles. Pakistani artillery deployed across the LOC supported these elements operating inside the Indian controlled areas. By the time the Indian army got its act together and started an operation to push back the Pak forces across the LOC, the Pakistani forces were already well entrenched and had stocked ammunition and ration there for months. The tough task of evicting well entrenched troops from the barren mountaintops fell on the shoulders of the Indian Army.

Because the heights were occupied by Pakistani infiltrators, the Indian Army indulged in the tactics of sending a troop to the top and from them finding the whereabouts of enemy troops (Kargil Review Committee., 1999). This resulted in a heavy loss of personnel for Indian army. At this stage, the Indian media highly criticized army's 'suicidal mission' (Sawant, 2000). Indian army through sheer will power and the determination to safeguard the sanctity of the Indian Territory fought against all odds, claimed one peak after another and at last forced the Pakistani troops to retreat. The Indian media, too, glorified the Army and publicized all their heroic victories. They kept on passing every details of valour of Indian soldiers to their respective news headquarters. It was only through media that, the whole nation indeed, the whole world could get to know about the valour and grit of the Indian Army and Indian Air Force.

Sacrifice

India is a nation well known for sacrifice, saffron itself being included in the National Flag for stressing on sacrifice. In the Kargil

War, more than 500 army personnel have sacrificed their lives for the nation. The injured soldiers, who lost parts of their body, did not care about their loss, but were bothered that they became disabled to take part in the war and teach the enemy a lesson. Not just the soldiers but their whole family parents, wife, children took great pride in sacrificing for the nation.

All the television channels in India, whenever they showed body bags of soldiers being carried to their homes, interviewed their parents. All the parents took great pride in their son's sacrifice. A letter written by the widow of a Kargil martyr from Punjab after losing her husband in the war to the Prime Minister was much publicized through all forms of Indian media. In the letter she had written that she was ready to send all her three sons to Kargil, if the country needed them. A bride from Rajasthan had donated all her jewellery, worth in lakhs, to the Kargil Relief Fund. A retired soldier wrote to the Chief of Army Staff that though he could not fight in the war, he could be used as a human bomb by the Army. When media brought to light all these facts, the viewing public appreciated the sacrifices and in turn, they were preparing themselves for such sacrifices (Verma, 1999).

Forgiveness

Forgiving the enemy is one of the features of Indian culture. The dead bodies of Captain Sourabh Khalia and five other were returned in a mutilated states by Pakistan. The Indian army forgave such a cruel and inhuman act and buried the leftover bodies of Pakistanis, according to Islamic customs (Sawant, 2000). When the international media highlighted these contrasting attitudes of two warring countries, the international community appreciated the gesture of Indian Army and its superior culture.

Commercialization

Of late, commercialization has become an inseparable part of Indian culture. The hype and hoopla given to war news by media resembled the up-market, commercial approach. The corporate sector also made use of the national sentiment for boosting their sales. They came out with schemes of donating money to the defense funds against purchase of their products. Some of the multinational corporations also followed this tactics. One of the newspapers published the picture of a rifle with actress Sonali Bendre's photograph affixed on it (Sachdev, 2000). The advertising lingo has also caught the soldiers fighting the war. In his interview to New Delhi Television (NDTV), one caption narrated how after capturing a peak, they decided 'Yeh Dil Mange More' (Pepsi's mission statement) and went on to capture another peak.

Conclusions

It has often been said that communication and culture are inseparable. Godwin C. Chu observed that every culture pattern and every single act of social behavior involves communication. To be understood, the two must be studied together. Culture cannot be known without a study of communication, and communication can only be understood with an understanding of the culture it supports (Jandt, 2000). This is quite true in the context of the Kargil War. A study of communication by media during the Kargil War explains the efficacy of Indian culture. Likewise, only those who know the basic values of Indian culture such as tolerance, forgiveness, non-violence, self-control, humanity etc. could appreciate the role played by Indian media during Kargil War.

References

Bahl, Y. (2000), 'Kargil Blunder Pakistan's Plight, India's Victory', Manas Publications, New Delhi, pp 34-35.

Fred E. Jandt (2000), 'International Communication - An Introduction', Sage publication, California, p 28.

Kargil Review Committee (1999), Report : 'From Surprise to Reckoning', Sage Publications, New Delhi, pp 91-92.

Sachdev, A.K. (1999), 'Media Related Lessons from Kargil', Strategic Analysis, IDSA website. www.idsa.org.

Sawant, G.C (2000), 'Dateline Kargil' Macmillan India Limited, New Delhi, pp 36-37.

Sidhu. S (2001, Sept 6-8), 'Media coverage of operation Vijay and Its Impact on The civilians' conference paper read at IAMCR section psychology and public opinion Budapest.

Verma, P.K. (1999, Sept), 'How much was Hype, How Much Real concern', Humanscape

website, <www.humanscaplndia.net/humanscape/hs0999/hs99991 t.htm>

Chapter **63**

Methods of Communication for General Science Education

N. Natessan, Mrs. S. Girija Rani

Effective communication is the most important function required for a successful teacher. Effective teaching motivates the student interests by kindling their abilities. Effective teaching, in any subject, depends upon the introduction of innovative methods of instruction. There is a growing need for trying out new methods of instruction and establishing their effectiveness in teaching. Now-a-days a teacher cannot depend on any single method of teaching. Multimedia has come into stay for ever because of its tremendous influence in the teaching learning process. Multimedia base modular approach is the 'latest trend' in teaching.

Module is a self-contained, suitable tested, auto instructional material. Modular approach provides strategy for the teacher and the learner to mutually share the responsibility for learning. The teacher becomes a facilitator of learning.

Multimedia or media mix means "combination of many media". The term multimedia refers to the use of appropriate and carefully selected varieties of learning experiences which are presented to the learner through selected teaching strategies. The components of multi-media based modules include modular lecture. Group discussion, role play, workshop, individual study, film audio-video cassettes etc. This will stimulate the student interest in learning. It economizes time and effort, reduces verbal teaching and imparts knowledge to students.

There is a need to experiment the effectiveness of multimedia based modules and to assess their advantage over the traditional lecture method. In the Indian context, studies have been made by Vardhini (1983), Krishnan (1983), Menon (1984) and Basu (1983) to prove the effectiveness of multimedia, Dhamiji (1985).

Hopper (1983) and Sahajahan (1980) have conducted experimental studies to prove and measure the effectiveness of modular approach.

But no attempt has been made to study on the effectiveness of multimedia based modular approach with special references to low, normal and high achievers. Hence, the objectives of the study are :

1. To prepare modules for selected topics in General Science of VIII Standard.
2. To prepare multimedia teaching methods for selected topics in general Science of VIII Standard.
3. To study the effectiveness of multimedia based modular approach.
4. To compare the academic performance of low, normal and high achievers of experimental and control groups.

The following hypotheses were made with reference to the above objectives :

1. The achievement in the post-test performance of the experimental group will be significantly greater than the post-test performance of the control group of low, normal and high achievers.
2. The post-test performance of the experimental and control groups will be significantly greater than their pre-test performance of low, normal and high achievers.
3. There will be significant difference between the mean gain score of the experimental and control groups of low, normal and high achievers.

Methodology

The investigator selected 150 students of VIII Standard of Sri Rao Bahadur A.K.D Dharmarja Girls Higher Secondary School, Rajapalayam during the academic year 1995-96 as the sample of the study.

The investigator constructed and validated the entry behaviour test and administered it to) 50 students of VIII Standard. Based on their performance and matching technique, 25 students were selected for each of low (marks below 40), normal (41 to 59 marks) and high achievers (60 marks and above) of control and experimental groups. Investigator also prepared multimedia based modules for the selected twenty four topics of VIII Standard General Science Text Book. The validated pre-test was administered to the control and experimental groups, to test the student's knowledge on the twenty four topics to be taught before the application of treatments. Then twenty four topics were taught through multimedia based modular approach to the experimental group and through traditional lecture method to the control, group. Finally post-test was administered to both the experimental and control groups after the respective treatment.

Results and Discussion

Pre -Test Performance

Before treatment the pre-test was administered to low, normal and high achievers. The mean score and standard deviation of the test scores of the control and experimental groups were calculated. By computing 't' test, the significance of difference between two groups were compared.

Table-1. Control vs. experimental – pre-test

Category	Group	N	Mean	S.D.	't'	Level of significance
Low	Expt.	25	12.6	5.08		0.22 N.S.
Achiever	Cont.	25	12.3	4.48		
Normal	Expt.	25	17.12	3.60		0.46 N.S.
Achiever	Cont.	25	17.84	6.36		
Higher	Expt.	25	21.8	7.40		0.11 N.S.
Achiever	Cont.	25	21.6	5.08		

N.S. – Not Significant, Expt. – Experimental, Cont. - Control

From the above table it is inferred that there is no significance difference between the experimental and control groups in their mean score and standard deviation. The low, normal and high achievers stand in the same academic achievement. It is also proved that the two groups are matched before the post-test performance.

Post-Test Performance

The post-test was administered to low, normal and high achievers after the treatment of multimedia based modularapproach. The mean and standard deviation of the test-scores of the control and experimental groups were calculated. By computing 't' test, the significance of difference between the two groups were compared.

Table – 2. Control vs. experimental – post test

Category	Group	N	Mean	S.D.	't'	Level of significance
Low	Expt.	25	52.20	8.88	3.45	0.01
Achiever	Cont.	25	43.31	9.32		
Normal	Expt.	25	65.10	14.96	5.36	0.001
Achiever	Cont.	25	46.34	14.89		
Higher	Expt.	25	81.20	13.39	3.32	0.01
Achiever	Cont.	25	68.70	13.24		

Expt. – Experimental, Cont - Control

From the above Table 2, it is inferred that there is a significant difference between the experimental and control groups of low, normal and high achievers in the post-test mean score. The significant difference is at 0.01 levels in low and high achievers and 0.001 levels in normal group. The experimental group student's achievements are higher than the control group students of low, normal and high achievers.

Pre-Test Vs. Post-Test - Control Group

The following table furnishes the details about pre-test and post-test scores of control group of low, normal and high achievers.

Table - 3. Pre-test vs. post-test control group

Category	Group	N	Mean	S.D.	't'	Level of significance
Low	Expt.	25	12.30	4.48	15.05	0.01
Achiever	Cont.	25	43.32	9.32		
Normal	Expt.	25	17.84	6.36	8.81	0.001
Achiever	Cont.	25	46.30	14.89		
Higher	Expt.	25	21.60	5.08	16.64	0.001
Achiever	Cont.	25	68.70	13.34		

Expt. – Experimental, Cont. - Control

From the table 3, it is inferred that the obtained V value is significant. The pre-test performance of low, normal and high achievers differs from the post-test performance. The significant variation is due to the control treatment given to the control group.

PRE-TEST Vs, POST-TEST - EXPERIMENT GROUP :

The following table furnishes the details of pre-test and post-test scores of experiment group of low, normal and high achievers.

Table – 4. Pre-test vs. post-test experimental group

Category	Group	N	Mean	S.D.	't'	Level of significance
Low	Expt.	25	12.60	5.08	19.40	.001
Achiever	Cont.	25	52.20	8.88		
Normal	Expt.	25	17.12	3.60	15.62	0.001
Achiever	Cont.	25	65.10	14.96		
Higher	Expt.	25	21.80	7.40	19.47	0.001
Achiever	Cont.	25	81.20	13.39		

Expt. – Experimental, Cont - Control

It is observed from the above table the experimental groups of low, normal and High achievers differ significantly between the pre-test and the post-test. It indicates that the achievement was enriched after teaching through multimedia based modular approach.

Gap Closure

The following table shows the gap closure of experimental and control group of low, normal and high achievers.

Table – 5. Gap closure

(In percentage)

S. No.	Achievers	Experimental Group	Control Group
1.	Low achievers	43.30	35.37
2.	Normal achievers	57.89	34.86
3.	High achievers	75.95	60.087

From the Table 5, it is concluded that the gap closure of the experimental group is greater than that of control group. This further confirms the superiority of the effectiveness of multimedia based modular approach in teaching of general science during the application of treatment.

The data of above tables indicate that there is no significant difference between the means of the experimental and control groups of low, normal and high achievers in the pre-test. This implies that the two groups were in the same level of academic performance of the topics prior to treatment.

The post-test score of the experimental group and control group of low, normal and high achievers differ significantly. The performance of the experimental group is greater than that of the control group. This implies that the improvement in the level of academic achievement is due to the application of experimental treatment. In other words, the multimedia based modular approach is more effective in teaching general science to the students of 8th standard than the traditional method.

The Gap closure also confirms the effectiveness of multimedia based modular approach over the traditional method in teaching of General Science during the application of treatment.

Conclusions

The traditional teacher depends on verbal exposition. In a fast developing world, where knowledge explosion is taking place in every sphere, it is unreasonable to expect that the spoken or written words alone could convey the volume of relevant information to the learner. Tape recorders and Television play a pivotal role in the teaching learning process. Multimedia packages help to overcome communication barriers. The dual effect of audio and video strengthens and enriches the understanding and mastery of the concept. Thus, the present study proves that the multimedia based modular approach is more effective in teaching general science at the VIII standard level than the traditional method.

References

Best, J. W., (ed.) 1977, Research in Education, *Prentice Hall of India, New Delhi.*

Borg. W.R. & Gall, M.D., 1924 Educational Research, *Mac Millan Co. New Delhi.*

Buch, M.B. (ed.), 1987, Third Survey of Research in Education, *NCERT, New Delhi.*

Cebeceinas & James, 1981. The Multimedia Library, Materials, Selection and Use, Academic Press, New York.

Chandran, M. R., 1990, Science Standard VIII, Tamilnadu Text Book Society, Madras.

Garrett, H.E., 1979, Statistics in Psychology and Education, Vakils Feffer and Simons Ltd., Bombay.

Chapter **64**

Impact of Mass Media on Regional Language

V. Renuga Devi

Today's world is world of modern technology communication field is of no exception. In our day-to-day life we come across so many information/ communication instrument one such is television. Whether it is a radio, television, pager, or e-mail we communicate through a code. That code is language. Without that code we cannot communicate. So, language plays an important role in communication. This paper attempts an analysis of language use in television because nowadays one cannot image a house without a television a powerful media which brings everything to our drawing room. The changes, whether a cultural change or language change, which are broadcast through television reaches the masses easily and the impact on masses is also in higher degree.

Moreover, the investment of time and money in the telecast media is obvious. It is this immense investment and the hotly debated effects of this investment that demonstrate the importance of the telecast media in our lives.

The audience in telecasting is not a single individual receiving message meant for him alone. 'It is meant for masses. Telecast audiences are in Head's word 'relatively large and undifferentiated[1]". Large is difficult to define. One might say that this means any audience too large for a speaker to communicate with in person in the amount of time available to him. By, 'undifferentiated' we mean that the audience is composed of a mixed group. It is heterogeneous and it hard to keep up with the hoursper day that a family watch television, the number of TV sets in the country.

A number of programmes are delivered through TV throughout the day. These programmes may be a talk programme, interview, discussion, drama etc. Even the advertisements which excel the programmes form part of the TV telecast. Thus, every channel puts out different things. The language with which you express ideas can build understanding and even create delight.

Tamil is diglossic language. InTV programmes both the varieties that is the written and the spoken variety are exploited to communicate to its various audiences.

Generally the written variety or the standard spoken variety is used in news reading and in some formal speeches. In most of the programmes the spoken varieties i.e. dialects are used. Moreover, in using the spoken variety also, the linguistic techniques such as code switching, code mixing, loan translation, loan blend, borrowing are extensively used in certain programmes.

The written variety is used in news reading on the other hand extensive use of code mixing and code swathing is used in Tele Shopping Network (TSN) programmes. In most of the programmes a mixture of Tamil and English is used. We can see the influence of English to a larger extent.

Of the major south Asian languages Tamil is least influenced by Sanskrit and infact has a strong persisting tradition of maintaining its distinctness from Sanskrit language and literature. Now the reverse is the case, of all the Dravidian languages, Tamil is mostly influenced by English. Like manipraval style, a style Tamil angilam style may come into existence in future.

Lexical Level

New experiences, new objects and practices bring new words into language. It does not matter whether the new objects and practices come to the community, by way of diffusion, or the community goes to the new objects and practices by way of migration the result is same. As Hockett (1970:405) rightly points out whenever the need filling motive plays a part the borrower is being confronted with some new object or practice for which he needs words.

Language is not frozen. It is alive and dynamic. Language is flexible, undergoing constant change unless the language is truly a dead language.

Tamil is a dynamic and very flexible language. We come across so many English Words in Tamil programmes. We are importing not only the technology from America with that we are importing language also. In TV programmes a large number of borrowed words are used.

Technology	Palm-Oil	Cricket
Formula	Coffee	Company
Family	Natural	Agency
Horlicks	Super	Audio
Colgate	Cumulative	Video
Bajaj	Special	Episode

These are the few words which are out rightly borrowed from English and used as such in Tamil programmes.

Phonological Level

Phonological system of Tamil is mostly violated due to the influence of English.

In Tamil, the voiced stops never occur in the word initial position. Because of this characteristic feature Tamil don't have any grapheme to denote voiced stops.

But due to borrowing of words like dappaa, andku, geettu etc. voiced sounds came into existence in Tamil phonological system. This change in the phonological system creates a need for the creation of graphemes for voiced phonemes in Tamil. Tamil doesn't have the fricative sound /f/ and so Tamil doesn't have the grapheme to denote this fricative sound. But, due to borrowing the phoneme /f/ came into existence in Tamil and it is denoted by the grapheme to denote this fricative sound. But, due to borrowing the phoneme /f/ came into existence in Tamil and it is denoted by the grapheme.

@ Toou[1]   @ **Toom**
@©Wv   @TooBXo

In Tamil, only vowels and vowelled constants can occur in the word initial position. But, due to borrowing the consonants are used in the word initial position.

KriimdAm   KriimardAUo

Tru   hU   tril   j p   praapalm **IWOTim**

The use of Tamil graphemes such as i (X), t (R), r (W), r(\), n (Q), n (e) etc., in word initial position is prohibited in writing by the grammars of Tamil. But these sounds are commonly used in wiring in the TV programmes.

raakam	WOLm	laari	XO
laa Tari	XOhP	dappaa	PITO
kraapar	dWoTo	Tru	hi

Grammatical Level

In grammatical level also English stems or English suffixes are borrowed. Tamil grammatical markers like case markers, question particles, plural marjkerare used and suffixes with English stems. English stem + Tamil case markers

Centre + il (Cp)	centre-il
Horlicks + in (Cu)	Horlicks-in
Technology + ai (I)	technology-ai
Agency + kku (di)	agency-kku

English doesn't have much of case markers in practice as in Tamil. Instead of case markers prepositions are used in English. But in TV language English stems are coined with the Tamil case markers.

The interrogative particle-aa is used with English stems,

wife-aa?
company-yaa?
anta agency-kaa?
Particle-um is also used with the English stems
milk shake-um
base-um
bate - um ball - um

The adverbial and adjectival markers - anna and -aaka is also used with English stems.

Chinese - aaka
program -aaka
super - aana
light - aana

The plural marker - kal in Tamil is coined with English stems to form plurals.

pumset - Kal
lattari - kal

Not only Tamil Gramatical suffixes are coined with English stems, the vice versa is also practiced in TV language.

The English possessive marker-s is used with Tamil nouns

Maleciya vaasu teevan's son
Velan's silk
English plural suffix -s is also used with Tamil nouns.

vidvaan-s
raakam-s

Syntactic Level

In any language, phonological system and grammatical system are susceptible fo influence of other languages. But syntax is comparatively rigid. In syntactic leave we come across extensive use of linguistic techniques called code-switching and code-mixing.

Code Switching

Code switching is the alternation of two languages within the single discourse, enkaloota best seller roti king atil imported

the most at-um irukku u Tanee order ceyyumka
innum oru free gift kitaikkum cooking and
surving-il murrilum atu unkalukku utavi ceyyum
niinkal protect-ai order ceyavatarkku inta
address-il to Tararpu koL Lalaam
namma problem solve akanmnutann naan India pooReen.

Code Mixing

Code mixing refers to the use of one or more language for consistent transfer of linguistic unit from one language to another. Code mixing takes place intrasententially.

inta ULLam ke Tkuinoo more
unka quality-ai eppa Ti namparatu
inta maatiri enta ka Taiyiliim checking ki Taiyaatu

Apart from code fixing and switching dialects are used in dramas to bring out the real colours of the characters.

Brahmin Dialect

Citta iruntaa cattamutu tayaarayiNTrukku

avaalellaam enkee?

nii turttam koNTu vaa

aattukkup pooRoom

Madras Dialect

olunka cooRu tuNRiyaa?
ellaam on annaattaiyaale vatatu

Any language need to be planned and should be standardized. If it is not standardized confusion will cause. For example, the word make-up is used as oppanai in certain instances and in certain instances it is transliterated as meekkap. The following words are the few examples for inconsistent usage.

Protection

A few words are completely translated into Tamil and they are consistently used in all their occurrences.

vacant tokuppu	**YNJj iiRoiix**
makkal to Tarpu	**uDls uRoPox**
vaakan oTtunar	**YoLi jhoSo**
tiraik **katai**	**giiWdLiiK**
vacanam	**YNIm**

A few words are transliterated and used as it in all their occurrences,

kiraapiks	iWoOdv
Tai Ttils	uPhYpv
cavuNT epek Ts	Niiih GuTdhv
aan lain e TiTTin	Bu uxu Gyhye
aap lain eTiTTin	BI uXu GyhYe

Change in Cultural Language

Kinship is a means of social control which demands precise and appropriate words. Each society or culture has a vocabulary which represents its important features and relationships. In Tamil culture father-in-law and mother-in-law are addressed as maamannar and maamiyaar.Bul due to the impact of TV, people started addressing their in -laws as uncle and aunti.

In yesteryears wife never address her husband. She may refer him as avaru or avanka. Now a days in the dramas which are telecasted in TV it is shown that a wife is addressing her husband by name and also by the pronoun nii.

This shows that due to the impact of TV, one can notice the changes in cultural level also.

Conclusion

This paper though, it is not a full-fledged study it stresses the need for the study of language use in Television. Though, the living language is flexible and dynamic, Tamil is more flexible. One can come across extensive use of English in Television programmes. Private channels are the promoters of this kind of language use. In this context, there is a possibility of formation of pidgin.

In one side, Government is trying to evolve Tamil as the administrative language in all its transactions. On the other hand, in a powerful media like television, English is extensively used in all the levels i.e. phonlogical, grammatical and syntactic level.

Borrowing may enrich a language. At the same time there should be a control over borrowing. If we borrow a term, it can be transliterated and used in the beginning. After sometime, if there is a possibility that word should translated into native word and then it should be standardized. The standardization process should be done on the following basis.

1. Adaptability
2. Economy / simplicity
3. Uniformity
4. Inter disciplinary approach
5. Language Purity

Mass media form a part of the society. We say it is a mirror, which reflects what is there in the society. So, steps should be taken to standardize the language use in mass media especially in the Television.

The above analysis reveals that the non-native elements have entered the native system and is enjoyed by the Tamil people, though the government is talking about pure Tamil or Tamilisation or nativization.

REFERENCES

Hocket,C.F, *1970. A* course in Modern Linguistics, *Oxford & IBH Publishing Co., New Delhi.*

Chapter **65**

Meaning Lag in Development Communication

Anita Jhamlani, Premlata Singh

Development depends heavily on communication however, the term Development Communication has come to mean different things to different persons at different times. Nora Quebral (1988) has aptly defined this term as Development Communication is the art and science of human communication applied to the speedy transformation of a country frompoverty to a dynamic state of economic growth that makes possible greater social and economic equality. In this light, communicating effectively in developing countries is a challenging, uphill task. Communicators have been very preoccupied with evaluating the relative effectiveness of different media. A basic premise is that, whatever the medium, it must be accessible for the use of all, and all must be able to participate for real development communication to take place. Aside from this need, it is essential that the source of information be not only comprehensible but also credible. This point may seem very obvious but it has often been overlooked. It may be even more important than being equipped with sophisticated media.

It is now important to understand why the fast hare of information is out stepping the slow tortoise of meaning formation. This crisis in meaning has to be carefully identified and analysed. There is no value in a flood of informationif it is just a heap of broken images. (T. S. Eliot). The crisis in meaning formation can be examined, in terms of information overload in context of what is happening to the amount and quality of information, by the well known concept of lag.

Ogburn's basic idea of cultural lag was that material culture change faster than non-material culture resulting in maladjustment because of old habits and ideas cannot keep up with new realities. Similarly meaning lag occurs between mere information outpour and meaning formation. Whereas facts accumulate rapidly, meaning emerges slowly, perhaps never at times. Not only have the biological limitations of the human brain been to be considered but also the noise. Noise is not merely auditory but has a broad spectrum, including semantic, informational and stylistic forms (Klopp, 1978).

One of the hypothesis for slow meaning formation is that poodering is inherently slow as suggested by synonyms like brooding, meditating, contemplating, deliberating etc. The rapid pace of modern life coupied with rapid media messages allow little time for pondering. According to Rosnow et. al. (1976) 'Talk has always been a major process of meaning formation whatever its form discussion, debate, brainstorming, counseling, psychotherapy, encountering, group therapy, chat, gossip and rumour. Slowness characterizes most talk compared with mere information transmission since it gives time for the other's thought and answers. The ever increasing mountain of information gives little time to gain meaning. Along with this the social networks are becoming sparse where people are unable to interact with neighbours and friends. In this era of rapid research, and technology generation, especially in agriculture, technologies are generated by the scientists and usually passed on through extension agencies to the farmers. This trend was followed throughout the sixties (one way information flow; soon after it was realized that technology generation without feedback from the farmers could not be as useful as it could be with input in terms of feedback from farmers. Thereafter in the seventies and eighties a lot of emphasis was placed on feedback in communication. Today, when the pace of technology generation is outstripping its consumption and traditional knowledge is once again looked upon as traditional wisdom for sustainability the age old tradition of farmer to farmer interaction is being revived by the extension agencies. This goes a long way in conserving the traditional wisdom and diffusing it among the new generation. Farmer to farmer interactions is once again being considered very valuable for sustainability in agriculture and ultimately sustainability in development. In the light of the declaration of the year 1993, as the 'Year of the Family' it is advisable for communication scientists and workers to lap emphasis on direct person to person interaction. In this fast pace of modern living and the television boom in rural as well as urban societies family members hardly spend enough time in talks and discussions with each other, leave alone with neighbours and friends. This busy life style further aggravates the meaning lag.

Further, brain research (Ferguson, 1980) reveals that our culture overloads the left hemisphere with calculative, verbal, digital information while inadvertently neglecting pattern recognition and intution by the right. If there is such a difference in hemispheric function, then much information is bound to seem noise like when unsuited to the hemisphere trying to process it. Further more the sea of information without feedback adds to the difficulty of meaning formation. Today media bombard, they do not answer us.

Hence in the modern society information is becoming more and more noise like because of not only the overload but due to it, being disconnected. There is a heavy burden in decoding such information.

This can be further analyzed, in terms of meaning lag in (a) pictorial conventions in development communication and (b) effects of ability on comprehension of television.

(a) Meaning lag in pictorial conventions :

In our country where the literacy rates are low and the written word fails to communicate effectively, pictures are often used in campaigns on family planning, nutrition, health aid hygiene; and agriculture. One can not miss these posters and pictures when one visits the Health Centre of the school or the block office or even on the roadsides.

However, the villagers may not be familiar with the use of pictures as a communication device or communicators may not be aware of a villager's ability to interpret learned pictorial conventions. Field workers and scientists in a variety of situations have found that pictures are not recognized as 'representations'. Visual messages are misinterpreted and message content is culturally taboo or offensive. Further, even visual messages that effectively communicate to one population may prove meaningless or offensive to another. This is another factor that widens the information lag. In fact the development of pictorial literacy skills can be viewed as a continuum that parallels a population's exposure to visual messages. Hence, to apply visual communication effectively it is especially important to assess the specific population's common visual conventions and their perception of pictures used in rural development programmes. More effective visual communication can be achieved by testing and using feedback to modify visual messages so that they are understood by the target populations. Efforts must be made to teach pictorial conventions in order to avoid noise, confusion or misinterpretation.

(b) Meaning lag in comprehension of television :

Very few studies have probed the effects of ability on comprehension of television. Fezdex et. al. (1987) focused on cognitive skills they considered unique to television comprehension. In adults, a facility in manipulating visual/spatial information (as measured by a sentence picture verification task) was found to be related to better comprehension of television. Cavanaugh (1983) found that adults free recall, probed recall and recognition of a television programme were related to their verbal ability. He hypothesized that low verbal individuals may encode less information about the programme than high verbal individuals. Likewise in children, Eckhardt, Jacobvits et. al. (1991) found that verbal skills accounted for forty two percent of the variance in children's comprehension of central events. High verbal children appeared to have more selective memory for central over non-central events, compared with average verbal children of the same age.

Encoding occurs in the context of knowledge (Hunt, 1985), Ferfitti (1985) Patts et. al. (1989). Background knowledge is necessary to encode, integrate and store information in meaningful ways. Knowledge is also important at the time of retrieval Knowledge cues can influence the type of content that is remembered and the type of errors that are made (Johnson, 1982). According to Graesser (1981) memory is reconstructive over time, and scheme knowledge, which guides the comprehender's interpretations and allows information to be inferred to fill in the missing gaps, becomes more important in guiding retrieval as the retrieval interval increases. So in this context, the meaning lag in communication can be reduced if due importance is given to verbal skills and background knowledge.

Hence, communicating effectively for development is a challenge. Communication methods and media should be used meaningfully and wisely to help in meaning formation and in speeding up this process. Probing and indepth studies are needed for education, for meaning rather than education by and for mere facts.

References

1. Cavanaugh, J. C. (1983). Comprehension and retention of television programmes be twenty and sixty year olds. Journal of Gerontology. 38 : 190-196

2. Eckhardt, Jacobitz, B. B. Wood M and Albin (1991). Verbal ability and prior knowledge : Contributions to adults comprehension of television,

Communication Research 18 : 5

3. Ferguson, Marilyn (1980) The Aquarian Conslpiracy.; Personal and Social Transformations in the 1980s. Los Angeles.
4. Graesser A. C. (1981) Prose comprehension beyond the word. New York, Springer Verlog.
5. Hunt, E (1985). Verbal ability in R. J. Stennberg (Ed) Human abilities. An information processing approach. Freeman New York, 31-58.
6. Johnson, RE (1982). Retrieval cues and the remembering of prose. A Review in a Flammer and D, Kintsch (Ed), Discourse processing. Amsterdam.
7. Klapp, Orrin E (1976). Opening and closing : Strategies of information Adaptation in Society. New York : Cambridge University Press.
8. Nora C. Quebral (1988). In Abhilasha Kumari. Development Communication : Need for meaningful reference. Communicator July.
9. Perfetti C. A. (1985), Reading ability. In R. J. Sternberg (Ed) Human abilities : An information processing approach Freeman. New York. 21-35
10. Pezdex, K., Simon S., Stoeckert, J. and Kiely, J. (1997), Individual differences in televisiòn comprehension : Memory and cognition. 15 : 428-435.
11. Paths J. R., John M. F. and Kirson D. (1989), Incorporating new information into existing world knowledge, Cognitive Psychology, 21 : 303-333,
12. Rosnow, Ralph. L and G, A, Fine (1976), Rumour and Gossip, The So.ial Psychology of Mearsay Review, New York.

Chapter **66**

Communication Policy Imperatives for Research

K. Stevenson

"Communication can no longer be regarded merely as an incidental service and its development left to chance. Recognition of its potential warrants the formulation by all nations, and particularly developing nations, of comprehensive communication policies linked to overall social, cultural, economic and political goals. Such policies should be based on inter-ministerial and inter-disciplinary consultations with broad public participation. The object must be to utilize the unique capacities of each form of communication from interpersonal and traditional to the most modern, to make men and societies aware of their rights, harmonise unity in diversity and foster the growth of individuals and communities within the wider framework of national development in an inter-dependent world.

"Within the framework of national development policies, each country will have to work out its own set of priorities, bearing in mind that it will not be possible to move, in all directions at the same time. But as far as resources allow, communication should aim at stimulating and encouraging ail means of communication."

Has India complied with the recommendation of the Mac Bride Commission in formulating a national communication policy? No. In fact, the Lok Sabha Committee on Assurances rapped the Government of India recently for not framing a national communication policy as assured eight years ago, as reported in The Hindu. A few attempts were still born and today we do not have a resolution listing the objectives of communication and its functions like in the other sectors. This paper examines the need for policy formulation and identifies a few areas of research which would help in framing it.

An expert body at a meeting convened by the UNESCO in Manila in 1977 considered the question of a communication policy which may be considered as a synonym for a National media policy. The body observed that "A national communication policy means an integrated and reasonably durable set of principles, rules and guidelines to assist the communication systems of a country to operate efficiently to help and to attain national goals. It provides a frame of reference for the elaboration of national strategies and plans for communication, infrastructures and activities making the best use of available resources". With these aims, objects and principles enunciated in such general terms, there may not be any controversy. Any broad policy stated in general terms, may not attract any legitimate objections, as these terms are not likely to interfere with the independence of any media. Every medium may subscribe to these broad propositions which will be conducive to the welfare of the country. However, the economic model, the political system in India and a perfunctory peek into the mass media tradition pose problems in framing a precisely defined communication policy.

Mass Media Tradition

The press and the film industry were born during the thick of the freedom struggle. The press identified itself with the aspirations of the nationalist movement and found expression in lofty ideals like freedom, liberty and patriotism. The radio, which, also has its roots in the colonial era and patronized by the alien rulers in its formative stage, adopted a different stance. The press and cinema remained in the private domain relatively free from State control and manipulation before independence. But, the Television which began in 1959, and the radio have been under State control with the market component incorporated into it in the latter phase. The State market media relationship hints at the contradictory lines of thought socialist and democratic. The socialist angle demands containing the market control over the media and the democratic facet implies media freedom from Government control. "The diffidence and the hesitancy in framing a policy stems out of the conflict between these two thoughts."

In this context, it would be apt to recall the Former Press Council Chairman, Justice A. N. Sen's remarks that there may not be any difficulty in arriving at the general consensus as to certain norms to be observed by the media and lists the following :

1. Combating illiteracy, superstition and poverty.
2. Removal of casteism, communalism and untouchability and social prejudices of every kind.
3. Fighting all kinds of divisive forces and promoting national unity and integration.

4. Encouraging all round economic, social, cultural and scientific development of the country.
5. Promoting and protecting the right of the individual to liberty and privacy and not to indulge in any kind of character assassination except on justifiable grounds and in public interest.

Currently our so-called communication policies (if we have any) are framed without a research base, if not fully based on arm chair thinking. For instance, the Television came into India as an accident having started on an experimental basis in September 1959 to impart social education to the citizens of Delhi. But the experience of TV over the last three decades indicates that it will reach millions of masses with considerable momentum owing to Government's patronage. But, the boom will never become a boom unless the software potential is monitored and geared accordingly. The technology push rather than a strong programming pull may achieve, to quote the late Iqbal Malik journalist, 'proliferation of the Central message as massage'. Lack of a clear cut policy on traditional media and folk forms of communication which are more potent in rural areas has undermined their importance and most of them are on the wane.

The necessity to have a dynamic integrated communication policy and planning cannot be disputed. A four fold function of communication was even specified for the communication policy formulation. India also participated in this exercise at a ministerial level held in Kaula Lumpur in 1947. The UNESCO has spelt out steps in formulation of communication policies and planning and stated that policies should.

a) serve to marshal national resources
(b) facilitate rational choices with regard to means
(c) strengthen the coordination of existing or planned infrastructure
(d) help to satisfy the needs of the most disadvantaged and to eliminate flagrant imbalances.
(e) emphasize universal and continuing education
(f) help in strengthening cultural identity and national independence
(g) enable all countries and cultures to play a prominent

Role in the international scene

Research inputs into these and allied areas changing media scene and its implications will go a long way in giving shape to a policy.

Print medium and Market Forces

The Indian economy offers a fairly large scope for State intervention which has led to State maneuvering and market manipulation. A scrutiny of the print tradition reveals that the market forces had a clear edge over the Government. Sometimes it acted as an ally and at times as an adversary. This makes it imperative to study the free market model of mass media in a developing society.

Except for the First Press Commission (1954) and the Second Press Commission (1980) there has not been any coordinated and concerted effort to study the concentration of ownership and economics of the newspaper industry. Such research is necessitated as the industry has undergone sweeping changes during the past one decade. In other words, there is an urgent need to probe into the aspect of who controls the media medium? Who are the 'mind managers'? Who is manipulating whom for what purpose? The functioning of the press even the electronic media calls for closer examination in view of the various pressures both from within and with out.

All this leads us to the debate on internal information order in the country. No nation with an imbalance in the information order can hope to achieve development. This imbalance throws up several imperatives for policy oriented research.

Communication Technology

There have been rapid developments in the realm of communication technology and India has adopted these new technologies at an equal pace for information and various other purposes. It is obvious that the innovative technologies will have an indelible social impact. Communication scholars can research on the extent these technologies would change our life styles, values, inter-personal relationships. Also research could focus on the implications of the technology for interpersonal and mass communication, and the imbalance and access to technology and information. Today, India is wired by 10,000 cable connections in the urban and the rural sector beaming programmes which know no national boundaries. Research can probe into the relevance of programmes originating from alien countries and the impact of the foreign media contents. The impact could be in terms of increase in communications leading to conflicts, tensions and deleterious tendencies.

Calturali Milieu

Communication occurs in a given culture and in India we see a trend towards a mass culture. India is a land of diverse ethnic, religious groups with innumerable dialects/languages. This poses a challenge to effective communication. Studies can emphasize on the quantum of distortion of messages due to varied cultural backgrounds of individuals. This would include verbal and non-verbal communication and the range of understanding or misunderstanding. To what extent would this interaction lead to a modified form of communication cross culturally? Also studies should be done on demographic profiles of potential audiences, their attitudes, and beliefs values cultural considerations influencing media production, dissemination and use.

Media and Literacy

Literacy for self-reliance is the slogan of the National Literacy Mission which envisages making the country literate in a phased manner. Various State Governments have embarked on the literacy drive and some of them have started using various media, especially the electronic media for instructional purposes (both formal and non-formal). Open Universities have started using the media. Technology is applied for instructional and demonstration purposes in development campaigns at the village level. Projects which may be studied in this area are 'Analysis of Educational Television' system relating to the utilization, and feedback, principles and techniques of distance education, and feasibility and application of low-cost media for education.

Message Assimilation

Another major theme for communication researchers would take the form of studying communication process beyond the mass media content and retention of messages. In other words to inquire into the manner in which the messages are absorbed and the utilization by different publishers. The reason for the failure of some of our ambitious development programmes has been the neglect of the crucial component of communication. While it is true that communication by itself cannot bring about a development it is also true that inadequate communication renders development slower and more difficult as well as impeding participation. Hence the need for cooperative research in various fields likes health, agriculture, science and technology. The priority under this theme would be evolving micro-level integrated communication support for development plans after a need assessment.

Evaluation Studies

The phrase 'evaluation studies' is deliberately mentioned as it denotes something more than the effect studies. What is needed is a new look at this theme by undertaking researches as economic evaluation of communication in view of the scarce resources in the country. Some research questions which may be asked can be : Which of the media or media-mix are cost effective for different purposes? What is the optimal investment strategy in the financing of communication technology?

Coming to impact studies, there remains a need to take a critical look at the studies in this are to identify the strengths and weaknesses. They need to be collated, classified and coded so that a theory adaptable to the Indian context can be evolved. At the same time efforts should be made to fill gaps in existing research and examine new aspects of impact studies.

A foremost area of communication research following Macbride's thinking falls under a very general theme-free flow of information. Macbride posed four connected questions focusing on information flow, a new order, the right to communicate and the media. The theme suggested here touches on the rights and responsibilities of journalists and communicators, free flow of information, legal environment and development of media, studies into the modus of media self-regulation and so forth.

Summing up

The themes mentioned here are neither exhaustive nor mutually exclusive. New issues could be taken up even as they are thrown up as a challenge to communicators while the field changes rapidly. These researches could gain the proportion of a national; or it could be at the regional or micro level. Lastly, the efforts will have to be the responsibility of a number of institutional teams, national institutes and agencies, media units and individual researches.

References

1. Bhaskara Rao, N. (1974) "Research in Newspaper Industry : Scope in India", Indian Press, March, pp. 21-24.
2. Malhan, P. N. (1986) "National Media Policy : Dimensions and

Perspectives", Vidura, February, p. 3-7.

3. Ministry of Information and Broadcasting, Government of India (1963) "Report of the Mass Communication Study Team" sponsored by the Ford Foundation, New Delhi.
4. Sen, A. N. (1986) "National Media Policy : A Debate", The Press Council of India Review, January, p. 1.
5. Sean MacBride (1980) "Communication Policies", Many Voices one World, UNESCO, Paris, p. 207.
6. Shukla, Y. V. P. (1966) "Communication Research an Imperative", Alpha, March, pp. 12-13,
7. Sinha, P. R. R. (1970) "Communication Research in India-Trends and Perspective", presented at AMIC Conference, Colombo, October, 12-17.
8. The Hindu (1992), July 15, p. 6.
9. UNESCO (1970) "Mass Media in Society - The Need for Research". Reports and Papers on Mass Communication, No. 59, Paris, UNESCO.
10. UNESCO (1976) "Towards Realistic Communication policies, Recent trends and ideas compiled and analyzed". Reports and Papers on Mass Communication, No. 76, Paris, UNESCO.
11. UNESCO (1979) "National Communication Policy Councils-Principles and Experiences", Report and Papers on Mass Communication, No. 83, Paris, UNESCO.

Index